Emerging Trends and Future Directions in Artificial Intelligence, Machine Learning, and Internet of Things Innovations

Emerging Trends and Future Directions in Artificial Intelligence, Machine Learning, and Internet of Things Innovations

A proceeding of NEIAIS – 2025

Edited by

Khumukcham Robindro Singh

Nazrul Hoque

Arnab Kumar Maji

Sabyasachi Mondal

Jyoti Sekhar Banerjee

Siddhartha Bhattacharyya

Panagiotis Sarigiannidis

CRC Press
Taylor & Francis Group
Boca Raton London New York

CRC Press is an imprint of the
Taylor & Francis Group, an **informa** business

First edition published 2026
by CRC Press
44 Park Square, Milton Park, Abingdon, Oxon, OX14 4RN

and by CRC Press
2385 NW Executive Center Drive, Suite 320, Boca Raton FL 33431

British Library Cataloguing-in-Publication Data
A catalogue record for this book is available from the British Library

ISBN: 9781041146087 (pbk)
ISBN: 9781041145387 (hbk)
ISBN: 9781003675235 (ebk)

DOI: 10.1201/9781003675235

Typeset in Times New Roman
by HBK Digital

Conference Proceedings Series on Futuristic Intelligent and Smart Technologies

Series Editors

Prof. (Dr.) Jyoti Sekhar Banerjee,
Department of Computer Science and Engineering (AI & ML),
Bengal Institute of Technology, Kolkata, India and
Remote Researcher, Internet of THings&AppliCAtions Lab (ITHACA),
University of Western Macedonia, Greece

Prof. (Dr.) Siddhartha Bhattacharyya,
Senior Researcher, Faculty of Electrical Engineering and Computer Science,
VSB Technical University of Ostrava, Czech Republic and
Algebra Bernays University, Zagreb, Croatia

About the Series

This series serves as a beacon for scholars, researchers, and innovators navigating the dynamic landscape of technological evolution. This series stands at the forefront of interdisciplinary discourse, providing a platform for the exchange of ideas, discoveries, and insights that propel society towards a smarter, more intelligent future. Encompassing a diverse array of topics, this series delves into the realms of artificial intelligence, machine learning, robotics, Internet of Things (IoT), smart cities, and beyond. Each volume within the series is a testament to the relentless pursuit of innovation and the relentless quest for solutions to the complex challenges of the modern world.

Contributions to this series emanate from the minds of visionaries across the globe, representing a rich tapestry of perspectives and expertise. From pioneering research papers to visionary keynote addresses, each piece of content within the series reflects the cutting-edge advancements and transformative potential of futuristic intelligent and smart technologies.

Aim and Scope of this Series

The proceedings serve not only as a repository of knowledge but also as a catalyst for collaboration and networking. By fostering connections among researchers, practitioners, industry leaders, and policymakers, the series cultivates a vibrant ecosystem of innovation, where ideas are shared, synergies are discovered, and partnerships are forged.

The integration of traditional and modern intelligent and smart techniques continues to play an increasingly important role in various fields, shaping our city and society through data analysis, optimization, decision-making, and system evaluation and analysis. Artificial intelligence, Machine Learning, Big Data, and the Internet of Things, etc., are examples of futuristic, intelligent, and smart technologies that are enabling practically every area in the future.

As the pace of technological innovation accelerates, the Conference Proceedings Series on Futuristic Intelligent and Smart Technologies stands as a beacon guiding society towards a future imbued with intelligence, efficiency, and sustainability. Through its commitment to excellence, collaboration, and forward-thinking, the series remains an indispensable resource for those shaping the course of technological evolution.

Dedication

This book of conference proceedings is a testament to a shared vision—a vision fuelled by curiosity, creativity, and the unwavering pursuit of knowledge. We dedicate the proceedings of NEIAS-2025 to all the researchers, practitioners, reviewers, and visionaries whose unwavering commitment for knowledge continues to inspire innovation and progress. Together, we have advanced the frontiers of artificial intelligence, machine learning, deep learning, and their transformative applications.

We extend our heartfelt gratitude to the Honourable Vice-Chancellor, Prof. N. Lokendra Singh, for his steadfast support and visionary leadership, which have been instrumental to the success of this conference. We also sincerely acknowledge the North Eastern Council (NEC), Shillong, Government of India, for their generous financial support, which helped create a platform for scholars and practitioners to exchange ideas and inspire future innovations.

In the years to come, may this collection act as an anchor of wisdom, encouraging more research and thought-provoking conversations on their strengths. We hope that these achievements inspire us to carry forward together—to transform knowledge into action, challenges into opportunities, and contribute to the continued advancement of knowledge in the field.

Contents

List of Figures   x

List of Tables   xiii

Preface   xv

Editors Biography   xvi

Reviewers   xix

Committee Members   xx

Chapter 1   Leveraging machine learning techniques to analyse soil organic carbon dynamics in the Eastern Himalayan Region   1
Puja Saha and Amitabha Nath

Chapter 2   Investigation into a suitable voice pathology detection and classification using glottal inverse filtering   8
Ranita Khumukcham and Kishorjit Nongmeikapam

Chapter 3   A hybrid approach for accurate fish segmentation: integrating Vision Transformer and MobileNetV2   13
Pradumn Kumar and Praveen Kumar Shukla

Chapter 4   Predictive efficacy assessment of deep learning models for apple leaf disease detection and classification   18
Vimal Kumar Singh, Deepak Gupta, Vaishnavi Pandey and Satyasundara Mahapatra

Chapter 5   An investigation of machine learning algorithms for lung disease detection   25
Rakesh Kumar Roshan and Devarani Devi Ningombam

Chapter 6   Beyond characters: a machine learning approach to password strength analysis   31
Mrinmoy Borah and Afsana Laskar

Chapter 7   Comparative analysis of machine learning-based specific capacitance prediction and experimental validation of supercapacitive performance in cerium-based electrodes   36
Mahatim Singh, Neeraj Sharma, Pangambam Sendash Singh and Preetam Singh

Chapter 8   Incremental learning using CART and Genetic Programming for handling dynamic training data   42
Ibadonbok Syiemlieh, Ymphaidien Sutong and Sufal Das

Chapter 9   Exploring deep learning approaches for hate speech detection in Assamese social media text   49
Tulika Chutia, Rituraj Phukan and Nomi Baruah

Chapter 10   Using sentiment analysis to enhance digital forensic evidence identification   56
Vishwavijay kumar and Vikas Pareek

Chapter 11   Machine learning methods for improving stock price prediction accuracy   61
Priya Rani and Devarani Devi Ningombam

Chapter 12   Effect of stop words in Assamese language for identifying toxic comments   66
Mandira Neog and Nomi Baruah

Chapter 13   Optimising low-resource Khasi NER through transfer learning   74
Ransly Hoojon

Chapter 14   Localised detection of rice disease using YOLO 11: an advanced approach for precision agriculture   81
Gitanjali Patowary and Dr Ridip Dev Choudhury

Chapter 15   Assamese handwritten script segmentation using deep learning approach   85
              Saurabh Sutradhar and Dr. Ridip Dev Choudhury

Chapter 16   Clustering of foods based on nutritional profile using principal component analysis and k-nearest
              neighbor classification   93
              Laishram Maria Devi, Chingriyo Raihing and Satishchandra Salam

Chapter 17   Estimating crop yields using machine learning   99
              Shrungashri Chaudhary and N Kishorjit

Chapter 18   Detection of sensitive information using CRF in unstructured text   104
              Longjam Velentina Devi and Navanath Saharia

Chapter 19   A home-based skill improvement tool utilising smartphone sensors   110
              Sudipta Saha, Saikat Basu and Koushik Majumder

Chapter 20   Offline handwritten character recognition of Manipuri Script using convolutional neural
              network (CNN) and vision transformer (ViT)   117
              Elangbam Binoy Singh and Thokchom Tangkeshwar Singh

Chapter 21   Denoising in Terahertz images   122
              Phibansabeth Nongkseh and Debdatta Kandar

Chapter 22   A novel ensemble rank aggregation feature selection algorithm for classification 128
              Nur Alom and Rubul Kumar Bania

Chapter 23   Analysing the impact of Word2Vec on stance detection in low-resource languages:
              a case study on Manipuri editorial articles   134
              Pebam Binodini, Sunita Sarkar and Kishorjit Nongmeikapam

Chapter 24   Robust MRI brain tumour segmentation using integrated mean shift clustering and
              adaptive thresholding   140
              Maya Pawar, Rubul Kumar Bania and Dibya Jyoti Bora

Chapter 25   Feature-driven classification of AI and human text: insights from traditional and neural models 144
              Potshangbam Kirankumar Singh,, Ksh Nareshkumar Singh and H. Mamata Devi

Chapter 26   A novel stacking ensemble machine learning approach for landslide susceptibility mapping 149
              Moziihrii Ado, Nongmaithem Kane and Khwairakpam Amitab

Chapter 27   Advancements in network security threat detection systems for wireless sensor networks:
              a review of machine learning and deep learning techniques   154
              Bikash Kalita and Satyajit Sarmah

Chapter 28   Comparison of machine learning, ensemble learning, and deep learning model for predicting
              ovarian cancer: a data-driven study   163
              Naba Jyoti Sarmah, Dwipen Laskar and Kaushik Kumar Bharadwaj

Chapter 29   Efficient crack classification through transfer learning with pre-trained convolutional
              neural networks   168
              Thokchom Chittaranjan Singh, Romesh Laishram and N. Basanta Singh

Chapter 30   Enhancing medical imaging diagnostics with federated learning: overcoming data heterogeneity 174
              Soumyaranjan Panda, Prayash Tah, Vikas Pareek and Sanjay Saxena

Chapter 31   A hybrid deep learning model for accurate brain tumour segmentation and classification
              using ResNet101 and U-Net   179
              Wu Min Fen, Yang Jin, Liao Shi Chin and Haobijam Basanta

Chapter 32   Evaluating machine learning models for precision disease prediction   186
Pallabi Patowary and Dhruba K Bhattacharyya

Chapter 33   Modelling stock prices using XGBoost, Prophet, and ARIMA: a comprehensive evaluation   191
K. B. Vaisshnavi and A Padmavathi

Chapter 34   Advancements in Sugarcane Disease Detection and Prediction   197
Ankita Goyal Agarwala and Mirzanur Rahman

Chapter 35   Impact of machine translation evaluation metrics on low resource languages   202
Sonithoi Ningombam, N Donald Jefferson Thabah, Arindam Roy and Bipul Syam Purkayastha

Chapter 36   Artificial intelligence and entrepreneurship in the context of current socioeconomic growth in India – an examination of the literature   207
Th. John Lerphangam Monsang and Sanasam Somokanta Singh

Chapter 37   An extended SMOTE-based class balancing technique to improve classification accuracy on imbalanced datasets   213
P. Sendash Singh, Ksh. Nilakanta Singh, Bishambhor Th, M. Surchand Singh, Joychandra N. and Khumukcham Robindro Singh

Chapter 38   AI-powered personalised learning and intelligent recommendation systems   218
Sri Sai Harshita Gadavarthi, Sanya Gulati, Sanika Jagiasi, Anushka Jaint, Kashish Jindal and Soni Sweta

Chapter 39   Detection of diabetic retinopathy using deep-learning based image classification   225
Ksh. Dayvason Singh, Leishangthem Sashikumar Singh, Ksh. Nilakanta Singh, N. Joychandra Singh and Kh. Robindro Singh

Chapter 40   In depth comparative analysis of the combine effect of ML models and feature selection methods in crop yield prediction   230
Mayanglambam Surchand Singh, Khumukcham. Robindro Singh, Ksh Nilkanta Singh, Nongdombam Joychandra Singh, Leishangthem Sashikumar Singh and M. Rajeshwar Singh

Chapter 41   Evaluating LLMs and pre-trained models for text summarisation across diverse datasets   235
Tohida Rehman, Soumabha Ghosh, Kuntal Das, Souvik Bhattacharjee, Debarshi Kumar Sanyal and Samiran Chattopadhyay

Chapter 42   Breast cancer classification using machine learning and deep learning: a systematic review of WBCD-based research and future directions   241
Shafiq Ahamed and Amitabh Wahi

Chapter 43   Sentiment analysis on code-mixed data: a comprehensive, qualitative and quantitative literature survey   245
Afsana Laskar and Shikhar Kumar Sarma

Chapter 44   Advanced fraud detection in financial transaction using machine learning   250
Pooja Baravkar, Sayali Jagdale, Vedant Salunke, Anuja Kanade and Nikhil Surse

Chapter 45   Efficient detection of offensive social media comments in Assamese language using LSTM   257
Tulika Chutia, Sanjib Bora, Nomi Baruah[b], Debajani Baruah, Swarnangka Barman, Joy Anupol Neog and Bikokhita Dutta

List of Figures

Figure 1.1	Data flow diagram	2
Figure 1.2	Study area map	2
Figure 1.3	Observed vs predicted graph for XGB, RF and AdaBoost model	4
Figure 1.4	Sensitivity analysis plot showing the impact of different input parameters	4
Figure 1.5	Taylor plot illustrating the performance of different models	4
Figure 1.6	Taylor plot illustrating the performance of different models	5
Figure 1.7	Pearson's correlation for different parameter	5
Figure 2.1	Block diagram of the proposed method	9
Figure 2.2	Confusion matrix showing the per result with kNN method using GIFEKF	9
Figure 2.3	Training progress of kNN with GIFEKF	9
Figure 2.4	Confusion matrix showing the per class result with SVM method using GIFEKF	10
Figure 2.5	Training progress of SVM with GIFEKF	10
Figure 2.6	Confusion matrix showing the per class result with kNN method using GIFQCP	10
Figure 2.7	Training progress of kNN with GIFQCP	10
Figure 2.8	Confusion matrix showing the per class result with SVM method using GIFQCP	10
Figure 2.9	Training progress of SVM with GIFQCP	11
Figure 3.1	Architecture for proposed methodology	15
Figure 3.2	Training and validation accuracy	15
Figure 3.3	Confusion matrix of segmentation	16
Figure 3.4	Training and validation loss	16
Figure 3.5	Precision, recall, and F1-score	16
Figure 4.1	Proposed methodology	20
Figure 4.2	(a) Original image and (b) Histogram equalised image	20
Figure 4.3	(a) Original image and (b) CLAHE image	20
Figure 4.4	(a) Original Image and (b) HSV Image	21
Figure 4.5	Confusion matrix for CLAHE processed dataset: a) MobileNet, b) ResNet50, c) VGG16, and d) VGG19	23
Figure 5.1	(a) shows a chest X-ray image with radiopaque areas showing fluid accumulation consolidations. (b) Demonstrates the cavities and unification of the right lung's top lobe	26
Figure 6.1	Dataset before pre-processing	32
Figure 6.2	Dataset after pre-processing	32
Figure 6.3	Confusion matrices of ML and DL algorithms	34
Figure 7.1	Original class distribution and class distribution after SMOTE	37
Figure 7.2	Heatmap of the correlation matrix illustrating the relationships among features	37
Figure 7.3	Comparative analysis of different machine learning algorithms	39
Figure 7.4	ROC curve for class A, B, C and D	39
Figure 8.1	Block diagram of the proposed method	44
Figure 8.2	Comparison with mean accuracy (%)	46
Figure 8.3	Comparison with mean precision (%)	46
Figure 8.4	Comparison with recall (%)	47
Figure 8.5	Comparison with F1-score (%)	47
Figure 9.1	Hierarchy of hate speech concepts	50
Figure 9.2	Pictorial diagram of proposed methodology	52
Figure 9.3	Sample of the proposed dataset	52
Figure 9.4	Model accuracy and model loss of LSTM	53
Figure 9.5	Model accuracy and model loss of Bi-LSTM	53
Figure 10.1	Research flowchart for sentiment analysis in digital forensics to the utilisation of all models	58
Figure 10.2	NB and CNN are machine learning models performing well on the same data set	58
Figure 11.1	The flowchart of the model	63
Figure 11.2	Performance comparison of stock price prediction and actual price	64

Figure 12.1	Demonstration of punctuation removal and tokenisation	70
Figure 12.2	Demonstration of domain specific Stopword creation	70
Figure 12.3	Demonstration of complete approach of Stopword removal with machine learning implementation	71
Figure 12.4	Demonstration of common and domain-specific Stopword removal	71
Figure 12.5	Demonstration of classifier (SVM, NB) performance comparison based on domain specific Stopwords removal and common Stopword removal	72
Figure 13.1	Dataset learning curve analysis	76
Figure 13.2	Model sample output	79
Figure 13.3	Comparison of models	79
Figure 14.1	F1, precision-confidence, recall-confidence and precision-recall curve	83
Figure 14.2	Results on training & validation loss	83
Figure 15.1	Propose methodology for this study	87
Figure 15.2	Sample Images of collected dataset	88
Figure 15.3	Samples of segmented images a. SegNet segmentation & b. U-Net segmentation	89
Figure 15.4	Statistical analysis of our study (a) Accuracy, (b) Dice coefficient, (c) Jaccard, and (d) Sensitivity	90
Figure 16.1	Correlation matrix of the nutrient attributes	95
Figure 16.2	Variance explained by the principal components	95
Figure 16.3	PCA cluster plotted against the first two principal components	96
Figure 16.4	Food similarity matrix	96
Figure 17.1	Decision tree	101
Figure 17.2	Decision tree	102
Figure 17.3	Random forest	102
Figure 17.4	Linear regression	102
Figure 18.1	Overall data flow of the proposed method	106
Figure 18.2	Visualization of performance based on different regularisation values	108
Figure 18.3	ROC curve For S class	108
Figure 18.4	ROC curve For N class	108
Figure 19.1	RADB practice scores of two elderly persons	113
Figure 19.2	Pre-test and post-test scores of two elderly persons in TM game	113
Figure 19.3	Pre-test and post-test scores of two elderly persons in BT and DS games	113
Figure 20.1	Manipuri Script	118
Figure 20.2	(a) A scanned image and (b) Some segmented handwritten characters images of Manipuri Script	119
Figure 20.3	Proposed CNN model	120
Figure 21.1	Workflow diagram of the proposed methodology	125
Figure 21.2	Confusion matrix for YOLOv8 without wavelet denoising, showing the classification performance across different object classes	125
Figure 21.3	Confusion matrix for YOLOv8 with wavelet denoising, showing classification performance across classes	125
Figure 21.4	mAP@05 score of the detection method	125
Figure 21.5	Training loss curve of YOLOv8 with wavelet denoising	126
Figure 21.6	Detected object using YOLOv8 with wavelet denoising	126
Figure 22.1	Block diagram of the proposed ensemble FS mechanism	129
Figure 23.1	Proposed architecture model	135
Figure 23.2	Confusion matrix for 1,000 combinations across all four classifiers	139
Figure 24.1	Flowchart of the proposed approach	141
Figure 24.2	Original image and the result obtained using the proposed approach	142
Figure 26.1	Proposed novel stacking model	152
Figure 26.2	Landslide susceptibility map for Meghalaya generated using the novel stacking model	152
Figure 27.1	A general architecture of WSN	155
Figure 27.2	A hierarchical taxonomy of network	158
Figure 28.1	A schematic representation of the studied methodology, indicating each step and the flow between them through arrows	165
Figure 29.1	Sample of the datasets	169
Figure 29.2	Flowchart of the proposed model	170
Figure 29.3	Confusion metrics for MobileNetV2	171
Figure 29.4	Confusion metrics for ResNet50	172

Figure 29.5	Confusion metrics for InceptionV3	172
Figure 29.6	Confusion metrics for Xception	172
Figure 29.7	Training and validation accuracy for MobileNetV2	172
Figure 29.8	Training and validation loss for MobileNetV2	172
Figure 30.1	The flow of the federated learning training	176
Figure 30.2	Heterogeneity data distribution over 10 clients	177
Figure 30.3	t-SNE visualisation for different federated learning models	177
Figure 31.1	Datasets used in this study. (a) Sample MRI scans from the segmentation dataset with ground truth tumour regions. (b) Representative images from the classification dataset	181
Figure 31.2	Overview of the hybrid model architecture	181
Figure 31.3	Architectural overview of the hybrid deep learning model	182
Figure 31.4	Confusion matrix for brain tumour classification	183
Figure 31.5	Brain tumour segmentation results	183
Figure 31.6	Brain tumour prediction results	183
Figure 32.1	Proposed framework for disease prediction	187
Figure 32.2	AUC-ROC curves for various machine learning. Models on symptom-based Dataset (Dataset 1) [The legend specifies the model name followed by its corresponding AUC score]	189
Figure 32.3	AUC-ROC curves for various machine learning models on biomarker-based dataset (Dataset 2)	189
Figure 33.1	Stock price prediction	193
Figure 33.2	Graph representation of the comparison of algorithms	194
Figure 34.1	Visual representation of sugarcane diseases	198
Figure 34.2	Healthy sugarcane plants	198
Figure 34.3	Comparative evaluation of accuracy and loss	200
Figure 34.4	Proposed CNN-SVM architecture	200
Figure 34.5	Proposed CNN-Vit architecture	200
Figure 35.1	Sentence embedding aggregation	203
Figure 35.2	2D Visualization of example 1 Word Embedding from using Meiteilon-BERT using t-SNE	205
Figure 35.3	2D Visualization of example 2 Word Embedding from using Meiteilon-RoBERTa using t-SNE	205
Figure 35.4	2D Visualization of example 3 Word Embedding from using Khasi-BERT using t-SNE	205
Figure 35.5	2D Visualization of example 3 Word Embedding from using Khasi-RoBERTa using t-SNE	205
Figure 37.1	Conceptual framework for class balancing technique	215
Figure 38.1	Mind map of the personalised model	219
Figure 38.2	Personalised learning framework model	221
Figure 38.3	Histogram for the distribution of scores	222
Figure 38.4	Distribution of quiz scores before and after personalised content	222
Figure 38.5	Correlation analysis and feature importance	223
Figure 39.1	Workflow diagram	228
Figure 40.1	Performance of ML models in different FS models in performance matric composite	232
Figure 40.2	Performance of RF, ANN, ETR in VIF, RFE and RF-EMB (R2 vs number of features selected)	233
Figure 40.3	Time taken by ML-FS models against number of features selected	233
Figure 41.1	Example from CNN/DM test set. Input taken from https://edition.cnn.com/2015/04/03/us/kentucky-ge-fire/index.html	238
Figure 41.2	Example from Gigaword test set. Input taken from https://www.cfr.org/backgrounder/sri-lankan-conflict	239
Figure 43.1	Sentiment analysis general workflow diagram	246
Figure 43.2	ACE-Mix: A Dataset for Assamese-English Code-Mixed Language Processing	247
Figure 43.3	Sample dataset with labelling	247
Figure 44.1	System architecture	251
Figure 44.2	Fraud detection visualizations: Dashboard	253
Figure 44.3	Fraud detection visualizations: Distribution chart	253
Figure 44.4	Original distribution of fraud and non-fraud	254
Figure 44.5	Original distribution of fraud and non-fraud	254
Figure 44.6	Fraud detection visualizations: Confusion matrix	255
Figure 45.1	Proposed methodology of the work	259
Figure 45.2	LSTM architecture	259
Figure 45.3	Confusion matrix for the proposed approach	261

List of Tables

Table 1.1	Description of different parameters used in the study.	3
Table 2.1	Comparison of the accuracy results using GIFEKF and GIFQCP.	11
Table 4.1	Performance evaluation with HSV pre-processing.	22
Table 4.2	Performance evaluation with histogram equalization.	22
Table 4.3	Performance evaluation with CLAHE	23
Table 5.1	Comparative findings of performance of literature work using various datasets.	29
Table 5.2	Comparative findings for Youden's index-based papers.	29
Table 6.1	Structural analysis of a password.	33
Table 6.2	Performance metrics of machine and deep learning classifiers.	33
Table 7.1	Precision, recall, F1 score, and accuracy of various algorithms.	38
Table 8.1	Description about datasets used.	46
Table 8.2	Comparison with mean accuracy (%).	46
Table 8.3	Comparison with mean precision (%).	46
Table 8.4	Comparison with recall (%).	46
Table 8.5	Comparison with F1-score (%).	47
Table 9.1	Comparison of results with proposed approach and existing work.	53
Table 10.1	Comparative table of all machine learning models with mean ± std deviation of the accuracy, precision, recall, and F1-score	59
Table 12.1	Stopwords with example.	69
Table 12.2	Output of processed input data with SVM and NB Classifier	72
Table 12.3	Result comparison with other languages	72
Table 12.4	Comparison of the two classifiers with and without using domain-specific Stopwords corpus	72
Table 13.1	Annotation classes.	76
Table 13.2	Total count of each entity type.	77
Table 13.3	Similar words to thiah which translate to sleep.	78
Table 13.4	Summary of Khasi NER model performance.	79
Table 14.1	Label counts for rice disease dataset.	82
Table 17.1	Optimal parameter resulted from grid search.	102
Table 18.1	Optimal parameter resulted from grid search.	107
Table 18.2	Models performance on imbalanced class.	107
Table 18.3	Models performance after Randomsampling.	107
Table 18.4	Performance based on different regularisation values.	107
Table 19.1	Descriptive statistics of games' scores.	114
Table 19.2	Correlation coefficients of games' parameters.	114
Table 19.3	Regression results.	114
Table 19.4	Elderly persons' practice score on Roll and Drop Ball.	115
Table 19.5	Elderly persons' pre-test and post-test scores on standard games.	115
Table 20.1	Manipuri script's works in HCR.	119
Table 20.2	Description of dataset	120
Table 20.3	Comparative results of CNN and ViT-B/16.	121
Table 21.1	Description of the concealed items and quantity of each object of the dataset.	123
Table 21.2	Performance evaluation of denoising techniques.	125
Table 22.1	Dataset details.	130
Table 22.2	Classification result of original feature set of different datasets.	131
Table 22.3	Comparison of classification result in hepatitis dataset.	131
Table 22.4	Comparison of classification result in dermatology dataset.	131
Table 22.5	Comparison of classification result in BCWD dataset.	131
Table 22.6	Comparison of classification result in lung cancer dataset.	132
Table 22.7	Comparison of classification result in arrhythmia dataset.	132

Table 23.1	Dataset distribution.	135
Table 23.2	Sample dataset annotations.	135
Table 23.3	Support vector machines.	137
Table 23.4	k-Nearest neighbor.	137
Table 23.5	Logistic regression.	137
Table 23.6	Random forest.	137
Table 23.7	F1 score comparison without and with SMOTETomek.	138
Table 23.8	Similar words for the word 'Sarkar'.	138
Table 24.1	Experimental results.	142
Table 25.1	Algorithms performance result on Dataset 1.	146
Table 25.2	Algorithms performance result on Dataset 2.	146
Table 25.3	Algorithms performance result on Dataset 3.	146
Table 25.4	Algorithms performance result on Dataset 4.	146
Table 26.1	Landslide causative factors.	151
Table 26.2	Landslide causative factors.	151
Table 27.1	Comprehensive overview of WSNs, covering technological advancements and limitation.	159
Table 28.1	Accuracy, precision, recall, F1-score, and ROC-AUC score of different classification methods.	166
Table 29.1	Evaluation metrics of pre-trained CNN models used.	171
Table 29.2	Comparative analysis of model architectures.	172
Table 30.1	Top1 accuracy achieved by various federated learning algorithm.	177
Table 30.2	Performance across various rounds during training.	177
Table 30.3	Various quantitative measures for COVID-19 tasks.	178
Table 31.1	U-Net + ResNet101 performance vs. leading models.	184
Table 32.1	Performance metrics for various ML models on symptom-based dataset (Dataset 1).	188
Table 32.2	Performance metrics for various ML models on biomarker-based dataset (Dataset 2).	188
Table 34.1	Common sugarcane diseases and their symptoms.	198
Table 35.1	Indic-English and English – Indic metrics.	204
Table 35.2	Example 1 sentence similarity and metrics score for BERT, S-BERT, RoBERTa, chrf and BLEU (Meiteilon language).	204
Table 35.3	Example 2 sentence similarity and metrics score for BERT, S-BERT, RoBERTa, chrf and BLEU (Khasi language).	204
Table 37.1	Description of the dataset.	216
Table 37.2	Result obtained on smotted minority sample.	216
Table 37.3	Result obtained on smotted majority sample.	217
Table 38.1	Comparative analysis of personalised learning systems and technologies.	219
Table 38.2	Paired t-test results for quiz score improvement.	222
Table 39.1	Performance comparison with existing literature work.	228
Table 40.1	Performance of best model among the combination of models of ML and FS.	232
Table 41.1	Performance comparison of models on CNN/DM.	237
Table 41.2	Performance comparison of models on Gigaword.	237
Table 41.3	Performance comparison of models on News Summary.	238
Table 41.4	Performance comparison of models on XSum.	238
Table 41.5	Performance comparison of models on BBC News.	238
Table 42.1	Performance-metrics of ML-models.	242
Table 42.2	Performance-metrics of DL-models.	242
Table 42.3	Performance-metrics of HL-models.	242
Table 44.1	Prediction results of ML model.	255
Table 45.1	Example of sentences with numerical values.	259
Table 45.2	Some examples of planning and control of offensive comments detection in Assamese text.	260
Table 45.3	Comparison with previous studies.	260

Preface

The "North East India AI Summit: Unravelling Trends (NEIAIS 2025)" served as a vibrant platform for the exchange of cutting-edge ideas and research in the field of Artificial Intelligence, with a strong emphasis on both foundational theories and real-world applications. The summit brought together experts, researchers, and enthusiasts to explore critical areas including Machine Learning, Deep Learning, Computer Vision, Natural Language Processing, Smart Systems, IoT Security, Network Technology, and Artificial Intelligence in Healthcare and Biomedical Applications. Discussions also delved into emerging trends and computational techniques, highlighting the transformative potential of AI in addressing complex, real-world challenges. The conference received an overwhelming response, attracting more than 120 research paper submissions from various regions of India and abroad. After a rigorous review process, 55 high-quality papers were accepted, out of which over 44 papers were registered for presentation at the summit. By fostering interdisciplinary collaboration and showcasing impactful innovations, NEIAIS 2025 aims to inspire sustained research, technological growth, and broader societal benefits.

May this collection of proceedings serve as a beacon of insight, fostering new discoveries, innovative thinking, and meaningful discussions for years to come.

The Department of Computer Science, Manipur University organised the first international conference on "North East India AI Summit: Unravelling Trends (NEIAIS 2025) with financial assistance from the North Eastern Council (NEC), Shillong at its campus from February 10–12, 2025. NEIAIS 2025 offered a vibrant platform for discussing recent advancements across a diverse array of subjects, with a particular focus on the core areas of Artificial Intelligence and emerging trends in Computer Science and Engineering. Envisioned as an open forum, NEIAIS encourages dialogues on regional and global challenges through the lens of technological innovation, particularly in the fields of engineering and information technology. With impactful research driving forward-thinking solutions, the summit aims to inspire innovations in modern science and technology that contribute meaningfully to solving pressing societal problems.

The organising committee of NEIAIS 2025 gratefully acknowledges the exceptional efforts of the Scientific, Advisory, and Technical Program Committee members, whose commitment and expertise were instrumental to the success of the event held from 10–12 February 2025. We extend our sincere appreciation to the Keynote Speakers for their insightful contributions that enriched the summit. Our heartfelt thanks also go to the reviewers for their valuable time, meticulous evaluations, and unwavering dedication in upholding the quality of the submissions. Furthermore, we would like to thank the Session Chairs for their pivotal role in ensuring the smooth conduct and coordination of the sessions throughout the conference.

We extend our heartfelt gratitude to all attendees of NEIAIS 2025, including presenters and participants, whose active engagement and contributions enriched the conference. We deeply appreciate everyone who supported and contributed to the success of this event. Our profound thanks and admiration go to the Organising Committee, Program Committee, volunteers, and support staff for their dedication, hard work, and unwavering commitment. We are also grateful to our Chief Patron, Patron, Honorary Chair, and the esteemed members of the various conference committees for their invaluable guidance and support throughout the planning and execution of the summit.

Finally, we would like to express our gratitude to Mrs. Shatakshi Mishra, the Editorial Manager of CRC Press, for her kind support in publishing this book.

Imphal	Prof. Dr. Khumukcham Robindro Singh
Imphal	Prof. Dr. Nazrul Hoque
Shillong	Prof. Dr. Arnab Kumar Maji
Shillong	Prof. Dr. Sabyasachi Mondal
Kolkata	Prof. Dr. Jyoti Sekhar Banerjee
Ostrava	Prof. Dr. Siddhartha Bhattacharyya
Kozani	Prof. Dr. Panagiotis Sarigiannidis

Editors Biography

Dr. Khumukcham Robindro Singh received his Master's degree in Computer Application from Manipur University, Imphal, in 2007, and was awarded his Ph.D. from Gauhati University, Guwahati, Assam, in 2013. With approximately 12 years of professional experience, he is currently serving as Assistant Professor and Head of the Department of Computer Science at Manipur University, Imphal—a Central University of India. Dr. Singh is also the Chairman of the eOffice Implementation Committee, where he is actively spearheading the university's digital transformation initiatives. Additionally, he has frequently served on the College Inspection Team of the university and has been appointed as the Nodal Officer of the Ek Bharat Shreshtha Bharat (EBSB) Cell, Manipur University. A sought-after speaker, Dr. Singh has been invited by several departments within and outside the university to deliver expert talks on Artificial Intelligence and its applications. He has published around 15 research articles in reputed SCIE/Scopus-indexed international journals, over 10 book chapters, and 12 papers in conference proceedings. Under his guidance, three Ph.D. scholars have successfully completed their research, and he has mentored 30 MCA major project theses. Dr. Singh has completed a major research project in collaboration with IIT Guwahati and Tezpur University, funded by the Ministry of Electronics and Information Technology (MeitY), Government of India. He has also organised two National Workshops on High Performance Computing in collaboration with CDAC Pune and IIT Kharagpur under the National Supercomputing Mission. He serves as a reviewer for several reputed international journals, and his primary research interests include Artificial Intelligence, Machine Learning, and Deep Learning.

Dr. Nazrul Hoque received his Ph.D. degree in Computer Science and Engineering from Tezpur University, Tezpur, India, in 2017. He is currently working as an Assistant Professor in the Department of Computer Science, Manipur University, Canchipur, Imphal, India. He has published more than 40 papers in international journals and referred to conference proceedings. Dr. Hoque has published papers in IEEE Communication Surveys and Tutorials, IEEE Access, Expert Systems with Applications, Computer Communications, Complex and Intelligent Systems, Security and Privacy, and Cybersecurity. He is a member of the editorial team of the Journal Scientific Reports. His research interests include Machine Learning, Network Security, and the Internet of Things.

Dr. Arnab Kumar Maji received his B.E. degree in Information Science and Engineering from Visvesvaraya Technological University (VTU) in 2003 and M.Tech in Information Technology from Bengal Engineering and Science University, Shibpur (currently IIEST, Shibpur) in 2006. He received his Ph.D. from Assam University, Silchar (a Central University of India) in 2016. He has approximately 20 years of professional experience. He is currently working as Associate Professor in the Department of Information Technology, North-Eastern Hill University, Shillong (a Central University of India). He has published around 40 articles in different reputed SCIE/SCOPUS-Indexed International Journals, more than 12 articles as book chapter, 30 papers as conference proceedings, and authored 03 books with several international Publishers like Elsevier, Springer, IEEE, MDPI, IGI Global, and McMilan International. 08 PhD scholars are successfully guided by him. He has also guided successfully 18 M.Tech theses. He is also reviewer of several reputed international journals and a guest editor for a Springer journal. 02 patents and 01 copyright are granted in his name. His Research interests include Computer Vision and Natural Language Processing.

Dr. Sabyasachi Mondal is an Associate Professor in the Department of Mathematics at North-Eastern Hill University, Shillong, Meghalaya, India. He obtained his B.Sc. (Hons.) in Mathematics, M.Sc. in Applied Mathematics, and Ph.D. in Computational Fluid Dynamics from Visva-Bharati Santiniketan, West Bengal, India. He worked as a 'Postdoctoral Research Fellow' at the University of KwaZulu-Natal, South Africa. His areas of research interest include Computational Fluid Dynamics, Heat and Mass Transport, and Nanofluid flow. With 13 years of teaching experience, he has more than 95 (SCOPUS 78) International Publications in different reputed journals, books, and conferences in his credit. Also, Dr. Mondal has published 01 Indian patent and filed 01 Indian patent recently. Dr. Mondal delivered 06 invited talks in different universities and colleges. He is an editorial board member in 06 international journals and is a reviewer for more than 50 reputed journals which include Elsevier, Springer, Wiley, and Taylor & Fransis. Dr. Mondal has received various prestigious funds and fellowships like UKZN PDF fellowship, Claude Leone Postdoctoral fellowship, and NRF travel grants. Dr. Mondal has research collaborations with various eminent scientists in India and abroad, like South Africa, Botswana, Malaysia, and Iran. His research has led to him being selected as 'Top 2% Most Influential Scientists' list published by 'Standford University and SCOPUS for the year 2022–2023 in the domain of 'Mechanical Engineering and Transports' and 'Applied Mathematics'.

Dr. Jyoti Sekhar Banerjee is currently serving as the Head of the Department of the Computer Science and Engineering (AI & ML) at the Bengal Institute of Technology, Kolkata, India. He also serves as the Professor-in-Charge, R&D and Consultancy Cell, and as the Nodal Officer of the IPR Cell at BIT. Since 2024, he also works as a Remote Researcher in the Internet of THings & AppliCAtions Lab (ITHACA) at the Department of Electrical and Computer Engineering, University of Western Macedonia, Greece. He is the former Remote Research Fellow of the Cognitive Computing and Brain Informatics Research Group (CCBI) at Nottingham Trent University (NTU), UK. Dr. Banerjee holds his Post-Doctoral Fellowship at Nottingham Trent University, UK, in the Department of Computer Science. He also completed the Post Graduate Diploma in IPR & TBM from MAKAUT, WB. He has teaching and research experience spanning 20 years and completed one IEI funded project. He is a member of the CSI, IEEE, ISTE, IEI, ISOC, IAENG, and a fellow of IETE. He presently serves as Secretary-cum-Treasurer of the ISTE WB Section and Secretary of the IETE, Kolkata Centre. He is the Immediate Past Secretary of the Computer Society of India, Kolkata Chapter. Dr. Banerjee is also elected as the Vice Chairman Cum Chairman Elect in Computer Society of India, Kolkata Chapter, for 2025–2027.

Dr. Siddhartha Bhattacharyya [FRSA, FIET (UK), FIEI, FIETE, FSCRS, LFOSI, SMIEEE, SMACM, SMAAIA, SMIETI, LMCSI, LMISTE] is currently a senior researcher at VSB – Technical University of Ostrava, Ostrava, Czech Republic. He is also a Scientific Advisor at Algebra University, Zagreb, Croatia. Prior to this, he was the Principal of Rajnagar Mahavidyalaya, Birbhum, India. Before this, he was a professor at CHRIST (Deemed to be University), Bangalore, India. He also served as the Principal of RCC Institute of Information Technology, Kolkata, India. He has also served as a Senior Research Scientist at VSB Technical University of Ostrava, Czech Republic. He is the recipient of several coveted national and international awards. He has received the Honorary Doctorate Award (D. Litt.) from the University of South America and the SEARCC International Digital Award ICT Educator in 2017. He was appointed as the ACM Distinguished Speaker for the tenure of 2018–2020. He has been appointed as the IEEE Computer Society Distinguished Visitor for 2021–2024. He is a co-author of 6 books and the co-editor of 106 books and has more than 400 research publications in international journals and conference proceedings to his credit.

Dr. Panagiotis Sarigiannidis is the Director of the ITHACA lab, co-founder of the first spin-off of the University of Western Macedonia: MetaMind Innovations P.C., and a Full Professor in the Department of Electrical and Computer Engineering at the University of Western Macedonia, Kozani, Greece. He received his B.Sc. and Ph.D. degrees in Computer Science from Aristotle University of Thessaloniki, Thessaloniki, Greece in 2001 and 2007, respectively. He has published over 360 papers in international journals, conferences, and book chapters, including IEEE Communications Surveys and Tutorials, IEEE Transactions on Communications, IEEE Internet of Things, IEEE Transactions on Broadcasting, IEEE Systems Journal, IEEE Wireless Communications Magazine, IEEE Open Journal of the Communications Society, IEEE/OSA Journal of Lightwave Technology, IEEE Transactions on Industrial Informatics, and IEEE Access and Computer Networks. He has received 6 best paper awards and the IEEE SMC TCHS Research and Innovation Award in 2023. He has been involved in several national, European and international projects, coordinating and technically leading numerous national and European projects including H2020, Horizon Europe, Erasmus+, and operational programs. His research interests include Telecommunication Networks, Internet of Things, and Network Security. He is an IEEE member and participates in the Editorial Boards of various journals, including IEEE Transactions on Communications, IET Networks, International Journal of Communication Systems, and International Journal of Information Security.

Advisory Committee

Sl. No.	Name	Affiliation
1	Prof. N. Basanta Singh	Dean, School of Engineering, Manipur University
2	Prof. Anjana Kakoti Mahanta	Gauhati University
3	Prof. Dhruba Kumar Bhattacharyya	Tezpur University
4	Dr. Jugal Kalita	College of Engineering and Applied Science, University of Colorado
5	Prof. Parmanand Astya	Hon'ble Pro Vice-Chancellor, Sharda University, Greater Noida
6	Prof. Kh. Manglem Singh	NIT Manipur
7	Yo-Ping Huang	Ph.D., FIEEE, FIET, FCACS, FTFSA, President, National Penghu University of Science and Technology, Penghu, Taiwan
8	Dr. Cheng-Ming Huang	National Taipei University of Technology, Taiwan
9	Prof. Chang-Yun Yang	National Taipei University, Taiwan
10	Prof. Yue-Shan Chang	National Taipei University, Taiwan
11	Prof. Subhash Chandra Yadav	Central University of Jharkhand
12	Prof. Sanjoy Das	IGNTU, RC Manipur
13	Prof. Jemal Hussein	Deakin University, Australia
14	Prof. Iftekhar Hussain	NEHU Shillong
15	Prof. Seokjoo Shin	Professor, Chosun University, South Korea
16	Prof. Ashutosh Sharma	Professor, Henan University of Science and Technology, China
17	Prof. T.P Singh	Professor, Bennett University, Greater Noida, India
18	Dr. Somnath Roy	IIT Kharagpur
19	Dr. Pralay Mitra	IIT Kharagpur
20	Dr. Michal Jasinski	Wroclaw University of Science and Technology, Wroclaw
21	Dr. Monowar Bhuyan	Umea University, Sweden
22	Dr. Arnab Kumar Majhi	NEHU Shillong
23	Dr. Prakash Chauhan	Cotton University
24	Dr. Hasin Ahmed	Gauhati University
25	Dr. Ganapathy Sannasi	NITTTR, Bhopal
26	Dr. Haobijam Basanta	National Taipei University, Taiwan
28	Dr. Devarani Ningombam	NIT Patna
29	Dr. Pangambam Sendash Singh	UPES Dehradun
30	Dr. Kishorjit Nongmeikapam	Indian Institute of Information Technology (IIIT) Manipur
31	Dr. Navanath Saharia	Indian Institute of Information Technology (IIIT) Manipur
32	Dr. Priyanka Singh	UPES, Dehradun
33	Dr. Achala Sakhya	UPES Dehradun
34	Arifa Ferdousi	Varendra University, Bangladesh
35	Ayesha Akter Lata	Chosun University, South Korea

Organizing Committee

Chief Patron	**Prof. N. Lokendra Singh, Hon'ble Vice-Chancellor, Manipur University**
Patron	Prof. Sumitra Phanjoubam, Dean, School of Mathematical & Physical Sciences, Manipur University
Chairman	Prof. W. Chandbabu Singh, Registrar, Manipur University
Honorary Chair	Prof. Shikhar Kumar Sarma, Dean of Technology, Department of IT, Gauhati University
General Chair(s)	Prof. Haobam Mamata Devi, Department of Computer Science, Manipur University
	Prof. Dr. Siddhartha Bhattacharyya, VSB Technical University, Ostrava, Czech Republic and Algebra University College, Zagreb, Croatia
	Prof. Dr. Jyoti Sekhar Banerjee, Bengal Institute of Technology, Kolkata, India
General Co-Chair	Dr. Th. Tangkeshwar Singh, Department of Computer Science, Manipur University
Convenor	Tejmani Sinam, Department of Computer Science, Manipur University
Organising Chair	Dr. Khumukcham Robindro Singh, Head, Department of Computer Science, Manipur University
Organising Co-Chair	Dr. Nazrul Hoque, Department of Computer Science, Manipur University
TPC Chair(s)	Prof. Dr. Siddhartha Bhattacharyya, VSB Technical University of Ostrava, Ostrava, Czech Republic and Algebra University College, Zagreb, Croatia
	Prof. Dr. Jyoti Sekhar Banerjee, Bengal Institute of Technology, Kolkata, India
	Dr. Arnab Kumar Maji, Department of Information Technology, NEHU Shillong
TPC Co-Chair	Dr. Th. Rupachandra Singh, Department of Computer Science, Manipur University
Program Chair	Dr. Nazrul Hoque, Department of Computer Science, Manipur University
Track Chair(s)	Dr. Devarani Ningombam, Department of Computer Science & Engineering, NIT Patna
	Dr. Pangambam Sendash Singh, School of Computer Science, UPES, Dehradun
	Dr. T. Sonamani Singh, Department of Physics, NIT Trichy
	Dr. Y. Surajkanta Singh, Department of Computer Science & Engineering, NIT Manipur
	Dr. Ksh. Nareshkumar Singh, Department of Computer Science, Manipur University
Publication Chair(s)	Dr. Navanath Saharia, Department of Computer Science & Engineering, IIIT Manipur
	Dr. Sabyasachi Mondal, Department of Mathematics, NEHU Shillong
Publicity & Sponsorship Chair	Dr. T. Romen Singh, Department of Computer Science, Manipur University
Publicity & Sponsorship Co-Chair(s)	Ksh. Nilakanta Singh, Department of Computer Science, Manipur University
	Nongthombam Joychandra Singh, Department of Computer Science, Manipur University
	A. Dorendro Singh, Department of Computer Science, Manipur University
Web Chair	Dr. S. Somorjeet Singh, Department of Computer Science, Manipur University
Web Co-Chair(s)	U. Boby Clinton, Department of Computer Science, Manipur University
	Salam Vivek, Department of Computer Science, Manipur University
Registration Chair	G. Rajeev Sharma, Department of Computer Science, Dhanamanjuri University

Registration Co-Chair(s)	M. Surchand Singh, Department of Computer Science, Manipur University
	L. Sashikumar Singh, Department of Computer Science, Manipur University
	Linthoingambi Takhellambam, Department of Computer Science, Manipur University
Transportation & Accommodation Chair	Rajeev Rajkumar, Department of Computer Science & Engineering, MIT, Manipur University
Transportation & Accommodation Co-Chair(s)	Yambem Ranjan Singh, Department of Computer Science, Manipur University
	Chandam Chinglensana Singh, Department of Computer Science, Manipur University

Technical Program Committee

Sl. No.	Name	Affiliation
1	Ajoy Kumar Khan	Department of Computer Engineering, Mizoram University, India
2	Alessandro Bevilacqua	University of Bologna, Italy
3	Anandarup Mukherjee	University of Cambridge, United Kingdom
4	Anindita Ganguly	IISC Bangalore (Postdoc), Dept. of Computational Data Science, India
5	Anindya Halder	North-Eastern Hill University, Shillong, India
6	Aparajita Ojha	PDPM IIITDM Jabalpur, India
7	Apurba Sarkar	Dept. Of Computer Science & Technology, IIEST, Shibpur, India
8	Arindam Kar	Indian Statistical Institute, Kolkata, West Bengal, India
9	Asish Bera	Edge Hill University, St Helens Rd, Ormskirk L39 4QP, United Kingdom
10	Atanu Kundu	Heritage Inst. of Technology, Kolkata, West Bengal, India
11	Bibhash Sen	National Institute of Technology, Durgapur, West Bengal, India
12	Biswapati Jana	Dept. of Computer Science, Visyasagar University, India
13	Bubu Bhuyan	North-Eastern Hill University, Shillong, India
14	Chandan Giri	Department of Information Technology, IIEST Shibpur, India
15	Christian Kollmann	Medical University of Vienna, Austria
16	Claudio Zito	Technology Innovation Institute, UAE
17	Consuelo Gonzalo Martín	Universidad Politécnica de Madrid, Spain
18	Dakshina Ranjan Kisku	National Institute of Technology, Durgapur, West Bengal, India
19	Diego Alberto Oliva Navarro	Universidad de Guadalajara, Mexico
20	Ernestina Menasalvas	Universidad Politécnica de Madrid, Spain
21	Esssam H Houssein	Minia University, Egypt
22	Euisik Yoon	University of Michigan, USA
23	Gordon Chan	Department of Oncology, University of Alberta Edmonton, Alberta, Canada
24	Hiroaki Hanafusa	Hiroshima University, Japan
25	Ilora Maity	Aalto University, Helsinki, Finland
26	Imon Mukherjee	Indian Institute of Information Technology, Kalyani, West Bengal, India
27	Indrajit Bhattacharjee	Dept. of MCA, Kalyani Govt. Engg. College, Kalyani, India
28	Indrajit Ghosh	Dept. of Computer Science, Ananda Chandra College, Jalpaiguri, India
29	Ioannis Pratikakis	Democritus University of Thrace, Greece
30	Jin Hee Yoon	Sejong University, Seoul, South Korea
31	Joanna Jaworek-Korjakowska	AGH University of Science and Technology, Poland
32	João Luís Garcia Rosa	Department of Computer Science, University of Sao Paulo (USP), Brazil
33	João Manuel R. S. Tavares	Faculdade de Engenharia, Universidade do Porto (FEUP), Portugal
34	Juan D. Velasquez	University of Chile, Chile

Sl. No.	Name	Affiliation
35	Jugal Kalita	University of Colorado, USA
36	Khwairakpam Amitab	North-Eastern Hill University, Shillong, India
37	Malay Kule	Dept. of Computer Science & Technology, IIEST, Shibpur, India
38	Michał Jasiński	Wroclaw University of Science and Technology, Poland
39	Mrinal Kanti Bhowmik	Tripura University, Tripura, India
40	Nanda Dulal Jana	Department of Computer Science & Engineering, NIT, Durgapur, India
41	Naoto Hori	University of Texas, Austin, USA
42	Narendra D. Londhe	National Institute of Technology Raipur, India
43	Nishatul Majid	Fort Lewis College, USA
44	Oishila Bandyopadhyay	IIIT Kalyani, Kalyani, West Bengal, India
45	Ozan Keysan	Middle East Technical University, Turkey
46	Paramartha Dutta	Visva-Bharati University, Shantiniketan, West Bengal, India
47	Partha Sarathi Paul	Research Associate, (III), IIIT Delhi, India
48	Partha Sarathi Roy	Lecturer, School of Computing and Information Technology, University of Wollongong, Australia
49	Paulo Quaresma	The University of Évora, Portugal
50	Pawan Kumar Singh	Jadavpur University, India
51	Pradyut Sarkar	Department of Computer Science & Engineering, Maulana Abul Kalam Azad University of Technology, West Bengal, India
52	Rafael Kleiman	McMaster University, Canada
53	Rajat Kumar Pal	University of Calcutta, Kolkata
54	Ranjit Ghoshal	Dept. of Information Technology, St. Thoma's College of Engineering & Technology, Kolkata, India
55	Ratna Mandal	Dept. of Computer Application, IEM Kolkata, India
56	Robert A. Taylor	University of New South Wales, Australia
57	Sakurai Kouichi	Kyushu University, Japan
58	Samarjeet Borah	SMIT, Sikkim Manipal University, India
59	Samir Malakar	Asutosh College, Kolkata, India
60	Samuelson W. Hong	Oriental Institute of Technology, Taiwan
61	Sandhya Arora	MKSSS's Cummins College of Engineering for Women, Pune, India
62	Sandip Rakshit	American University of Nigeria, Nigeria
63	Santanu Das	Dept. of Electrical Engineering, Jalpaiguri Govt. Engg. College, Jalpaiguri, West Bengal, India
64	Santanu Phadikar	Department of Computer Science & Engineering, Maulana Abul Kalam Azad University of Technology, West Bengal, India
65	Santanu Sarkar	IIT Madras, India
66	Sema Candemir	The Ohio State University, USA
67	Serestina Viriri	University of KwaZulu-Natal, South Africa
68	Shibaprasad Sen	Future Institute of Engineering & Management, India
69	Somnath Mukhopadhyay	Assam University, Silchar

Sl. No.	Name	Affiliation
70	Soumen Bag	Indian Institute of Technology (Indian School of Mines), Dhanbad, India
71	Soumen Kumar Pati	Department of Bioinformatics, Maulana Abul Kalam Azad University of Technology, West Bengal, India
72	Soumya Pandit	University of Calcutta, West Bengal, India
73	Soumyabrata Dey	Clarkson University, New York, USA
74	Sourav De	Department of Computer Science & Engineering, Cooch Behar Government Engineering College, India
75	Subhas Barman	Dept. Computer Science & Engineering, Jalpaiguri Govt. Engg. College, India
76	Subhas Chandra Sahana	North-Eastern Hill University, Shillong, India
77	Sufal Das	North-Eastern Hill University, Shillong, India
78	Sujatha Krishamoorthy	Wenzhou-Kean University, China
79	Sung-Yun Park	University of Michigan, USA
80	Sungmin Eum	Booz Allen Hamilton/U.S. Army Research Laboratory, USA
81	Swarup Roy	Central University of Sikkim, Gangtok, India
82	Tamghana Ojha	National Research Council, Italy
83	Teresa Goncalves	University of Evora, Portugal
84	Tetsushi Koide	Hiroshima University, Japan
85	Tomas Klingström	Swedish University of Agricultural Sciences, Sweden
86	Vijay Mago	Lakehead University, Canada
87	Vijayalakshmi Saravanan	University at Buffalo, The State University of New York, USA
88	Dr. Takhellambam Sonamani Singh	Manipur University
89	Dr. Rajeev Rajkumar	Manipur University
90	Dr. Romesh Laishram	Manipur University
91	G. Rajeev Sharma	Dhanamajuri University
92	Dr. Usham Robinchandra Singh	SEMCO Komlathabi
93	Dr. Yumnam Surajkanta	NIT Manipur
94	Thokchom Aken Singh	Moirang College
95	Khumukcham Angamba Singh	Don Bosco College Maram
96	Ningmathan Jajo	Don Bosco College, Maram
97	Khaidem Anandkumar Singh	Thoubal College
98	Dr. Sunita Ningthoujam	Mangolnganbi College
99	Dr. Hussain Ahmed Chodhury	VIT Chennai
100	Dr. Kishore Medhi	North Eastern Hill University
101	Dr. Satyajit Sarma	Gauhati University
102	Dr. Abdul Hannan	Assam Skill University
103	Dr. Bornali Phukan	University of Illinois Urbana Champaign
104	Dr. Chayanika Deka Nath	University of Science & Technology, Meghalaya
105	Dr. Tasher Ali Ahmed	Golaghat Engineering College

Sl. No.	Name	Affiliation
106	Dr. Nabajyoti Sarmah	Gauhati University
107	Dr. Rita Chakraborty	Gauhati University
108	Dr. Chandan Kalita	Gauhati University
109	Dr. Sikdar MD Askari	Rajiv Gandhi University, Itanagar
110	Mr. Romesh Laishram	Manipur University
111	Mr. Rajeev Gurumayum	Dhanamanjuri University
112	Dr. Ravi Yadav	UPES Dehradun
113	Dr. Sounak Sadukhan	Bennet University
114	Dr. Ksh. Nilakanta Singh	Manipur University
115	Ahongsangbam Dorendro	Manipur University
116	Leishangthem Sashikumar Singh	Manipur University
117	Nongthombam Joychandra Singh	Manipur University
118	Mayanglambam Surchand	Manipur University
119	Ranjan Yambem	Manipur University
120	Urikhimbam Boby Clinton	Manipur University
121	Linthoingambi Takhellambam	Manipur University
122	Chandam Chinglensana Singh	Manipur University
123	Salam Vivek	Manipur University

Manipur University

Canchipur, Imphal-795003
Manipur, India

Vice-Chancellor

MESSAGE

I am happy to learn that the Department of Computer Science, Manipur University is organizing the North East India AI Summit (NEIAIS) 2025 under the sponsorship of North East Council (NEC) on Monday, the 10th February, 2025 at Manipur University.

In the era of digital transformation, AI is playing a crucial role in diverse applications such as Smart Cities, Healthcare, Agriculture, and Cyber security. By bringing together experts, researchers and students, this summit will serve as a platform for knowledge exchange and skill development, fostering AI-driven solutions for real-world challenges.

The Summit, thus, will be a significant initiative in advancing research, innovation, and collaboration in Artificial Intelligence and its applications across various domains. It will also make a meaningful impact on the technological landscape in the North East region and beyond.

I commend the Department for its dedication and commitment for organizing the Summit.

I wish the event a great success.

Prof. N. Lokendra Singh
Vice Chancellor, Manipur University

---

1 Leveraging machine learning techniques to analyse soil organic carbon dynamics in the Eastern Himalayan Region

Puja Saha[1,a] and Amitabha Nath[2,b]

[1]Research Scholar, Department of Information Technology, North Eastern Hill University, Shillong, India

[2]Assistant Professor, Department of Information Technology, North Eastern Hill University, Shillong, India

Abstract

Soil organic carbon (SOC) is an indispensable parameter of soil. It is a primary indicator of soil fertility and health. It provides the necessary support for a plant to grow and thrive. Therefore, understanding its spatial distribution and controlling factors is crucial for soil management practices. In this study, ensemble machine learning techniques—Extreme Gradient Boost (XGB), Random Forest (RF), and Adaptive Boost (AdaBoost)—were implemented and tested for estimating SOC content. Multiple parameters, including bulk density (BD), above-ground biomass (AGB), below-ground biomass (BGB), digital elevation model (DEM), sand, clay, NDVI, moisture, temperature, rainfall, and pH, were used as independent parameters. AdaBoost emerged as the best-performing model, with considerable R-square and RMSE values of 0.873 and 0.047, respectively. Sensitivity analysis suggests BD, NDVI, moisture, AGB, DEM, and temperature as the most influential attributes in SOC prediction.

Keywords: Soil fertility, soil organic carbon, ensemble model, Eastern Himalayan Region, remote sensing

Introduction

Soil organic carbon (SOC) is a crucial component of soil. Its management is necessary for maintaining soil fertility and boosting crop productivity [10]. Soil acts as a reservoir for organic carbon and therefore plays an important role in maintaining the global carbon budget [30]. During photosynthesis, the plant absorbs CO_2 from the environment and converts it into organic matter. When a plant dies and decomposes, a portion of this organic matter gets transformed and retained as SOC in the soil. This makes SOC a crucial component in maintaining the global carbon balance and a catalyst for mitigating climate change [23]. Besides this, SOC content regulates soil structure, soil fertility, water retention, and the welfare of the ecosystem [22]. Studies have shown that it not only influences soil fertility and crop productivity but also helps with SOC sequestration [23]. The Eastern Himalayan Region (EHR) is an ecologically rich and biodiverse region. The livelihood of most people in the region depends on agriculture [5]. We have chosen the EHR for our study to ensure sustainable development by monitoring and managing its SOC content. We intend to identify and quantify the influence of various climatic and soil parameters responsible for determining SOC distribution through a machine learning-based modelling technique. Manual interpretation of SOC distribution is labour-intensive and time-consuming. However, advancements in the fields of remote sensing and machine learning algorithms have simplified this process. These methods enable us to study and analyse large

areas with relative ease and better accuracy. This article presents a case study of the EHR, renowned for its diverse agricultural practices. However, this region faces challenges due to soil degradation, land use change, or climate change. To address this issue, ensemble machine learning techniques with remote sensing data were employed for SOC prediction and identification of the most influential attributes. The implications of this work are far-reaching, with significant impacts on crop production, climate change mitigation, and sustainable development. The findings of this study are valuable for improving agricultural productivity and guiding climate change mitigation efforts by offering valuable insights into soil health and carbon sequestration potential. This article is structured into six different sections. The Introduction and the Literature Review sections provide background to the problem. The Materials and Methods section narrates the acquisition of data and the methodologies employed in the study. The Results section discusses the outcomes of the applied algorithms, while the Discussion section offers a concise scrutiny of these results. In the end, the Conclusion section provides an overall summary of the work.

Literature review

Recognizing the role of SOC in biodiversity preservation, soil fertility, climate change mitigation, and sustainable development, researchers have shown growing interest in this field. SOC monitoring plays a crucial role in food security and environmental sustainability [27]. Remote

---

[a]pooja12saha@gmail.com; [b]amitabha.me@gmail.com

DOI: 10.1201/9781003675235-1

sensing data, along with machine learning techniques, are being primarily used in modelling SOC. These models have proven to be crucial in understanding the complex relationship of SOC with other environmental and soil attributes [30]. Heil et al. predicted SOM from the top layer of soil in cropland using Random Forest [16]. Similarly, Pavlovic et al. used different deep learning techniques for prediction [28]. Neural networks, linear regression, support vector machines, decision trees, and Random Forest are some of the commonly used machine learning techniques employed in this field [1,23,33]. So far, except for Random Forest (RF), we could not find references to any other ensemble learning methods used in estimating SOC. This is encouraging, given the fact that ensemble models have performed significantly well

in other allied fields [31]. This gives us the necessary motivation to explore and analyse alternative ensemble techniques like Gradient Boosting, AdaBoost, or stacking models to improve prediction accuracy. This study presents three popular ensemble learning techniques (XGB, Random Forest, and AdaBoost) for modelling SOC based on climatic data (rainfall, temperature), soil attributes (above-ground biomass, below-ground biomass, bulk density, clay, sand, moisture, pH), vegetation (Normalised Difference Vegetation Index), and elevation data. The choice of these parameters is mostly influenced by their importance as reported in the literature [23,30].

Materials and methods

The outline of the paper is presented with the data flow diagram (Figure 1.1). The first step is data collection, followed by data pre-processing (combining data and removing missing values). After that, the data is split into a 70:30 ratio.

Then, use 70% data to train the models. Finally, use 30% data to test the model's performance.

Study site

The Eastern Himalayan Region consists of eight states: Arunachal Pradesh, Tripura, Meghalaya, Mizoram, Nagaland, Manipur, Assam, and Sikkim (Figure 1.2) [9]. It covers approximately 273,326 km², accounting for about 8.3% of India's total area [8,9]. This region features an ecologically diverse and fragile ecosystem. The subtropical climate makes it favourable for many plant and animal species. The region contains rich, fertile soil due

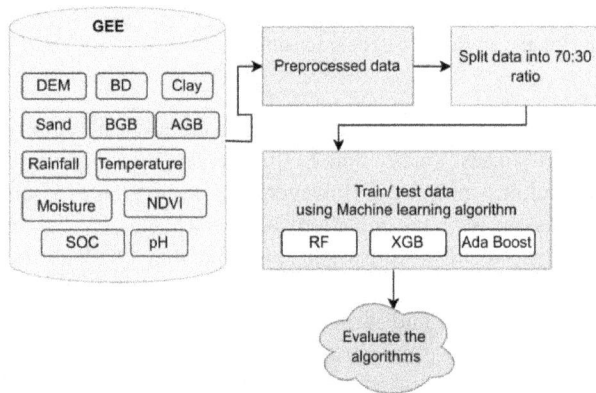

Figure 1.1 Data flow diagram
Source: Author

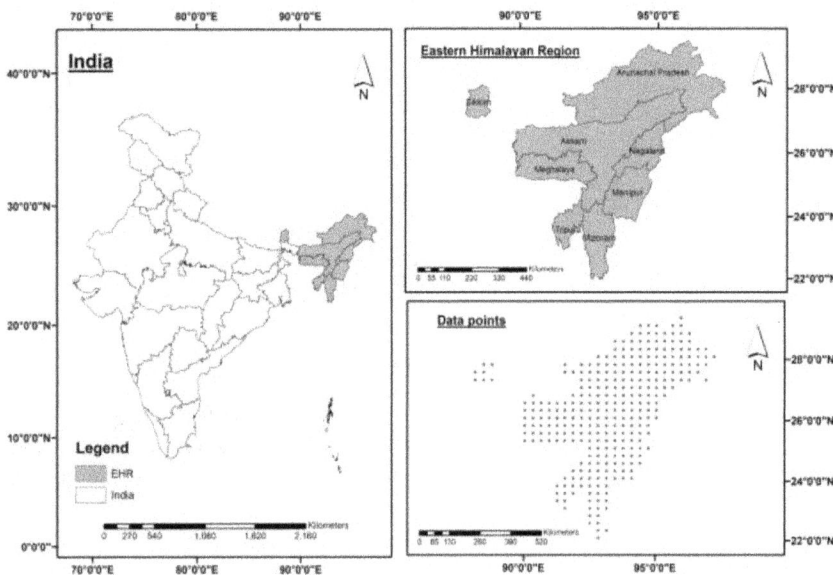

Figure 1.2 Study area map
Source: Author

Table 1.1 Description of different parameters used in the study.

Data	Standard deviation	Unit
Above ground biomass (AGB) [32]	36.297	M g/ha
Bellow ground biomass (BGB) [32]	1.484	M g/ha
Soil organic carbon (SOC) [21]	5.362	g/kg
Sand [17]	5.496	%(kg/kg)
Clay [18]	5.387	%(kg/kg)
Bulk density (BD) [19]	20.212	kg/m^3
Temperature 29]	5.630	K
Rainfall [13]	11.120	mm
Moisture [11]	2.557	mm
pH [20]	7.120	
Digital elevation model (DEM) [4]	82.932	m
Normalised Difference Vegetation Index (NDVI) [26]	0.199	

Source: Author

to dense forests, continuous litter production, and a high rainfall pattern [3]. However, the growing population and deforestation are some issues that make these rich habitats vulnerable [8].

Dataset acquisition
Google Earth Engine (GEE) is an online cloud-based data warehouse. It provides multiple functionalities such as cloud masking, mosaic raster, clip raster, and many more [14,15]. All the datasets used in this study were collected using the GEE platform. The description of the datasets is presented in Table 1.1. Subsequently, 300 data points were generated using the Fishnet tool in ArcGIS 10.3 software (Figure 1.2). Using these points, multiple temporal information were extracted from multiple raster datasets using ArcGIS. The data were merged into a single CSV file without any missing values. The dataset was then split into a 70:30 ratio, with 70% data used for training the models and the remaining 30% used for testing.

Methods
- Extreme Gradient Boost (XGB): It is a sequential ensemble learning technique where multiple decision trees are created in sequence. Each tree tries to reduce the errors made by the previous ones [7]. It employs a gradient descent algorithm to mitigate the loss function.
- Random Forest (RF): It is a parallel ensemble learning technique. RF randomly constructs multiple decision trees at the same time. The final model output is a weighted average of all individual tree outputs [6].

- Adaptive Boost (AdaBoost): It is also known as AdaBoost. It is a sequential prediction technique in which weak learners are given more weightage compared to strong learners [12]. This process continues and during each iteration, it corrects misclassified instances.

Results

SOC is a strong indicator of soil health. Through this analysis, we aim to analyse the SOC distribution and the most influential factors responsible for determining the SOC content.

Multiple parameters, viz. BD, AGB, BGB, DEM, sand, clay, NDVI, moisture, rainfall, temperature, pH, and elevation, are used as descriptive attributes. Three ML models (XGB, RF, and AdaBoost) are used to predict the relationship between the explanatory variables and the SOC. The performance of these models is shown in the form of an observed versus predicted graph (Figure 1.3). The red diagonal line in the plot represents the best-fit line. The x- and y-axes represent the observed and predicted values, respectively. It is evident from the figure that AdaBoost performed better compared to all other models in terms of R-square and RMSE value. A residual plot is also presented in Figure 1.4 to provide insight into the relative prediction error of all three models.

Additionally, to understand the contribution of each variable in SOC prediction, we have plotted the feature sensitivity graph, as shown in Figure 1.5. The key observation that we can make from the sensitivity graph is that

BD, moisture, and NDVI emerge as the most dominant factors in determining SOC content.

Discussion

Model performance

The performance of the predictive models—XGB, RF, and AdaBoost—was evaluated using R-squared and RMSE values (Figure 1.3). It can be seen from the figure

that AdaBoost's performance is the best, with an R-square and RMSE values of 0.873 and 0.047, respectively.

In contrast, the XGB model performed the worst, with R-square and RMSE values of 0.775 and 0.063, respectively. However, RF showed average performance with an R-square value of 0.801. The residual plot (Figure 1.4) also supports AdaBoost's better performance, with the lowest MSE value of 0.002. It can also be observed that the XGB model (blue dots) deviated from the zero residual line, while AdaBoost's (red dot) prediction points are much closer to the zero residual line, indicating lower prediction error. Furthermore, the Taylor graph (Figure 1.6) shows the centered root mean square error (CRMSE), standard deviation, and correlation coefficient of all the three models. AdaBoost's closeness to the x-axis indicates a high positive correlation, which means it can closely follow the actual data. Similar to Matinfar et al., our findings also suggest that two-step machine learning algorithms are more suitable for SOC prediction [25].

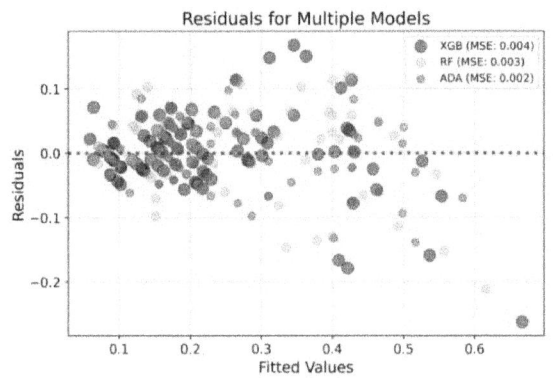

Figure 1.4 Sensitivity analysis plot showing the impact of different input parameters
Source: Author

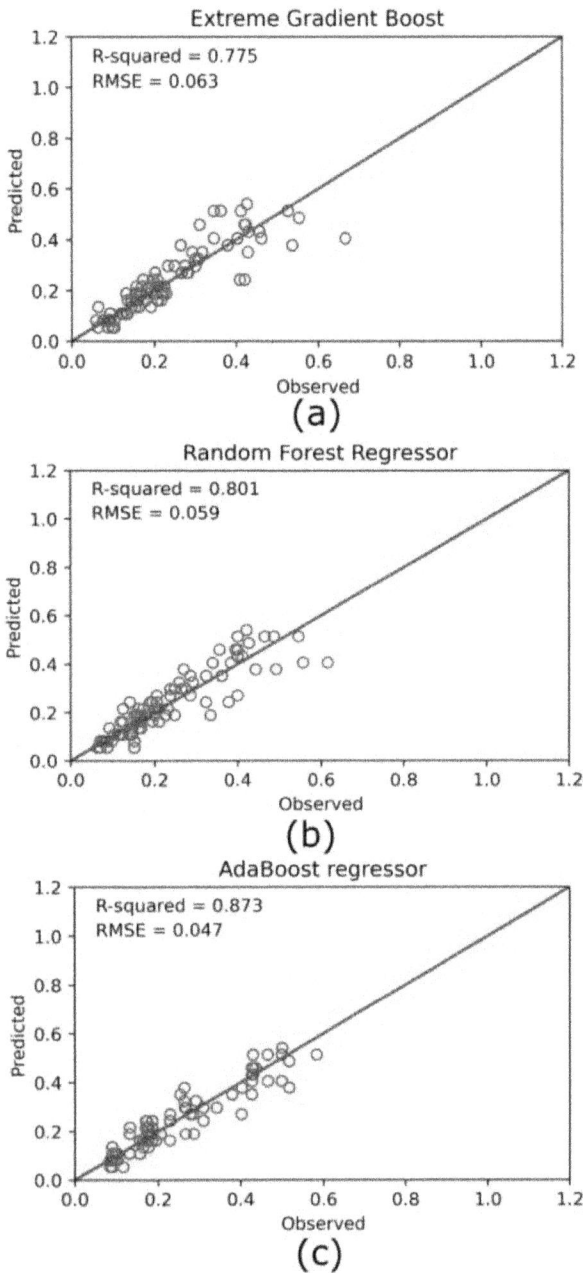

Figure 1.3 Observed vs predicted graph for XGB, RF and AdaBoost model
Source: Author

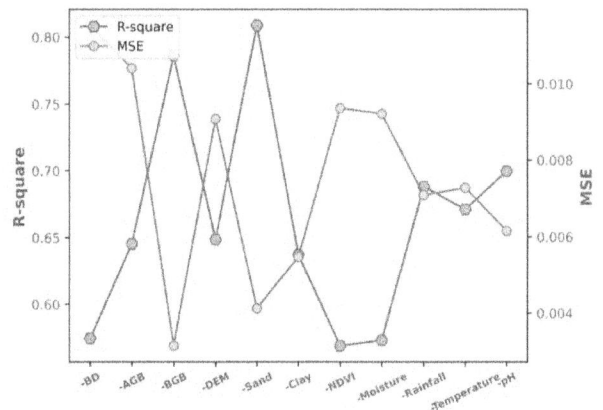

Figure 1.5 Taylor plot illustrating the performance of different models
Source: Author

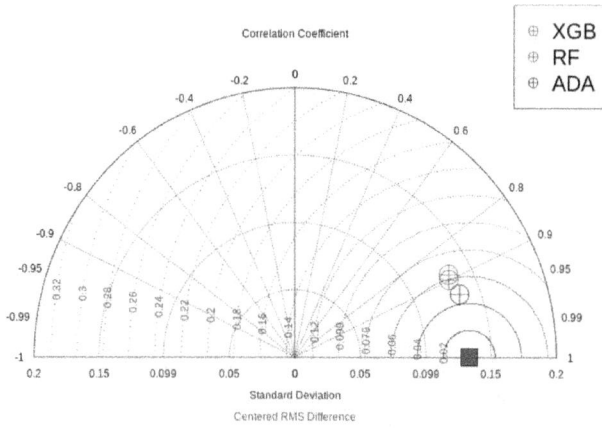

Figure 1.6 Taylor plot illustrating the performance of different models
Source: Author

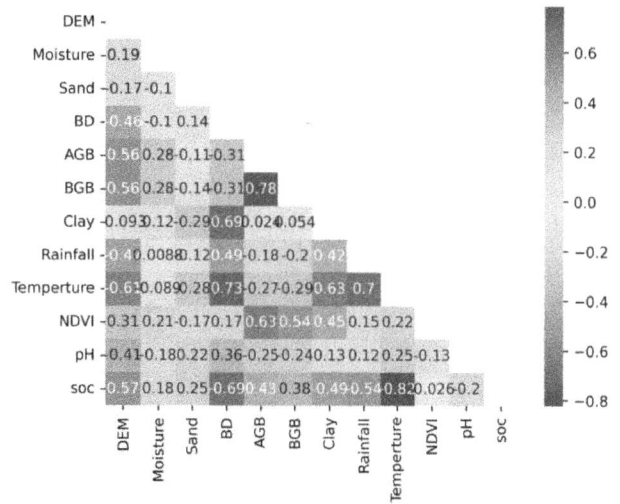

Figure 1.7 Pearson's correlation for different parameter
Source: Author

Parameter sensitivity analysis

A sensitivity test was conducted to understand the relative importance of each descriptive attribute on the SOC value (Figure 1.5). The test involves sequentially excluding variables and observing the impact on the R-square and RMSE values. The graph shows that the exclusion of BD, NDVI, moisture, DEM, AGB, temperature, rainfall, pH, and BGB significantly impacts model performance. These findings are congruous with those of Adhikari et al. [2], Mahmoudzadeh et al. [24], and Matinfar et al. [25], where BD, soil moisture, and NDVI were reported to be the primary contributing factors in SOC prediction. Pearson's correlation map (Figure 1.7) also suggests a positive correlation of SOC with DEM, moisture, NDVI, clay, BGB, AGB, and sand variables, thereby revalidating the model findings. Both outcomes are consistent with [3], suggesting AGB, temperature, and rainfall make a substantial contribution to the SOC management. From the discussion, it is apparent that soil properties, vegetation cover, and climatic factors play a strong role in determining the SOC content for a region. Predictive models of SOC can play an important role in managing agriculture, ecology, and mitigating climate change. This observation can be useful for environmentalists or climatologists in better policy-making concerning sustainable development and food security.

Conclusion

SOC is an essential component of soil and plays a crucial role in nutrient cycling and crop production. This study emphasised on the inclusion of ensemble-based machine learning models, with vegetation index, elevation, and soil properties, for accurate SOC estimation.

Among all predictor variables, BD, DEM, moisture, NDVI, clay, BGB, AGB, temperature, pH and sand were found to be the most significant attributes for determining SOC distribution of the region. Findings suggest that machine learning techniques combine with remote sensing information can form a convenient model for precise SOC prediction. The findings of this study have significant implications for global SOC management and sustainable development. Further research can introduce refine models by incorporating additional soil properties and land use pattern to enhance global SOC monitoring and carbon sequestration strategies.

References

[1]   Adeniyi, O. D., Brenning, A., Maerker, M. (2024). Spatial prediction of soil organic carbon: combining machine learning with residual kriging in an agricultural lowland area (Lombardy region, Italy). *Geoderma*, 448, 116953. https://doi.org/10.1016/j.geoderma.2024.116953

[2]   Adhikari, K., Hartemink, A. E., Minasny, B., Bou Kheir, R., Greve, M. B., Greve, M. H. (2014). Digital mapping of soil organic carbon contents and stocks in Denmark. *PLoS One*, 9, e105519. https://doi.org/10.1371/journal.pone.0105519

[3]   Ahirwal, J., Saha, P., Nath, A., Nath, A. J., Deb, S., Sahoo, U. K. (2021). Forests litter dynamics and environmental patterns in the Indian Himalayan region. *Forest Ecology and Management*. https://doi.org/10.1016/j.foreco.2021.119612

[4]   Airbus Defence and Space GmbH, G. (n.d.). Copernicus DEM GLO-30: Global 30m Digital Elevation Model [WWW Document]. https://developers.google.com/earth-engine/datasets/catalog/COPERNICUS_DEM_GLO30 (accessed 1.3.24).

[5]   Bhadwal, S., Sharma, G., Gorti, G., Sen, S. M. (2019). Livelihoods, gender and climate change in the Eastern Himalayas. *Environmental Development*, 31, 68–77. https://doi.org/10.1016/j.envdev.2019.04.008

[6]   Breiman, L. (2001). Random Forests. Random Forests, 1–122. *Machine Learning*, 45, 5–32.

[7]   Chen, T., Guestrin, C. (2016). XGBoost: a scalable tree boosting system. In Proceedings of the ACM SIGKDD International Conference on Knowledge Discovery and Data Mining, 785–794. https://doi.org/10.1145/2939672.2939785

[8]   Das, M., Das, A., Pandey, R. (2023a). A socio-ecological and geospatial approach for evaluation of ecosystem services to communities of the Eastern Himalayan Region, India. *Environmental Science and Pollution Research*, 30, 116860–116875. https://doi.org/10.1007/s11356-023-25746-7

[9]   Das, M., Mandal, A., Das, A., Inácio, M., Pereira, P. (2023b). Mapping and assessment of carbon sequestration potential and its drivers in the Eastern Himalayan Region (India). *Case Studies in Chemical and Environmental Engineering*, 7, 100344. https://doi.org/10.1016/j.cscee.2023.100344

[10]   Dinesh, G. K., Sinduja, M., Priyanka, B., Sathya, V., Karthika, S., Meena, R. S., Prasad, S. (2022). Enhancing soil organic carbon sequestration in agriculture: plans and policies. In Plans and Policies for Soil Organic Carbon Management in Agriculture, 95–121. https://doi.org/10.1007/978-981-19-6179-3_4

[11]   Entekhabi, D., Njoku, E., O'Neill, P., Kellogg, K., Crow, W., Edelstein, W., Entin, J., Goodman, S., Jackson, T., Johnson, J., Kimball, J., Piepmeier, J., Koster, R., Martin, N., McDonald, K. C., Moghaddam, M., Moran, S., Reichle, R., Shi, J., Spencer, M. W., Thurman, J. (2010). The soil moisture active passive (SMAP) mission. *Proceedings of the IEEE*, 98, 704–716. https://doi.org/doi:10.1109/JPROC.2010.2043918Article

[12]   Freund, Y., Schapire, R. E. (1995). A decision-theoretic generalization of on-line learning and an application to boosting. In Lecture Notes in Computer Science (Including Subseries Lecture Notes in Artificial Intelligence and Lecture Notes in Bioinformatics). Elsevier, 23–37. https://doi.org/10.1007/3-540-59119-2_166

[13]   Funk, C., Peterson, P., Landsfeld, M., Pedreros, D., Verdin, J., Shukla, S., Husak, G., Rowland, J., Harrison, L., Hoell, A., Michaelsen, J. (2015). The climate hazards infrared precipitation with stations - a new environmental record for monitoring extremes. *Scientific Data*, 2, 1–21. https://doi.org/10.1038/sdata.2015.66

[14]   Gorelick, N. (2013). Google earth engine. In EGU General Assembly Conference Abstracts. American Geophysical Union Vienna, Austria, 11997.

[15]   Gorelick, N., Hancher, M., Dixon, M., Ilyushchenko, S., Thau, D., Moore, R. (2017). Google Earth Engine: planetary-scale geospatial analysis for everyone. *Remote Sensing of Environment*, 202, 18–27. https://doi.org/10.1016/j.rse.2017.06.031

[16]   Heil, J., Jörges, C., Stumpe, B. (2022). Fine-scale mapping of soil organic matter in agricultural soils using UAVs and machine learning. *Remote Sensing*, 14, 3349. https://doi.org/10.3390/rs14143349

[17]   Hengl, T. (2018a). Sand Content in % (kg/kg) at 6 Standard Depths (0, 10, 30, 60, 100 and 200 cm) at 250 m Resolution (Version v0.2) [WWW Document]. Zenodo. https://developers.google.com/earth-engine/datasets/catalog/OpenLandMap_SOL_SOL_SAND-WFRACTION_USDA-3A1A1A_M_v02#citations (accessed 2.19.24).

[18]   Hengl, T. (2018b). Clay Content in % (kg/kg) at 6 Standard Depths (0, 10, 30, 60, 100 and 200 cm) at 250 m Resolution (Version v02) [Data set] [WWW Document]. https://doi.org/10.5281/zenodo.1476854

[19]   Hengl, T. (2018c). Soil Bulk Density (Fine Earth) 10 × kg m-cubic at 6 Standard Depths (0, 10, 30, 60, 100 and 200 cm) at 250 m Resolution (Version v02) [data set] [WWW Document]. Zenodo. https//doi.org/10.5281/zenodo.1475970 (accessed 2.11.24).

[20]   Hengl, T. (2018d). Soil pH in H2O at 6 Standard Depths (0, 10, 30, 60, 100 and 200 cm) at 250 m Resolution (Version v02) [WWW Document]. Zenodo. https://doi.org/10.5281/zenodo.1475459

[21]   Hengl, T., Wheeler, I. (2018). Soil Organic Carbon Content in × 5 g/kg at 6 Standard Depths (0, 10, 30, 60, 100 and 200 cm) at 250 m Resolution. Zenodo.

[22]   Koorneef, G. J., Pulleman, M. M., Comans, R. N., van Rijssel, S. Q., Barré, P., Baudin, F., de Goede, R. G. (2024). Assessing soil functioning: what is the added value of soil organic carbon quality measurements alongside total organic carbon content? *Soil Biology and Biochemistry*, 196, 109507. https://doi.org/10.1016/j.soilbio.2024.109507

[23]   Li, T., Cui, L., Wu, Y., McLaren, Timothy I. Xia, A., Pandey, R., Liu, H., Wang, W., Xu, Z., Song, X., Dalal, R. C., Dang, Y. P. (2024). Soil organic carbon estimation via remote sensing and machine learning techniques: global topic modeling and research trend exploration. *Remote Sensing*, 16, 3168. https://doi.org/https://doi.org/10.3390/rs16173168

[24]   Mahmoudzadeh, H., Matinfar, H. R., Taghizadeh-Mehrjardi, R., Kerry, R. (2020). Spatial prediction of soil organic carbon using machine learning techniques in western Iran. *Geoderma Regional*, 21, e00260. https://doi.org/10.1016/j.geodrs.2020.e00260

[25]   Matinfar, H. R., Maghsodi, Z., Mousavi, S. R., Rahmani, A. (2021). Evaluation and prediction of topsoil organic carbon using machine learning and hybrid models at a field-scale. *Catena*, 202, 105258. https://doi.org/10.1016/j.catena.2021.105258

[26]   MODIS (n.d.). Modis Combined 16-day ndvi [WWW Document]. Google. https://developers.google.com/earth-engine/datasets/catalog/MODIS_MCD43A4_006_NDVI#bands (accessed 2.4.24).

[27]   Neofytou, E., Neophytides, S. P., Eliades, M., Papoutsa, C., Tzouvaras, M., Hadjimitsis, D. G. (2024). A review of soil organic carbon (SOC) prediction techniques in agricultural lands using remote sensing. In IGARSS 2024-2024 IEEE International Geoscience and Remote Sensing Symposium. IEEE, 1273–1279. https://doi.org/10.1109/igarss53475.2024.10641028

[28]   Pavlovic, M., Ilic, S., Ralevic, N., Antonic, N., Raffa, D. W., Bandecchi, M., Culibrk, D. (2024). A deep learning approach to estimate soil organic carbon from remote sensing. *Remote Sensing*, 16, 655. https://doi.org/10.3390/rs16040655

[29]   Sabater, J. M. (n.d.). ERA5-Land Monthly Averaged Data From 1981 to Present. Copernicus Climate Change

Service (C3S) Climate Data Store (CDS) [WWW Document]. 2019. https://doi.org/https://doi.org/10.24381/cds.68d2bb30

[30] Segura, C., Neal, A., Castro-Sardiňa, L., Harris, P., Rivero, M., Cardenas, L., Irisarri, J. (2024). Comparison of direct and indirect soil organic carbon prediction at farm field scale. *Journal of Environmental Management*, 365, 121573. https://doi.org/https://doi.org/10.1016/j.jenvman.2024.121573

[31] Shahri, N. H. N. B. M., Lai, S. B. S., Mohamad, M. B., Rahman, H. A. B. A., Rambli, A. Bin. (2021). Comparing the performance of AdaBoost, XGboost, and logistic regression for imbalanced data. *Mathematical Statistics*, 9, 379–385. https://doi.org/10.13189/ms.2021.090320

[32] Spawn, S. A., Sullivan, C. C., Lark, T. J., Gibbs, H. K. (2020). Harmonized global maps of above and belowground biomass carbon density in the year 2010. *Scientific Data*, 7, 112. https://doi.org/10.1038/s41597-020-0444-4

[33] Wang, B., Waters, C., Orgill, S., Gray, J., Cowie, A., Clark, A., Liu, D. L. (2018). High resolution mapping of soil organic carbon stocks using remote sensing variables in the semi-arid rangelands of eastern Australia. *Science of the Total Environment*, 630, 367–378. https://doi.org/10.1016/j.scitotenv.2018.02.204

2 Investigation into a suitable voice pathology detection and classification using glottal inverse filtering

Ranita Khumukcham[1,a] and Kishorjit Nongmeikapam[2,b]

[1]Department of Electronics and Communication Engineering, Indian Institute of Information Technology, Manipur, India

[2]Department of Computer Science and Engineering, Indian Institute of Information Technology, Manipur, India

Abstract

The paper presents a methods on the pathological voice detection and classification using glottal inverse filtering. It is based on the glottal inverse filtering with extended Kalman filter and glottal inverse filtering with quasi-closed phase. It classifies the database into healthy and pathological categories. With the help of glottal inverse filtering, the effects of vocal tract and lip radiation are cancelled from the output of the speech signal. The method of glottal inverse filtering with the extended Kalman filter provides better results compared to glottal inverse filtering with the quasi-closed phase. Mel-frequency Cepstral Coefficients (MFCCs) and pitch characteristics are extracted from the input signal. Support Vector Machine (SVM) and k-nearest neighbor (kNN) are the two machine learning classifiers used for classifying voice signal into healthy and pathological categories with high accuracy.

Keywords: Glottal inverse filtering, vocal tract, Kalman filter, MFCC, SVM, kNN

Introduction

Voice pathology classification is mostly done by machine learning techniques in classifying voice signal into binary and multiclass classification [1,2]. Machine learning techniques are applied in most of the medical applications [3]. Detection of voice pathology is also one of the applications of these techniques. A voice that causes discomfort or fatigue while producing sound is called voice pathology. Researchers have found the voice pathology classification in many different ways. Traditional classification models, such as Support Vector Machine (SVM), k-Nearest Neighbor (kNN), Naïve Bayes, Decision trees, and Discriminating Analysis (DA), have been widely used by many researchers. However, classification using filtering methods, such as glottal inverse filtering with extended Kalman filter (GIFEKF) based on first-order derivative waveforms, is hardly used. Glottal inverse filtering is used by many researchers due to its functioning such as speech recognition [4], speech synthesis [5], emotion and stress analysis [6,7], and speaker identification [8]. Glottal inverse filtering (GIF) is a method of estimating the input voice signal, the glottal excitation, when the output is known. It investigates the functioning of the human speech production mechanism. Three mechanisms that are used in speech producing in human voice are the energy source, the sound source, and the filter. The source of energy is provided by the lungs. The vocal fold within the larynx is the source of sound. And the filter is done by vocal tract. The glottal excitation is the excitation of human voice speech which match with the air flow from

the lungs. The rate at which the vocal folds open and close is determined, to a large extent, by subglottal pressure. Finally, the glottal excitation is filtered by the vocal tract. According to Paavo Alku [9], GIF is mostly used in the analysis of voice production because it is a non-invasive method. Analysis can be conducted with only one input (the speech recorded in a microphone in a normal environment), it is at least in principle possible to implement both the inverse filtering and the parameterisation stage in a completely automatic manner, and lastly, the output of GIF can directly be used in speech production technology. The technique described in Refs. [10,11] is based on the linear prediction (LP) technique, in which the values of the output signal are reproduced from the past samples, except at the place of input pulses. Iterative adaptive inverse filtering, as discussed in Ref. [12], is based on linear prediction analysis. In this method, glottal flow and vocal tract contributions are eliminated in an iterative manner to get the estimated glottal signal. GIFEKF, introduced in Ref. [13], estimates glottal air-volume-velocity as well as intermediate pressure values within the vocal tract using the EKF. This method calculates the glottal signal which is more applicable to high pitch speech signal when vowel is spoken. Glottal source feature is studied in Ref. [14], where it is extracted using glottal flows estimated with the quasi-closed phase (QCP) glottal inverse filtering method, using approximate glottal source signals computed with the zero-frequency filtering (ZFF) method, and using acoustic voice signals directly. In addition, they propose to derive Mel-frequency Cepstral Coefficients (MFCCs) from the glottal source waveforms computed by QCP and

---

[a]ranitakh89@gmail.com, [b]kishorjit@iiitmanipur.ac.in

DOI: 10.1201/9781003675235-2

ZFF to effectively capture the variations in glottal source spectra of pathological voices.

Method

The block diagram of the proposed method is shown in Figure 2.1. The voice is collected at the initial stage of the method. The voice is used from the available dataset in Ref. [15]. The dataset is used as an input of the proposed method. Glottal inverse filtering with extended Kalman filter (GIFEKF) and glottal inverse filtering with quasi-closed phase (GIFQCP) are the two filtration methods used here. The Mel-frequency Cepstral Coefficients (MFCC) and pitch characteristics are extracted from the input signal which is used in training. Then, the final stage involves classification using machine learning techniques, such as Support Vector Machine (SVM) and k-Nearest Neighbor (kNN). It classifies the voice pathology into healthy and pathological categories.

Dataset

The dataset contains 151 diseased and 55 healthy speech samples, respectively. Each sample has a length of 4.76 seconds with the vowel sound 'a' at an 8 kHz sampling rate. The samples are segmented into 10 equal lengths of 0.476 seconds and 3808 sampling points. After segmentation, 550 and 1510 of samples are achieved for healthy and pathological data, respectively. Since the segmented number of healthy class is 550, in order to keep balance number for both the classes, 550 segmented number is selected from both healthy and pathological classes. To improve accuracy, the training and testing samples are divided into 80:20 ratio for both classes which results in the formation of 440 training and 110 testing samples, respectively.

Filtering method

Two techniques of filtering are used for the proposed method, such as glottal inverse filtering with extended Kalman Filter (GIFEKF) [13] and glottal inverse filtering with quasi-closed phase (GIFQCP) [14]. GIFEKF is employed in this method because it increased the pitch level and simultaneously reduced signal loss. So, in this

work, the glottal flow derivative from the input signal is used. GIFQCP is used for the analysis of closed phase which is the estimation of vocal tract during the closed phase of glottal flow. It also gives more importance on the samples which are located in the closed phase.

Experimental and evaluation results

Evaluation metrics

The proposed work is evaluated using the confusion metrics of accuracy. The confusion metrics is essential for classification model performance. Here, TP, TN, FP, and FN stand for true positive, true negative, false positive, and false negative, respectively. When there is an uneven class distribution in a dataset, this matrix is especially helpful in evaluating a model's performance beyond basic

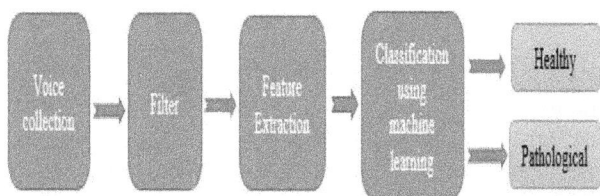

	Healthy	Pathological		
Healthy	106	4	96.36 %	3.63 %
Pathological	9	101	91.81 %	8.18 %
	92.17%	96.19%		
	7.82%	3.77%		

Figure 2.2 Confusion matrix showing the per result with kNN method using GIFEKF

Source: Author

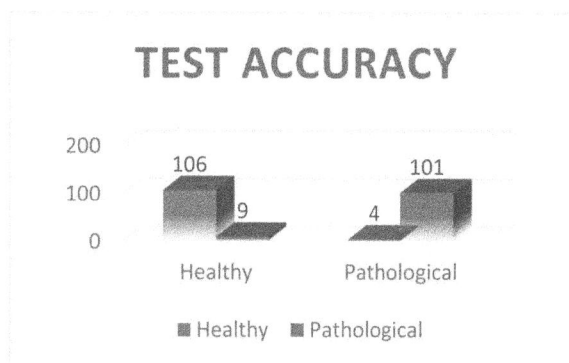

Figure 2.3 Training progress of kNN with GIFEKF

Source: Author

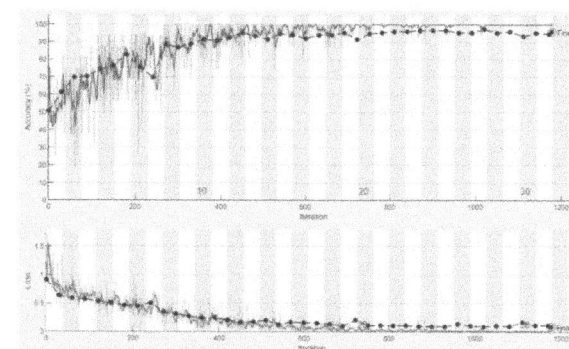

Figure 2.1 Block diagram of the proposed method

Source: Author

	Healthy	Pathological		
Healthy	108	2	98.18 %	1.8 %
Pathological	8	102	92.72 %	7.27 %
	93.10%	98.07%		
	6.89%	1.92%		

Figure 2.4 Confusion matrix showing the per class result with SVM method using GIFEKF
Source: Author

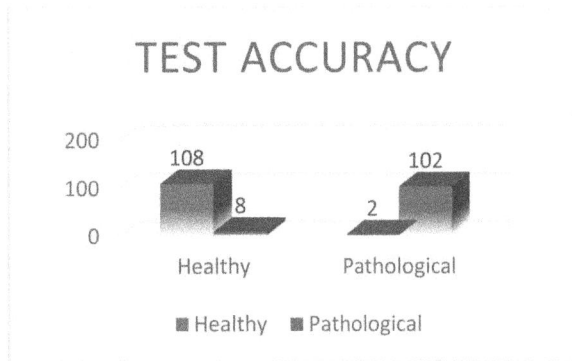

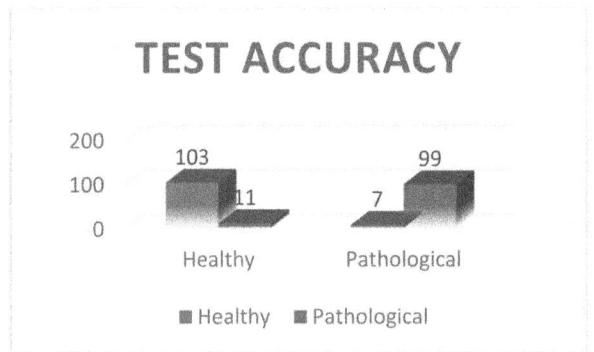

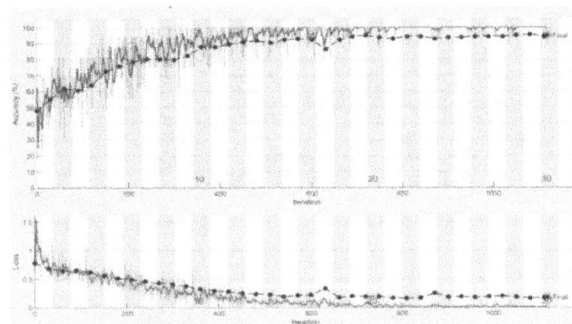

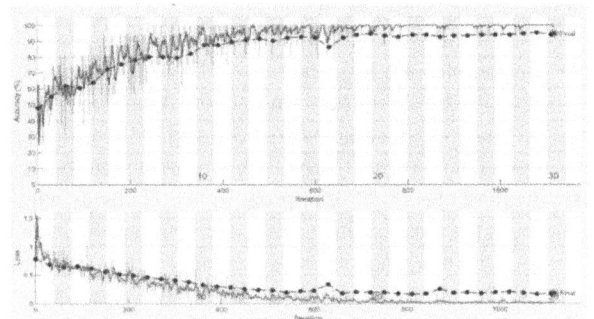

	Healthy	Pathological		
Healthy	103	7	96.63 %	6.36 %
Pathological	11	99	90.00 %	10.00 %
	90.35%	93.39%		
	9.64%	6.61%		

Figure 2.6 Confusion matrix showing the per class result with kNN method using GIFQCP
Source: Author

Figure 2.5 Training progress of SVM with GIFEKF
Source: Author

Figure 2.7 Training progress of kNN with GIFQCP
Source: Author

	Healthy	Pathological		
Healthy	102	8	92.72 %	7.27 %
Pathological	9	101	91.81 %	8.19 %
	92.72%	92.66%		
	7.28%	7.34%		

Figure 2.8 Confusion matrix showing the per class result with SVM method using GIFQCP
Source: Author

accuracy metrics. Accuracy (Acc) is used to measure performance, which is calculated as:

$$Acc = \frac{TP + TN}{TP + TN + FP + FN}$$

The performance of classifying the dataset using the kNN method with GIFEKF is shown in Figure 2.2 through the confusion matrix. It classifies the 110 samples into 106 healthy and 101 pathological, with an accuracy of 96.36% healthy and 91.81% pathological using GIFEKF. The training progress of kNN with GIFEKF is shown in Figure 2.3.

Figure 2.4 shows the performance of classifying the dataset using the SVM method with GIFEKF, through the confusion matrix. It classifies the 110 samples into 108 healthy and 102 pathological, with an accuracy of 98.18%

healthy and 92.72% pathological using GIFEKF. The training progress of SVM method with GIFEKF is shown in Figure 2.5.

The performance of classifying the dataset using kNN method with GIFQCP is shown in Figure 2.6 through the confusion matrix. It classifies the 110 samples into 103 healthy and 99 pathological, with an accuracy of 93.63%

healthy and 90% pathological using GIFQCP. The training progress of kNN with GIFQCP is shown in Figure 2.7.

The performance of classifying the dataset using SVM method with GIFQCP is shown in Figure 2.8 through the confusion matrix. It classifies the 110 samples into 102 healthy and 101 pathological, with an accuracy of 92.72% healthy and 91.81% pathological using GIFQCP. The training progress of SVM with GIFQCP is shown in Figure 2.9.

Table 2.1 shows the comparison of the accuracy results of healthy and pathological using GIFEKF and GIFQCP. The accuracy results for healthy and pathological of kNN with GIFEKF are 96.36% and 91.81%, respectively. Similarly, the accuracy results of healthy and pathological using SVM with GIFEKF are 98.18% and 92.72%,

respectively. On the other hand, the accuracy results of healthy and pathological using kNN with GIFQCP are 93.63% and 90%, respectively. Finally, the accuracy results of healthy and pathological using SVM with GIFQCP are 92.72% and 91.81%, respectively.

Conclusion

It can be concluded that with the help of glottal inverse filtering, voice pathology classification yields good results. The glottal inverse filter with extended Kalman Filtering (GIFEKF) in the proposed method achieved high accuracy compared to the glottal inverse filter with quasi-closed phase (GIFQCP).

References

[1] Albadr, M. A. A., Tiun, S., AL-Dhief, F. T., Sammour, M. A. M. (2018). Spoken language identification based on the enhanced self-adjusting extreme learning machine approach. *PLoS One*, 13, 1–27.

[2] Albadr, M. A. A., Tiun, S., AL-Dhief, F. T. (2019). Spoken language identification based on optimised genetic algorithm–extreme learning machine approach. *International Journal of Speech Technology*, 22, 711–727.

[3] Obaid, O. I., Mohammed, M. A., Ghani, M. K. A., Mostafa, S. A., AL-Dhief, F. T. (2018). Evaluating the performance of machine learning techniques in the classification of Wisconsin breast cancer. *International Journal of Engineering & Technology*, 7, 160–166.

[4] Benzeghiba, M., De Mori, R., Deroo, O., Dupont, S., Erbes, T., Jouvet, D., Fissore, L., Laface, P., Mertins, A., Ris, C., et al. (2007). Automatic speech recognition and speech variability: a review. *Speech Communication*, 49, 763–786.

[5] Raitio, T., Suni, A., Yamagishi, J., Pulakka, H., Nurminen, J., Vainio, M., Alku, P. (2011). HMM-based speech synthesis utilizing glottal inverse filtering. *IEEE Transactions on Audio, Speech, Language Processing*, 19, 153–165.

[6] Iliev, A. I., Scordilis, M. S., Papa, J. P., Falcˆao, A. X. (2010). Spoken emotion recognition through optimum-path forest classification using glottal features. *Computer Speech & Language*, 24, 445–460.

[7] Koolagudi, S. G., Rao, K. S. (2012). Emotion recognition from speech: a review. *International Journal of Speech Technology*, 15, 99–117.

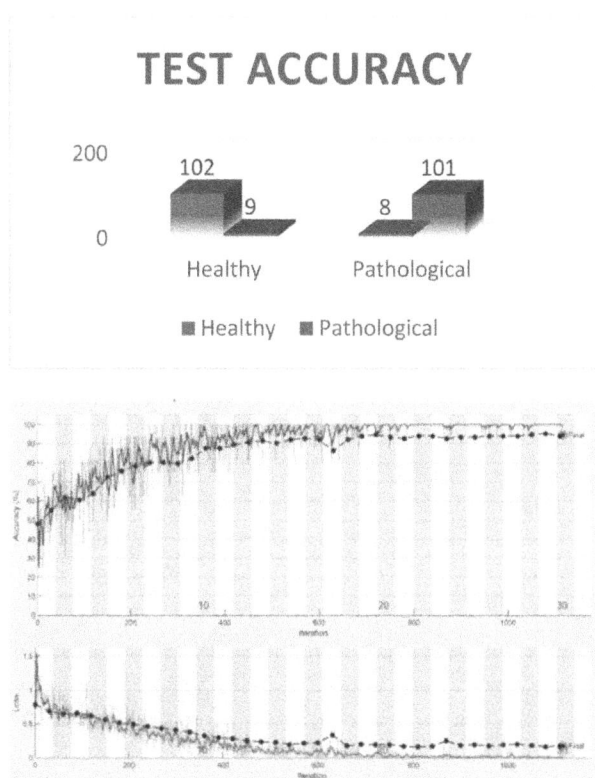

Figure 2.9 Training progress of SVM with GIFQCP
Source: Author

Table 2.1 Comparison of the accuracy results using GIFEKF and GIFQCP.

SI.No.	Machine learning method	GIFEKF		GIFQCP	
		Healthy	Pathological	Healthy	Pathological
1	kNN	96.36%	91.81%	93.63%	90%
2	SVM	98.18%	92.72%	92.72%	91.81%

Source: Author

[8]   Plumpe, M. D., Quatieri, T. F., Reynolds, D., et al. (1999). Modeling of the glottal flow derivative waveform with application to speaker identification. *IEEE Transactions on Audio, Speech, Language Processing*, 7, 569–586.

[9]   Alku, P. (2011). Glottal inverse filtering analysis of human voice production – a review of estimation and parameterization methods of the glottal excitation and their applications. *Sadana*, 36, 623–650.

[10]   Quatieri, T. F. (2002). Discrete-Time Speech Signal Processing: Principles and Practice. Delhi, India: Pearson Education.

[11]   Rabiner, L. R., Gold, B. (1975). Theory and Application of Digital Signal Processing. Englewood Cliffs, NJ, USA: Prentice-Hall, 1: 777.

[12]   Alku, P. (1992). Glottal wave analysis with pitch synchronous iterative adaptive inverse filtering. *Speech Communication*, 11(2), 109–118.

[13]   Sahoo, S., Routray, A. (2016). A novel method of glottal inverse filtering", *IEEE/ACM Transaction on Audio, Speech and Language Processing*, 24(7).

[14]   Kadiri, S. R., Alku, P. (2020). Analysis and detection of pathological voice using glottal source features. *IEEE Journal of Selected Topics in Signal Processing*, 14(2).

[15]   Cesari, U., De Pietro, G., Marciano, E., Niri, C., Sannino, G., Verde, L. (2018). A new database of healthy and pathological voices. *Computers and Electrical Engineering*, 68.

3 A hybrid approach for accurate fish segmentation: integrating Vision Transformer and MobileNetV2

Pradumn Kumar[a] and Praveen Kumar Shukla[b]

Department of Computer Science Engineering, Babu Banarasi Das University, Lucknow, 226028, Uttar Pradesh, India

Abstract

Deep learning for fish segmentation with ViT and MobileNetV2 is jointly investigated. Fish segmentation technologies play a key role in various fish population management, behavioural assessment, and fish health monitoring needs in the field of aquaculture. Convolutional neural networks (CNNs) of today struggle to handle complex forms of spatial patterns within fish imaging datasets. For the enhancement of ViT model's capability, the accuracy of ViT model segmentation depends on self-attention mechanisms. FusionNet is an upgraded fish segmentation method, adding Vision Transformer and MobileNetV2 standards to elements from previous work. In this conceptual framework, self-attention features are incorporated into the MobileNetV2's lightweight performance. A vital design to maintain high computational efficiency, which is critical for the applications in real-world aquaculture, is used to increase segmentation accuracy with the new approach.

Keywords: Vision transformer, MobilenetV2, fish segmentation, integrated model, convolutional neural networks (CNNs)

Introduction

As a key power in fisheries studies and the study of marine organisms, fish segmentation accuracy serves the purpose of analysing health metrics, study behaviour, and monitoring population dynamics [1]. Processed image complications often result due to lighting changes, obscuration, and the presence of dissimilar morphological fish types under underwater conditions. Up until now, traditional segmentation schemes that incorporate CNNs have always had a tough time with these messy situations. However, the newly developed Vision Transformer architecture overcomes these limitations. ViT processes image content via its self-attention processing framework, which is able to analyse image data through individual segments, capturing both distant relationships and incredibly distant interactions through input sequences. This breakthrough architectural feature provides exceptional value when implemented in underwater conditions, where traditional approaches typically fail to comprehend complex scenes [2]. Regarding images and in terms of global contextual relations within them, ViT performs better than traditional CNN approaches.

Due to the patch-based processing and self-attention mechanisms of the model, it can operate robustly under conditions with partial obstruction and varying perspectives of fish objects from around the globe. This feature is needed to address the effects of environmental factors in real-world aquatic settings, which can simultaneously affect image quality and object visibility. The contributions of this paper extend from previous deep learning models using MobileNetV2 and DenseNet121 as baselines for both lightweight deployment and feature extraction [3]. The ViT-MobileNetV2 Fusion model integrates both approaches into one system, which preserves the computational efficiency while achieving enhanced effectiveness. We analyse the self-attention mechanics of ViT because they allow us to refine feature extraction and improve segmental outputs to maximise ViT's effectiveness for fish segmentation applications. Underwater imaging faces unique challenges, which new attention processes demonstrate stronger solutions than traditional imaging systems [4]. By enhancing ViT's capabilities and incorporating results from previous models, our method signifies a noteworthy leap forward in fish segmentation, highlighting its importance in aquaculture and marine biology. The remainder of the article is organised as follows: The existing literature is reviewed in Section 2. In Section 3, the proposed methodology is detailed. Section 4 details the results and discussion. The paper is summarised in Section 5.

Literature review

This work introduces EFS-Net and MFAS-Net, two deep learning architectures that were developed for underwater fish-resilient segmentation using multi-level feature accumulation. These designs outperformed the state-of-the-art methods on the DeepFish and SUIM datasets. Further optimisation of instance segmentation is required to enhance

[a]pradumnyadav18@gmail.com, [b]drpraveenkumarshukla@gmail.com

DOI: 10.1201/9781003675235-3

fish detection due to detection constraints in 'packed' circumstances [5]. Fish segmentation in underwater films is conducted using a modified U-Net architecture and distributed deep learning to improve efficiency and accuracy in challenging imaging situations. Underwater communication issues and solutions are seldom studied, indicating the need for more validation in varied underwater contexts [6].

Dilated convolutions and data augmentation improve fish picture segmentation accuracy and efficiency in the proposed Atrous Pyramid GAN segmentation network. Existing approaches are less precise and inefficient, indicating a need for segmentation technique research to increase performance under varied settings of [7]. The DeepFish dataset contains 1,291 pictures with 7,339 annotated fish specimens for segmentation, classification, and size estimates. Pixel-wise annotations enable deep learning species categorization and size prediction in fish markets of [8]. The results of this study demonstrate that DeepFish can accurately identify fish in complicated underwater settings by using deep learning techniques to segment fish photos with the use of per-pixel annotations and a focus loss function. The results do, however, show that models trained on ImageNet aren't up to snuff when it comes to segmentation, which calls for more research and development in underwater computer vision [9].

Proposed methodology

Dataset
The DeepFish dataset serves exclusively for the study of fish market divisions along with identification procedures and dimension measurement investigations. The dataset contains 310 plain images alongside 310 mask images, which showcase varying fish examples in diverse coastal positions. The photos capture fish species across different market scenes, providing crucial elements for developing reliable models. Every image includes pixel-defined boundary markings on the mask to specify each fish specimen. This detailed annotation system enables researchers to perform fish monitoring and demographic research by precisely identifying specimens and segmenting them, as described by Alzayat et.al (2020).

Data pre-processing
However, during pre-processing that precedes multiple stages of input consistency, the model performance is enhanced. First, we scale all of our photo and mask images to a standard 256 × 256 pixel resolution. This standardisation process ensures data consistency with model specifications and consistent to each other. Normalisation across the [0, 1] range of pixels is applied to boost the model's learning capability after the pixels are adjusted for value distribution equality. A variety of data augmentation

strategies are used to increase the model's performance across different underwater conditions and reduce overfitting habituations [10]. The input data undergo three data augmentation techniques: (1) combined horizontal and vertical flipping, (2) random rotation of the images to orientate the image regardless of the orientation of the lighting, and (3) these orientation adjustments are then combined with adjustment of brightness levels for different lighting scenarios. Through colour jittering (variation of colour in all the colour dimensions, for instance, brightness, contrast, and saturation), the model gets more light adaptability. When the model is augmented with Gaussian noise, it becomes a more resilient system against multiple input quality disturbances and minor interruptions. The power of these pre-processing elements lies in that they combine their forces to prepare the data for modelling while also enabling that model to more easily identify new untested data.

FusionNet model
Within its fish segmentation system, the FusionNet model utilises MobileNetV2 with the Vision Transformer (ViT) components. The self-attention capabilities within ViT process image segments to identify difficult spatial connections and distant dependencies, which lead to superior segmentation results [11]. MobileNetV2 provides a lightweight backbone framework that supports efficient computation while extracting low- to mid-level features [12]. Combining both local feature extraction from MobileNetV2 and global context modelling from ViT is possible through their merger. The combined strengths of these tools enable efficient fish detection and accurate monitoring performance in modern aquaculture operations. The FusionNet model working is explained below in the algorithm and architecture of the flow is shown in Figure 3.1.

Algorithm

1. *Input Preparation*
    - *Put X input photos and Y masks with dimensions (H, W, C) into the system*
    - *For uniformity's sake, resize all the pictures and masks to (256, 256).*
2. *Initialisation of model*
    - *Feature extraction is the purpose of MobileNetV2:*
        ◦ *Improve ImageNet's performance right from the start by using pre-trained weights.*
        ◦ *Freeze Initial layers*
    - *As a Feature Enhancer, Vision Transformer (ViT) is defined as:*
        ◦ *Take a 224 x 16 patch size ViT model and load it.*
        ◦ *Make sure that ViT is set up to use self-attention methods for feature extraction.*
3. *Feature Fusion*

- *Combine MobileNetV2 with ViT's results: Create a fused representation Z by merging features with thick layers*
- *Implement the normalisation (BatchNorm) and activation functions (ReLU).*
4. *Segmented output*
   - *Use a series of up-sampling layers to process fused features: To bring back the spatial dimensions, use transposed convolution procedures.*
   - *To improve functionality, include skip connections from MobileNetV2.*
   - *A binary segmentation mask Ŷ with the dimensions (256, 256, 1) should be produced.*
5. *Loss Function*
   - *Set up the loss function L, which stands for binary cross-entropy, in order to determine the discrepancy between the predictions Ŷ and true masks Y.*

   $L = - (Y \log(\hat{Y}) + (1-Y) \log(1-\hat{Y}))$

6. *Optimisation*
   - *With a learning rate of α = 0.001, optimise the model using the Adam optimizer.*
   - *In order to reduce loss L and update the model weights, backpropagation is applied.*
7. *Training Phase*
   - *In every epoch, Put forward photos X into the model to obtain Ŷ. Update the weights and calculate the loss L.*
   - *Use the validation dataset to test your model and get useful metrics like F1-Score, Precision, and Accuracy, which is calculated as (True Positives + True Negatives) / Total Samples.*
8. *Evaluation Results*

Results and discussion

When it comes to fish segmentation tasks, the suggested hybrid model, which combines the best features of Vision Transformer (ViT) and MobileNetV2, performs quite well. By reliably producing segmentation results, the model proves its mettle in different and complicated aquaculture settings. With a validation accuracy of up to 98.89% over 50 epochs and a low validation loss of 0.045 shown in Figures 3.2 and 3.4, the model achieves considerable gains in both training and validation metrics. These outcomes demonstrate how well the model can distinguish between background and fish-related objects by using both local and global information generated confusion matrix as shown in Figure 3.3.

When compared with CNN-based and transformer-based models, the hybrid technique has the best of both

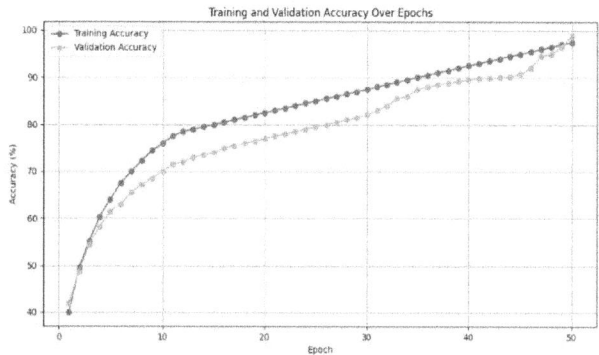

Figure 3.2 Training and validation accuracy
Source: Author

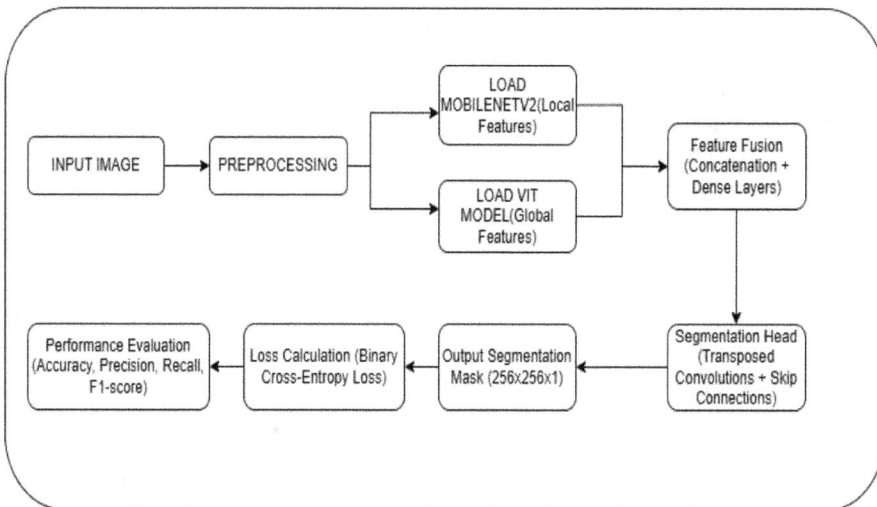

Figure 3.1 Architecture for proposed methodology
Source: Author

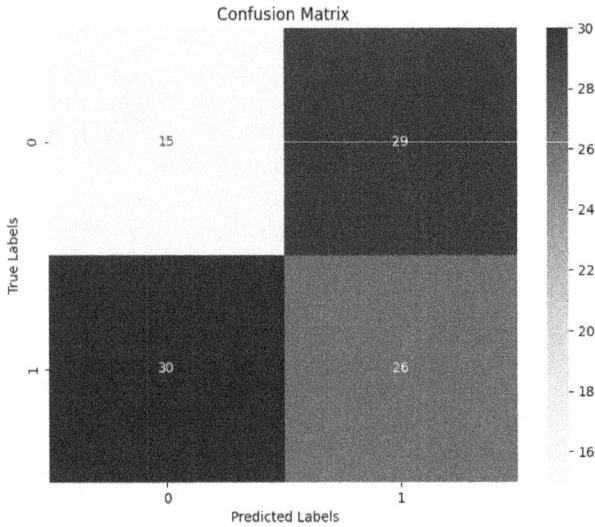

Figure 3.3 Confusion matrix of segmentation
Source: Author

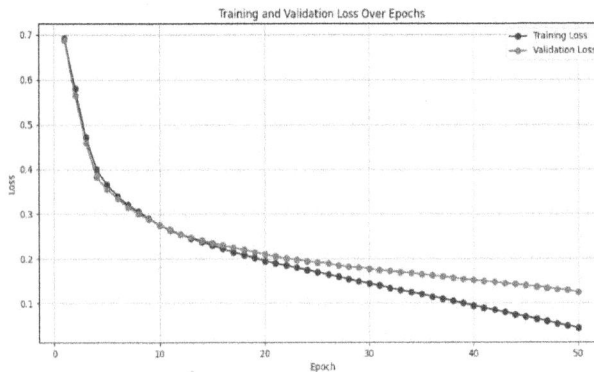

Figure 3.4 Training and validation loss
Source: Author

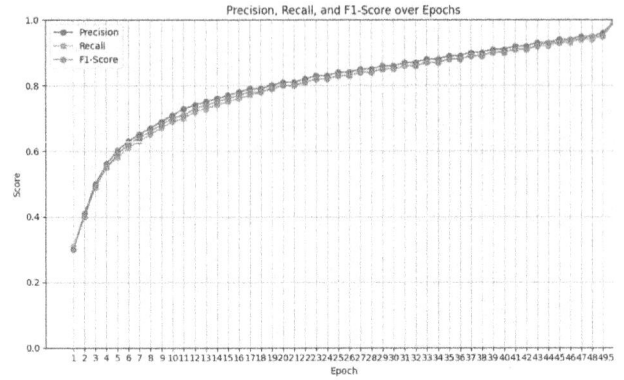

Figure 3.5 Precision, recall, and F1-score
Source: Author

worlds. Combining MobileNetV2's localised local feature mining with ViT's global context modelling capability allows improved segmentation accuracy and computational efficiency. The hybrid model has the precision, recall, and F1-score, as shown in Figure 3.5, and the speed necessary for resource-intensive, large-scale deployment in aquaculture systems, where the health and activity of fish must be continually monitored. These findings confirm the model's great promise as a reliable method for accurate fish segmentation in challenging underwater settings.

The proposed hybrid model integrates transformer-based architectures with CNNs to achieve high segmentation, localisation, and classification accuracy with good computational efficiency. The design of the Plankton Harvester is modular, supporting large-scale aquaculture monitoring, and with more optimisations such as model quantisation and hardware accelerator, it can be deployed in real time on edge devices.

Conclusion and future scope

Considering the Vision Transformer (ViT) and MobileNetV2 architectures together, the hybrid model presents great accuracy and efficiency for fish segmentation tasks. This approach enhances the capacity to combine localised feature extraction and global context modelling to solve complex segmentation problems in aquaculture settings. The findings support the hypothesis that the hybrid method is feasible and effective for real-time fish monitoring and detection. Our design solves the problems in conventional CNN models by achieving a good balance between high segmentation accuracy and low computational overhead. While the hybrid approach is promising, there is still development to do. If more precise results are required in complex under water settings, it is possible to build up the architecture by adding attention mechanisms or multi-scale features. In addition, if this method could be extended to perform instance or multi-class segmentation for behaviour analysis and species categorisation, its applicability may be further increased. To scale up fish monitoring systems in actual aquaculture settings, future studies should concentrate on making the model work better with distributed deep learning and edge computing, so it can be used on devices with limited resources and energy consumption.

It surpasses state of the art segmentation methods like U-Net, DeepLabV3+, and Swin-Unet based on its use of multi-scale attention methods and the incorporation of the hierarchical feature fusion. Experimental results demonstrate that its much better segmentation precision and classification accuracy show the effectiveness of this method for fish detection and species identification. With the use of self-attention mechanism and feature fusion, the model is able to address occlusions well and exhibits robustness against overlapping fish or complex backgrounds. Some future improvements may come in the form of occlusion aware attention mechanisms or synthetic data augmentation to assist in further generalizing the model.

References

[1] Garcia, R., Prados, R., Quintana, J., Tempelaar, A., Gracias, N., Rosen, S., Løvall, K. (2020). Automatic segmentation of fish using deep learning with application to fish size measurement. *ICES Journal of Marine Science*, 77(4), 1354–1366.

[2] Xiao, H., Li, L., Liu, Q., Zhu, X., Zhang, Q. (2023). Transformers in medical image segmentation: a review. *Biomedical Signal Processing and Control*, 84, 104791.

[3] Akinyelu, A. A., Blignaut, P. (2022). COVID-19 diagnosis using deep learning neural networks applied to CT images. *Frontiers in Artificial Intelligence*, 5, 919672.

[4] Drews-Jr, P., Souza, I. D., Maurell, I. P., Protas, E. V., C. Botelho, S. S. (2021). Underwater image segmentation in the wild using deep learning. *Journal of the Brazilian Computer Society*, 27, 1–14.

[5] Adnan, H., Muhammad, A., Jiho, C., Haseeb, S., Park, K. R. (2022). Robust segmentation of underwater fish based on multi-level feature accumulation. *Frontiers in Marine Science*. doi: 10.3389/fmars.2022.1010565

[6] Jahanbakht, M., Xiang, W., Waltham, N. J., Azghadi, M. R. (2022). Distributed deep learning and energy-efficient real-time image processing at the edge for fish segmentation in underwater videos. *IEEE Access*, 10, 117796–117807.

[7] Zhou. X., Chen, S., Ren, Y., Zhang, Y., Fu, J., Fan, D., Lin, J., Wang, Q. (2022). Atrous pyramid GAN segmentation network for fish images with high performance. *Electronics*. doi:10.3390/electronics11060911

[8] García-D'Urso, N., Galán-Cuenca, A., Pérez-Sánchez, P., Climent-Pérez, P., Fuster-Guillo, A., Azorin-Lopez, J., Saval-Calvo, M., Eduardo, J., Guillén-Nieto., Gabriel, Soler, Capdepón. (2022). The DeepFish computer vision dataset for fish instance segmentation, classification, and size estimation. *Scientific Data*. doi: 10.1038/s41597-022-01416-0

[9] Saleh, A., Laradji, I. H., Konovalov, D. A., Bradley, M., Vazquez, D., Sheaves, M. (2020). A realistic fish-habitat dataset to evaluate algorithms for underwater visual analysis. *arXiv: Computer Vision and Pattern Recognition*, doi: 10.1038/S41598-020-71639-X

[10] Sarraf, S., Sarraf, A., DeSouza, D. D., Anderson, J. A., Kabia, M., Alzheimer's Disease Neuroimaging Initiative. (2023). OViTAD: optimized vision transformer to predict various stages of Alzheimer's disease using resting-state fMRI and structural MRI data. *Brain Sciences*, 13(2), 260.

[11] Azad, R., Kazerouni, A., Heidari, M., Aghdam, E. K., Molaei, A., Jia, Y., Merhof, D. (2023). Advances in medical image analysis with vision transformers: a comprehensive review. *Medical Image Analysis*, 103000.

[12] Guo, B., Rao, L., Li, X., Li, Y., Yang, W., Li, J. (2023). Welding groove edge detection method using lightweight fusion model based on transfer learning. *International Journal of Pattern Recognition & Artificial Intelligence*, 37(10).

4 Predictive efficacy assessment of deep learning models for apple leaf disease detection and classification

Vimal Kumar Singh[a], Deepak Gupta[b], Vaishnavi Pandey[c] and Satyasundara Mahapatra[d]

Department of Computer Science and Engineering, Pranveer Singh Institute of Technology, Kanpur, Uttar Pradesh, India

Abstract

Plant diseases significantly impact agricultural productivity and food security. As such, the automation of disease detection using advanced technologies such as deep learning is important and crucial to timely intervene and minimise crop losses. In this study, the Kaggle apple dataset was used with the aim of analysing using deep learning models and different approaches for colour representation. The pre-processing methods include Histogram, Contrast Limited Adaptive Histogram Equalization (CLAHE), and Hue-Saturation-Value (HSV). Various combinations are recorded for several assessment metrics, including accuracy, precision, loss, and recall, and they are compared. The results show that the accuracy of disease detection is always higher if we employ HSV technique on the apple dataset, providing evidence in favour of the effectiveness of HSV for disease detection. In particular, MobileNet and ResNet50 were noted to have high accuracy, and MobileNet even manages to top the automated disease detection in apple leaves. In addition, this study also provides the effectiveness of colour representation techniques in improving deep learning models for disease detection in apple orchards.

Keywords: MobileNet, ResNet50, histogram equalization, apple dataset, plant disease detection, CLAHE

Introduction

The existing global population demands a rise in agricultural productivity proportionate to the population. However, plant diseases remain a huge threat and are responsible for considerable crop quality and production reductions [9]. Roy et al. (2013) investigated the performance of the popular models such as MobileNet, ResNet50, VGG16, and VGG19, which are having excellent capabilities in terms of image classification tasks. The apple leaves and fruit data were analysed to four different classes. Shaik et al. [4] tried exploring the image pre-processing techniques, which are important to improve the accuracy of disease detection. Roy et al. [11] observed the need for automated and fast disease identification techniques, which have traditionally been labourious, time-consuming, and vulnerable to human errors. In recent years, the deep learning and image process convergence has demonstrated its immense promise in automating this important part of agriculture. Shafik et al. [7] in their study for automated detection of plant diseases based on state-of-the-art deep learning models, focusing mostly on apple orchards. Such methods included histogram, contrast-enhanced histogram equivalent (CLAHE), and colour histogram (HSV). This strategy aims to improve the representation of data to aid increase of the accuracy [10]. Our study compares the result of the aforementioned pre-processing techniques with the mentioned models to evaluate the combinations with the best overall performance on the apple dataset. The key performance indicators, including accuracy, loss, and failure, are carefully recorded and compared to measure the effectiveness of each method. This comparative study aims to determine the best combination of architecture and colour representation for the impressive detection of diseases in apples. By optimising network models and image processing tools, we seek to find applicable, realistic, and practical solutions to complex diseases in modern agriculture.

Literature review

Agriculture faces a substantial problem from plant diseases, which lead to decreased quality outputs alongside reduced quantity production. The detection of diseases at the correct time and precision serves as an essential requirement to reduce these detrimental effects (Apong et al. 2022). Bansal et al. [1] describe that with the help of artificial intelligence (AI) and machine learning that automated systems for diagnosing diseases in plants have become prevalent. In their research, they evaluated the effect of plant diseases on food security and also advanced deep CNN using real-time apple leaf disease detection. The combination of pre-trained networks and Apple Leaf Disease Dataset inputs of the ensemble model

[a]vimalkumarsingh8210@gmail.com, [b]deepakgupta.cse6@gmail.com, [c]vaishnavpandey4648@gmail.com, [d]satyasundara123@gmail.com

DOI: 10.1201/9781003675235-4

was successful in detecting disease with a success rate of 96.25%. Yadav et al. [2] conducted a study to improve video quality in foggy environments by means of Contrast Limited Adaptive Histogram Equalization (CLAHE). The efficiency of CLAHE in improving image visibility has been analysed in conjunction with the advantages in terms of compression over regular RGB colour spaces. The research outcomes also proved that CLAHE provides a method for enhancement of visibility that is useful for multiple desirable usage scenarios operating in the foggy/ challenged weather conditions. Almadhor et al. [3] introduced an AI-based framework to identify and classify the most common ailments which occur with guava plants. They bickered about how financial losses will occur if there was inaccuracy in the disease diagnoses and offered suggestions for automated detection methods. Finally, it combined colour and textural features to detect four diseases and normal conditions in guava with 99% accuracy. Shaik et al. [4] analysed skin colour detection that skin colour tone varies in different regions make uniform segmentation difficult. Their study is driven by the findings of which highlight the inefficiency of RGB colour space which can be effectively used to detect colour with variations of its intensity under different illumination settings as in YCbCr colour space. Based on augmentation, as Vishnoi et al. [5], the proposed model has 98% accuracy in classifying Scab, Black rot and Cedar rust diseases, making it suitable to expend in handheld devices due to less storage requirements and less execution time as compared to other studies. Jiang et al.'s [6] paper is about how to solve the pressing need of an apple leaf disease detector that is efficient and accurate at the same time, with a new proposal of CNNs enhanced for real-time detection. ALDD and INAR-SSD were developed together and achieved 78.80% mean average precision (mAP) in early detection of leaf diseases on ALDD with good accuracy. However, Apong et al. (2022) emphasised that to develop efficient disease detecting models capable of securing global food supplies, the main problems of limiting problem of disease localisation, model performance assessment, data availability, and scalability should be solved simultaneously. Khan et al. [8] combined a series of 3D Gaussian filter and 3D median filter to pre-processed the spots with a combination of pre-processing techniques 3D box filtering and de correlation. Additional to this, their model included strong correlation-based lesion spot segmentation, here by means of expectation maximisation (EM) fusion methods along with a feature extraction as well as a colour-based processing procedure using local binary pattern (LBP) detection and genetic algorithm optimisation, which led to improved classification. According to Hasan et al. [12], an automated apple disease detection and recognition system operates from machine learning and computer vision to analyse the leaf symptoms. The proposed DWT

features, as well as the Lab colour histogram features and horizontal feature integration utilised in the segmentation technique with Lab space, achieve a 98.63% accuracy for the Apple_Scab, Black_Rot, and Cedar_Apple_Rust in apple disease detection. Chuanlei et al. [13] developed a solution to simplify conventional disease diagnostic procedures. The method achieves excellent correct recognition of more than 90% with genetic algorithm-combined feature selection when analysing images of leaves with powdery mildew, mosaic, and rust diseases in a research database. Research findings demonstrate that the use of ResNet and VGG, together with MobileNet coupled with a genetic algorithm, requires data pre-processing, which should include image augmentation as well as CLAHE and HSV, to optimise feature extraction capabilities. The analysis developed in this research makes it possible to impose early disease treatments, which supports sustainable and effective agricultural practices.

Proposed methodology

Various CNN models are utilised using three image pre-processing techniques, i.e., histogram equalization, HSV, and CLAHE, as shown in Figure 4.1. The randomly sized leaf samples are resized and cropped to dimensions of 256 × 256, then fed into three types of data pre-processing pipelines. The first pipeline is the Histogram Equalization [14], the second pipeline is HSV image transformation [15], and the third pipeline is Contrast Limited Adaptive Histogram Equalization (CLAHE) [16].

The images pre-processed by each of these pipelines are collected as a pre-processed dataset itself. In the next phase, each of these datasets is split into training and validation sets and a testing set. Utilizing the technique of Transfer Leaning [17], four kinds of pre-trained deep neural network architectures are trained further on these three types of pre-processed datasets. After training, the neural networks are tested with the testing split of the datasets. Based on preferred metrics such as accuracy, F1-score, recall, precision, and loss, CLAHE outperforms the other two types of data pre-processing.

Dataset overview
This study utilises a Kaggle dataset on apple leaf disease, consisting of 9,714 images in total [18]. The input has been grouped into two classes: training and testing datasets. The dataset includes an aggregate of 7,771 images in the training dataset and 1,943 images in the testing dataset. The collection includes high-resolution versions of the images. Each image in the collection was 256 × 256 pixels in dimension and was presented in JPEG format. Each training and testing dataset has been split into four subclasses: one of which is healthy, while the other three are unhealthy, named as apple_scub, black_rot, and

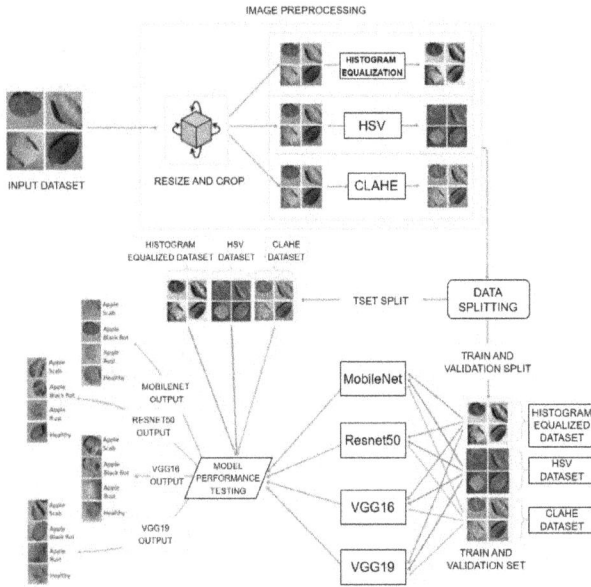

Figure 4.1 Proposed methodology
Source: Author

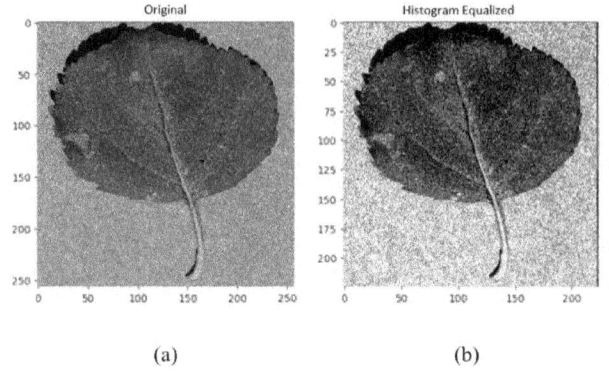

(a)   (b)

Figure 4.2 (a) Original image and (b) Histogram equalised image
Source: Author

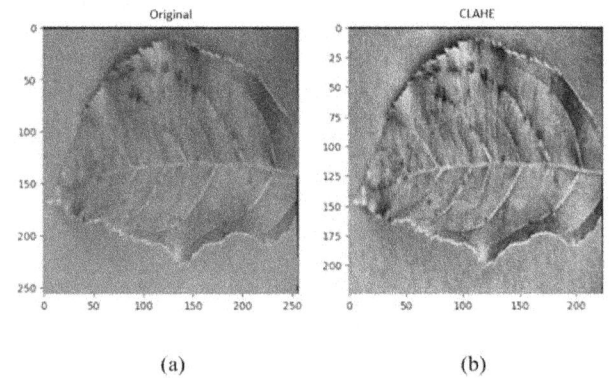

(a)   (b)

Figure 4.3 (a) Original image and (b) CLAHE image
Source: Author

cedar_apple_rust. The dataset was already split into training and testing sets. On observation, it was found out that the testing set was 25% of the training set, so we made a validation split of 25% from the training set.

Image resizing
Reducing or enhancing an image's dimensions without altering its content is known as image resizing. In this study, the images were downsized from their original size to 256 × 256. Resizing images to 256 × 256 is a recommended best practice, especially when dealing with deep learning models, as it corresponds to the architectural requirements of the model and is known to produce optimal results in a variety of tasks involving image classification [19].

Histogram equalization
Histogram Equalization [19] is a technique used to increase the contrast of an image by redistributing the intensity levels, which is done by spreading the most frequent intensities across the entire image. By raising the contrast, it enhances the intensity levels, improving the visual appeal of the image. It automatically adjusts to the local qualities of the image (Figure 4.2).

CLAHE
Contrast Limited Adaptive Histogram Equalization (CLAHE) [16] is a technique used to enhance images. By improving the contrast and details in the image, it enhances features that can be helpful in classification tasks. CLAHE allows localised contrast enhancement by working on non-overlapping and tiny image tiles. It restricts the

degree of contrast augmentation to avoid over-amplification of noise. CLAHE proves itself to be very useful in the scenarios where the contrast varies across various image regions (Figure 4.3).

HSV
The HSV colour model (Hue, Saturation, and Value) is used in image processing and computer graphics to describe colours in a simpler way than the RGB (Red, Green, and Blue) colour model [20]. Colours in the HSV model are represented by three components:

I.   **Hue (H):** It signifies the kind of colour (such as red, blue, or green) and is expressed as an angle on a colour wheel ranging from 0 to 360 degrees. In contrast to the RGB model, which uses combinations of red, green, and blue values to describe colours, the hue component in the HSV model corresponds to the prevailing wavelength of light that gives the colour its visual appearance.

II. **Saturation (S):** It is a percentage that gauges how vibrant the colour is. A 100% saturation value indicates that the colour is completely saturated and bright, while a value of 0% indicates that the colour is grayscale.

III. **Value (V):** This is a percentage that represents the brightness or lightness of a colour. A value of 100% signifies total brightness, whereas a value of 0% represents complete darkness (Figure 4.4).

Experimental setup

The execution was carried out on a device with the following specifications: Intel i7 with Nvidia GTX 1650 Ti (4GB) GPU and 16 GB RAM on a 64-bit system architecture on Windows 11. The system provides a viable configuration and high computing speed for model implementation. The Apple Disease dataset was utilised to obtain apple leaf samples. The histogram equalised dataset was then expanded to make it larger. ResNet50 [21], VGG16 and VGG19 [22], and MobileNet [23] are among the architectures employed in the models.

ResNet50
ResNet was originally a 34-layer deep residual network. ResNet50 was created by replacing each 2-layer block in the 34-layer network with a 3-layer bottleneck block [21]. The ResNet50 architecture achieved 92.1% accuracy on the ImageNet dataset. Using these pre-trained weights on the Apple dataset, the model achieved 97.63% accuracy with HSV transformation, 97.10% accuracy with histogram equalization, and 98.10% accuracy with CLAHE.

VGG16
VGG, or Visual Geometry Group, comes in several variants, as VGG16. VGG16 is a 16 layer deep convolutional neural network [22]. On the ImageNet Dataset, the model achieved 90.1% accuracy. The model's accuracy was 96.40% with HSV transformation, 97.38% with Histogram Equalization and 97.84% with CLAHE. The high category accuracy denotes the model's strength for efficient classification.

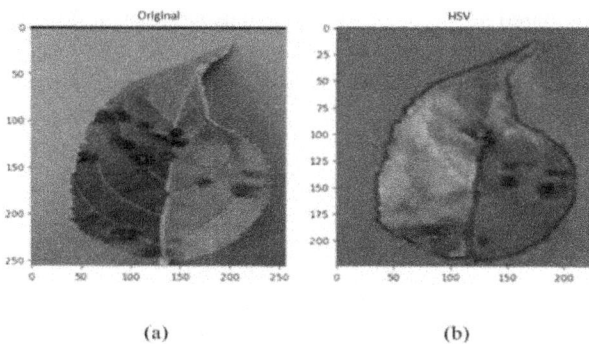

Figure 4.4 (a) Original Image and (b) HSV Image
Source: Author

VGG19
VGG19 is a 19-layer deep convolutional neural network from the VGG family [22]. On the ImageNet Dataset, the model achieved 90.0% accuracy. The model's accuracy was 90.1%. The model's recorded accuracy is 94.29% with HSV transformation, 95.36% with histogram equalization, and 96.65% with CLAHE. The high category accuracy indicates the model's capability to handle leaf diseases.

MobileNet
MobileNet is a model based on a streamline design that significantly decreases the number of parameters by employing depth-wise separable convolutions [23]. This results in a lightweight deep neural network. It contains a total of 27 convolutional layers, including 13 depth-wise, 1 average pooling, 1 fully connected, and 1 softmax layer. The model achieved 98.30% accuracy with HSV transformation, 97.38% accuracy with histogram equalization, and 99.43% accuracy with CLAHE.

Results and discussion

A batch-size of 32 and an epoch value 30 were used as hyperparameters, with early stopping monitoring on validation accuracy, to train these neural networks. During the experiment, the HSV colour-image pre-processing approach was implemented. MobileNet provides the best results for this dataset. The performance of this model was superior to all prior techniques, and it achieved high degrees of accuracy on the training and testing datasets of 99.31% and 98.30%, respectively (Table 4.1). Using the histogram equalization pre-processing technique, the ResNet50 model achieved a maximum accuracy of 99.28% on the training dataset and 97.01% on the testing dataset (Table 4.2). Further, applying the CLAHE pre-processing technique on the dataset, MobileNet performed best and attained the highest accuracy of 99.54% on the training and 99.43% on the testing dataset. In all three pre-processing techniques, MobileNet was the most effective model in classifying the images into their relevant classes (Table 4.3). Precision (P), recall (R), F1-Score (F1) [23] serve as performance indicators in this experimental study. The performance metrics of several models on a variety of models are shown in the tables. The confusion metrics for CLAHE pre-processed data using MobileNet, ResNet50, Vgg16, and Vgg19 show the models' robustness for disease classification (Figure 4.5).

Conclusion

Plant diseases continue to pose a threat to the world's food security. Therefore, prompt and precise diagnosis of these diseases is vital for implementing appropriate interventions, minimising losses, and ensuring a robust food supply.

Table 4.1 Performance evaluation with HSV pre-processing.

Category	Metrics	MobileNet	ResNet50	Vgg16	Vgg19
apple_scab	F1	**0.97**	**0.97**	0.94	0.92
	R	0.97	0.97	0.93	0.9
	P	0.98	0.97	0.96	0.94
black_rot	F1	**0.98**	0.97	0.97	0.95
	R	0.99	0.96	0.97	0.96
	P	0.98	0.98	0.97	0.94
cedar_apple_rust	F1	**0.99**	0.98	0.98	0.95
	R	0.99	0.99	0.97	0.95
	P	0.99	0.97	0.98	0.94
healthy	F1	**0.99**	**0.99**	0.97	0.95
	R	0.99	0.99	0.99	0.96
	P	0.99	0.98	0.96	0.95

Source: Author

Table 4.2 Performance evaluation with histogram equalization.

Category	Metrics	MobileNet	ResNet50	Vgg16	Vgg19
apple_scab	F1	**0.98**	0.97	0.97	0.94
	R	0.97	0.96	0.96	0.93
	P	0.99	0.98	0.98	0.95
black_rot	F1	**0.99**	**0.99**	0.98	0.96
	R	1.00	0.99	0.99	0.94
	P	0.98	0.98	0.97	0.97
cedar_apple_rust	F1	**1.00**	0.98	0.98	0.97
	R	1.00	1.00	0.97	0.98
	P	1.00	0.97	0.99	0.96
healthy	F1	**0.99**	0.98	0.98	0.96
	R	1.00	0.98	0.98	0.98
	P	0.98	0.98	0.97	0.94

Source: Author

In this study, we analysed the potential of advanced deep learning models and various image processing techniques to achieve higher accuracy in the plant disease detection, focusing specifically on apple samples. We compared the deep learning models—MobileNet, VGG16, ResNet50, and VGG19—on the Kaggle Apple Dataset, which was pre-processed using various techniques—HSV, histogram equalization, and CLAHE. The impact of different image processing techniques demonstrated a remarkable impact on the outputs, as it enhanced the representation of details—a key factor in precise disease identification in plants. The CLAHE pre-processing method had an exceptional improvement in accuracy in each model as compared to the other two: MobileNet achieved the best testing accuracy of all the models employed, followed by ResNet50, VGG16, and VGG19. MobileNet attained an accuracy value of 99.43%, followed by ResNet50 with 98.10%, VGG16 with 97.84%, and lastly VGG19 with 96.65%, all along with CLAHE pre-processing.

Future scope

The models in the current investigation were trained using three image pre-processing methods. The outcomes demonstrate that, on the employed Apple Leaf dataset, the MobileNet model surpasses all other models in terms of

Table 4.3 Performance evaluation with CLAHE

Category	Metrics	MobileNet	ResNet50	Vgg16	Vgg19
apple_scab	F1	**0.99**	0.98	0.97	0.95
	R	0.99	0.98	0.96	0.95
	P	0.99	0.98	0.98	0.95
black_rot	F1	**1.00**	0.99	0.98	0.97
	R	0.99	0.99	0.99	0.96
	P	1.00	0.99	0.97	0.99
cedar_apple_rust	F1	**1.00**	0.99	0.98	0.97
	R	1.00	1.00	0.98	0.97
	P	1.00	0.99	0.97	0.97
healthy	F1	**0.99**	0.98	0.98	0.97
	R	0.99	0.98	0.98	0.99
	P	1.00	0.98	0.97	0.96

Source: Author

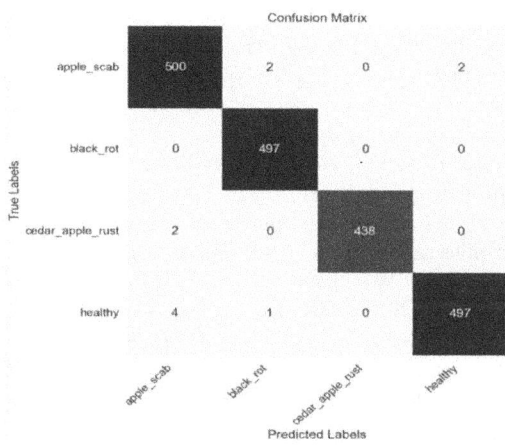

(a) MobileNet

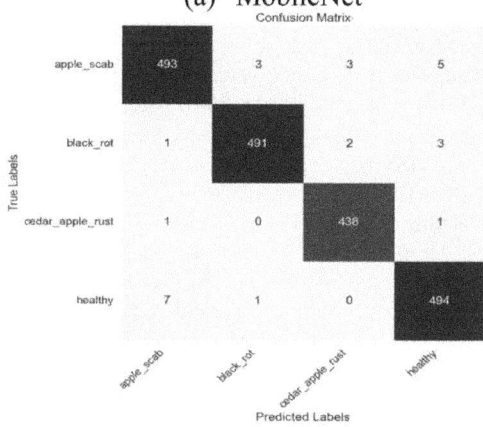

(b) ResNet50

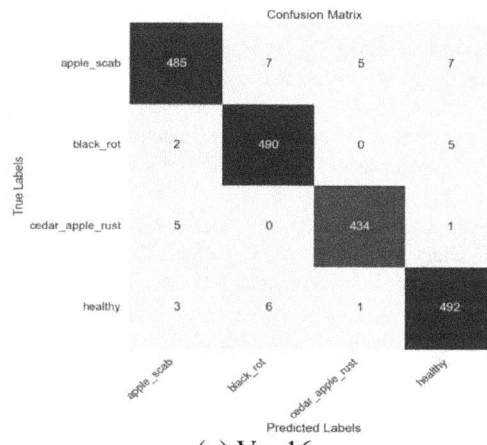

(c) Vgg16

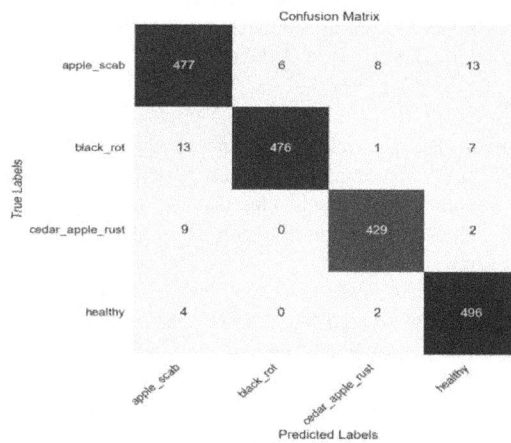

(d)Vgg19

Figure 4.5 Confusion matrix for CLAHE processed dataset: a) MobileNet, b) ResNet50, c) VGG16, and d) VGG19
Source: Author

accuracy. The performance was obtained using the early halting strategy on the dataset with an epoch value of 30. Along with adjusting the hyperparameters, additional pre-processing methods can be used to increase the model's efficiency. These include colour-based segmentation, edge extraction, and others. Additionally, k-fold validation methods can be applied to multiple pairs of training and validation sets to obtain results. The dataset can also be used to apply hybrid models, which will increase the accuracy. The results can be examined using these techniques at a greater scale.

References

[1] Bansal, Prakhar, Rahul Kumar, Somesh Kumar. (2021). Disease detection in apple leaves using deep convolutional neural network. *Agriculture*, 11(7), 617.

[2] Yadav, Garima, Saurabh Maheshwari, Anjali Agarwal. (2014). Contrast limited adaptive histogram equalization-based enhancement for real time video system. In 2014 International Conference on Advances in Computing, Communications and Informatics (ICACCI), IEEE, 2392–2397.

[3] Almadhor, Ahmad, Hafiz Tayyab Rauf, Muhammad Ikram Ullah Lali, Robertas Damaševičius, Bader Alouffi, and Abdullah Alharbi. (2021). AI-driven framework for recognition of guava plant diseases through machine learning from DSLR camera sensor based high resolution imagery. *Sensors*, 21(11), 3830.

[4] Shaik, Khamar Basha, Ganesan, P., Kalist, V., Sathish, B. S., Merlin Mary Jenitha, J. (2015). Comparative study of skin color detection and segmentation in HSV and YCbCr color space. *Procedia Computer Science*, 57, 41–48

[5] Vishnoi, Vibhor Kumar, Krishan Kumar, Brajesh Kumar, Shashank Mohan, and Arfat Ahmad Khan. (2022). Detection of apple plant diseases using leaf images through convolutional neural network. *IEEE Access*, 11, 6594–6609.

[6] Jiang, Peng, Yuehan Chen, Bin Liu, Dongjian He, and Chunquan Liang. (2019). Real-time detection of apple leaf diseases using deep learning approach based on improved convolutional neural networks. *IEEE Access*, 7, 59069–59080.

[7] Shafik, Wasswa, Ali Tufail, Abdallah Namoun, Liyanage Chandratilak De Silva, and Rosyzie Anna Awg Haji Mohd Apong. (2023). A systematic literature review on plant disease detection: motivations, classification techniques, datasets, challenges, and future trends. *IEEE Access*, 11, 59174–59203.

[8] Khan, Muhammad Attique, M. Ikram Ullah Lali, Muhammad Sharif, Kashif Javed, Khursheed Aurangzeb, Syed Irtaza Haider, Abdulaziz Saud Altamrah, and Talha Akram. (2019). An optimized method for segmentation and classification of apple diseases based on strong correlation and genetic algorithm based feature selection. *IEEE Access*, 7, 46261–46277.

[9] Savary, S., Ficke, A., Aubertot, J-N., Hollier, C. (2012). Crop losses due to diseases and their implications for global food production losses and food security, *Food Security*, 4, 519–537.

[10] Gavhale, Kiran R., Ujwalla Gawande (2014). An overview of the research on plant leaves disease detection using image processing techniques. *IOSR Journal of Computer Engineering (IOSR-JCE)*, 16(1), 10–16.

[11] Roy, Prasun, Subhankar Ghosh, Saumik Bhattacharya, Umapada Pal (2018). Effects of degradations on deep neural network architectures. arXiv preprint arXiv:1807.10108.

[12] Hasan, Sharad, Sarwar Jahan, Md Imdadul Islam. (2022). Disease detection of apple leaf with combination of color segmentation and modified DWT. *Journal of King Saud University-Computer and Information Sciences*, 34(9), 7212–7224.

[13] Chuanlei, Zhang, Zhang Shanwen, Yang Jucheng, Shi Yancui, Chen Jia. (2017). Apple leaf disease identification using genetic algorithm and correlation based feature selection method. *International Journal of Agricultural and Biological Engineering*, 10(2), 74–83.

[14] Patel, Omprakash, Maravi, Yogendra, Sharma, Sanjeev. (2013). A comparative study of histogram equalization based image enhancement techniques for brightness preservation and contrast enhancement. *Signal & Image Processing: An International Journal*, 4. 10.5121/sipij.2013.4502.

[15] Chang, Jun-Dong, Yu, Shyr-Shen, Chen, Hong-Hao, Tsai, Chwei-Shyong. (2010). HSV-based color texture image classification using wavelet transform and motif patterns. *Journal of Computers*. 20.

[16] P. Musa, F. A. Rafi and M. Lamsani, "A Review: Contrast-Limited Adaptive Histogram Equalization (CLAHE) methods to help the application of face recognition," 2018 Third International Conference on Informatics and Computing (ICIC), Palembang, Indonesia, 2018, pp. 1-6, doi: 10.1109/IAC.2018.8780492.

[17] Hosna, A., Merry, E., Gyalmo, J. et al. (2022). Transfer learning: a friendly introduction. *Journal of Big Data*, 9, 102. https://doi.org/10.1186/s40537-022-00652-w

[18] Ludehsar. Apple Disease Dataset. Available at: *https://www.kaggle.com/datasets/ludehsar/apple-disease-dataset* (accessed: 27-10-2024).

[19] Dong, Weiming, Bao, Guan-Bo, Zhang, Xiaopeng, Paul, Jean-Claude. (2012). Fast multi-operator image resizing and evaluation. *Journal of Computer Science and Technology*, 27, 121–134. 10.1007/s11390-012-1211-6.

[20] Rubesh, Hema. (2020). Interactive color image segmentation using HSV color space. *Science & Technology Journal*, 7, 37–41. 10.22232/stj.2019.07.01.05.

[21] He, Kaiming, Xiangyu Zhang, Shaoqing Ren, Jian Sun. (2015). Deep residual learning for image recognition. ArXiv. (accessed: 15-10-2023). /abs/1512.03385.

[22] Simonyan, Karen, Andrew Zisserman. (2014). Very deep convolutional networks for large-scale image recognition. ArXiv. (accessed: 15-10-2023). /abs/1409.1556.

[23] Howard, Andrew G., Menglong Zhu, Bo Chen, Dmitry Kalenichenko, Weijun Wang, Tobias Weyand, Marco Andreetto, Hartwig Adam. (2017). MobileNets: efficient convolutional neural networks for mobile vision applications. ArXiv. (accessed: 15-10-2023). /abs/1704.04861.

5 An investigation of machine learning algorithms for lung disease detection

Rakesh Kumar Roshan[a] and Devarani Devi Ningombam[b]

National Institute of technology Patna, Bihar, India

Abstract

Respiratory diseases are a major health concern globally, impacting millions of people. Conditions such as lung cancer, tuberculosis, and pneumonia pose serious risks to public health. Effective treatment and management of lung disorders depend on early and precise identification. Artificial intelligence has demonstrated encouraging outcomes in a number of medical applications, including the identification of lung diseases. Lung diseases, which affect the deepest layers of the lungs, can cause breathing difficulties and various other respiratory problems. An extensive review of the most recent artificial intelligence methods for lung disease detection is presented in this study. This investigation aims to emphasise significant advancements in the identification and evaluation of lung diseases, as well as to highlight existing issues and promising directions for the future. We have analysed various data sources, including all type of radiographic images, and discussed the challenges in diagnosing respiratory illnesses such as chronic airflow limitation, lung infection, and pulmonary tuberculosis. We also present the performance indicators and advantages and disadvantages of various methods. Lastly, we point out areas of unfinished business and offer potential future paths for enhancing lung disease diagnosis.

Keywords: Machine learning, deep learning, pneumonia, tuberculosis, radiographic image

Introduction

Lung disease is a type of respiratory disorder that is closely linked to issues with oxygen circulation, pulmonary tissue infection, and breathing difficulties in humans. The majority of respiratory illnesses are caused by abnormal cell formation, air pollution, and smoking. The pulmonary loop, which consists of the heart and lungs working together, ensures that the blood has the appropriate quantity of oxygen for every region of the body. According to the latest report published by the World Health Organization (WHO), tuberculosis (TB) ranks fifth among the leading causes of mortality in low-income countries within the category of communicable diseases.

The *Mycobacterium tuberculosis* bacteria cause TB, a dangerous lung illness that affects human pulmonary cells [7]. A chest X-ray of the human body can indicate TB infection, although primary tuberculosis cannot be diagnosed based solely on radiological evidence. The radiological example that shows pneumonia & tuberculosis is shown in Figure 5.1. The most typical signs and symptoms of TB include: more than three weeks of persistent coughing, blood or phlegm may come from the cough, chest pain especially when breathing or coughing, chest pain or discomfort fatigue, weakness, loss of appetite, and unexplained weight loss can all occur; these symptoms are frequently associated by low energy levels. Pneumonia is the second most common lung disease and can be caused by a virus, bacterium, or fungus.

As per the 2023 WHO report on the prevailing global status of tuberculosis (TB), the number of TB-related fatalities in 2022 was 1.3 million, indicating a modest decline from 1.4 million in 2021. Still, the most common cause of mortality for HIV-positive individuals is tuberculosis. Due to setbacks brought on by the COVID-19 epidemic, between 2020 and 2022, there were about 500,000 more TB-related fatalities. In 2022, India accounted for around 2.8 million cases of TB, or 27% of the global TB burden. The country has the greatest burden of TB cases worldwide. This fits with a broader global trend; in the same year, 7.5 million cases of tuberculosis were reported globally, the greatest number since the WHO started keeping count of infections in 1995.

Various machine learning (ML) methods have been utilised to support computer-assisted diagnostics and lung condition identification. ML techniques may be divided into two categories: those that accept a raw image as input and those that ask the user to select input parameters directly. Researchers frequently utilise deep learning (DL) approaches, including residual networks (ResNets), such as ResNet18 and ResNet50, have proven effective for feature extraction in medical imaging tasks [6], ResNet101, ChexNet, InceptionV3, VGG19, DenseNet201, SqueezeNet, and MobileNet, to analyse medical images.

For conditions like pneumonia, imaging techniques such as radiographs and computed tomography (CT) scans are commonly used. Pneumonia is a lung infection that causes

[a]rakeshr.ph22.cs@nitp.ac.in, [b]devarani.cs@nitp.ac.in

DOI: 10.1201/9781003675235-5

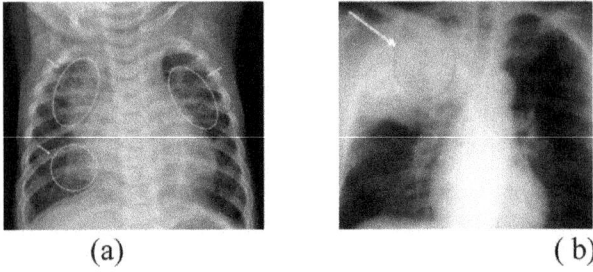

Figure 5.1 (a) shows a chest X-ray image with radiopaque areas showing fluid accumulation consolidations [12]. (b) Demonstrates the cavities and unification of the right lung's top lobe [7]

Source: Author

inflammation in the alveoli, leading to fluid or pus buildup. This condition can result in symptoms such as coughing, fever, chills, breathing difficulties, chest discomfort, and fatigue. It remains a major cause of death in children under five years old. Deep learning (DL) frameworks have been designed to automatically detect pneumonia from chest X-ray images. Numerous researchers have put up other frameworks for the diagnosis or prediction of pneumonia, such as the 121-layer dense convolutional network (DenseNet), which produces better results than logistic regression [1].

This research provides a summary of various methods for utilizing CT-scan and chest X-ray images for diagnosing pneumonia and tuberculosis in the lungs. The aim of this study is to learn about various AI techniques including ML and DL for tuberculosis and pneumonia disease detection and classification.

The structure of this paper is as follows: Section 2 explores related research; Section 3 examines model comparisons; Section 4 explores current methodologies; and Section 5 summarises the findings.

Literature study

Iqbal et al. [7] suggest two phases of the computer aided detection and diagnosis system that emphasises lung area segmentation and categorisation. The study explores TB-DenseNet for classifying tuberculosis in the lungs using X-ray images. After the input of chest radiographic images of dimension of 224 × 224, the architecture starts with dual convolution blocks for feature extraction, followed by a pre-trained DenseNet-169 block for deeper feature learning, then a feature fusion step, and finally a classification block that predicts the medical condition from the chest radiographic image. The proposed TB-DenseNet utilises Gradient-based Class Activation Mapping (Grad-CAM) to generate heatmaps over input chest X-ray images, enabling accurate localisation of lung-affected regions and providing insights into network learning from a medical imaging perspective. Densely connected networks, such as DenseNet, have been widely

adopted for medical screening applications due to their feature reuse and gradient propagation benefits [17].

The U-Net architecture, unveiled by Ronneberger et al. [13], employs an advanced segmentation technique designed specifically for segmentation tasks. Widely used in image segmentation, particularly in the medical field, U-Net consists of two main components: the down-sampling path (contracting path/encoder) and the up-sampling path (expanding path/decoder). This study introduces a U-Net architecture enhanced with a dilated fusion (DF) block and an attention block. The purpose of the dilated fusion block (DF) is to blend feature maps produced by several convolutional procedures and adaptively modify the network's response field, therefore efficiently overcoming the issues posed by the wide fluctuations in lung image quality. An attention coefficient is calculated using the sigmoid function for all feature maps from the dilated fusion blocks. Spatial detail vanishes because of cascaded convolution and non-linearity at deeper level of encoding from the DF block. To solve this problem, an attention block (AB) is included to improve network's sensitivity and accuracy for foreground pixels. By using Eq. (1), one may obtain the attention block's linear attention coefficient.

$$q_{attn}^t = \psi^T \left(\sigma \left(w_X^T F_{x_i}^L + w_G^T F_G + b_g \right) \right) + b\psi \qquad (1)$$

where, q_{attn}^t = attention map, $\emptyset^T$ = the linear transformation (learnable parameters), σ = ReLU activation function, w_x^T and w_G^T = Weights for feature maps x and g, respectively, F = Feature map operations (convolutions) and b_g = bias term. The attention block's function is to emphasise pertinent aspects by integrating activation maps from many levels. To guarantee non-linearity and efficient feature selection, it employs both ReLU and Sigmoid activations. Prior to calculating an attention coefficient, which aids in feature map refinement, the procedure combines both spatial and contextual data.

Moreover, reinforcement learning (RL) has played a key role in categorising diseases like tuberculosis (TB) and pneumonia [10]. RL helps to develop adaptive, self-learning classifiers by allowing systems to learn the best strategies through interaction and feedback, without needing mathematical models or predefined outcomes. The Modified Fuzzy Q-Learning algorithm (MFQL) is applied to improve learning by using fuzzy logic, which handles uncertainty and continuous data. Wavelets are used to reduce data complexity, thereby speeding up the process. The RL-based approach achieves classification success rates of 90.85% for TB and 96.75% for pneumonia.

Pneumonia is a serious respiratory illness that poses a major risk to both children and older adults, often leading to severe lung infections [4]. This research introduces a deep learning framework that utilises four different models to detect and classify pneumonia using chest X-ray images.

This two-tier process involves:

1. **First tier: pre-processing**
   - Chest X-ray images are imported, resized, augmented, normalised, and split into training, validation, and testing sets.
2. **Second tier: feature extraction and classification**
   - MobileNetV2 and ResNet152V2, two pre-trained models, are used for feature extraction and classification. MobileNetV2 provides efficient and fast feature extraction, while ResNet152V2 performs deeper analysis to capture subtle signs of pneumonia.
   - Custom models, CNN and LSTM-CNN, are also used. CNN focuses on traditional image classification, while LSTM-CNN combines CNN for spatial features and LSTM to capture more complex spatial dependencies.

Results show high accuracy in pneumonia detection: 99.22% for ResNet152V2, 96.48% for MobileNetV2, 92.19% for CNN, and 91.80% for LSTM-CNN. The authors also proposed a DL framework for pneumonia detection using transfer learning. This research utilises image pre-processing, data augmentation, and pre-trained convolutional neural network (CNN) models such as AlexNet, DenseNet121, InceptionV3, ResNet18, and GoogLeNet. These models, originally trained on the ImageNet dataset, assist in extracting features from X-ray images, improving both training efficiency and classification accuracy compared to training from scratch. The ensemble approach, which combines multiple CNN architectures, achieves a leading accuracy of 96.39%, outperforming individual models: ResNet18 (94.23%), GoogLeNet (93.12%), AlexNet (92.86%), DenseNet121 (92.62%), and InceptionV3 (92.01%).

Model comparison

The TB-DenseNet model presented by Ronneberger et al. [13] was trained on four datasets and evaluated using chest X-ray images. Its performance was compared with other models based on specificity, sensitivity, F1 metric, and accuracy. These measures evaluate the reliability of positive predictions, the model's responsiveness to true positives, the equilibrium between specificity and sensitivity, and the overall accuracy. TB-DenseNet's results were compared to models like ResNet50 [16], VGG16 [5], EfficientNet (Scaling et al., 2019), and others across different datasets. The findings show that TB-DenseNet outperforms all other models.

To emphasise the benefits of the proposed Xception Network model, Lujan-García et al. [12] evaluated its effectiveness in relation to other cutting-edge benchmark models, including VGG-16, ResNet50v2 [8], DenseNet121, and the high-performing LZNet2019 [11]. For performance comparison, authors used X-ray images of size 224 × 224 as the image input for VGG-16 and DenseNet121. Additionally, the LZNet2019 consider images of size 150 × 150. Conversely, 299 × 299 is the X-ray image input form in the proposed model, which also incorporates VGG-16, ResNet50v2, and DenseNet121. The model's performance is assessed using Gradient-weighted Class Activation Mapping (Grad-WCAM), which produces a heatmap to highlight areas potentially affected by pneumonia. This study evaluates the performance metrics of different models for pneumonia classification on a test dataset. The proposed approach surpasses other models in nearly all evaluated metrics.

Kukker et al. [10] proposed a Modified Fuzzy Q-Learning (MFQL) classifier for diagnosing pneumonia and tuberculosis, incorporating wavelet-based feature extraction. This classifier identifies cases as no pneumonia, mild pneumonia, severe pneumonia, tuberculosis present, or tuberculosis absent. By utilizing wavelet-derived features from X-ray images, MFQL enhances the accuracy of classification. Compared to Support Vector Machines (SVM) and K-Nearest Neighbour (KNN), MFQL performs better, especially with wavelet-based features. It achieves 91.6% accuracy for detecting tuberculosis and 90.1% for ruling it out. For pneumonia, MFQL records 95.7% accuracy for moderate cases, 96.17% for severe cases, and an overall accuracy of 98.4%. On average, MFQL achieves 96.75% accuracy for pneumonia and 90.85% for tuberculosis, outperforming KNN (92%) and SVM (94%) in pneumonia classification.

The enhanced decision-making capabilities of Q-learning in conjunction with fuzzy logic in MFQL may be responsible for this increased accuracy. By obtaining more precise image characteristics, wavelet-based feature extraction is also probably responsible for the enhanced classification performance. A comparative finding of performance of literature work using various datasets is presented in Table 5.1.

The proposed ensemble model by Dasanayaka et al. [3] is based on the UNet architecture for semantic segmentation of image data, specifically for chest X-rays to detect TB infections. It uses the Dice loss function for training, which is common in medical segmentation. The model's performance is evaluated using Youden's Index, accuracy, specificity, and sensitivity, and is compared to VGG16 and InceptionV3.

- **Youden's Index** combines sensitivity and specificity to assess diagnostic effectiveness.
- **Accuracy** reflects the proportion of correct results.
- **Specificity** evaluates the accurate recognition of negative cases.

- **Sensitivity** evaluates the accurate recognition of positive cases.

The proposed model outperforms VGG16 and InceptionV3, achieving 94.1% for Youden's Index, an accuracy of 97.1%, specificity of 96.3%, and sensitivity of 97.9%. In comparison, VGG16 scores 87.5%, 93.8%, 93.0%, and 94.5%, and InceptionV3 scores 92.5%, 96.3%, 95.3%, and 97.2% for these metrics. A comparative analysis based on Youden's Index is presented in Table 5.2.

The pneumonia detection model proposed by Elshennawy et al. [4] uses CNN and LSTM-CNN models for evaluation, and ResNet152V2 and MobileNetV2 for pre-training. Stochastic Gradient Descent (SGD) with a learning rate of 0.0001 is utilised for pre-trained models, whereas the Adamax optimizer with learning rates of 0.00003 and 0.00006 is employed for the proposed models. Key performance findings are as follows:

- **ResNet152V2** (pre-trained) achieved the best results across all metrics: Accuracy (99.22%), Precision (99.44%), AUC (99.77%), F1-Score (99.44%), and Recall (99.43%).
- **MobileNetV2** (pre-trained) performed slightly lower but still well, particularly in Recall (99.44%), with lower scores in Accuracy (96.48%) and F1-Score (97.52%).
- **CNN** (proposed model) showed moderate performance: Accuracy (92.19%), Precision (95.57%), and Recall (92.07%).
- **LSTM-CNN** (proposed model) had the lowest performance: Accuracy (91.80%), F1-Score (92.29%), but performed decently in Recall (92.62%).

The transfer learning technique is utilised to identify pneumonia from chest radiographs by combining the outputs of five well-known deep learning architectures, namely AlexNet, DenseNet121, InceptionV3, GoogLeNet, and ResNet18. This method, known as ensemble learning, aggregates the strengths of each model to improve classification performance.

1. **Prediction aggregation**: The ensemble uses majority voting, where the final prediction is the class that most models agree on, improving the reliability of predictions.
2. **Performance improvement**: The ensemble outperformed individual models with:
   - **Test accuracy**: 96.39%, meaning 96.39% of test samples were correctly classified.
   - **AUC (Area under ROC Curve)**: 99.34%, showing excellent ability to distinguish between classes.

- **Sensitivity (recall)**: 99.62%, indicating that the model correctly identified 99.62% of true positive cases.

Discussion

This is an outline of various approaches for lung disease detection and classification, particularly TB and pneumonia, using DL models and computer-aided diagnosis (CAD) systems. That include:

Lung segmentation and classification
The lung segmentation and classification process involves using a modified UNet architecture for lung area segmentation, incorporating a Dilated Fusion (DF) block to address variations in image quality and an Attention Block (AB) to focus on important areas. The attention mechanism uses both ReLU and Sigmoid activations for better feature refinement. For TB classification, the TB-DenseNet model, based on DenseNet-169, was proposed. The Dice loss function is used for segmentation, and an ensemble of VGG-16 and InceptionV3 is employed for classification, with 97.10% accuracy.

DL framework for tuberculosis diagnosis
DConv-GAN is employed to overcome the data shortage for TB diagnosis by generating synthetic data.

DL framework for pneumonia detection
Chouhan et al. [2] proposed a transfer learning-based model for pneumonia detection using chest X-rays, demonstrating that pre-trained models can significantly enhance classification performance. A two-tier pneumonia detection system utilises ResNet152V2, MobileNetV2, CNN, and LSTM-CNN for data pre-processing and feature extraction, with the document highlights the use of pre-trained models combining these networks outperforms individual models with 96.39% accuracy, compared to the best individual result from ResNet18 at 94.23%.

The ensemble models like AlexNet [9], one of the earliest deep CNN architectures trained on ImageNet, is also employed in this study for pneumonia classification., DenseNet121, InceptionV3 has an ability to handle multiscale features [14] for early Inception advancements [15], ResNet18, and GoogLeNet are used for pneumonia detection through transfer ResNet152V2, achieving the highest accuracy of 99.22%.

Transfer learning for pneumonia detection
Overall, this study demonstrates significant progress in lung disease diagnosis using advanced DL techniques, improving both segmentation and classification accuracy across various models.

Table 5.1 Comparative findings of performance of literature work using various datasets.

Reference	Iqbal et al., 2023				García et al., 2020	Kukker et al., 2021	
Model used	**TB-DenseNet**				**Xception Network**	**Modified Fuzzy Q-Learning (MFQL)**	
Disease	**TB**				**Pneumonia**	**Pneumonia**	**TB**
Dataset used	Dataset A: Shenzhen Hospital (SH) and Montgomery County (MC) dataset	Dataset B: Ministry of Health, Republic of Belarus.	Dataset C: TB Portals – a publicly accessible web-based platform.	Dataset D: Kaggle public dataset	Dataset introduced by Kermany et al. in 2018.	Kaggle data base National Library of Medicine, Maryland, USA	
Accuracy (%)	98.98	97.23	96.06	95.1	‥	98.4	90.85 (TB)
F1 Score	99.01	97.22	96.08	95.38	0.912	‥	‥
Precision (%)	99.03	97.21	96.09	95.67	84.3	‥	‥
Recall (%)	98.98	97.23	96.06	95.1	99.2	96.75	90.85 (TB)
AUC (%)	‥	‥	‥	‥	96.8	‥	‥
Sensitivity (%)	‥	‥	‥	‥	99.2	‥	‥

Source: Author

Table 5.2 Comparative findings for Youden's index-based papers.

Reference	Dasanayaka et al., 2021	Elshennawy et al., 2020				Chouhan et al., 2020
Model used	**U-Net based**	**Ensemble model, ResNet152V2 (pre-trained), MobileNetV2 (pre-trained), LSTM-CNN**				**Ensemble Learning**
Disease	**TB**	**Pneumonia**				**Pneumonia**
Dataset used	Shenzhen No. 3 hospital in Shenzhen, Guangdong, China	Kaggle data base				Guangzhou Women and Children's Medical
		Ensemble model	**ResNet152V2**	**MobileNetV2**	**LSTM-CNN**	**LSTM-CNN**
Accuracy (%)	97.1	99.22	96.48	92.19	91.8	96.39
F1 Score	‥	99.44	97.52	92.29	‥	‥
Precision (%)	‥	99.44	‥	95.57	‥	‥
Recall (%)	97.9	99.43	99.44	92.07	92.62	99.62
AUC (%)	N/A	99.77	‥	‥	‥	99.34
Specificity (%)	96.3	‥	‥	‥	‥	‥
Youden's Index	94.1	‥	‥	‥	‥	‥

Source: Author

Reinforcement learning (RL) for disease classification

Reinforcement learning, specifically Modified Fuzzy Q-Learning (MFQL), is utilised for classifying tuberculosis and pneumonia, attaining 90.85% accuracy for TB and 96.75% for pneumonia. To reduce data size, wavelet transformation is employed, while MFQL effectively manages uncertainty and continuous data in decision-making. Additionally, Contrast Limited Adaptive Histogram Equalization (CLAHE) enhances image quality before model training.

Conclusion

This research emphasises the efficiency of advanced artificial intelligence techniques, including deep learning models and computer-aided diagnosis (CAD) systems, in accurately detecting lung diseases such as tuberculosis and pneumonia. By leveraging models such as UNet for segmentation, DenseNet for feature extraction, and ensemble techniques combining pre-trained networks, these approaches have achieved high accuracy rates in both classification and segmentation tasks. Techniques like Dilated Fusion, Attention Blocks, reinforcement learning with Modified Fuzzy Q-Learning, and transfer learning further enhance the model's performance. Overall, these innovations represent a significant advancement in the early detection of lung diseases, demonstrating the potential for widespread clinical applications with high precision and efficiency.

References

[1]   Antin, Benjamin, Joshua Kravitz, Emil Martayan. (2021). Detecting pneumonia in chest X-Rays with supervised learning.

[2]   Chouhan, V., et al. (Jan. 2020). A novel transfer learning based approach for pneumonia detection in chest X-ray images. *Applied Sciences (Switzerland)*, 10(2). doi: 10.3390/app10020559.

[3]   Dasanayaka, C., Dissanayake, M. B. (2021). Deep learning methods for screening pulmonary tuberculosis using chest X-rays. *Computer Methods in Biomechanics and Biomedical Engineering: Imaging & Visualization*, 9(1), 39–49. doi:10.1080/21681163.2020.1808532.

[4]   Elshennawy, N. M., Ibrahim, D. M. (Sep. 2020). Deep-pneumonia framework using deep learning models based on chest X-ray images. *Diagnostics*, 10(9). doi: 10.3390/diagnostics10090649.

[5]   Geng, Zhang, S., Tong, J., Xiao, Z. (Oct. 2019). Lung segmentation method with dilated convolution based on VGG-16 network. *Computer Assisted Surgery*, 24(suppl. 2), 27–33. doi: 10.1080/24699322.2019.1649071.

[6]   He, K., Zhang, X., Ren, S., Sun, J. (2015). Deep Residual Learning for Image Recognition. [Online]. Available: http://image-net.org/challenges/LSVRC/2015/

[7]   Iqbal, Ahmed, Muhammad Usman, Zohair Ahmed. (2023). Tuberculosis chest X-ray detection using CNN-based hybrid segmentation and classification approach. *Biomedical Signal Processing and Control*, 84 (July). Elsevier Ltd. doi: 10.1016/j.bspc.2023.104667.

[8]   Jung M., Chi, S. (Jun. 2020). Human activity classification based on sound recognition and residual convolutional neural network. *Automation in Construction*, 114. doi: 10.1016/j.autcon.2020.103177

[9]   Krizhevsky, A., Sutskever, I., Hinton, G. E. (2015). ImageNet Classification with Deep Convolutional Neural Networks. [Online]. Available: http://code.google.com/p/cuda-convnet/

[10] Kukker, A., Sharma, R. (Oct. 2021). Modified fuzzy Q learning based classifier for pneumonia and tuberculosis," *IRBM*, 42(5), 369–377. doi: 10.1016/j.irbm.2020.10.005.

[11] Liang, G., Zheng, L. (Apr. 2020). A transfer learning method with deep residual network for pediatric pneumonia diagnosis. *Computer Methods and Programs in Biomedicine*, 187. doi: 10.1016/j.cmpb.2019.06.023.

[12] Lujan-García, Juan Eduardo, Cornelio Yáñez-Márquez, Yenny Villuendas-Rey, Oscar Camacho-Nieto. (2020). A transfer learning method for pneumonia classification and visualization." *Applied Sciences (Switzerland)*, 10(8). MDPI AG.

[13] Ronneberger, O., Fischer, P., Brox, T. (2015). U-net: Convolutional networks for biomedical image segmentation. In Lecture Notes in Computer Science (including subseries Lecture Notes in Artificial Intelligence and Lecture Notes in Bioinformatics), Springer Verlag, 234–241.

[14] Szegedy, C., Vanhoucke, V., Ioffe, S. and Shlens. J. (2016). Rethinking the inception architecture for computer vision.

[15] Szegedy, C., et al. (2015). Going deeper with convolutions. In 2015 IEEE Conference on Computer Vision and Pattern Recognition (CVPR), Boston, MA, USA, 1–9. doi: 10.1109/CVPR.2015.7298594 14 L.

[16] Wang, F., et al. (2017). Residual attention network for image classification. In Proceedings – 30th IEEE Conference on Computer Vision and Pattern Recognition, CVPR 2017, Institute of Electrical and Electronics Engineers Inc., 6450–6458. doi: 10.1109/CVPR.217.683.

[17] Yao, Z., Li, J., Guan, Z., Ye, Y., Chen, Y. (Mar. 2020). Liver disease screening based on densely connected deep neural networks," *Neural Networks*, 123, 299–304. doi: 10.1016/j.neunet.2019.11.00

6 Beyond characters: a machine learning approach to password strength analysis

Mrinmoy Borah[a] and Afsana Laskar[b]

Royal School of Information Technology, The Assam Royal Global University, Guwahati, India

Abstract

With the ever-increasing reliance on digital systems, the importance of robust password security has become paramount. Passwords function as a barrier against unauthorised access, thwarting cybercriminals from commandeering our accounts and potentially causing financial harm or identity theft. This study investigates the strength of passwords using several classification machine learning techniques. We have implemented a hybrid methodology for feature engineering, integrating TF-IDF with a structural analysis of each password. Three novel algorithms—K Nearest Neighbors (KNN), Convolutional Neural Network (CNN), and Ensemble Neural Network (ENN)—alongside with conventional machine learning classification techniques, to assess password strength, have been introduced. Of the nine machine and deep learning models, seven demonstrated approximately 99% accuracy.

Keywords: Password strength analysis, KNN, ENN, CNN

Introduction

Passwords are combinations of characters, such as uppercase, lowercase, symbols, and numbers, used to secure access to digital systems. Password security is crucial in the contemporary digital landscape, as passwords serve as the principal mechanism for safeguarding personal, financial, and sensitive information across diverse online platforms; they are, in essence, the keys to our digital existence. According to Bitwarden's research, convenience often drives 85% of individuals to reuse identical passwords across several websites and accounts, resulting in the difficulty of recalling numerous unique passwords [3]. In such scenarios, having a strong password plays an important role. Weak passwords provide hackers simple entry points, allowing them to access our emails, bank accounts, and other sensitive information. Social engineering, dictionary attacks, reverse brute-force attacks, brute-force attacks, and password guessing are a few common ways through which hackers can access our digital information [1,7,16,17,18].

In our analysis of password strength, we applied a range of machine learning (ML) and deep learning (DL) models, acknowledging their extensive use in cybersecurity for purposes such as detecting fraud and spam. Machine learning represents a fusion of mathematics and statistics that empowers computers to learn from data and make predictions autonomously, without the need for explicit programming [6]. Such algorithms can recognise patterns on their own from data, thereby eliminating the necessity for human involvement in numerous classification tasks. Deep learning uses neural networks with multiple hidden layers to discern intricate relationships within data. The architecture and function of the human brain, which processes information through interconnected neuronal layers, serve as a model for these neural networks. This study examines the classification of password strength using the subsequent methodologies. First, we use a dataset containing passwords classified as weak, medium, and strong. Next, after pre-processing, feature engineering, and partitioning the data, we train nine ML and DL learning classifiers on the dataset. Finally, we evaluate the results using various performance metrics.

The structure of this paper is as follows: Section 2 reviews the existing literature. Section 3 provides an outline of the dataset. Section 4 details the research methodology. Section 5 discusses the findings. Section 6 concludes with a summary of the paper.

Literature review

Researchers have performed comprehensive studies on password strength analysis and password guessing using ML and DL techniques. Some studies looked at deep learning algorithms like LSTM, ANN, RNN, and GAN in more detail [2,6,8,12], while others employed machine learning methods such as Decision Trees, Support Vector Machines, Logistic Regression, Naive Bayes, XGBoost, and Random Forests to analyse password strength [6,9,11]. Researchers have also conducted studies on passwords from minor languages, such as Lithuanian [5]. Most studies employed TF-IDF to extract features from individual passwords, often neglecting structural analysis of the passwords.

[a]mrinmoyborahdev@gmail.com, [b]laskar.afs@gmail.com

DOI: 10.1201/9781003675235-6

There are several password strength assessment tools that adhere to traditional LUDS criteria (lowercase and uppercase letters, digits, and special characters) widely used by tech giants to assess passwords. However, Dropbox deviates from this convention, employing zxcvbn to evaluate a password's strength by considering its commonality across multiple sources [4,5,8]. Claude Shannon initially introduced the concept of entropy in his information theory. Researchers have examined passwords with an entropy-based combinatorial approach and an enhanced method that incorporates more sophisticated elements to address the deficiencies of conventional password strength assessments [4,15]. Password guessing extensively relies on the concept of entropy [10,19].

This research builds upon earlier investigations in the domain by utilising a hybrid feature engineering methodology that integrates TF-IDF with structural password analysis. It aims to examine the dataset using both machine and deep learning algorithms to assess password strength.

Data and variables

We utilised a dataset from one of the most significant password breaches: 000webhost, which is accessible on Kaggle. Georgia Tech University developed the PARS tool, which categorised the passwords into three classifications: 0 for weak, 1 for medium, and 2 for strong [2]. The dataset comprises 669,640 passwords, categorised as follows: 89,702 are classified as weak (0), 496,801 as medium (1), and 83,137 as strong (2). There is an imbalance in the dataset: 74.18% of the passwords are in class 1, 13.39% are in class 0, and 12.41% are in class 2. Figure 6.1 illustrates the imbalance between different classes of passwords in the dataset.

Dependent variables

The dependent variable is the target we seek to predict or explain; in simpler terms, it represents the output. In this dataset, the strength column serves as the dependent variable and consists of three classes: 0 (weak), 1 (medium), and 2 (strong).

Independent variables

The independent variable is the one we alter or choose to assess its impact on the dependent variable; input values represent the independent variables. The passwords in the password column of the dataset denote independent variables.

Methodology

Pre-processing

Pre-processing entails converting unrefined data into a polished, organised version appropriate for modelling by removing missing values, eliminating duplicates, and guaranteeing an equitable representation of all classes within the dataset to mitigate bias.

We undersampled the class 1 passwords to 86,419, the mean of the class 0 and class 2 passwords, to mitigate biases in the dataset, as outlined in the Data and variables section. Upon eliminating all missing values and duplicates from the dataset, the total count of passwords is 259,182. We categorise 89,662 passwords as class 0, 86,405 as class 1, and 83,115 as class 2. Figure 6.2 illustrates the dataset after pre-processing.

Feature extraction

The process of feature extraction consists of recognising and choosing the most pertinent characteristics from the dataset that are related to the specific problem being tackled. This study utilises a hybrid methodology that integrates TF-IDF with structural analysis of passwords applied to the dataset, a technique not previously implemented.

*   *Term frequency-inverse text frequency (TF-IDF)*: TF-IDF is a popular method in natural language process-

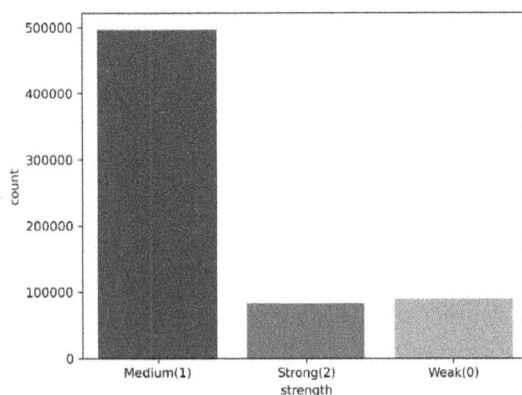

Figure 6.1 Dataset before pre-processing
Source: Author

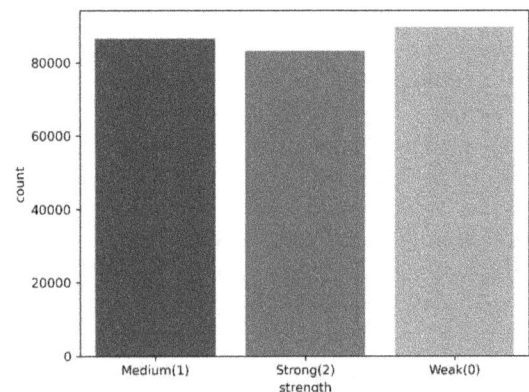

Figure 6.2 Dataset after pre-processing
Source: Author

ing and information retrieval that assesses the significance of a word or term within a text in relation to a corpus of documents. Mathematically, it can be defined as follows:

*TF-IDF (t, d) = TF (t, d) * IDF (t)*

In this context, *t* represents the term (word) being analysed, and *d* denotes the document in which the term appears. *TF(t, d)* stands for Term Frequency, indicating how often the term *t* appears in the document *d*. *IDF(t)* refers to Inverse Document Frequency, which measures the rarity of the term *t* across the entire document set [2,6,9].

- *Structural analysis*: Structural analysis of passwords refers to the quantifiable attributes that delineate a password's composition, encompassing the counts of uppercase letters, lowercase letters, symbols, digits, and the total length of the password. Table 6.1 illustrates the structural analysis of a password.

Training and testing the model

We utilised Scikit-learn to perform standard scaling on the input columns and partition the dataset into two sets: 80% designated for training and 20% allocated to evaluate our models.

Machine learning classifiers

We utilised Scikit-learn and TensorFlow to construct nine ML and DL algorithms: K-Nearest Neighbors (KNN), Random Forest (RF), Decision Tree (DT), Naive Bayes (NB), Multilayer Perceptron (MLP), Logistic Regression (LR), Neural Network (NN), Convolutional Neural Network (CNN), and Ensemble Neural Network (ENN) [5,13,14,20]. Prior studies in this area did not employ the following algorithms:

- *K-Nearest Neighbors (KNN)*: KNN is a supervised ML algorithm that classifies a data point based on the configuration of its neighbours. It depends on distance measures, including Euclidean, Manhattan, Minkowski, and Hamming distances, along with data similarities.
- *Ensemble Neural Network (ENN)*: An ensemble neural network integrates many neural networks to enhance forecast accuracy and resilience. We employed the bagging ensemble learning technique, wherein we randomly sampled the dataset into smaller subgroups. We additionally trained 100 models utilising each

bootstrap sample. This approach reduces variability and alleviates overfitting by aggregating the predictions of multiple models, resulting in improved accuracy of results.

- *Convolutional Neural Network (CNNs)*: They are specialised neural networks designed to process structured, grid-like data, such as images. By employing convolutional layers, CNNs extract hierarchical features from input data, facilitating pattern recognition and spatial hierarchy analysis. They play an important role in computer vision tasks, including image classification, object detection, and image segmentation.

Results

Our research indicates that the suggested combination of machine and deep learning models can effectively estimate and classify the strength of any password. Among all the algorithms, RF achieved the highest accuracy of 0.9998 and NB recorded the lowest accuracy of 0.6143. The findings demonstrate that the accuracy recorded for the KNN model was 0.8451, whereas the ENN model demonstrated an accuracy of 0.9984, and the CNN model achieved an accuracy of 0.9988. Table 6.2 below details the performance metrics for each model, which include accuracy, recall, precision, and F1 score. Figure 6.3 refers to confusion matrices of all ML and DL algorithms

Conclusion

This study aimed to assess password strength using ML and DL algorithms. The process involved data collection, pre-processing, feature engineering, model selection, and evaluation. For feature engineering, a hybrid approach was employed, integrating TF-IDF with structural analysis

Table 6.1 Structural analysis of a password.

Password	Uppercase	Lowercase	Symbol	Numbers
abc123#	0	3	1	3

Source: Author

Table 6.2 Performance metrics of machine and deep learning classifiers.

	KNN	DT	RF	NB	LR
Accuracy	0.8451	0.9997	0.9998	0.6143	0.9995
Precision	0.8590	0.9998	0.9998	0.7338	0.9995
Recall	0.8456	0.9998	0.9998	0.6262	0.9995
F1 Score	0.8441	0.9998	0.9998	0.5188	0.9995

	MLP	NN	ENN	CNN
Accuracy	0.9995	0.9991	0.9984	0.9988
Precision	0.9995	0.9991	0.9984	0.9988
Recall	0.9995	0.9991	0.9984	0.9988
F1 Score	0.9995	0.9991	0.9984	0.9988

Source: Author

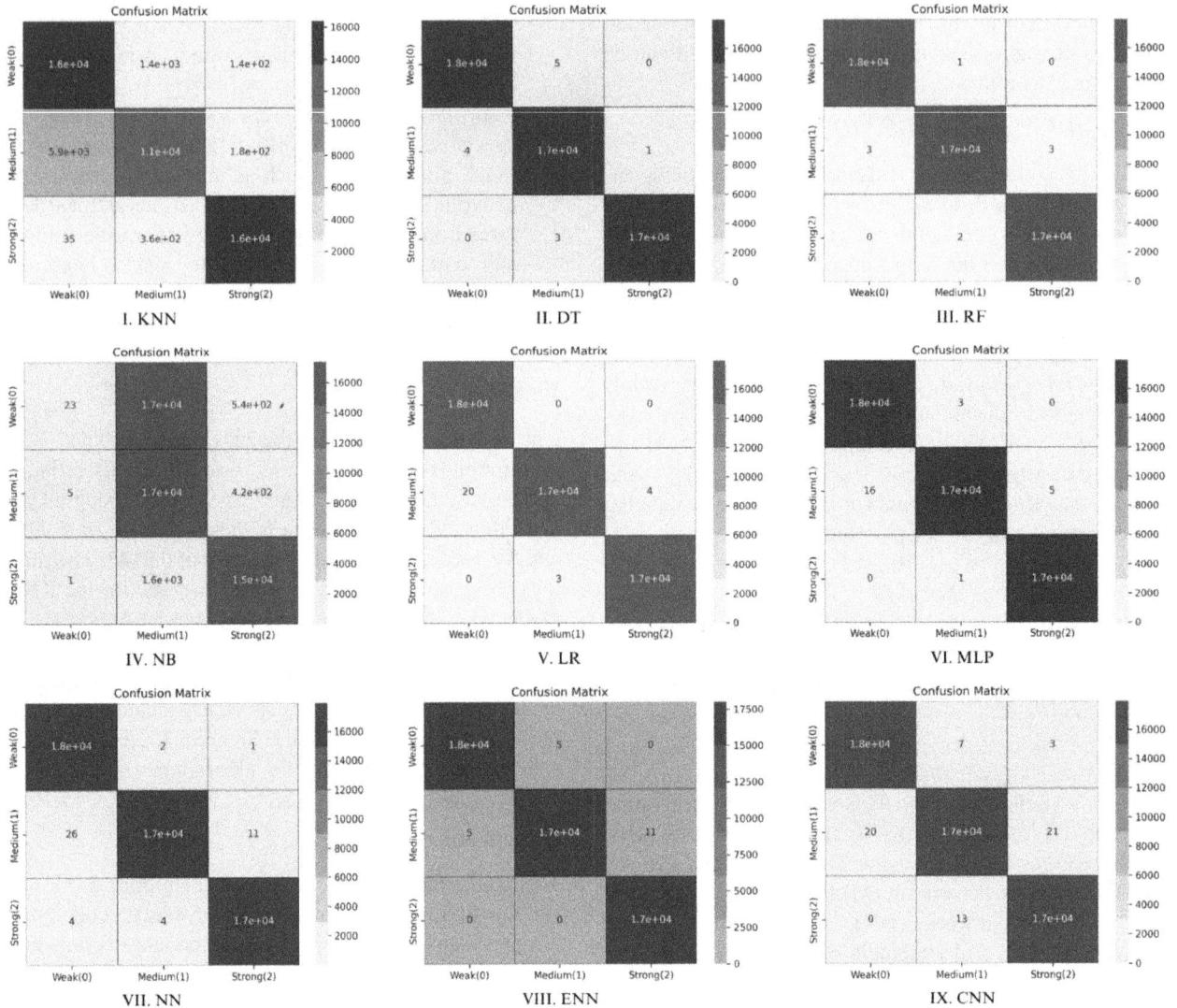

Figure 6.3 Confusion matrices of ML and DL algorithms
Source: Author

of each password. Three novel algorithms—KNN, ENN, and CNN—were implemented alongside other classification algorithms. Among these, Random Forest achieved the highest performance with an accuracy of 0.9998.

The results indicate that both ML and DL methodologies possess the capacity to transform password strength analysis, offering more accurate instruments for bolstering security in digital systems. Subsequent investigations may examine a broader array of password datasets and integrate real-time analytical techniques. Future studies could explore the use of regional languages as passwords instead of English, and a detailed analysis of the relationship between entropy and password length could improve prediction models. The findings of this study underscore the critical role of ML and DL algorithms in the assessment of password strength, which in turn facilitates the development of more robust authentication systems.

References

[1]   Asaduzzaman, Abu, Declan D'souza, Md Raihan Uddin, Yoel Woldeyes. (2024). Increase security by analyzing password strength using machine learning. In 2024 Joint International Conference on Digital Arts, Media and Technology with ECTI Northern Section Conference on Electrical, Electronics, Computer and Telecommunications Engineering (ECTI DAMT & NCON). Chiang-mai, Thailand: IEEE, 5.

[2]   Belikov, Vladimir V., Prokuronov, I. A. (2023). Password strength verification based on machine learning algorithms and LSTM recurrent neural networks. *Russian Technological Journal*, 9.

[3]   Bitwarden. (Aug. 8, 2023). Security habits around the world: a closer look at password security statistics. (accessed 22-10-2024). https://bitwarden.com/blog/a-closer-look-at-password- statistics/.

[4]   Chowdhury, Naem Azam. (2024). Analyzing password strength: a combinatorial entropy approach. In The 22nd International Conference on Computer and Information Technology, 9.

[5]   Darbutait, Ema, Stefanovi, Pavel, Ramanauskait, Simona. (2023). Machine-learning-based password-strength-estimation approach for passwords of Lithuanian context. *Applied Sciences*, 15.

[6]   Farooq, Umar. (2020). Real time password strength analysis on a web application using multiple machine learning approaches. *International Journal of Engineering Research and Technology*, 6.

[7]   Hatice, Serkan, Mehmet Ali. (2021). Password attack analysis over honeypot using machine learning password attack analysis. *Turkish Journal of Mathematics and Computer Science*, 15.

[8]   Kim, Seok Jun, Lee, Byung Mun. (2023). Multi-class classification prediction model for password strength based on deep learning. *Journal of Multimedia Information System*, 8.

[9]   Sarkar, Sakya, Nandan, Mauparna. (2022). Password strength analysis and its classification by applying machine learning based techniques. In 2022 Second International Conference on Computer Science, Engineering and Applications (ICCSEA). Gunupur, India: IEEE, 1–5.

[10] Ur, Blase, Noma, Fumiko, Bees, Jonathan, Segreti, Sean M., Shay, Richard, Bauer, Lujo, Christin, Nicolas, Cranor, Lorrie Faith. 2015. "I Added'!'at the End to Make It Secure": Observing Password Creation in the Lab. In Eleventh Symposium on Usable Privacy and Security (SOUPS 2015). Ottawa, Canada: USENIX, 123–140.

[11] Wang, Ding, Zou, Yunkai, Zhang, Zijian, Xiu, Kedong. (2023). Password guessing using random forest. In 32nd USENIX Security Symposium (USENIX Security 23). N/A: USENIX, 965–982.

[12] Zhang, Tao, Cheng, Zelei, Qin, Yi, Li, Qiang, Shi, Lin. (2020). Deep learning for password guessing and password strength evaluation, a survey. In 19th International Conference on Trust, Security and Privacy in Computing and Communications (TrustCom). Guangzhou, China: IEEE, 5.

[13] Divya, R., Shridhar B. Devamane, Dharshini V., Deepika, S. (2023). Performance analysis of machine learning algorithms for password strength check. In 2023 International Conference on Computational Intelligence for Information, Security and Communication Applications (CIISCA). IEEE, 338–343.

[14] Kuriakose, Sony, Krishna Teja, G., Sravan Duggi, Harshel Srivatsava, A., Venkat Jonnalagadda. Machine learning based password strength analysis. *International Journal of Innovative Technology and Exploring Engineering*, 11(8), 5–8. *Blue Eyes Intelligence Engineering and Sciences Engineering and Sciences* , 2022.

[15] Taneski, Viktor, Marko Kompara, Marjan Heričko, Boštjan Brumen (2021). Strength analysis of real-life passwords using Markov models. *Applied Sciences*, 11(20), 9406.

[16] Yıldırım, M., Ian Mackie. (2019). Encouraging users to improve password security and memorability. *International Journal of Information Security*, 18, 741–759.

[17] Hingmire, Amruta, Sinju Saliya. (2017). A multimodal metric for password strength estimation. *International Journal of Recent Trends in Engineering & Research*, 3, 21–30.

[18] Melicher, William, Blase Ur, Sean M. Segreti, Saranga Komanduri, Lujo Bauer, Nicolas Christin, Lorrie Faith Cranor. (2016). Fast, lean, and accurate: modeling password guessability using neural networks. In 25th USENIX Security Symposium (USENIX Security 16), 175–191.

[19] Kelley, Patrick Gage, Saranga Komanduri, Michelle L. Mazurek, Richard Shay, Timothy Vidas, Lujo Bauer, Nicolas Christin, Lorrie Faith Cranor, Julio Lopez. (2012). Guess again (and again and again): measuring password strength by simulating password-cracking algorithms. In 2012 IEEE Symposium on Security and Privacy. IEEE, 523–537.

[20] Suganya, G., Karpgavalli, S., Christina, V. (2010). Proactive password strength analyzer using filters and machine learning techniques. *International Journal of Computer Applications*, 7(14), 1–5.

7 Comparative analysis of machine learning-based specific capacitance prediction and experimental validation of supercapacitive performance in cerium-based electrodes

Mahatim Singh[1,a], Neeraj Sharma[2,b], Pangambam Sendash Singh[3,c] and Preetam Singh[4,d]

[1]Research Scholar, School of Biomedical Engineering, Indian Institute of Technology (BHU), Varanasi, UP, India

[2]Professor, School of Biomedical Engineering, Indian Institute of Technology (BHU), Varanasi, UP, India

[3]Assistant Professor, Department of Computer Science & Applications, Dr. Harisingh Gour Vishwavidyalaya Sagar, M.P., India

[4]Associate Professor, Department of Ceramic Engineering, IIT (BHU), Varanasi, UP, India

Abstract

Specific capacitance is regarded as the most vital performance-related characteristic for supercapacitor electrodes, making its prediction essential for assessing the suitability of a material for supercapacitor-electrode applications. The unique properties of cerium position it as a highly promising candidate for supercapacitors, affecting the functionality of these devices. This study sought to examine the essential aspects of cerium's properties in relation to its applications in supercapacitors. Machine learning techniques were utilised to forecast the capacitance of cerium-based supercapacitors. A comprehensive collection of experimental datasets from various publications was assembled to train the ML models, enabling the assessment of the significance of various electrode characteristics. This technique may be instrumental in predicting and discovering improved cerium electrode materials.

Keywords: Supercapacitor, specific capacitance, electrode

Introduction

The electrochemical kinetics of electrode materials is essential for advancing a range of energy storage technologies, including batteries, supercapacitors, and hybrid supercapacitors. Battery-type hybrid supercapacitors are considered promising solutions to address the performance disparity between supercapacitors and batteries [16]. The electrode is a fundamental component in the research and development of supercapacitors. Over the years, numerous material [5,9] classes have been explored for the creation of electrodes used in supercapacitors. Cerium oxide (CeO2) is classified as a ceramic material that exhibits advantageous electrochemical characteristics relative to a range of widely employed materials. Its potential as an active material for supercapacitor applications, especially when combined with nanostructured composites, is currently the focus of significant research efforts [7].

A data-driven approach, known as the machine learning algorithm, has been utilised as an alternative solution to address various real-world challenges [15]. This method involves developing models and techniques that allow computers to learn from data, enabling them to make predictions or decisions without the need for explicit programming, which is a fundamental aspect of the artificial intelligence domain of machine learning [2,14]. The application of machine learning (ML) techniques to enhance energy storage performance has shown significant promise, resulting in its utilisation across multiple disciplines, including physics, chemistry, biology, and engineering [5,6,13]. Notably, integrating machine learning into the design and management of energy storage systems has demonstrated encouraging results, leading to substantial improvements in prediction accuracy and computational efficiency.

This paper applies machine learning methodologies to analyse the structural properties of cerium electrodes that are likely to have the most considerable impact on capacitance. Additionally, experimental validation was conducted using real experimental data. The prediction-validation process reveals that the observed results are firmly situated within the acceptable range of the predicted grades.

Materials and methods

Dataset description

The dataset used in this work consists of multiple attributes that describe various properties about the material's

[a]mahatimsingh.rs.bme19@iitbhu.ac.in, [b]neeraj.bme@iitbhu.ac.in, [c]sendashpangambam@gmail.com, [d]preetamsingh.cer@iitbhu.ac.in

DOI: 10.1201/9781003675235-7

structure, electrochemical performance, and experimental conditions of cerium-based composite materials used in electrochemical applications, as sourced from the published literature [4]. The attributes are as follows:

Code: Besides Cerium, the database also includes oxides, nitrides, oxynitrides, and composites based on vanadium, titanium, and tungsten, which are assigned the codes C, V, T, and W, respectively.

Composite or not: A binary attribute indicating whether the material is a composite or not.

Code morphology: Categorical attributes illustrate various morphological classifications of materials. These classifications include 1 - 1D structures (like quantum dots), 2 - 2D forms (for example, rods and tube-like shapes), 3 - 3D nanoparticles (which lack a primarily defined morphology), 4 - combinational structures (such as core-shell configurations that exhibit multiple morphologies, particularly in composites), 5 - highly porous structures, and 6 - 3D nanoparticles with unique morphologies (including flower-like and spherical shapes).

Surface area: A real-valued metric that determines the specific surface area of the material, articulated in square meters per gram.

Potential window: A real-valued attribute representing the electrochemical potential window, measured in volts.

Current density: A real-valued attribute that measures the current density of the material in amperes per gram.

Substrate/current collector: A categorical attribute representing different types of substrates or current collectors used in the experiments. 1 - Ni-based, 2 - Stainless steel, 3 - Ti-based, 4 - Carbon based substrate, 5 - Glassy Carbon, 6 - Graphite, 7 - Vanadium-based, and 8 - Unknown

Molar solution: A real attribute representing the molarity of the solution used.

Electrolyte: A categorical attribute indicating the type of electrolyte used, such as $CaCl_2$, H_2SO_4, KOH, K_2SO_4, etc.

Specific capacitance: A real-valued attribute measuring the specific capacitance of the material in farads per gram.

Prediction protocols

The specific capacitance prediction experiment is conducted using a Capacitance Grade Prediction approach. To do this, the data is classified into four distinct grades based on specific capacitance values:

Grade A for values below 200 F/g, **Grade B** for values ranging from 200 to 500 F/g, **Grade C** for values between 501 and 1000 F/g, and **Grade D** for values exceeding 1000 F/g.

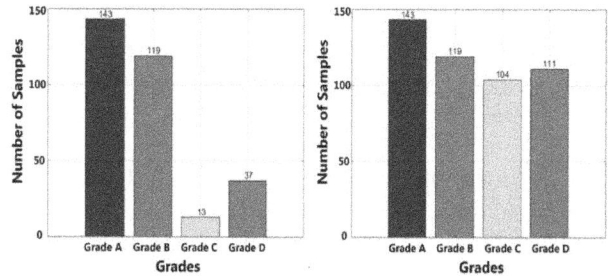

Figure 7.1 Original class distribution and class distribution after SMOTE

Source: Author

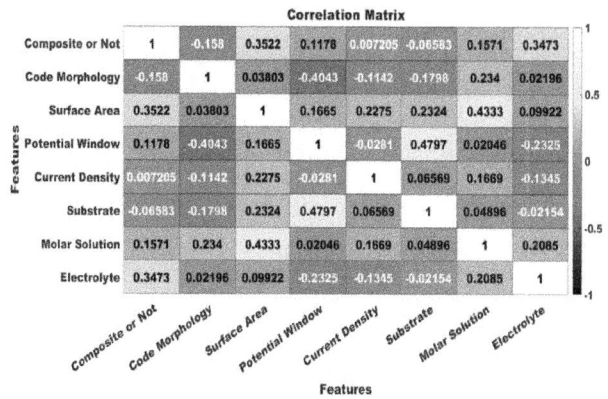

Figure 7.2 Heatmap of the correlation matrix illustrating the relationships among features

Source: Author

The distribution of samples of different grades/classes (as per the categorisation described above) in the dataset is shown in Figure 7.1 (Left). The class distribution reveals an imbalanced dataset, where Grade A dominates with 143 instances, followed by 119 instances of Grade B samples, with minority samples of Grade C and D with only 37 and 13 instances, respectively. To address this, Synthetic Minority Over-sampling Technique (SMOTE) [11] was applied as part of pre-processing to rebalance the datasets. SMOTE helps by generating synthetic samples for the minority classes, thus mitigating the bias introduced by class imbalance and ensuring that the classification models are trained on a more balanced dataset [1,8]. The SMOTE-rebalanced distribution of samples for different classes is shown in Figure 7.1 (Right).

Before applying machine learning algorithms, a correlation analysis was performed to explore the relationships among features in the dataset. The correlation analysis is represented in the heatmap shown in Figure 7.2. Although the correlation coefficients among features were generally low, this does not rule out the potential importance of all features for prediction tasks. As modern machine learning algorithms can capture complex and non-linear relationships; subsequently, all features were retained for model training.

After rebalancing the dataset, different machine learning models were implemented to classify specific capacitance grades and evaluate their predictive performance. The purpose of this analysis was to identify the most effective models for accurately predicting material grades based on their specific capacitance. The machine learning algorithms utilised in this study include K-nearest neighbors (KNN) classifier, Decision tree, Classification and regression tree (CART), Reduced Error Pruning (REP) Tree, Random Tree, Naive Bayes classifier, Logistic regression, Support Vector Machine (SVM) with various kernels, Multilayer perceptron (MLP), and Ensemble classifiers such as the Voting classifier, Bagging, and Boosting. The primary classification metrics used in our evaluation are precision, recall, F1-score, and classification accuracy. These performance metrics were computed through a 10-fold cross-validation process, with results for each grade (A, B, C, D) aggregated into weighted averages to present the overall performance for each algorithm.

Electrochemical experiment

The machine learning-based prediction result is validated with experimental data. The protocol for the preparation of electrodes for this purpose is described in this section.

Material synthesis

The highly porous compound Ce2(C2O4)3·10H2O was synthesised via a precipitation method. The process began with the dissolution of Ce2(C2O4)3·10H2O in 200 ml of deionised water, which was continuously stirred on a hot plate magnetic stirrer. In a stoichiometric ratio, 1.27 g of oxalic acid dihydrate (H2C2O4·2H2O) was subsequently added to the solution. The mixture was stirred vigorously at 80°C for a total of 3 hours. At the end of this stirring period, a white precipitate of Ce2(C2O4)3·10H2O was produced [12].

Electrode preparation

Hydrated $Ce_2(C_2O_4)_3 \cdot 10H_2O$ working electrodes were fabricated by combining active material, Activated Carbon, and binder (polyvinylidene fluoride, PVDF) in a ratio of 7:2:1, utilising N-methyl-2-pyrrolidone (NMP) as the solvent. A uniform slurry was prepared in a mortar, and a portion of this slurry (B1 mg) was applied to a 1 cm² area of Toray carbon paper. The resulting coated electrode was subsequently dried at 80°C for 12 hours.

Electrochemical testing

The electrochemical experiments were conducted in a 2M KOH electrolyte, using Ce2(C2O4)3·10H2O and CeO2, which were obtained from TGA investigations of cerium oxalate, as the working electrodes. A three-electrode system was implemented, featuring a saturated Hg/HgO (1 M KOH) reference electrode and a platinum wire as the counter electrode. A significant specific capacitance of 78 mAhg⁻¹ (corresponding to 401 Fg⁻¹) was achieved at a current density of 1 Ag⁻¹ within the potential range of −0.3 to 0.5 V in the aqueous 2 M KOH electrolyte.

Results and discussions

The machine learning work was executed using WEKA 3.8.6 software [3,10]. Table 7.1 and Figure 7.3 present the

Table 7.1 Precision, recall, F1 score, and accuracy of various algorithms.

Algorithm	Precision	Recall	F1 Score	Accuracy
KNN	80.5	80.5	80.5	80.503
Decision Tree	80.9	81.1	80.9	81.132
Logistic Regression	61.9	62.5	61.1	62.473
Naïve Bayes	77.8	78.0	77.8	77.987
CART	79.9	79.9	79.9	79.874
SVM(Linear)	63.0	62.7	61.0	62.683
SVM (RBF)	70.3	69.6	69.8	69.607
SVM (Poly Kernel = 2)	72.7	66.7	66.4	66.667
Random Tree	80.6	80.5	80.5	80.503
REP Tree	71.6	72.1	71.5	72.117
MLP	73.3	74.0	73.5	74.004
Voting	85.8	85.7	85.7	85.744
Bagging	84.4	84.3	84.2	84.276
AdaBoost	85.6	85.5	85.5	85.534

Source: Author

comparative analysis results of different machine learning algorithms based on different performance measures. The results highlight significant variability in the performance of the evaluated algorithms. The Voting classifier stands out as the most effective model in this study, achieving a precision of 85.80, recall of 85.70, F_1-score of 85.70, and an accuracy of 85.744. By integrating the predictions of multiple classifiers, it harnesses the strengths of each model, thereby enhancing overall performance. This ensemble method effectively reduces bias and variance, leading to improved reliability in predictions. The

consistent high scores in different metrics demonstrate its robustness in handling various complexities within the dataset.

The receiver operating characteristic (ROC) curves for the Voting classifier for each grade are shown in Figure 7.4. The area under the curve (AUC) values are notably high: 0.9806 for Grade A, 0.9413 for Grade B, 0.9503 for Grade C, and an exceptional 0.9954 for Grade D. These results indicate the model's effectiveness in distinguishing between grades, with particularly robust classification for Grades A and D. Overall, the high AUC values validate the decision to utilise all features for training the Voting classifier, showcasing its capability to handle class imbalances and optimise predictive accuracy.

The Decision Tree algorithm employs a tree-like model of decisions, enabling it to classify data by splitting it based on feature values. Its structure allows for clear interpretation of decision pathways, which is critical in applications like material grading, where understanding the rationale behind predictions is necessary. The strong precision and recall indicate its ability to balance between correctly identifying grades and minimising misclassifications. Random Tree, a variant of decision trees, uses random feature selection at each split. This approach reduces overfitting and enhances generalisation by promoting

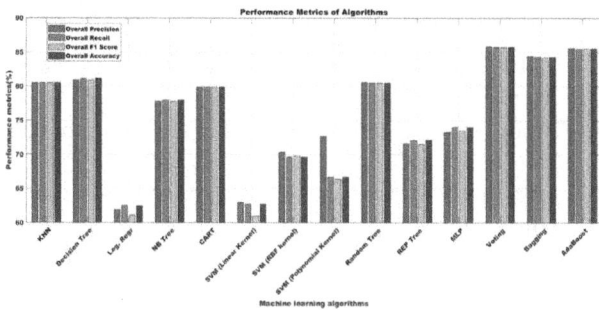

Figure 7.3 Comparative analysis of different machine learning algorithms

Source: Author

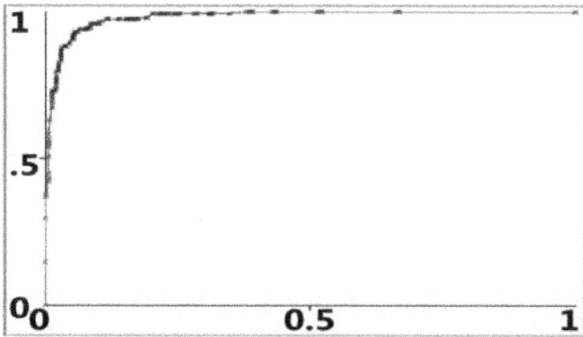

Plot (Area under ROC = 0.9806)

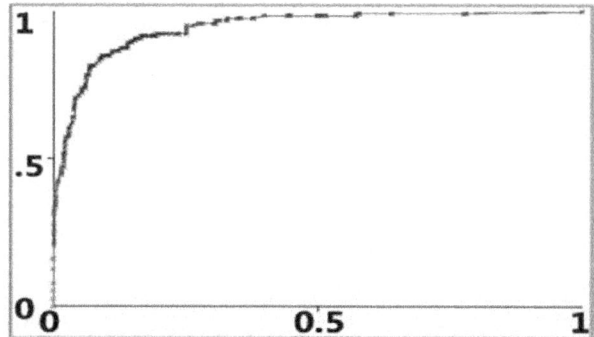

Plot (Area under ROC = 0.9413)

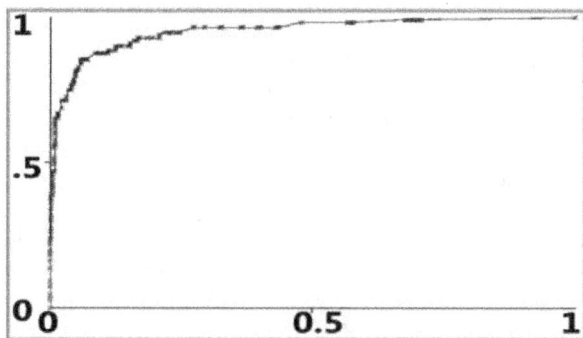

Plot (Area under ROC = 0.9503)

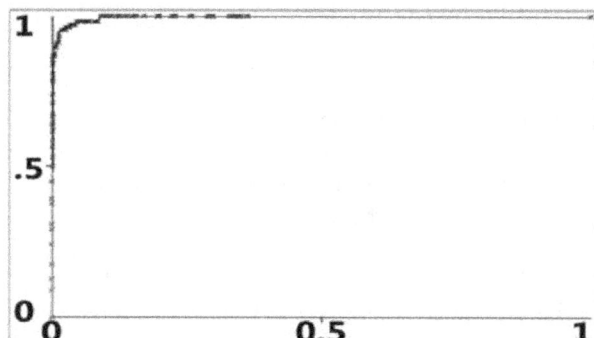

Plot (Area under ROC = 0.9954)

Figure 7.4 ROC curve for class A, B, C and D

Source: Author

diversity among individual trees, contributing to its solid performance metrics.

K-Nearest Neighbors (KNN) classifies instances based on the majority label among the closest data points in the feature space. This instance-based learning captures local patterns effectively, leading to good performance in distinguishing between grades. However, KNN may become computationally intensive with larger datasets due to its reliance on distance calculations for all instances.

Ensemble methods, such as Bagging and AdaBoost, enhance model performance by leveraging the strengths of multiple classifiers. Bagging (Bootstrap Aggregating) improves stability and accuracy by creating multiple versions of a training dataset through resampling and averaging the predictions from separate models, which is particularly beneficial for decision trees that are prone to overfitting. In contrast, AdaBoost (Adaptive Boosting) focuses on misclassified instances by adjusting their weights in subsequent iterations, allowing the model to progressively improve upon weaker classifiers and achieve high precision and recall scores.

Lower performing models include Logistic Regression, Support Vector Machines (SVM), Multilayer Perceptron (MLP), and Naive Bayes. Logistic Regression's dependence on linear relationships limits its effectiveness in datasets with complex, non-linear dynamics, resulting in subpar performance metrics despite its simplicity and interpretability. The SVM with linear kernel struggles with non-linear data due to its linear decision boundary, while with the polynomial kernel with degree = 2, despite using a polynomial kernel, fails to significantly improve performance, highlighting the need for appropriate kernel selection based on data characteristics. The MLP may benefit from further tuning of hyperparameters and larger datasets to enhance its ability to learn complex patterns, whereas the Naive Bayes Tree's assumption of feature independence may hinder its effectiveness, suggesting that incorporating techniques to account for feature dependencies could improve its performance.

Conclusion and future work

This work emphasises the importance of selecting suitable machine learning algorithms for predicting specific capacitance grades based on the unique characteristics of the materials, through an extensive comparative analysis of different machine learning algorithms. The ensemble methods, particularly the Voting classifier, demonstrated superior performance, highlighting their effectiveness in accurately classifying materials into grades A, B, C, and D. The experimental results (~401 F g^{-1}, Grade B) considerably validate the predictive approach presented here. This suggests that further investigation into ensemble techniques may yield enhanced predictive capabilities,

especially when combined with hyperparameter tuning and cross-validation strategies.

Future work could also explore the following areas:

- Feature engineering: Improving the input features could lead to better model performance in predicting specific capacitance grades. Techniques such as feature scaling, normalisation, and dimensionality reduction may enhance the algorithms' learning capabilities.
- Hyperparameter optimisation: Employing different hyperparameter optimisation methods to refine model parameters could improve performance, particularly for algorithms like Support Vector Machines (SVM) and Multilayer Perceptron (MLP), which may benefit from tailored settings.
- Model interpretability: As machine learning models gain importance in critical applications like capacitance grading, focusing on interpretability becomes crucial. Investigating techniques to elucidate model predictions (such as SHAP values) will be particularly beneficial in materials science, where understanding decision pathways is essential.
- Application in real-world scenarios: Validating these models in practical settings to assess their effectiveness in real-world specific capacitance grading applications could provide valuable insights into their utility, limitations, and potential areas for improvement.

References

[1]   Alex, S. A. (2025). Imbalanced data learning using SMOTE and deep learning architecture with optimized features. *Neural Computing and Applications*, 37(2), 967–984.

[2]   Chong, S. S., Ng, Y. S., Wang, H.-Q., Zheng, J.-C. (2024). Advances of machine learning in materials science: ideas and techniques. *Frontiers in Physics*, 19(1), 13501.

[3]   Frank, E., Hall, M. A., Witten, I. H. (2016). The WEKA Workbench. Online Appendix for "Data Mining: Practical Machine Learning Tools and Techniques".

[4]   Ghosh, S., Rao, G. R., Thomas, T. (2021). Machine learning-based prediction of supercapacitor performance for a novel electrode material: cerium oxynitride. *Energy Storage Materials*, 40, 426–438.

[5]   Ghosh, S., Santhosh, R., Jeniffer, S., Raghavan, V., Jacob, G., Nanaji, K., Kollu, P., Jeong, S. K., Grace, A. N. (2019). Natural biomass derived hard carbon and activated carbons as electrochemical supercapacitor electrodes. *Scientific Reports*, 9(1), 16315.

[6]   Jamaluddin, A., Nursanti, A. D., Nur'aini, A., Putri, R. R. M., Arshad, M. U. (2023). Graphene as an active material for supercapacitors: a machine learning approach. *Indonesian Journal of Applied Physics*, 13(2), 305–312.

[7]   Kowsuki, K., Nirmala, R., Ra, Y.-H., Navamathavan, R. (2023). Recent advances in cerium oxide-based nanocomposites in synthesis, characterization, and energy storage

applications: a comprehensive review. *Results in Chemistry*, 5, 100877.

[8] Liaw, L. C. M., Tan, S. C., Goh, P. Y., Lim, C. P. (2025). A histogram SMOTE-based sampling algorithm with incremental learning for imbalanced data classification. Information Sciences, 686, 121193.

[9] Liu, Y., Xu, X., Shao, Z., et al. (2020). Metal-organic frameworks derived porous carbon, metal oxides and metal sulfides-based compounds for supercapacitors application. *Energy Storage Materials*, 26, 1–22.

[10] Markov, Z., Russell, I. (2006). An introduction to the WEKA data mining system. ACM SIGCSE Bulletin, 38(3), 367–368.

[11] Nitesh, V.C. (2002). SMOTE: synthetic minority over-sampling technique. *Journal of Artificial Intelligence Research*, 16(1), 321.

[12] Singh, M., Mondal, R., Singh, P., Sharma, N. (2023). Framework structured Ce 2 (C 2 O 4) 3·10H 2 O as a pseudocapacitive electrode of a hybrid (asymmetric) supercapacitor

(HSC) for large scale energy storage applications. Physical Chemistry Chemical Physics, 25(16), 11429–11441.

[13] Su, H., Lin, S., Deng, S., Lian, C., Shang, Y., Liu, H. (2019). Predicting the capacitance of carbon-based electric double layer capacitors by machine learning. *Nanoscale Advances*, 1(6), 2162–2166.

[14] Thakkar, P., Khatri, S., Dobariya, D., Patel, D., Dey, B., Singh, A. K. (2024). Advances in materials and machine learning techniques for energy storage devices: a comprehensive review. *Journal of Energy Storage*, 81, 110452.

[15] Wei, J., Chu, X., Sun, X.-Y., Xu, K., Deng, H.-X., Chen, J., Wei, Z., Lei, M. (2019). Machine learning in materials science. *InfoMat*, 1(3), 338–358.

[16] Zhao, X., Sánchez, B. M., Dobson, P. J., Grant, P.S. (2011). The role of nanomaterials in redox-based supercapacitors for next generation energy storage devices. *Nanoscale*, 3(3), 839–855.

8 Incremental learning using CART and Genetic Programming for handling dynamic training data

Ibadonbok Syiemlieh[a], Ymphaidien Sutong[b] and Sufal Das[c]

Department of Information Technology, North-Eastern Hill University, Meghalaya, India

Abstract

Genetic Programming (GP) is a technique that focuses on evolving a program from a population of unfit programs to fit a specific task by applying operations similar to natural genetic processes on the program population. GP can be used to dynamically adapt tree classifiers to concept drift in data streams. The sudden changes in data may lead to concept drift, which is one of the significant challenges in machine learning. This paper presents a new technique for classification with dynamic data through the hybridisation of GP and the CART algorithm. CART is a well-established decision-tree-based classifier. This proposed method efficiently handles concept drift, in which the pattern of data changes over time and affects the model's performance. The proposed approach improves decision tree adaptability using Selection, Crossover, and Mutation to iteratively make model populations. Several experiments were conducted with breakneck-mark datasets. We have also compared the developed decision tree to baselines models and other methods. The paper helps deepen the understanding of adaptive machine learning approaches in dynamic data, where continuous learning is essential. Metrics such as accuracy, precision, recall, and F1-score are used for evaluating performance.

Keywords: Genetic programming, decision tree, classification, incremental learning, adaptive classifier

Introduction

Data stream classification is a process of continuously categorizing the arriving data points in actual time where the classifier needs to adapt with concept drift and shortage of memory [11, 21]. The study of adaptive machine learning models for data stream classification is motivated since streaming data is used in many different applications [3,14]. Data streams require real-time decision-making to handle constant information influx, but concept drift, a phenomenon where data distribution evolves over time, poses significant challenges [28,26]. Continuous transformation in data streams necessitates adaptive models to maintain accurate predictions, as traditional static models struggle with shifting patterns. This mismatch compromises the accuracy and reliability of predictions, rendering conventional approaches unsuitable for the exigencies of dynamic environments [29, 33]. Concept drift, categorized into abrupt, incremental, and gradual forms, is influenced by factors like abruptness, incrementality, and change magnitude, each requiring unique adaptation mechanisms. Concept drift [12] is a challenge that requires timely adaptation and detection. It is essential for an adaptive model to adapt with the changing pattern using active or passive methods.

This paper has presented an incremental , model using Genetic Programming (GP) to make adaptive decision-tree based classifiers for dynamic training dataset. The proposed method overcomes the concept drift which is a big challenge for traditional classification models. The work's aim is to obtain efficiency in accuracy, and other metricslike precision, f1-score and recall for classifiers. The paper provides an integration of Genetic Programming with the CART classifier to make a flexible classifier that can dynamically adapt with dynamic training dataset [22]. A variety of datasets and measurement parameters with real world applications are used to evaluate the proposed method. The proposed method has achieved high predictive accuracy, precision, recall, F1-score, and computational efficiency, contributing to the advancement of adaptive machine learning models.

This is how the rest of the paper is organized.Section 2: Review of Literature.In Section 3, the proposed method is explained. Section 4 describes the experimental setup.The findings and analysis are covered in Section 5.The paper is concluded with a summary in Section 6.

Literature review

This literature review explores the advancements and contributions in adaptive machine learning, highlighting various approaches, methodologies, and applications that have emerged in the field. Adaptive machine learning techniques, introduced by Widmer and Kubat (1996), handle evolving data distributions and concept drift, demonstrating model accuracy over time without retraining from scratch [2,9,19]. Concept drift by Xiang and Qiuyan, et al. (2023) under deep learning help decision maker to

[a]ibadonbok32@gmail.com, [b]ymphaidiensutong99@gmail.com, [c]sufal.das@gmail.com

DOI: 10.1201/9781003675235-8

predict accurate data through constant refining the learning model. This method has the Concept drift detection which contains both Active (update model when concept drift is detected) and Passive modes (update model continuously) and Model update consist of structure updates and parameter updates [34]. Concept drift by Yuan, Liheng, et al. (2022) divided the concept drift into 2parts (1) model parameter updating and (2) model structure updating (adjust network structure) [32, 35]. Jameel et al. (2020) have shown better solution that can handle stream data for gradual using Deep learning in Big Data Stream but the accuracy degraded during online learning [29,36]. Mehmood et al. (2021) Ensemble based approach are suitable for non stationaryenvironment due to their accuracy and are flexible to incorporate with new data. Using dynamic selection and voting technique the old classifier can remove the irrelevant old data [37]. Ensemble methods, including the AdaBoost algorithm by Freund and Schapire (1997), have been extensively studied for adaptive learning, combining weak classifiers to form a strong classifier.[13]. Techniques like Online Bagging and Online Boosting, utilized ensemble learning strengths, offer robust performance in the presence of concept drift. [10,26]. Detecting concept drift is crucial for adaptive learning systems. Gama et al. (2004) presented a framework for detecting concept drift by monitoring statistical properties of the data stream and employing change detection tests [15]. This work laid the foundation for variousdrift detection methods, including the DDM and EDDM[16]. Adaptive machine learning is effectively used in financial forecasting to accurately predict stock prices and market trends, despite the volatile nature of financial data. [30]. In cybersecurity, adaptive learning systems detect and respond to new and evolving threats, enhancing the robustness of intrusion detection systems [2]. Healthcare is another critical area where adaptive machine learning has shown promise. Models that adapt to new patient data and evolving medical knowledge are crucial for personalized medicine and early disease detection [25]. In industrial settings, adaptive learning is used for predictive maintenance, where models predict equipment failures based on real-time sensor data, reducing downtime and maintenance costs [33].

Representative adaptive ensemble methods
Adaptive ensemble methods are effective for classifying data streams with concept drifts by weighting, evaluating, and updating component classifiers based on incoming data. AWEby Wang et al. [31], works by training a new classifier on each data chunk and replaces the worst performing component. Accuracy Updated Ensemble (AUE) proposed by Brzezinski et al. [8], works by incrementally updating the component classifiers and adjusting

their weights according to their accuracy. Kolter et al [27] Dynamic weighted majority, an ensemble method that adjusts online learners' weights based on concept drift, adding, removing, and decreasing weights for mistakes and underperforming learners. Online Bagging by Oza et al. [16], a method that enhances generalization by combining multiple base models, compared to batch bagging. It adapts to streaming data by dispatching K copies of each example, updating each base model. Bagging with a Change Detector by Bifet et al. [4] is an ensemble method uses ADWIN[3] ADWIN bagging is a change detector that adds a new classifier to an ensemble after removing the weakest one, adapting to different change rates. Oza and Russell [16] Online Boosting is an online version of AdaBoost algorithm that generates base models from weighted training examples. It updates each model based on previous misclassifications and normalized error. Online boosting performs similarly to AdaBoost on most datasets and improves with increasing examples.

Adaptive Random Forest is a data stream classification method that creates decision trees with features and drift detection methods. Streaming Random Patches introduced by Gomes et al. [17, 20] an ensemble method that combines random subspaces [23] and online bagging [16]. SRP employs global subspace randomization and sampling with replacement to train base models on random features, enhancing accuracy and efficiency by reducing feature space and variance.SRP uses the ADWIN [3] algorithm to monitor the performance of each base model and detect concept drifts. When a drift is detected, SRP replaces the affected baseline model with a new one trained on a new random subspace. This allows SRP to adapt to changing data distributions over time. Hovakimyan et al. (2024) enhance the accuracy using the method Text to text Transfer Transformer (T5) model and highlight different approach to handle drift detection such as Online learning Theory used incremental model upon arival data by adjusting the new data and maintaining prior knowledge [38].

Proposed method

To simulate batches of data streams, the training data is split into several parts. The decision tree is trained and evolved using these batches, adapting to concept drift. Initially, a population of decision trees is created using the CART algorithm, with two decision trees trained on the first two parts of the data. These trees form the foundation for the subsequent operations of the Genetic Programming (GP) algorithm. For every batch of the subsequent simulated stream, a decision tree is created. These trees are then fed as inputs to the Genetic Programming algorithm, along with the new population, as the process continues.

Decision Tree Generation Using CART Algorithm
The decision trees are constructed using the Classification and Regression Trees (CART) algorithm, a widely-used method for generating decision trees.

Initial Population Creation
Initial Decision Trees Population: Using the CART technique, 2 decision trees were trained on the first two segments of the input stream to form the first decision tree population. Each of these parts acts as an initial batch of the simulated streaming data, providing a foundation for the decision trees. This process ensures that the initial trees are well-grounded in the underlying data distribution.

Genetic Programming Algorithm for Decision Tree Evolution
The GP algorithm aims to evolve the decision tree population over a series of generations to enhance model performance and adapt to changes. This is achieved through a combination of selection, crossover, and mutation operations.

Design of Selection, Crossover and Mutation Operations for Binary Trees

1. **Selection:** The selection process employs tournament selection to choose parent models for reproduction. A subset of the population is evaluated in each tournament, , and the model with the highest fitness, as measured by accuracy on the next batch of sequential data partitions, is selected. This guarantees that the models that perform the best have a higher chance of being selected as parents.Consequently, this helps

the decision tree classifier to adapt and evolve in response to the changing nature of the data.

2. **Crossover:** In the context of binary trees, the crossover operation involves selecting random crossover points within two parent trees. These points are chosen such that each selected point is an internal node with at least one child. Two parent trees are then exchanged of the subtrees. To ensure that the offspring trees maintain the binary tree structure, while potentially combining beneficial traits from both parents.

3. **Mutation**: For binary trees, the mutation operation is designed as a sub- tree exchange within a single tree. A subtree exchange mutation selects two random subtrees within the same tree and swaps them. This mutation introduces diversity without disrupting the binary tree structure, as the overall tree maintains its balance and integrity.

Evaluation
Each model in the population is evaluated for fitness using its accuracy on the next batch of sequential data partitions. This metric offers a quantitative assessment of each model's performance and guides the selection, crossover, and mutation processes throughout the evolution as new data becomes available.

The GA runs iteratively for a predefined number of generations taking a tree created from that batch and populations generated at that instance as input.A new population is createdin each generation,by repeatedly performing selection, crossover, and mutation until the population size is restored. This iterative process keeps going until the maximum number of generations or the convergence-criteria are met.

Experimental Setup

All experiments were carried out utilizing the Python language to implement the algorithms. The comprehensive experiments were executed on five distinct synthetic datasets sourced from the UCI repository [1] and some generated with MOA [5] using appropriate parameters, serving as the recognized benchmarks for the analysis of data streams, each designed to exhibit concept drift of different form.

Evaluated Algorithms
In this study we evaluated 3 algorithms along with the proposed method utilized in conducting our analysis on different forms of concept drifts. They are listed below and their details have been given in previous chapters.

1. CART
2. Online Bagging [26].
3. Streaming Random Patches [20].

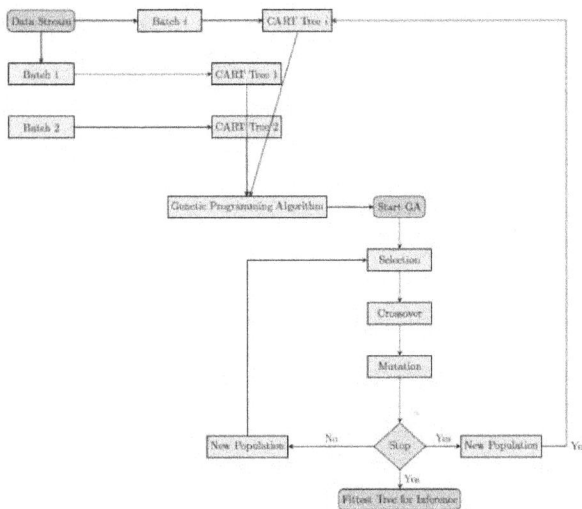

Figure 8.1 Block diagram of the proposed method
Source: Author

The decision trees were generated using the CART algorithm implemented in Scikit-Learn an open-source Python package for data science and machine learning. Below, we provide details of each dataset used during the experiments.

Computational Complexity

CART worst case time complexity is $O(n^2 \log n)$ where m represent the number of training samples. Incremental reduce the cost by updating the affected nodes. GP complexity is $O (G \times P \times F)$ where G is the number of generations or how many times the population evolves, P is the number of individuals per generation or the Population size, and F is the Fitness evaluation cost per individual. In worst case tree evaluation can reach $O(P \times n \log n)$

Dataset Used

We have considered both real world dataset from Openml an open source platform and popular synthetic datasets used by researchers as benchmarks, using the MOA framework with different parameters, ensuring that each dataset incorporates one of the concept drift typesrecurring, sudden, incremental, gradual, or mixed drift.This deliberate selection methodology was employed to facilitate a comprehensive analysis of diverse data stream mining algorithms.

1. Hyperplane: In [24], the CVFDT and VFDT were compared using the hyperplane problem as a test. A hyperplane dynamically simulates evolving concepts by smoothly adjusting orientation and position through changes in weights. Mathematically expressed as $w \cdot x + b = 0$, where w stands for weights, x is the input vector, and b is the bias term, this equation characterizes the hyperplane's discriminative boundary.
2. Radial Basis Function: The function (RBF) introduces a drifting centroid by computing the similarity between an input and fixed basis functions. This dynamic adjustment of centroids allows RBF networks to effectively simulate evolving concepts, making them adaptable to changes in data distribution and providing robustness in modeling complex, dynamic systems.
3. SEA: This dataset is obtained by using the SEA generator proposed in [27] to study reaction to abrupt concept drift, it uses three attributes, with the first two being relevant, divided into four blocks with distinct concepts based on specific threshold values.
4. LED: The LED dataset generates a challenge in predicting the digit presented on a 7segment LED display, sourced from the CART book[6]. Here, the objective is to forecast the digit on the display, considering a certain possibility of inversion for each

attribute. We created mixed drifts using MOA's LED feature.

Table 8.1 show the different dataset generate from MOA (Massive Online Analysis) framework using built in stream generators such as Hyperplane , LED, SEA and RBF each simulating various kind of drift such as recurrent, abrupt, gradual or mixed. (Bifet, Albert, et al. [5], Frank, Andrew [1])

Real Dataset

1. Electricity Dataset has 45,312 rows (data points), 8 features per sample and The target (y) has 45,312 values. Used for time-series forecasting and detecting concept drift (when data patterns change over time).
2. The Covertype Dataset contains seven distinct forest cover types as Target Classes, 581,012 instances, and 54 features. The Purpose is to predict forest cover type using cartographic factors obtained from remote sensing data.

Evaluation Method

In this study we used Interleaved Test Then Train approach to measure the performance of the algorithms. Interleaved Test Then Train evaluates a classifier's performance on a stream of data by testing and training it one instance at a time. This method doesn't have a forgetting mechanism, thereby assigning equal importance to all instances during the evaluation process. A modified version of this technique could optionally include a fading factor or a sliding window as forgetting mechanisms to adaptivelyprioritize recent instances and accommodate evolving patterns in the data stream.

Result Analysis

We provide the findings and analysis of our experiments in this part. The primary focus of these experiments was to analyze the performance of decision trees created using the CART algorithm, both independently as a baseline model and in combination with a Genetic Programming (GP) algorithm compared to other methods. The aim was to observe how these approaches cope with concept drift in data streams.

The results of different approaches was evaluated in terms of mean accuracy across the datasets. As shown in Table 6.1, we compared the accuracy of decision trees trained on the whole dataset, Online Bagging, Streaming Random Patches and the proposed method.

Table 8.2 shows the mean accuracy of each algorithm across the datasets. The proposed method achieves higher accuracy compared to the baseline CART, Online Bagging, and SRP methods. This suggests that the genetic programming's evolutionary operations enhance the

Table 8.1 Description about datasets used.

Generator	Type of concept drift	No. of attributes	No. of attributes with drift	No. of class
HYP(R)	Recurrent	10	5	2
HYP(G)	Gradual	10	5	2
RBF	Gradual	10	4	2
SEA	Abrupt	3	9	4
LED	Mixed	7	2	10

Source: Author

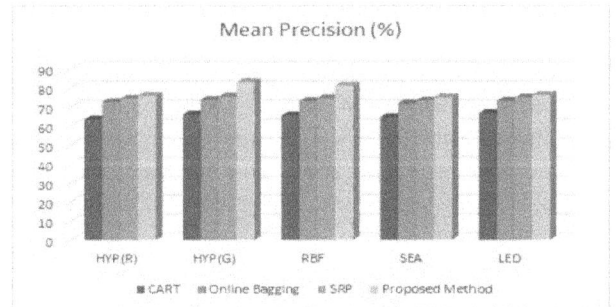

Figure 8.2 Comparison with mean accuracy (%)
Source: Author

Table 8.2 Comparison with mean accuracy (%).

Algorithm/ Dataset	HYP(R)	HYP(G)	RBF	SEA	LED
CART	63.34	66.12	65.45	64.78	66.89
Online Bagging	72.34	73.56	72.89	71.45	73.00
SRP	74.12	75.23	74.45	73.00	74.56
Proposed Method	**75.56**	**83.62**	**81.88**	**74.5**	**75.67**

Source: Author

Table 8.3 Comparison with mean precision (%).

Algorithm/ Dataset	HYP(R)	HYP(G)	RBF	SEA	LED
CART	63.34	66.00	65.34	64.56	66.67
Online Bagging	72.45	73.67	73.00	71.78	73.12
SRP	74.23	75.34	74.56	73.23	74.89
Proposed Method	75.78	82.9	81.02	74.90	76.13

Source: Author

decision tree's ability to adapt to changing data streams, leading to better classification performance.

Table 8.3 presents the mean precision values for the algorithms. The proposed method consistently shows higher precision, indicating that it generates fewer false positives compared to other methods. This improvement is crucial in applications where precision is a key performance indicator, as it reduces the likelihood of incorrect positive classifications.

Table 8.4 shows the mean recall scores. The proposed method demonstrates superior recall, reflecting its ability to correctly identify positive instances more effectively than the other models. High recall is particularly beneficial in scenarios where capturing all relevant instances is critical, even at the expense of increased false positives.

Table 8.5 provides the mean F1-scores for each algorithm. The F1-score is the harmonic mean of precision and recall, offering a balance between the two metrics. The proposed method consistently outperforms the baseline models, highlighting its overall effectiveness in managing the trade-off between precision and recall.

The experiments conducted in this study underscore the effectiveness of Genetic Programming (GP) in enhancing decision tree classifiers to address concept drift within data streams. Evaluation across five bench-mark datasets i.e. Rotating Hyperplane (Recurrent and Gradual), RBF, SEA, and LED. It can be concluded that integrating

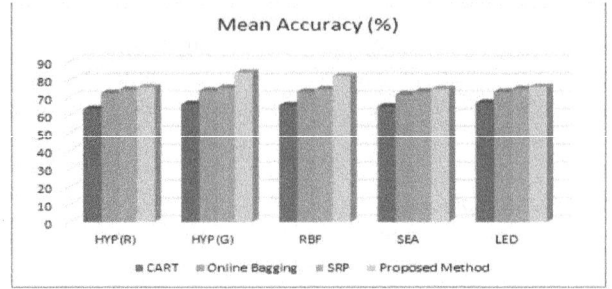

Figure 8.3 Comparison with mean precision (%)
Source: Author

Table 8.4 Comparison with recall (%).

Algorithm/ Dataset	HYP(R)	HYP(G)	RBF	SEA	LED
CART	63.34	66.12	65.45	64.89	66.78
Online Bagging	72.34	73.56	72.89	71.56	72.98
SRP	74.12	75.23	74.45	73.12	74.56
Proposed Method	**75.67**	**83.15**	**81.9**	**74.67**	**75.78**

Source: Author

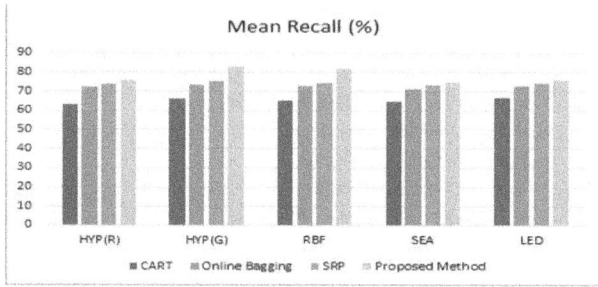

Figure 8.4 Comparison with recall (%)
Source: Author

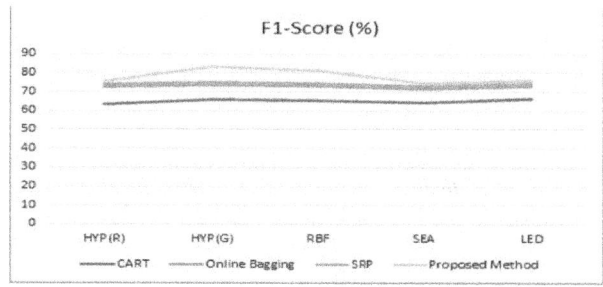

Figure 8.5 Comparison with F1-score (%)
Source: Author

Table 8.5 Comparison with F1-score (%).

Algorithm/ Dataset	HYP(R)	HYP(G)	RBF	SEA	LED
CART	63.34	66.06	65.39	64.72	66.72
Online Bagging	72.39	73.61	72.94	71.67	73.05
SRP	74.17	75.28	74.50	73.17	74.72
Proposed Method	**75.72**	**83.02**	**81.46**	**74.78**	**75.95**

Source: Author

Genetic Programming with the CART algorithm significantly improves the efficiency compared to decision trees trained on the entire dataset or other existing methods.

Statistical test has been performed using T-test and shown promising results as the Proposed Method has continuously Outperform CART and Online bagging with 0.00017 and 0.0187 respectively. If $p < 0.05$ means the difference is statistical significance.

Conclusion

Adaptive machine learning faces challenges in balancing complexity and adaptability, with complex models potentially overfitting data and simplistic models failing to capture essential patterns, resulting in poor performance. This study offers a novel and effective technique for classification with incremental learning. Thus why this proposed method is suitable for classification with dynamic training dataset. Genetic programming is applied to achieve search optimality so that an effective decision-tree based classifier is generated. Moreover, through the iterative session of genetic programming, a new dataset is being considered so the classifier can adopt it and accordingly enhance the learning. Thus why the proposed method is efficient in time complexity as no re-apply of the classier for future data. Extensive experiment has been implemented with branch-marked datasets. Several comparison tales/figures are using to establish the effectiveness of the proposed

method. Future research will be focused on hybrid adaptive learning approaches, integrating domain knowledge and human-in-the-loop systems for improved interpretability. Exploring the interplay between adaptive learning and emerging fields like reinforcement learning and deep learning is also promising. Continuous innovation is crucial for maximizing adaptive learning potential.

Reference

[1] Arthur Asuncion and David Newman. Uci machine learning repository, 2007.
[2] Dimitris Bertsimas and Matthew Peroni. Policy trees for prediction: Interpretable and adaptive model selection for machine learning. *arXivonlinebagg preprint arXiv:2405.20486*, 2024.
[3] Albert Bifet and RicardonlinebaggGavalda. Learning from time-changing data with adaptive windowing. In *Proceedings of the 2007 SIAM international conference on data mining*, pages 443–448. SIAM, 2007.
[4] Albert Bifet, Geoff Holmes, Bernhard Pfahringer, Richard Kirkby, and Ricard Gavalda. New ensemble methods for evolving data streams. In *Proceedings of the 15th ACM SIGKDD international conference on Knowledge discovery and data mining*, pages 139–148, 2009.
[5] Albert Bifet, Geoff Holmes, Bernhard Pfahringer, Philipp Kranen, Hardy Kremer, Timm Jansen, and Thomas Seidl. Moa: Massive online analysis, a framework for stream classification and clustering. In *Proceedings of the first workshop on applications of pattern analysis*, pages 44–50. PMLR, 2010.
[6] Leo Breiman. *Classification and regression trees*. Routledge, 2017.
[7] Dariusz Brzezinski and Jerzy Stefanowski. Reacting to different types of concept drift: The accuracy updated ensemble algorithm. *IEEE transactions on neural networks and learning systems*, 25(1):81–94, 2013.
[8] K Deb. Multi-objective optimization techniques in engineering, 2001.
[9] Russell Eberhart and James Kennedy. Particle swarm optimization. In *Proceedings of the IEEE international conference on neural networks*, volume 4, pages 1942–1948. Citeseer, 1995.
[10] Yoav Freund and Robert E Schapire. A decision-theoretic generalization of on-line learning and an application

to boosting. *Journal of computer and system sciences*, 55(1):119–139, 1997.

[11] Joao Gama. *Knowledge discovery from data streams*. CRC Press, 2010.

[12] Joao Gama, Pedro Medas, Gladys Castillo, and Pedro Rodrigues. Learning with drift detection. In *Advances in Artificial Intelligence–SBIA 2004: 17th Brazilian Symposium on Artificial Intelligence, Sao Luis, Maranhao, Brazil, September 29-Ocotber 1, 2004. Proceedings 17*, pages 286–295. Springer, 2004.

[13] Joãͦo Gama, Indreˊ Zˇliobaiteˊ, Albert Bifet, Mykola Pechenizkiy, and Abdelhamid Bouchachia. A survey on concept drift adaptation. *ACM computing surveys (CSUR)*, 46(4):1–37, 2014.

[14] Heitor M Gomes, Albert Bifet, Jesse Read, Jean Paul Barddal, FabrˊıcioEnembreck, Bernhard Pfharinger, Geoff Holmes, and Talel Abdessalem. Adap- tive random forests for evolving data stream classification. *Machine Learning*, 106:1469–1495, 2017.

[15] Heitor Murilo Gomes, Jesse Read, and Albert Bifet Streaming random patches for evolving data stream classification. In *2019 IEEE international conference on data mining (ICDM)*, pages 240–249. IEEE, 2019.

[16] Vivekanand Gopalkrishnan, David Steier, Harvey Lewis, and James Guszcza. Big data, big business: bridging the gap. In *Proceedings of the 1st International Workshop on Big Data, Streams and Heterogeneous Source Mining: Algorithms, Systems, Programming Models and Applications*, pages 7–11, 2012.

[17] Tin Kam Ho. The random subspace method for constructing decision forests. *IEEE transactions on pattern analysis and machine intelligence*, 20(8):832– 844, 1998.

[18] Geoff Hulten, Laurie Spencer, and Pedro Domingos. Mining time-changing data streams. In *Proceedings of the seventh ACM SIGKDD international conference on Knowledge discovery and data mining*, pages 97–106, 2001.

[19] Alistair EW Johnson, Mohammad M Ghassemi, Shamim Nemati, Katherine E Niehaus, David A Clifton, and Gari D Clifford. Machine learning and decision support in critical care. *Proceedings of the IEEE*, 104(2):444–466, 2016.

[20] Ralf Klinkenberg. Learning drifting concepts: Example selection vs. example weighting. *Intelligent data analysis*, 8(3):281–300, 2004.

[21] J Zico Kolter and Marcus A Maloof. Dynamic weighted majority: An ensemble method for drifting concepts. *The Journal of Machine Learning Research*, 8:2755–2790, 2007.

[22] John R Koza. A genetic approach to finding a controller to back up a tractor- trailer truck. In *1992 American Control Conference*, pages 2307–2311. IEEE, 1992.

[23] Gourav Kumar, Sanjeev Jain, and Uday Pratap Singh. Stock market fore- casting using computational intelligence: A survey. *Archives of computational methods in engineering*, 28(3):1069–1101, 2021.

[24] Leandro L Minku, Allan P White, and Xin Yao. The impact of diversity on online ensemble learning in the presence of concept drift. *IEEE Transactions on knowledge and Data Engineering*, 22(5):730–742, 2009.

[25] R Keith Mobley. *An introduction to predictive maintenance*. Elsevier, 2002.

[26] Nikunj C Oza and Stuart J Russell. Online bagging and boosting. In *International Workshop on Artificial Intelligence and Statistics*, pages 229–236. PMLR, 2001.

[27] Parneeta Sidhu and MPS Bhatia. Empirical support for concept drifting approaches: Results based on new performance metrics. *International Journal of Intelligent Systems & Applications*, 7(6), 2015.

[28] Robin Sommer and Vern Paxson. Outside the closed world: On using machine learning for network intrusion detection. In *2010 IEEE symposium on security and privacy*, pages 305–316. IEEE, 2010.

[29] Rainer Storn and Kenneth Price. Differential evolution–a simple and efficient heuristic for global optimization over continuous spaces. *Journal of global optimization*, 11:341–359, 1997.

[30] W Nick Street and YongSeog Kim. A streaming ensemble algorithm (sea) for large-scale classification. In *Proceedings of the seventh ACM SIGKDD international conference on Knowledge discovery and data mining*, pages 377– 382, 2001.

[31] Li Su, Hong-yan Liu, and Zhen-Hui Song. A new classification algorithm for data stream. *IJ Modern Education and Computer Science*, 4:32–39, 2011.

[32] Haixun Wang, Wei Fan, Philip S Yu, and Jiawei Han. Mining concept-drifting data streams using ensemble classifiers. In *Proceedings of the ninth ACM SIGKDD international conference on Knowledge discovery and data mining*, pages 226–235, 2003.

[33] Gerhard Widmer and Miroslav Kubat. Learning in the presence of concept drift and hidden contexts. *Machine learning*, 23:69–101, 1996.

[34] Xiang, Qiuyan, et al. "Concept drift adaptation methods under the deep learning framework: A literature review." *Applied Sciences* 13.11 (2023): 6515.

[35] Yuan, Liheng, et al. "Recent Advances in Concept Drift Adaptation Methods for Deep Learning." *IJCAI*. 2022.

[36] Jameel, Syed Muslim, et al. "A critical review on adverse effects of concept drift over machine learning classification models." *International Journal of Advanced Computer Science and Applications (IJACSA)* 11.1 (2020): 2020.

[37] Mehmood, Hassan, et al. "Concept drift adaptation techniques in distributed environment for real-world data streams." *Smart Cities* 4.1 (2021): 349-371.

[38] Hovakimyan, Gurgen, and Jorge Miguel Bravo. "Evolving strategies in machine learning: A systematic review of concept drift detection." (2024).

9 Exploring deep learning approaches for hate speech detection in Assamese social media text

Tulika Chutia[a], Rituraj Phukan[b] and Nomi Baruah[c]

Dibrugarh University, Dibrugarh, India

Abstract

Hate speech detection is a crucial task in natural language processing for identifying and mitigating harmful content. Effective detection systems are imperative for safe and inclusive online environments. This paper addresses the application of Long-Short Term Memory (LSTM) and Bidirectional Long-Short Term Memory (Bi-LSTM) models for hate speech detection in Assamese text. Using a dataset of 387,488 words over 2,557 sentences, the LSTM model achieved 98.44% of accuracy and the Bi-LSTM model achieved 97.22%. The accuracy rates are very high, which shows that both models can be used to identify hate speech with good accuracy. Future work may further enhance these models by expanding datasets and using advanced pre-processing techniques to create more robust hate speech detection systems.

Keywords: Bi-LSTM, hate speech, LSTM, natural language processing, sentiment analysis

Introduction

In recent times, individuals have been using social media to interact and communicate with the world. The majority of work that is performed in sentiment analysis is done in social media domains such as Facebook, Twitter, Instagram, blogs, and consumer-generated content (CGC). For instance, on dark web online forums, terrorist and extremist organisations communicate with one another, exchange ideas, and implement the site as a groove for extremism [1]. As far as the current state is concerned, hate speech recognition model has predominantly been estimated using held-out performance on some limited sample of hate speech [2–5].

However, recent work has pointed out the weakness of the estimated paradigm. In the domain of sentiment analysis, various web discussions surrounded the evaluation of newsgroups, blogs, and web forums. However, this domain is often used nowadays as a tenet for remarks containing hateful words and fires [1]. About recognising hate speech in social media text, like comments on news reviews, blogs, and web forums. Different researchers have investigated the identification of brutal or terrorising texts and fires on social media as well as the diffusion of hateful content in the dark web forums [6,7]. Popular languages like English, Arabic, Russian, Mandarin, and also Indian languages like Hindi, Bengali, and Tamil have seen a considerable amount of work in this area [8]. Hate speech can open out through different websites processed by small intimate groups, but it may extend via popular social media sites like Facebook, Twitter, Yahoo, different blogs, etc. Although several researchers have got into the discipline of hate speech recognition in recent years, they typically only discuss supervised methods for recognising racist speech [9,10].

According to several studies [3,9,11], hate speech is a type of abuse against a protected group or its members for being part of that group. The protected group can be based on disability, gender, race, pregnancy, sexual orientation, religion, community, etc.

The people who generally post hateful content on social media are typically associated with a competing group, from different ethnic communities. Hate speech on social media speaks about writing that diminishes a loser or victim and has the potential to threaten or harm them. It is discussed that is antagonist, malicious, and biased against a person or a group of people due to some perceived or real intrinsic quality [12].

This type of speech is directed to make it obvious for the purpose of causing harm, hatred, or otherwise to encourage harmfulness. With the Industrial Revolution 4.0 and interactive Web 2.0 era, the social media and electronic marketing context offer an exceptionally simple one to produce, share, post, and pass on in the context of digital media platforms and social media. The opinions and sentiments are expressed on social media sites, digital marketing, news channel reviews, internet forums, and group discussions along with micro-blogging sites [13]. According to Chakravarthi [14] and Cohen R [18], 'Hate speech involves a direct attack on people based on what we call protected characteristics—race, ethnicity, national origin, religious affiliation, sexual orientation, caste, sex,

[a]tulikachutiadu24@gmail.com, [b]riturajphukan01@gmail.com, [c]baruahnomi@gmail.com

DOI: 10.1201/9781003675235-9

gender, gender identity, and serious disease or disability. We also offer some protections based on immigration status. The authors define attack as speech that is violent or dehumanising, or which proclaims inferiority, or which urges exclusion or segregation.'

The act of posting heinous or discriminatory material and information over computer-based communication media on social networking sites is referred to as Cyber-hate [20]. It is a unique form of Cyberbullying or Cyberstalking that is different in that it can prompt communications of hate [21], any verbal or written expression that condemns unjustly violence towards someone due to his or her sexual, ethnic, national, religious, or gender orientation. Ghosh et al. [23] stated that hate speech may have different linguistic forms, lack of respect, irritation, annoying, despicable, and aggressive behaviours.

Example: ফালটু জনি ৰিলিচ বনোৱা বাদ দে / Faltu Joni reels bonua baad de/ Useless girl stop making reels. Here, ফালটু / Faltu / Useless is used as a hate word and the sentence is considered as a hate speech/sentence. It is a challenging task to understand hate speech since various personal views and cultural differences can have varying suggestions or opinions regarding what is hate speech or not. Figure 9.1 illustrates the various dimensions of hate speech. Davidson et al. [19] asserted that the sharing of offensive features between the abusive and aggressive text layers has been observed in hate speech. The chart illustrates that the segmentation of the concept is complicated, a call that has been analysed in current research [24]. Gitari N [24] also noted that contentious definitions of hate speech involve the observation that homophobic and racist tweets have higher chances of being labelled as abusive or offensive content. Thus, it is impossible to conclude confidently whether controversial content sums up hate speech, in general, [15,25].

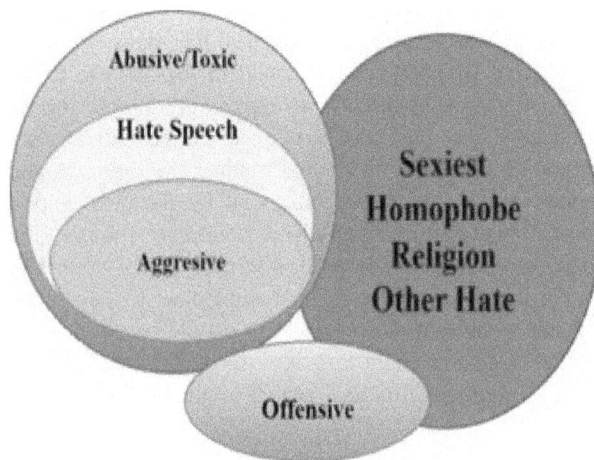

Figure 9.1 Hierarchy of hate speech concepts [19]
Source: Author

The goal of this experiment is to identify hate speech from the Assamese sentences that were collected from the different domains using LSTM and Bi-LSTM. Essentially, this study aims to address two research questions, which are as follows:

RQ1 How to design a deep learning model for Assamese text that automatically detects hate speech labelled as TRUE/FALSE in the social media text, without using any profanity dictionary?

RQ2 How to determine the effectiveness in detecting hate speech in Assamese Social media text?

This paper is organised as follows: Section 2 includes a review of related studies. The methodology of the proposed research work is described in Section 3. Section 4 describes the dataset description, experimental results, and analysis. Section 5 describes the conclusion and future scope of our study.

Literature Review

Al Hassan et al. [3] introduced a model to identify hate speech using machine learning methods like Logistic Regression, Nave Bayes classifier, Decision Tree, SVM, and Random Forest classifier. The authors achieved a total precision of 0.91%, recall of 0.90, and an F1 score of 0.90%. Cohen R [18] describes different theories of hate speech based on the *Encyclopaedia of the American Constitution*. According to Cohen R [18] and Kamble et al. [26], 'Hate speech is a speech that targets a person or group based on characteristics like race, religion, ethnic origin, national origin, sex, disability, sexual orientation, or gender identity'. As stated by Cohen R [18] and Karim et al. [27], 'Hate speech is a direct attack on an identifiable group of people motivated by some aspect of the group's identity'. Kulkarni et al. [28] and Kwok and Wang [29] developed the largest manually annotated corpus of hate speech identification in Malayalam, Tamil, and English, with 10,705, 28,451, and 20,198 sentences respectively. MacAvaney et al. [30] proposed a machine learning model with SVM and Random Forest Classifiers for hate speech identification and achieved an average F1 score and 81% using 10-fold cross-validation. Miro et al. [31] suggested a technique for Hindi–English code-mixed language to detect hate speech. Thus, they introduced their work by employing machine learning technique on code-mixed social media text. They further stated that 'The methodology they have used for Facebook's pre-trained word embedding library, fastText to represent 10,000 sentences that were accumulated from distinct domains as non-hate or hate'. Their proposed model's performance was compared to doc2vec with word2vec features that illustrates how fastText features yield the feature extraction more effectively using the SVM-RBF classifier. Mustafa et al. [32] contributed to the state-of-the-art hate speech detection for English–Hindi code-mixed tweets, in which they compare three deep learning models using domain-specific embeddings. They also found that domain-specific embeddings provided promising outputs for the target groups. Baruah et al. [8] provided a lot of classification rules for the Bengali language, where they took into account three hate speech topics: sentiment analysis, document classification, and detection on opinion mining. They have also shown the Bengali word embedding models, based on 250 million articles, named BengFastText. The authors conducted three experiments, i.e., sentiment analysis, hate speech detection, and document classification. He employed the Multichannel Convolutional LSTM (MC-LSTM) for his work. They have been successful in proving that BengFastText performs against other baseline embedding models, i.e., 90.45% F1-Score, 82.25% GloVe, and 92.30% Word-Vec for hate speech detection, sentiment analysis, and document classifications, at the time of 5-fold cross-validation.

In natural language processing, opinion mining and sentiment analysis are the basic tasks. Neog et al. [33] created a dataset in Marathi sentiment analysis called L3CubeMahaSent. They collected data from different publicly available social media platforms, in which the dataset L3CubeMahaSent contains 16,000 tweets labelled as positive, negative, and neutral. They provided baseline classification outputs utilising deep learning architectures like LSTM, C, ULMFit, and BERT-based models. Nockleby et al. [34] presented their work by utilising SVM and XLM-RoBERT in 'Dravidian-CodeMix-HASOC 2020'. In the code-mixed Malayalam dataset, the best performance was achieved using an SVM classifier with TF-IDF features and n-grams, in which they achieved an F1 score of 95% on YouTube data, 76% on Twitter data, and XLM-RoBERTa scored 87% of F1 score on Tamil Tweet data. Phukan et al. [35] presented sentiment analysis using BERT and ULMFit on Tanglish (Tamil+English) with an F1-score of 86%, and an accuracy of 62% [36].

Proposed Methodology

Below are the steps of the suggested methodology of this paper.

A. Methodology

In Figure 9.2, the input dataset is initially pre-processed using the methodologies outlined in the figure. Following pre-processing, the data enters the classification phase, employing both LSTM and Bi-LSTM approaches. After evaluation, the model predicts the desired output.

- Data pre-processing:
  Before applying the Deep Learning Classification algorithm to the input, we implement different pre-processing techniques:

- Removal of emojis:
  Since the data we collected from different social media domains contain different stickers or emojis, they will be removed [16,17,37].

- Data cleaning:
  In this phase, unpleasant portions of a sentence are deleted. For instance: mostly eliminating punctuation marks (e.g., Question mark (?) or প্রশ্নবোধক (?), Full stop (.) or দাঁৰি (।), Comma (,) or কমা (,), Semicolon (;) or ছেমিকলন (;), Exclamatory(!) or ভাৱবোধক (!), Digits (0,1,2,3,4,5,6,7,8,9) or সংখ্যা (০,১,২,৩,৪,৫,৬,৭,৮,৯), etc. was being removed from the collected texts [17].

- Tokenisation:
  It is a process that splits a sentence into words, which are known as tokens, and these tokens help to detect the output, improving the efficiency of our proposed algorithm [38].

- Word Embedding:
  It is an evolutionary technique process in various areas of sentiment analysis as well as natural language processing (NLP). For identifying the patterns and significant relations, and increasing the accuracy and efficiency of AI models, it can be done by representing words in the form of numerical vectors.
  For example: Before Tokenisation: 'আমি তেওঁলোকক পাগল বুলি কও ।' After Tokenisation: আমি তেওঁলোকক পাগল বুলি কও

B. Used approach

Long short-term memory (LSTM):

Long Short-Term Memory (LSTM) networks are a type of recurrent neural network (RNN) architecture that effectively captures sequential data by solving the vanishing gradient problem that is common with normal RNNs [39]. Developed by Hochreiter and Schmidhuber in 1997, LSTMs use various gating techniques, including input, output, and forget gates, to control the flow of information inside the cell state. This gating technique enables LSTMs to keep and update important information over extended sequences. It makes them especially good at retaining long-term dependencies in time series data, natural language processing jobs, and other sequential inputs [40]. The effectiveness of LSTM networks has been demonstrated in various disciplines, including speech recognition, machine translation, and sentiment analysis. Their architecture allows them to store crucial information over long sequences while selectively forgetting irrelevant data, minimising the problems associated with short-term memory restrictions [41]. LSTMs learn to perceive patterns and dependencies in data sequences by dynamically altering gate weights during training, resulting in improved performance in tasks that require context knowledge and temporal sensitivity. As a

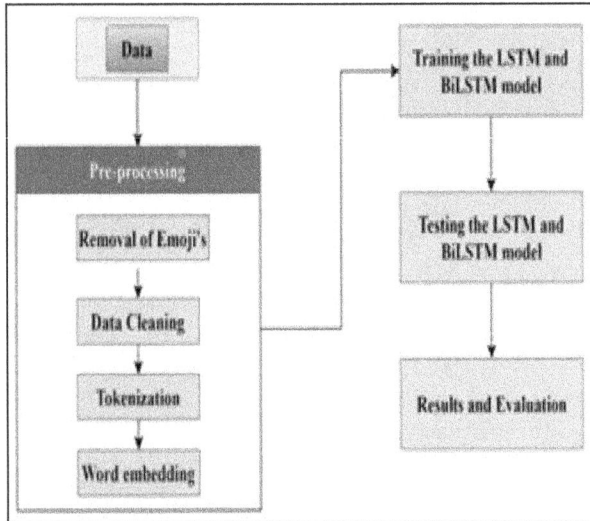

Figure 9.2 Pictorial diagram of proposed methodology
Source: Author

	Comments	label
0	পইছা নিদিয়া কাৰনে আন্দোলন ইয়াতকে লাজ লগা কি ...	1
1	যদি পইচা নাই উত্তৰৰ কিয় পাতিব লাগে বেলেগৰ পৰা...	1
2	কিছুমান ভিক্ষাৰীৰ কাৰণে অসমীয়া জাতীটোকে কলংকিত...	1
3	ই জন পাগল হয় ইয়াক দেখিলে ঘৃণা লাগে	1
4	তাৰ মুখ খন দেখিবলে গাহৰি টো নিছিনা	1

Figure 9.3 Sample of the proposed dataset
Source: Author

Dataset and experiment

A. Dataset

The corpus for training and testing of LSTM and Bi-LSTM has 387,488 words and 2,557 sentences. Every sentence is labelled for containing hate speech or not so that it becomes a very clear target for classification. This corpus has all kinds of examples of hate and non-hate speech. Thus, both classes are well-balanced in this dataset. In the given corpus, 1,400 non-hate sentences and 1,157 hate sentences are found. Apart from that, the dataset also has class labels, which enrich the data and give another feature the models can use while training. The dual labelling of the datasets helps the models understand and discern different types of speech even better. The figure 9.3 depicted the sample of collected dataset.

B. Implementation:

1) **LSTM**: Training an LSTM model involves preparing a dataset, designing the model architecture, and iteratively updating the model parameters to minimise a loss function. The dataset for the task of sentiment analysis or hate speech identification usually consists of sentences that are labelled to denote whether the text is hate speech. Before training, the text data is tokenised and padded to have a uniform input length, and labels and classes are encoded numerically. For instance, labels often are binary encoded, while classes could be one-hot encoded [6] [22] [42]. The LSTM model is then initialised with an embedding layer to map words into dense vectors, followed by LSTM layers to identify temporal relationships in the text. It will train through back-propagation through time (BPTT) of its model weights. It adjusts its parameters like learning rate, batch size, and epoch number. Early stopping is one typical way of preventing overfitting. Once trained, the model's performance is measured by testing it on another test set [15].

2) **Bi-LSTM**: The training and testing approach of a Bi-LSTM model is the same as that of an LSTM, but with more features to capture bidirectional context inside sequences. The dataset, like LSTM, consists of labelled and classified sentences. Sentences are tokenised and padded, and categorical data is nu-

result, LSTMs have emerged as a critical component in creating complex sequential data models, making substantial contributions to profound learning application progress [43].

Bidirectional long short-term memory (Bi-LSTM)
Bidirectional Long Short-Term Memory (Bi-LSTM) networks enhance the capabilities of regular LSTM networks by processing data in both forward and reverse directions [44]. This dual-processing approach allows Bi-LSTMs to capture dependencies and contextual information from both previous and future states in a sequence. Bi-LSTMs, introduced as an improvement over unidirectional LSTMs, comprise two parallel LSTM layers: one processing the input sequence from start to finish (forward layer) and the other from end to start (reverse layer). The outputs of these two layers are then concatenated, usually by concatenation, to create a complete representation of the input sequence [41].

The bidirectional feature of Bi-LSTMs provides considerable advantages in tasks that require context from both preceding and succeeding parts for good predictions (Lauritzen et al., 2020). In natural language processing, for example, understanding a specific word's context can considerably improve job performance through named entity recognition, part-of-speech tagging, and machine translation [41]. Bi-LSTMs can acquire a more sophisticated understanding of the data by combining input from both directions, enhancing model accuracy and resilience. This bidirectional technique has proven particularly useful in applications requiring fine-grained contextual analysis, cementing Bi-LSTMs as a potent extension of the LSTM architecture in deep learning [45].

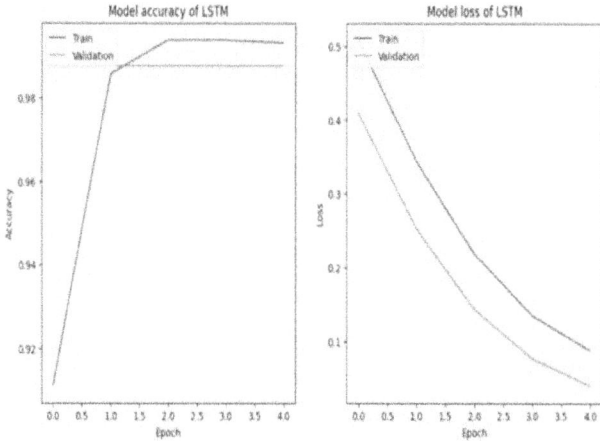

Figure 9.4 Model accuracy and model loss of LSTM
Source: Author

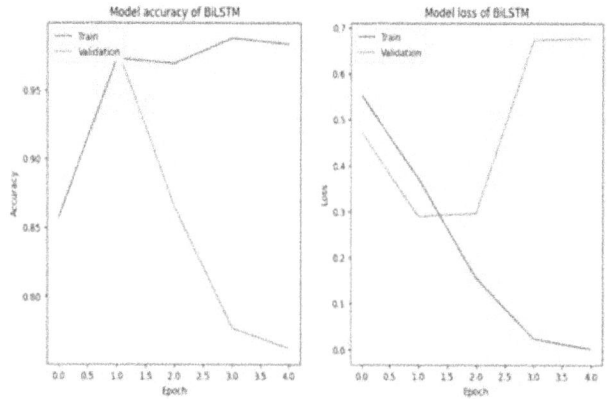

Figure 9.5 Model accuracy and model loss of Bi-LSTM
Source: Author

merically encoded. The distinction is in the model's architecture: a Bi-LSTM has two LSTM layers that pass the input sequence in both forward and backward directions [15].

This bidirectional structure allows the model to take context from both past and future situations, thereby enhancing its ability to perceive the intricacies of the text. The combined outputs of the forward and backward layers provide a more complete representation of the input sequence.

Training a Bi-LSTM requires modification of the same parameters as in an LSTM, including learning rate, batch size, and epochs, but with the additional complexity, tuning becomes important. The use of early halting and validation splits are essential to ensure monitoring of model performance and the prevention of overfitting. After the training procedure, the Bi-LSTM model is tested on an entirely independent set of samples while the accuracy, in most cases, is a measure for determining its efficacy in detecting hate speech. Bidirectionality often makes the Bi-LSTM perform better than standard models like regular LSTMs, mainly in tasks with comprehensive context dependency [15].

C. Result and comparison
The performance of Bi-LSTM and LSTM models was evaluated using an Assamese text dataset with hate speech labels and class categories. The test dataset accuracy of the LSTM model was 98.44%, which means that it can distinguish hate speech from non-hate speech texts. Bi-LSTM, although not that accurate at 97.22%, did fairly well in the classification task. Bi-LSTM, with bidirectional processing, could obtain the context of both past and future states, which is extremely important for the proper identification of hate speech. Figures 9.4 and 9.5 illustrate LSTM models' model accuracy and model loss, while Figure 9.4 illustrates

Table 9.1 Comparison of results with proposed approach and existing work.

Existing work	Language	Approach	Accuracy
Neog N. et al. 2023 [33]	Assamese	CNN + BiLSTM	88.43%
Baruah N 2023 [10]	Assamese	Linear SVC and RF	88.3%
Baruah N et al. 2023a [9]	Assamese	NB, SVM	86.9%
LSTM	Assamese	2,557 sentences	98.44%
Bi-LSTM	Assamese	2,557 sentences	97.22%

Source: Author

the model accuracy and model loss of the Bi-LSTM model. This work has far superior results to anything previously obtained in other research endeavours in detecting Assamese hate speech. Ghosh et al. suggested sentiment analysis using a Dataset of 50,000 sentences, the accuracy achieved was 90.51% [46]. Baruah et al. proposed research work on Assamese sentiment analysis with a dataset of 2,000 sentences with an accuracy of 88.3% [29].

In contrast, the proposed work attained 98.44% accuracy with the LSTM model and 97.22% with the Bi-LSTM model, showing significant performance. Neog et al. reported an accuracy of 88.43% with a dataset size of 2,000 sentences in their toxic comments detection study in the low-resource Assamese language, implementing LSTM and Bi-LSTM models [47]. The comparison is shown in Table 9.1.

Conclusion and future work

LSTM and Bi-LSTM models were experimented with to assess their performance on hate speech detection in Assamese. There is an intensive processing and testing of

387,488 words from 2,557 phrases for training and testing the models. The model attained 98.44% for the LSTM model, while the performance of the Bi-LSTM model was noted to be 97.22%. These results indicate that both models perform hate speech classification exceedingly well; the LSTM narrowly outperforms its bidirectional counterpart. The matrices presented here in full detail illustrate how each model classified instances with a breakdown of true positives, true negatives, false positives, and false negatives, providing more detailed information about each model.

Further research can enhance performance by increasing the dataset to comprise more representative samples, enhancing pre-processing approaches such as stemming and lemmatisation, and investigating idiomatic idioms. Ensemble models that combine LSTM and Bi-LSTM architectures may also achieve superior outcomes. Integrating attention mechanisms could help the models focus on relevant text sections, increasing detection accuracy. Transfer learning with pre-trained models such as BERT or GPT that have been fine-tuned for Assamese text could improve performance even further. Such models could be used for developing real-time hate speech detection systems that help in monitoring social media and online communities to reduce harmful content. The following research directions are based on the encouraging findings, aimed at developing more robust hate speech detection systems for Assamese and other languages.

References

[1] Abbasi, A., Chen, H. (2007). Affect intensity analysis of dark web forums. In 2007 IEEE Intelligence and Security Informatics. IEEE, 282–288.

[2] Abbasi, A., Chen, H., Salem, A. (2008). Sentiment analysis in multiple languages: feature selection for opinion classification in web forums. *ACM Transactions on Information Systems (TOIS)*, 26(3), 1–34.

[3] Al-Hassan, A., Al-Dossari, H. (2019). Detection of hate speech in social networks: a survey on multilingual corpus. In 6th International Conference on Computer Science and Information Technology, vol. 10. ACM, 10-5121.

[4] Alkomah, F., Ma, X. (2022). A literature review of textual hate speech detection methods and datasets. *Information*, 13(6), 273.

[5] Anis, M. Y. (2017). Hatespeech in Arabic language. In International Conference on Media Studies, 724.

[6] Ayo, F. E., Folorunso, O., Ibharalu, F. T., Osinuga, A. (2020). Machine learning techniques for hate speech classification of twitter data: state-of-the-art, future challenges and research directions. *Computer Science Review*, 38, 100311.

[7] Banko, M., MacKeen, B., Ray, L. (2020). A unified taxonomy of harmful content. In Proceedings of the Fourth Workshop on Online Abuse and Harms, 125–137.

[8] Baruah, A., Das, K. A., Barbhuiya, F. A., Dey, K. (2021). Iiitg-adbu@hasoc-dravidian-codemix-fire2020: offensive content detection in code-mixed Dravidian text. *arXiv preprint* arXiv:2107.14336.

[9] Baruah, N., Gogoi, A., Phukan, R., Goutom, P. J. (2023a). Hate speech detection for a low-level language (Assamese). In 2023 14th International Conference on Computing Communication and Networking Technologies (ICCCNT). IEEE, 1–6.

[10] Baruah, N., Kalita, A. J., Gogoi, A., Bezbaruah, M., Prasad, N., Das, V. P., Phukan, R. (2023b). Detection of explicit lyrics in Hindi music using LSTM. In 2023 14th International Conference on Computing Communication and Networking Technologies (ICCCNT). IEEE, 1–5.

[11] Biradar, S., Saumya, S. (2022). Iiitdwd@tamilnlp-acl2022: transformer-based approach to classify abusive content in Dravidian code-mixed text. In Proceedings of the Second Workshop on Speech and Language Technologies for Dravidian Languages, 100–104.

[12] Blaya, C. (2019). Cyberhate: a review and content analysis of intervention strategies. *Aggression and Violent Behavior*, 45, 163–172.

[13] Brown, A. (2017). What is hate speech? Part 1: The myth of hate. *Law and Philosophy*, 36, 419–468.

[14] Chakravarthi, B. R. (2020). Hopeedi: a multilingual hope speech detection dataset for equality, diversity, and inclusion. In Proceedings of the Third Workshop on Computational Modeling of People's Opinions, Personality, and Emotion's in Social Media, 41–53.

[15] Chutia, Tulika, Nomi Baruah. (2024). A review on emotion detection by using deep learning techniques. *Artificial Intelligence Review*, 57(8), 203.

[16] Chutia, Tulika, Nomi Baruah. (2024a). Text-based emotion detection: a review. *International Journal of Digital Technologies*, 3(I).

[17] Chutia, Tulika, Nomi Baruah (2024b). Emotion recognition in Assamese text using LSTM. 2024 15th International Conference on Computing Communication and Networking Technologies (ICCCNT). IEEE.

[18] Cohen-Almagor, R. (2011). Fighting hate and bigotry on the internet. *Policy & Internet*, 3(3), 1–26.

[19] Davidson, T., Warmsley, D., Macy, M., Weber, I. (2017). Automated hate speech detection and the problem of offensive language. In Proceedings of the International AAAI Conference on Web and Social Media, vol. 11, 512–515.

[20] De Gibert, O., Perez, N., García-Pablos, A., Cuadros, M. (2018). Hate speech dataset from a white supremacy forum. arXiv preprint arXiv:1809.04444.

[21] Dutta, S., Neog, M., Baruah, N. (2024). Assamese toxic comment detection on social media using machine learning methods. In 2024 Second International Conference on Emerging Trends in Information Technology and Engineering (ICETITE). IEEE, 1–8.

[22] Fortuna, P. Nunes, S. (2018). A survey on automatic detection of hate speech in text. *ACM Computing Surveys (CSUR)*, 51(4), 1–30.

[23] Ghosh, K., Senapati, A., Narzary, M., Brahma, M. (2023). Hate speech detection in low-resource Bodo and Assamese texts with ML-DL and BERT models. *Scalable Computing: Practice and Experience*, 24(4), 941–955.

[24] Gitari, N. D., Zuping, Z., Damien, H., Long, J. (2015). A lexicon-based approach for hate speech detection. *International Journal of Multimedia and Ubiquitous Engineering*, 10(4), 215–230.

[25] Goutom, P. J., Baruah, N., Sonowal, P. (2023). An abstractive text summarization using deep learning in Assamese. *International Journal of Information Technology*, 15(5), 2365–2372.

[26] Kamble, S., Joshi, A. (2018). Hate speech detection from code-mixed Hindi-English tweets using deep learning models. arXiv preprint arXiv:1811.05145.

[27] Karim, M. R., Chakravarthi, B. R., McCrae, J. P., Cochez, M. (2020). Classification benchmarks for under-resourced Bengali language based on multichannel convolutional-LSTM network. In 2020 IEEE 7th International Conference on Data Science and Advanced Analytics (DSAA). IEEE, 390–399.

[28] Kulkarni, A., Mandhane, M., Likhitkar, M., Kshirsagar, G., Joshi, R. (2021). L3cubemahasent: a Marathi tweet-based sentiment analysis dataset. arXiv preprint arXiv:2103.11408.

[29] Kwok, I., Wang, Y. (2013). Locate the hate: detecting tweets against blacks. In Proceedings of the AAAI Conference on Artificial Intelligence, vol. 27, 1621–1622.

[30] MacAvaney, S., Yao, H.-R., Yang, E., Russell, K., Goharian, N., Frieder, O. (2019). Hate speech detection: challenges and solutions. *PLoS One*, 14(8), e0221152.

[31] Miro-Llinares, F., Rodríguez-Sala, J. J. (2016). Cyber hate speech on twitter: analyzing disruptive events from social media to build a violent communication and hate speech taxonomy. *International Journal of Design & Nature and Ecodynamics*, 11(3), 406–415.

[32] Mustafa, R. U., Nawaz, M. S., Farzund, J., Lali, M. I., Shahzad, B., Viger, P. (2017). Early detection of controversial Urdu speeches from social media. *Data Science and Pattern Recognition*, 1(2), 26–42.

[33] Neog, M., Baruah, N. (2023). A deep learning framework for Assamese toxic comment detection: Leveraging LSTM and Bi-LSTM models with attention mechanism. In International Conference on Advances in Data-Driven Computing and Intelligent Systems. Springer, 485–497.

[34] Nockleby, J. T., Levy, L. W., Karst, K. L., Mahoney, D. J. (2000). Encyclopedia of the American constitution. Hate Speech; Levy, L., Karst, K., (eds.), 1277–1279.

[35] Phukan, R., Baruah, N., Sarma, S. K., Konwar, D. (2024a). Exploring character-level deep learning models for POS tagging in Assamese language. *Procedia Computer Science*, 235, 1467–1476.

[36] Phukan, R., Goutom, P. J., Baruah, N. (2024b). Assamese fake news detection: a comprehensive exploration of LSTM and Bi-LSTM techniques. *Procedia Computer Science*, 235, 2167–2177.

[37] Phukan, R., Neog, M., Baruah, N. (2023). A deep learning based approach for spelling error detection in the Assamese language. In 2023 14th International Conference on Computing Communication and Networking Technologies (ICCCNT). IEEE, 1–7.

[38] Phukan, R., Neog, M., Goutom, P. J., Baruah, N. (2024c). Automated spelling error detection in Assamese texts using deep learning approaches. *Procedia Computer Science*, 235, 1684–1694.

[39] Röttger, P., Vidgen, B., Nguyen, D., Waseem, Z., Margetts, H., Pierrehumbert, J. B. (2020). Hatecheck: functional tests for hate speech detection models. arXiv preprint arXiv:2012.15606.

[40] Sane, K. R., Kolla, S., Sane, S. R., Srirangam, V. K., Mamidi, R. (2019). Corpus and baseline system for hate speech detection in Telugu-English code-mixed tweets. In Proceedings of the 20th International Conference on Computational Linguistics and Intelligent Text Processing (CICLing'19).

[41] Saumya, S., Kumar, A., Singh, J. P. (2021). Offensive language identification in Dravidian code mixed social media text. In Proceedings of the First Workshop on Speech and Language Technologies for Dravidian Languages, 36–45.

[42] Spertus, E. et al. (1997). Smokey: automatic recognition of hostile messages. In AAAI/IAAI, 1058–1065.

[43] Sreelakshmi, K., Premjith, B., Soman, K. (2020). Detection of hate speech text in Hindi-English code-mixed data. *Procedia Computer Science*, 171, 737–744.

[44] Stjernfelt, F., Lauritzen, A. M., Stjernfelt, F., Lauritzen, A. M. (2020). Facebook's Handbook of Content Removal. Your Post has been Removed: Tech Giants and Freedom of Speech, 115–137.

[45] Strossen, N. (2016). Freedom of speech and equality: do we have to choose. *Journal of Law and Policy*, 25, 185.

[46] Warner, W., Hirschberg, J. (2012). Detecting hate speech on the world wide web. In Proceedings of the Second Workshop on Language in Social Media, 19–26.

[47] Waseem, Z., Thorne, J., Bingel, J. (2018). Bridging the gaps: multi task learning for domain transfer of hate speech detection. *Online Harassment*, 29–55.

10 Using sentiment analysis to enhance digital forensic evidence identi ication

Vishwavijay Kumar[a] and Vikas Pareek[b]

Department of CS&IT, Mahatma Gandhi Central University, Motihari, Bihar, India

Abstract

The rapid proliferation of digital communication has increased the complexity of digital forensic investigations. Identifying crucial evidence from vast amounts of data can be challenging. Sentiment analysis, a form of natural language processing (NLP), offers a promising solution by providing insight into the emotional tone of digital communications. This paper explores the application of sentiment analysis to enhance the identification and classification of digital forensic evidence. By analysing emails, social media posts, and messages, sentiment analysis can detect hostile or aggressive language, which may correlate with illicit activities or behaviour. Integrating this technique into forensic workflows allows investigators to prioritise critical evidence more effectively, reducing the time and effort required to sift through large datasets. A comparative analysis of multiple models, including Naive Bayes (NB), Logistic Regression (LR), SVM, Random Forest (RF), Decision Tree (DT), KNN, ANN, Convolutional Neural Networks (CNN), Long Short-Term Memory (LSTM), CatBoost, LightGBM, XGBoost, and AdaBoost was conducted. The NB and CNN models demonstrated high performance, both achieving an accuracy of 98.39 ± 3.20 on the same dataset. This approach also provides deeper insights into suspect intentions, which can assist in solving cases related to cyberbullying, fraud, harassment, and criminal conspiracies. The findings suggest that sentiment analysis can significantly enhance the efficiency and accuracy of digital forensic investigations.

Keywords: Behaviour; social media; sentiment analysis; machine learning; natural language processing

Introduction

Sentiment analysis analyses the text's positive, negative, or neutral sentiment. It is a field of natural language processing (NLP) that aims to identify and categorise the emotional tone of the text. Sentiment analysis is essential in various fields, such as social media monitoring, customer feedback analysis, healthcare, financial analysis, and political analysis. Digital forensics has become integral to modern criminal investigations, where analysts sift through enormous volumes of digital data—such as emails, social media posts, and text messages—to extract helpful evidence. However, with the proliferation of digital communication, identifying key pieces of evidence has become increasingly difficult [13]. Traditional methods of keyword-based searches may miss critical information hidden in the tone or sentiment of communications. In digital forensics, sentiment analysis can filter and prioritise digital communications, guiding investigators towards emotionally charged or contextually relevant conversations that may contain valuable evidence [4]. By identifying messages with emotional solid undertones, investigators can focus on digital artefacts more likely to contain motive, intent, or indicators of criminal behaviour [10]. Integrating sentiment analysis into digital forensics introduces a more nuanced approach to evidence identification, allowing for the detection of covert signals that might go unnoticed through traditional [12]. This method could improve the efficiency of forensic investigations and provide deeper insights into the behaviour, state of mind, and intentions of suspects involved in criminal activity. In the evolving landscape of criminal investigations, digital forensics plays a critical role by extracting and analysing evidence from digital devices, networks, and online communications [12]. As individuals' digital data grows exponentially, forensic analysts face increasing challenges in identifying relevant evidence efficiently. This overwhelming influx of data, often in the form of emails, text messages, social media posts, or forum discussions, can obscure critical information, slowing down investigations. Incorporating sentiment analysis into digital forensics provides several potential advantages. For instance, messages conveying strong negative emotions could signal frustration, threats, or intent, which may be crucial in criminal cases like cyberbullying, harassment, or fraud [8]. By using sentiment analysis, investigators can filter through vast volumes of digital communications and hone in on conversations that carry emotional weight, likely tied to the case under investigation [16]. This technology offers a more targeted approach than simple keyword searches, which might miss subtleties in context or emotional [3].

Literature review

Sentiment analysis can help investigators focus on communications that carry emotional weight. For instance,

---

[a]vishwavijaykumar123@gmail.com, [b]vikaspareek@mgcub.ac.in

DOI: 10.1201/9781003675235-10

negative sentiments like anger, frustration, or hostility may signal conversations relevant to criminal activity [2]. Instead of manually sorting through countless messages, sentiment analysis tools can flag emotionally charged messages, allowing investigators to prioritise those for further examination. By analysing the sentiment trends over time, investigators can identify patterns of behaviour or mood changes in suspects or victims [11]. For example, a person showing increasing levels of aggression or distress might be linked to premeditated crimes, harassment, or cyberstalking. Detecting these emotional shifts can help predict potential criminal actions or provide deeper insight into ongoing activities [9]. Traditional keyword searches often need help to grasp a message's context or underlying emotions. Sentiment analysis adds a layer of understanding by capturing the emotional tone, which can reveal intent, motive, or urgency [1]. For instance, a keyword search might find a threat, but sentiment analysis could reveal whether the threat was genuine anger or sarcastic humour, helping investigators better evaluate its relevance.

Dataset/case study, model, performance metrics, and identified research gaps

Krishnan, Sundar and Shashidhar, Narasimha and Varol, Cihan and Islam [13] applied supervised learning and neural network models on three fictional legal case datasets, achieving CNN accuracy of 81.4% and LSTM accuracy of 80.6%, highlighting a lack of focus on specific sentiment analysis applications. Budiman [4] analysed 300 tweets on sexual harassment using Naive Bayes and DBN, achieving 83% accuracy, and identified the need for better multi-language sentiment models. Hamdi [8] developed a forensic email classifier for 600,000 emails using K-Means, Naive Bayes, PSO, and SVM, with 92% accuracy, addressing gaps in model optimisation and integration. Studiawan et al. [17] used GRU for anomaly detection in system logs, achieving an F1-score of 99.84% and accuracy of 99.93%, noting limited research on sentiment analysis for anomaly detection. Ashraf et al. [2] analysed datasets, applied deep learning models, evaluated accuracy and F1-score, and highlighted gaps in multilingual sentiment analysis and real-time detection.

Methodology

The dataset comprises a total of 5,572 distinct messages. The columns include Label: A binary classification denoting whether the message is categorised as spam (1) or ham (0). Text: This column encompasses the substantive content of the messages. There exist a total of 5,169 unique entries of text, albeit with some occurrences of repetition. The class distribution reveals that approximately 13.4% of the messages are categorised as spam (designated by Label 1), whereas the remaining 86.6% are classified as ham (designated by Label 0). Naive Bayes is a probabilistic classification algorithm that relies on Bayes' Theorem, operating under the assumption that features are independent. It is particularly effective for text classification tasks such as spam filtering and sentiment analysis [7]. Logistic Regression is a linear model commonly used for binary classification. It estimates the probability that a given instance falls into a specific class based on input features. It applies a logistic (sigmoid) function to model the relationship between the dependent variable and one or more independent variables [20]. Support Vector Machine (SVM) is a supervised learning algorithm designed to determine an optimal hyperplane for effectively separating data into distinct classes. It is particularly well-suited for high-dimensional datasets and is widely used for both classification and regression tasks [18]. Random Forest is an ensemble learning technique that constructs multiple decision trees and combines their outputs to enhance prediction accuracy and stability. It effectively reduces overfitting and performs well in both classification and regression tasks [14]. A Decision Tree splits data into subsets based on feature values, using a tree structure to make predictions. It is easy to interpret but can overfit if not appropriately managed [6]. KNN is an instance-based, non-parametric algorithm that categorises data by considering the majority label among its closest neighbours. It is simple but can be computationally expensive with large datasets [17]. ANN is a computational model inspired by the human brain, composed of layers of interconnected neurons. It can model complex relationships in data and is commonly used for tasks like image recognition and natural language processing [19]. This makes them well-suited for handling sequential data, such as time series or natural language processing [7]. CatBoost is a gradient-boosting method that handles categorical data efficiently. It is robust, scalable, and often used in ranking, recommendation systems, and regression tasks generally [20]. LightGBM is an advanced gradient boosting framework that utilises tree-based learning techniques to efficiently handle large datasets. It supports parallel and distributed learning, achieving faster training with lower memory usage, and is often used for classification, regression, and ranking tasks [5]. XGBoost is an optimised method of gradient boosting that offers superior performance and speed. It is used in machine learning competitions and excels in handling large-scale datasets [15]. The methodology for this research focuses on developing, testing, and integrating sentiment analysis tools into the digital forensic process to enhance evidence identification (Figure 10.1). The approach will combine qualitative and quantitative methods to evaluate the effectiveness of sentiment analysis in a forensic context. The research will be conducted through several stages: data collection, model development, testing, evaluation, and integration. A large dataset of digital

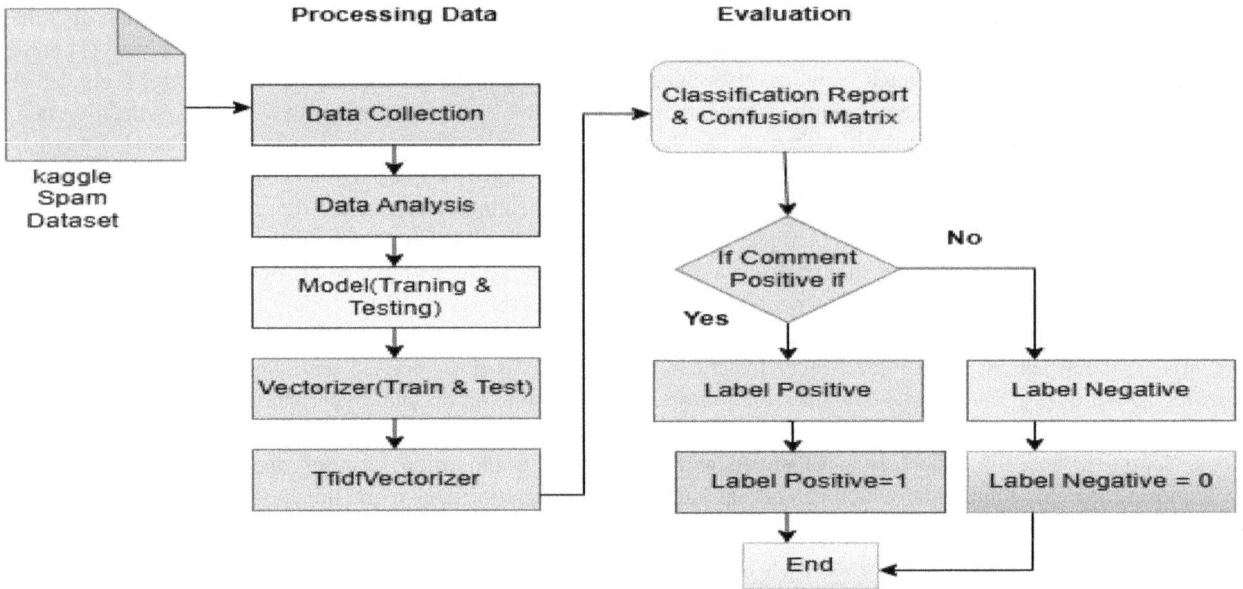

Figure 10.1 Research flowchart for sentiment analysis in digital forensics to the utilisation of all models
Source: Author

communications will be collected, including emails, text messages, social media posts, online forum discussions, and instant messaging conversations.

The dataset will be drawn from both public and anonymised forensic case data. Data types: Data will be categorised based on digital communication formats, such as text, audio transcripts, and chat logs. Where necessary, metadata (e.g., time, sender, and receiver information) will be included to add contextual relevance. The raw data will be cleaned to remove noise, including irrelevant symbols, broken text, and spam messages. Text will be tokenised, normalised, removing inconsistencies like special characters, and structured into a format compatible with sentiment analysis models. Various sentiment analysis algorithms will be explored, including lexicon-based approaches such as VADER and machine learning based models, e.g., Support Vector Machines, Naive Bayes, and Deep Learning models like LSTMs and Transformers. The models will be trained on labelled datasets where the emotional tones (positive, negative, neutral) and specific emotions (anger, fear, and joy) are already categorised. Part of the dataset will be reserved for testing and validation to evaluate the model's accuracy. Fine-tuning for forensic context: The sentiment analysis model will be refined specifically for the forensic context by including domain-specific keywords, phrases, and emotional cues commonly encountered in criminal investigations, such as threats, deception, and aggression. Contextual analysis will be incorporated to interpret better sarcasm, indirect speech, and cultural differences in emotional expressions. The trained model will be applied to digital

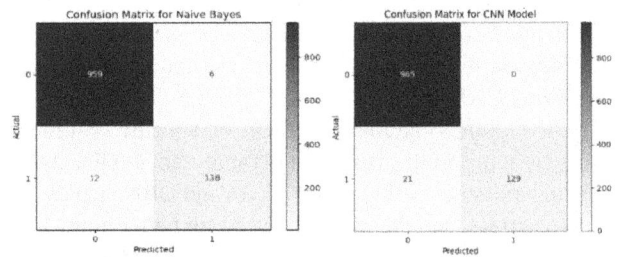

Figure 10.2 NB and CNN are machine learning models performing well on the same data set
Source: Author

communication datasets to categorise messages by emotional tone and detect spikes or shifts in sentiment over time. This will help identify communications indicating intent, distress, threats, or relevant emotional states. Using the sentiment scores generated by the model, messages with solid emotional and prioritised for further manual review by investigators. The performance of the sentiment analysis model will be measured through metrics such as accuracy, precision, recall, and F1-score. These metrics will evaluate how well the model correctly classifies sentiment in text, especially compared to traditional keyword-based forensic methods. Investigators and forensic experts will review the sentiment analysis results to assess its real-world applicability. Their feedback will focus on the model's usefulness in identifying critical evidence, the quality of flagged messages, and how well it integrates into their investigative workflows.

Table 10.1 Comparative table of all machine learning models with mean ± std deviation of the accuracy, precision, recall, and F1-score

Machine learning algorithms	Accuracy	Precision	Recall	F1-score
NB	**98.39 ± 3.20**	**0.99 ± 0.04**	0.99 ± 0.01	**0.99 ± 0.15**
LR	97.85 ± 3.20	0.98 ± 0.04	**1.00 ± 0.01**	**0.99 ± 0.15**
SVM	98.03 ± 3.20	0.98 ± 0.04	**1.00 ± 0.01**	**0.99 ± 0.15**
RF	97.67 ± 3.20	0.97 ± 0.04	**1.00 ± 0.01**	**0.99 ± 0.15**
DT	97.13 ± 3.20	0.98 ± 0.04	0.99 ± 0.01	0.98 ± 0.15
KNN	91.66 ± 3.20	0.91 ± 0.04	**1.00 ± 0.01**	0.65 ± 0.15
ANN	98.12 ± 3.20	0.98 ± 0.04	**1.00 ± 0.01**	**0.99 ± 0.15**
CNN	**98.39 ± 3.20**	**0.99 ± 0.04**	0.99 ± 0.01	**0.99 ± 0.15**
LSTM	86.55 ± 3.20	0.87 ± 0.04	**1.00 ± 0.01**	0.93 ± 0.15
CAT BOOST POOL	97.40 ± 3.20	0.97 ± 0.04	**1.00 ± 0.01**	**0.99 ± 0.15**
LIGHTGBM	97.58 ± 3.20	0.98 ± 0.04	**1.00 ± 0.01**	**0.99 ± 0.15**
XGBOOST	96.95 ± 3.20	0.97 ± 0.04	0.99 ± 0.01	0.98 ± 0.15
ADABOOST	97.40 ± 3.20	0.98 ± 0.04	0.99 ± 0.01	**0.99 ± 0.15**

Source: Author

Result analysis

A visual of the conduction matrix for all models classified into the binary classes 0 and 1. The performance of the NB model outperforms the class one (Figure 10.2). The CNN correctly classifies all zero-class instances (Figure 10.2). The results analysis compares the efficacy of various machine learning algorithms in sentiment analysis within digital forensic investigations. This scholarly work examines an array of machine learning algorithms, including Naive Bayes, LR, SVM, RF, DT, KNN, ANN, CNN, LSTM, CatBoost, LightGBM, XGBoost, and AdaBoost. Both Naive Bayes and CNN achieved the highest accuracy rates of 98.39% (Table 10.1). Logistic Regression, SVM, ANN, and CatBoost also demonstrated commendable performance, with accuracy metrics exceeding 97%. In contrast, the LSTM model exhibited a comparatively lower accuracy of 86.55%. The KNN algorithm recorded the least effective performance, particularly regarding the F1-score, which stood at 0.65. The models exhibiting the most favourable overall performance—NB, CNN, SVM—were found to accurately classify sentiment, thereby establishing their suitability as candidates for application in forensic contexts.

Conclusion

This study highlights the potential of sentiment analysis in enhancing digital forensic investigations by identifying and prioritising critical emotional cues in digital communications. The findings demonstrate that machine learning models, particularly Naïve Bayes (NB) and Convolutional Neural Networks (CNN), perform effectively in analysing sentiment-based forensic evidence, achieving an accuracy of 98.39 ± 3.20.

These results emphasise the theoretical significance of integrating sentiment analysis into forensic workflows and its practical advantages in improving investigative efficiency. The study contributes to digital forensics by proposing a structured approach to applying sentiment analysis, offering improved evidence detection and prioritisation. In practice, the integration of these models can help investigators process vast amounts of data efficiently, enhancing the accuracy of identifying illicit activities. The research also underscores the need for specialised models capable of interpreting complex emotions in various digital contexts. However, certain limitations remain. The study primarily focuses on specific datasets, which may not fully capture the diversity of online communication. Additionally, challenges in distinguishing sarcasm, irony, and context-dependent sentiments affect model interpretability. The dependency on pre-trained models may also limit adaptability to emerging forensic challenges. Future research should explore the integration of multimodal sentiment analysis, combining text, images, and voice data for a more comprehensive forensic approach.

References

[1] AbdulSattar, K., Obeidat, Q., Akour, M. (2020). Towards harnessing based learning algorithms for tweets sentiment analysis. 2020 International Conference on Innovation and

Intelligence for Informatics, Computing and Technologies (3ICT), 1–5.

[2] Ashraf, N., Mahmood, D., Obaidat, M. A., Ahmed, G., Akhunzada, A. (2022). Criminal behavior identification using social media forensics. *Electronics*, 11(19), 3162.

[3] Bokolo, B. G., Ogegbene-Ise, E., Chen, L., Liu, Q. (2023). Crime-intent sentiment detection on Twitter data using machine learning. 2023 8th International Conference on Automation, Control and Robotics Engineering (CACRE), 79–83.

[4] Budiman, K., Z. N., N. U., F. F. M. N. (2020). Analysis of sexual harassment tweet sentiment on Twitter in Indonesia using Naive Bayes method through national institute of standard and technology digital forensic acquisition approach. *Journal of Advances in Information Systems and Technology*, 2, 21–30.

[5] Cervantes, J., Garcia-Lamont, F., Rodríguez-Mazahua, L., Lopez, A. (2020). A comprehensive survey on support vector machine classification: applications, challenges and trends. *Neurocomputing*, 408, 189–215.

[6] Devi, G. D., Kamalakkannan, S. (2020). Literature review on sentiment analysis in social media: open challenges toward applications. *Test Engineering & Management*, 83(7), 2466–2474.

[7] Dey Sarkar, S., Goswami, S., Agarwal, A., Aktar, J. (2014). A novel feature selection technique for text classification using Naive Bayes. *International Scholarly Research Notices*, 2014(1), 717092.

[8] Hamdi, S. D., R. A. M. (2020). Digital cyber forensic email analysis and detection based on intelligent techniques investigation. *Iraqi Journal of Information and Communication Technology*, 11–25.

[9] Harris, M., Jacobson, J., Provetti, A. (2024). Sentiment and time-series analysis of direct-message conversations. *Forensic Science International: Digital Investigation*, 49, 301753.

[10] Hijan, A. I. (2023). Digital forensic investigation tools for cases related to social media and cybersecurity. *Ajrsp*, 5, 5–20.

[11] Hussain, M. M., Anandkumar, B. (2015). An approach for digital forensics using behavior analysis. *International Journal Of Engineering Development And Research*, 3(3), 1–9.

[12] Jain, A., G. D. K., G. K. K. (n.d.). A Review on Social Sentiment Based Predicting Cyber-Attacks.

[13] Krishnan, Sundar, Shashidhar, Narasimha, Varol, Cihan, Islam, A. R. (2022). Sentiment analysis of case suspects in digital forensics and legal analytics. *International Journal of Security (IJS)*, 13, 1–15.

[14] Mallick, C., Mishra, S., Giri, P. K., Paikaray, B. K. (2023). Machine learning approaches to sentiment analysis in online social networks. *International Journal of Work Innovation*, 3(4), 317–337.

[15] Mienye, I. D., Sun, Y. (2022). A survey of ensemble learning: concepts, algorithms, applications, and prospects. *IEEE Access*, 10, 99129–99149.

[16] Mukherjee, S., Bala, P. K. (2017). Sarcasm detection in microblogs using Naïve Bayes and fuzzy clustering. *Technology in Society*, 48, 19–27.

[17] Studiawan, H., Sohel, F., Payne, C. (2020). Sentiment analysis in a forensic timeline with deep learning. *IEEE Access*, 8, 60664–60675.

[18] Tabany, M., Gueffal, M. (2024). Sentiment analysis and fake amazon reviews classification using SVM supervised machine learning model. *Journal of Advances in Information Technology*, 15(1), 49–58.

[19] Waheed, A., Salam, A., Bangash, J.-I., Bangash, M. (2021). *Lexicon and Learn-Based Sentiment Analysis for Web Spam Detection*. IEEE-SEM.

[20] Zaidi, A., Al Luhayb, A. S. M. (2023). Two statistical approaches to justify the use of the logistic function in binary logistic regression. *Mathematical Problems in Engineering*, 2023(1), 5525675.

11 Machine learning methods for improving stock price prediction accuracy

Priya Rani[a] and Devarani Devi Ningombam[b]

Department of Computer Science and Engineering, National Institute of Technology Patna, Patna, Bihar, India

Abstract

The last few years have seen large-scale changes in the use of financial methods for stock selection, with the increased use of machine learning techniques. The objective of this study is to enhance the forecasting accuracy of stock prices by integrating market sentiment with traditional market metrics. This research focuses on the Indian stock market, utilising financial news from top news sources for sentiment analysis and the stock-related data from the National Stock Exchange (NSE). The independent variables used to determine the dependent variable—the closing price of the stock next trading session—are the Open price, High price, Low price, Close price, and volume of trade, along with quantified sentiment scores derived from financial news articles. The methodology used in the study includes pre-processing of data, sentiment score extraction, closing price prediction, and model evaluation. The results of the study show that using sentiment analysis along with the price and trade volume metrics improves the accuracy of stock price prediction, offering valuable insights for the investors. This research provides a useful contribution to the predictive modelling field in finance and highlights the importance of market sentiment in determining stock prices.

Keywords: Machine learning, sentiment analysis, predictive modelling, portfolio optimisation

Introduction

There has been growing interest in recent few years in the effect of integration of sentiment analysis with technical and fundamental indicators for predicting stock values. This has been caused by the entry of a large number of retail investors into the stock market in recent few years, especially in the Indian context, which makes the market sentiment more significant. It is further fuelled by the increased social media and internet sources accessibility to the masses.

Fundamental and technical indicators are proven beneficial tools for stock price prediction [9,13]. Past stock price data and trading volumes of stocks have been used to capture stock market trends. However, market sentiment—shaped by news articles, social media, and investor psychology—has been recognised as an important force which influences price movements [1,4]. The readily available digital communication and information channels pave way for easier assessment of investor sentiment.

This study proposes an integrated model for stock price movement prediction by using both the stock metrics—Open price, High price, Low price, Close price and volume of trade—and the sentiment score derived through natural language processing (NLP). The sentiment score captures the mood and attitude of investors towards a stock, which is an important factor driving the directionality and depth of price changes of the stocks. The sentiment factor, which represents the perception of a stock's strength and potential, among the investors, holds much importance along with the traditionally considered factors.

The aim of this study is to assess the effect of combining market sentiment factor with the stock price metrics and trading volume for improvement in the prediction accuracy, and to propose a model to do so. Moreover, the insights gained from this can be used by investors for optimising and strengthening their portfolios and make informed decisions in trading stocks.

The organisation of this paper is as follows: The literature on stock price prediction models, especially those that include sentiment analysis, is reviewed in Section 2. The variables and data sources are described in Section 3; and the methodology and model specifications, including metrics for model evaluation, are described in Section 4. The results are finally shown in Section 5, and the conclusion is covered in Section 6.

Literature review

The predictive modelling of stocks by including sentiment analysis along with the past data follows the demand of the field to consider the psychological and emotional state of the investors.

Theoretical framework

Behavioural finance, introduced by Kahneman et al. [10], provides a framework for understanding how cognitive biases and emotions influence investor behaviour,

[a]priyar.ph23.cs@nitp.ac.in, [b]devarani.cs@nitp.ac.in

DOI: 10.1201/9781003675235-11

resulting in market anomalies. Psychological factors, including investor sentiment, are important factors which can lead to price fluctuations.

Role of investor sentiment on stock prices
Research has shown that shifts in stock prices and market sentiment are correlated [4,15]. Negative market sentiment predicts a downward trend in stock prices, while positive market sentiment predicts a rise. Stock returns are highly impacted by market sentiment, especially in less liquid stocks [3,14]. Moreover, the effect of negative sentiment persists for longer period than the positive sentiment [1].

Methodology for sentiment analysis
Sentiment analysis has improved with the advancement in the field of machine learning and computational linguistics. The initial studies were keyword-based, which were limited by their ability of applying context and capturing the nuance in usage. However, the new developments in the Natural Language Processing (NLP) field have changed this, and now context etc. is also taken well into consideration [2,11].

Bidirectional Encoder Representations from Transformers (BERT), presented by Devlin et al. [7] shows a significant advancement in NLP capabilities. By understanding the context of a word in a sentence in relation to the rest of the words of the sentence, BERT improves sentiment classification accuracy. Its application in financial sentiment analysis has been promising, as it can effectively differentiate between positive, negative, and neutral sentiments in complex financial narratives with a score.

Comparative analysis with global markets
Various studies across several markets show the significance of sentiment analysis in prediction of stock prices. Zeitun et al. [16] demonstrate the effect of the prevailing sentiment around a stock on Twitter on the returns of various sectors in the U.S. market. Chen [6] reviews various research papers and concludes that the positive investor sentiment gives higher returns, while the negative sentiment leads to a decline in returns.

In spite of substantial proof backing the use of sentiment analysis in stock price prediction, there are some limitations that have been pointed out by some studies. For example, noise in social media sentiment can be misleading [12]. Thus, herd behaviour needs to be filtered out for getting positive returns by filtering out non-informative noise.

Sentiment analysis in Indian context
Although there have been numerous research on market sentiment in the context of Western markets, its effect on the Indian market is not fully explored. Agarwal et al. [1]

and Bhardwaj et al. [5] study the effect of market sentiment on emerging markets, particularly India.

Data and variables

Study period and datasets
The BERT model is trained on an Indian financial news article dataset spanning from 2003 to 2020, sourced from Kaggle [8]. Stock prices and trade volume data are sourced from the National Stock Exchange (NSE) of India. Financial news for sentiment analysis is sourced from top news sources.

Independent variables
This study uses stock price metrics—namely Open price, High price, Low price, Close price, volume of trade—and sentiment scores from financial news to predict stock movements. The Open price is the stock's initial price at the start of the trading day, while the High and Low prices indicate the highest and lowest values, respectively, reached during the trading session. The Close price, often the most important metric, reflects the stock's final value at market close and helps identify price trends. Additionally, the volume of trade shows the number of shares traded into the market that day, which provides insight into market activity and investor interest in any stock. Market sentiment is represented as a quantified sentiment score, representing positive, negative, or neutral sentiment around any stock. It is estimated through sentiment analysis of financial news, by capturing the prevailing attitudes and emotions surrounding a stock which influences trading decisions. By combining both market data and sentiment analysis, this approach aims to create a more complete understanding of stock price movements, leading to more accurate predictions and better insights into market trends.

The independent variables set for a given day t is denoted by:

$$X_t = [\text{Open}_t, \text{High}_t, \text{Low}_t, \text{Close}_t, \text{Volume}_t, S_t]$$

Dependent variable
The dependent variable is the closing value of the stock in the next trading session. The dependent variable, i.e., next day's closing price is Y_{t+1}.

Methodology and model specifications

Methodology
Sentiment analysis
The methodology of sentiment analysis in this study involves several steps to ensure accurate and reliable results. It includes data collection, data pre-processing, sentiment analysis, and feature extraction.

Data collection: The financial news related to stocks is collected from top news sources.

Data pre-processing: The initial step in the pre-processing phase is to filter out non-relevant content. This is achieved by using keyword-based filtering, removing all the extraneous characters, HTML tags, and any noise to standardise the data. Normalisation of text is done by lowercasing the characters so as to standardise the data for analysis.

Sentiment analysis: After the pre-processing of data is completed, sentiment analysis of the news is conducted using the specialised BERT model. It is the BERT model fine-tuned on a labelled dataset specific to financial sentiment, ensuring that it effectively distinguishes between positive, negative, and neutral sentiments and gives a quantified score.

Feature extraction: After sentiment scores are generated, they are quantified and converted into numerical features suitable for machine learning applications. All the news sentiment related to a particular stock for a day is aggregated, and its mean is calculated after taking the positive and negative news into consideration and ignoring the neutral ones to give a net sentiment score of a stock for a day.

The sentiment score S_t for a stock on day t is given by:

$$S_t = \frac{1}{N} \sum_{i=1}^{N} s_i$$

where, N is total number of positive and negative (neutral ones are dropped) news article about a particular stock on day t, and s_i is the sentiment score of the ith news article.

Stock price predictive analysis
The derived net sentiment scores of a stock for different days are then integrated with other independent variables—namely Open price, High price, Low price, Close price and volume of trade—to form a comprehensive dataset for predictive analysis using random forest regression.

The random forest regression model having M decision trees, which are indexed by k is given as:

$$Y_{t+1} = \frac{1}{M} \sum_{k=1}^{M} Y_{t+1}^k$$

here, Y_{t+1}^k is the predicted closing price by the kth decision tree.

This study uses both the market sentiments and the stock price data to improve future stock movement prediction. The stock price metrics capture the trends and trading activity, while market sentiment helps track investor feelings around a stock. This approach improves prediction

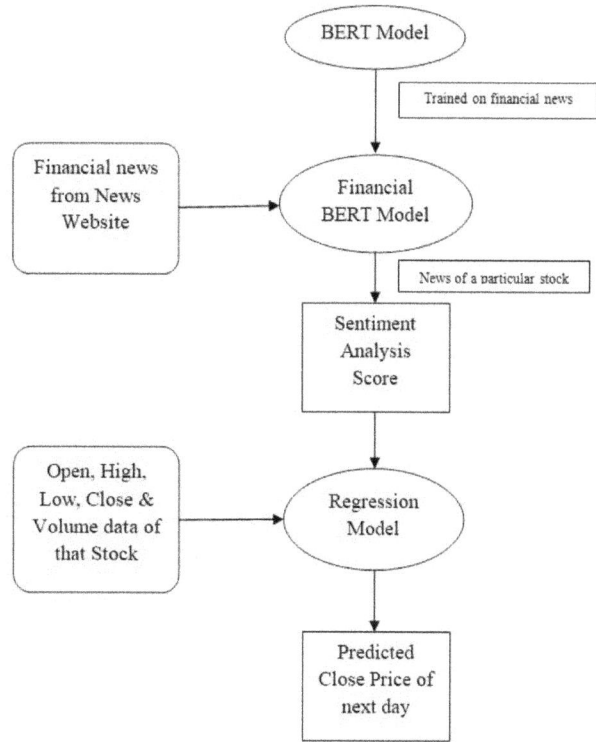

Figure 11.1 The flowchart of the model
Source: Author

accuracy; giving investors useful insights and helping them make smarter trading decisions.

Model specifications
The predictive model combines the daily price metrics, trading volume data, and the net market sentiment score for that stock for that particular day. The closing price of that stock for the next day is predicted using random forest regression. The temporal dependencies and momentum effect of a stock rally or downfall are taken into consideration by the various price metrics and the trading volume, while the effect of investor sentiment is reflected in the sentiment score.

The predictive model for stock price utilises the combined dataset of daily price metrics along with trade volume and market sentiment. A hybrid approach is employed which integrates both time series forecasting methods and machine learning techniques. The model framework consists of 'Time Series Component' to capture the temporal dependencies in the stock price, and the 'Sentiment Component' related to the sentiment scores derived from the NLP model.

The dataset is split into two subsets: a training set and a testing set. The training set comprises 80% of the data, while 20% data is used as the testing set. A cross-validation technique is employed to ensure the robustness of the model's performance.

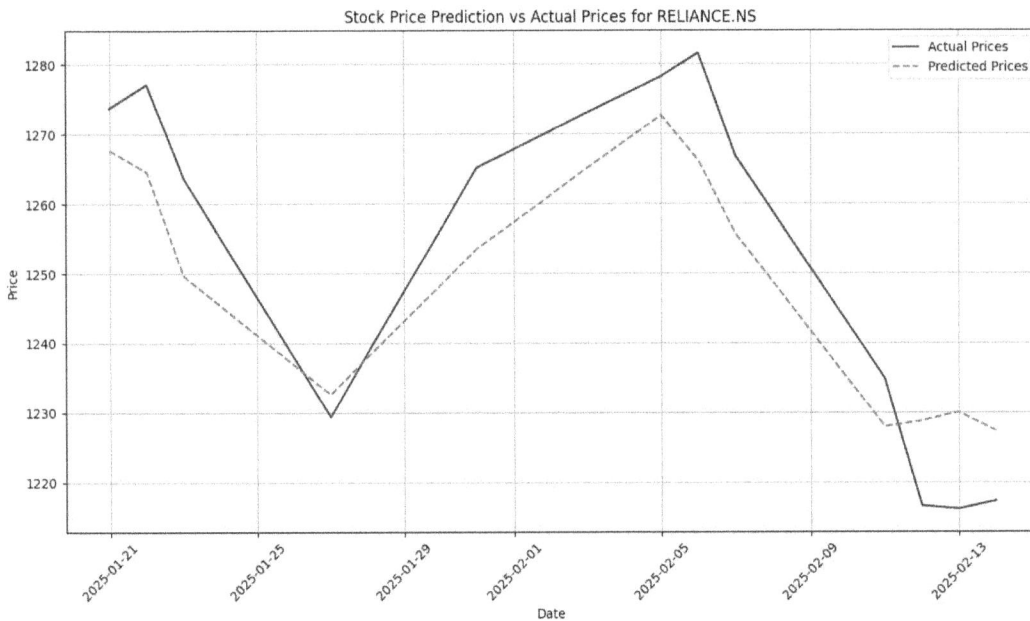

Figure 11.2 Performance comparison of stock price prediction and actual price
Source: Author

Model evaluation

The proposed predictive model, using random forest regression trained on integrated dataset comprising market sentiment with price and volume metrics, of stock is evaluated against a random forest regression model and a linear regression model on a dataset comprising only the price and volume metric and not the market sentiment. The model is evaluated by comparing the predicted and actual values. The evaluation metrics used are Mean Squared Error (MSE) and R^2.

Empirical results

The random forest regression model, trained on the integrated dataset comprising market sentiment along with price and volume metrics, performed best in around half of the cases when compared to the random forest regression model and linear regression model on a dataset comprising only the price and trade volume metrics based on MSE and R^2.

Figure 11.2 shows the performance comparison of the predicted and actual prices for one month stock data for a stock. It is seen that both the values have significant correlation, which shows that not only the time series data (comprising of Open, High, Low, Close, and volume values), but also sentiment values are also very important for stock price prediction.

Conclusion

As financial markets change with time, it is important to use the best performing analytical tools and methods such as sentiment analysis with traditional predictive modelling. This study can be used to add the existing knowledge in this field, and to provide valuable insights for financial professionals. It gives good correlation between actual and predicted values when news data for sentiment analysis is available in a complete sense. This study aspires to deepen the understanding of market dynamics and enhance the predictive capabilities of stock price models. It can be used in portfolio optimisation and improving the gains from the market for the investors. It can also be used in short-term day trading by investors to make profit.

References

[1] Agarwal, S., Kumar, S., Goel, U. (2021). Social media and the stock markets: an emerging market perspective. *Journal of Business Economics and Management*, 22(6), 1614–1632. https://doi.org/10.3846/jbem.2021.15619

[2] Araci, D. (2019). FinBERT: financial sentiment analysis with pre-trained language models. *arXiv* (Cornell University). https://doi.org/10.48550/arxiv.1908.10063

[3] Baker, M., Wurgler, J. (2006). Investor sentiment and the cross-section of stock returns. *The Journal of Finance*, 61(4), 1645–1680.

[4] Bharathi, S., Geetha, A. (2017). Sentiment analysis for effective stock market prediction. *International Journal of Intelligent Engineering and Systems*, 10(3), 146–154. https://doi.org/10.22266/ijies2017.0630.16

[5] Bhardwaj, A., Narayan, Y., Vanraj, Pawan, Dutta, M. (2015). Sentiment analysis for Indian stock market prediction using Sensex and Nifty. *Procedia Computer Science*, 70, 85–91. https://doi.org/10.1016/J.PROCS.2015.10.043

[6] Chen, X. (2023). Stock price prediction using machine learning strategies. *BCP Business & Management*, 36, 488–497. https://doi.org/10.54691/bcpbm.v36i.3507

[7] Devlin, J., Chang, M., Lee, K., Toutanova, K. (2018). BERT: Pre-training of deep bidirectional transformers for language understanding. *arXiv* (Cornell University). https://doi.org/10.48550/arxiv.1810.04805.

[8] Indian financial news articles (2003–2020). (2020b, May 26). Kaggle. https://www.kaggle.com/datasets/hkapoor/indian-financial-news-articles-20032020/code

[9] Jegadeesh, N., Titman, S. (1993). Returns to buying winners and selling losers: implications for stock market efficiency. *The Journal of Finance*, 48(1), 65–91. https://doi.org/10.1111/j.1540-6261.1993.tb04702.x

[10] Kahneman, D., Tversky, A. (1979). Prospect theory: an analysis of decision under risk. *Econometrica*, 47(2), 263. https://doi.org/10.2307/1914185

[11] Khurana, D., Koli, A., Khatter, K., Singh, S. (2022). Natural language processing: state of the art, current trends and challenges. *Multimedia Tools and Applications*, 82(3), 3713–3744. https://doi.org/10.1007/s11042-022-13428-4

[12] Li, T., Chen, H., Liu, W., Yu, G., Yu, Y. (2023). Understanding the role of social media sentiment in identifying irrational herding behavior in the stock market. *International Review of Economics & Finance*, 87, 163–179. https://doi.org/10.1016/j.iref.2023.04.016

[13] Malkiel, B. G., Fama, E. F. (1970). Efficient capital markets: a review of theory and empirical work*. *The Journal of Finance*, 25(2), 383–417. https://doi.org/10.1111/j.1540-zz6261.1970.tb00518.x.

[14] Nyakurukwa, K., Seetharam, Y. (2023). The evolution of studies on social media sentiment in the stock market: insights from bibliometric analysis. *Scientific African*, 20, e01596. https://doi.org/10.1016/j.sciaf.2023.e01596

[15] Tetlock, P. C. (2007). Giving content to investor sentiment: the role of media in the stock market. *The Journal of Finance*, 62(3), 1139–1168.

[16] Zeitun, R., Rehman, M. U., Ahmad, N., Vo, X. V. (2022). The impact of Twitter-based sentiment on US sectoral returns. *The North American Journal of Economics and Finance*, 64, 101847. https://doi.org/10.1016/j.najef.2022.101847

12 Effect of stop words in Assamese language for identifying toxic comments

Mandira Neog[a] and Nomi Baruah[b]

Dibrugarh University, Dibrugarh, India

Abstract

Stopwords are common words in natural language processing (NLP) that are frequently excluded from text data analysis because they are considered irrelevant or provide little value in understanding the meaning of a document. For the text processing methods like Stopwords removal, there are numerous natural language processing tools available. The linguistic tools used by these instruments are language-specific. Many languages lack such tools due to the intricacy of creating them. This paper presents an empirical study of Indian language Stopwords removal methods for information retrieval. The study evaluates the performance of different Stopwords removal methods, including frequency-based and clustering-based methods, automatic Stopwords generation methods, and domain-specific Stopwords removal, on a large Indian language dataset. Based on this analysis, we present a domain-specific Stopwords removal approach for Indian language, Assamese, in particular. This new method is used to create a list of Stopwords that are specific to the domain or Assamese language of the text corpus. Six domain-specific corpora and two classifiers (NB and SVM) were used to assess the impact of dimensionality reduction. The domain-specific method enhanced mutual information, reduced dimensionality, and required significantly less computational time.

Keywords: Stopwords, dimensionality reduction, Assamese, NB, SVM

Introduction

The rapid increase in the generation of raw text data across fields such as science, finance, social media, and medicine has required efficient text pre-processing methods. Unprocessed text data pose significant computational and analytical challenges mainly due to their high dimensionality [1]. In an attempt to overcome these challenges, text pre-processing has emerged as a critical step in preparing raw data to be used in text mining tasks. Among various techniques is Stopwords removal. That mainly refers to the deletion from a corpus of highly commonly occurring, non-informative words. By avoiding these words, it gets the text mining algorithm's focus on meaningful content thereby improving predictive accuracy and insight in the analysis.

Stopword removal for Indian languages is highly problematic since the degree of inflectional and derivational morphology is very high along with the complexity in script [2]. Methods of Stopword removal from Indian languages that have been proposed include rule-based approaches, frequency-based and clustering-based methods, automatic Stopword generating methods, and domain-specific Stopword removal. Some Stopword removal methods used by various researchers for text pre-processing in Indian languages are numerous. In a word, a particular remove Stopwords methodology may end up in another way if viewed according to some languages [3]. It states in the final that all of these work by completely varying with other languages. This paper researches and evaluates the existing Stopwords removal techniques in six Indian major languages: Hindi, Bengali, Telugu, Tamil, Gujarati, and Punjabi. The work analyses the usability of different techniques in terms of retrieval performance, computational complexity, and the preservation of relevant data. Based on this analysis, we have provided a domain-specific Stopword removal technique for Indian languages, specifically for Assamese. Similar to other Indian languages, Assamese includes a set of Stopwords which occur frequently but contribute to no meaning in the text. But, due to different spelling of Assamese words, finding these Stopwords is a tough task. For Assamese word লগ means, লগত, লগাম, লগালগিকৈ, লগালগ are also different. There are some difficulties involved with Stopword removal. They are as follows:

- Lack of comprehensive Stopword lists: There is no list of Stopwords for Assamese as is the case for English. Hence, it becomes a difficult job to find out all Stopwords that should be excluded from a document.
- Assamese is an inflectional language, which means that words can change according to their grammatical function. Stopwords are hard to find because their inflected forms may not be recognised as Stopwords [4].
- Domain-specific Stopwords: Assamese has specific domains-based Stopwords that do not exist in general

[a]mandira.neog@gmail.com, [b]baruahnomi@gmail.com

DOI: 10.1201/9781003675235-12

vocabulary. Most standard Stopword lists lack such terms and so it has to be discovered manually [5].

- Ambiguity: Some Assamese words have different meanings and are ambiguous about whether they are Stopwords or not. Researchers [6] have analysed these spelling differences and come up with some solutions to resolve them in natural language processing applications. In the Assamese repository 'Assamese Stopwords' by P. Sarma and S. Kr. Sarma, the total number of words is 264. This includes 143 words that are frequently used in everyday conversation, and 121 Assamese-specific words.

To justify and evaluate the success of our proposed methodology with respect to other phrase filtering techniques, we had implemented it on six domains: defence, medical, religious, educational, political, and sports. Deleting domain-specific Stopwords varied due to the type of corpus being tagged. We, therefore, applied the TF-IDF method to identify how important each phrase would be in the dataset. Finally, the frequent terms were checked against the domain-specific Stopword list to make sure no major terms were accidentally deleted. This method enables more meaningful and relevant results in natural language processing applications. We test the accuracy by passing the pre-processed corpus through two different classifiers, Naive Bayes and Support Vector Machine. Comparing the results of these tests helped us determine the contextual difference between our proposed approach and other methods.

In particular, we try to explore the following research questions (RQs):

RQ1: How does the use of different Stopword lists affect the performance of information retrieval systems in Indian languages?

RQ2: What impact does the removal of Stopwords have on the performance of text classification in Indian languages?

RQ3: What is the impact of using a combination of statistical and rule-based methods for Stopwords removal in the Assamese language?

Literature review

Stopword removal is one of the must-do pre-processing steps in any job related to natural language processing in Indian languages. The Indian languages have diverse features, including complex grammar rules, word inflection, and compound words, making the removal of Stopwords tough. It should be mentioned that the effectiveness of the mentioned strategies depends on the individual Indian language and its specific characteristics. As a consequence, researchers must devise language-specific approaches to the removal of Stopwords, accounting for the inherent

properties of the language. Tables 12.1 and 12.2 represent an exhaustive review of the research done in the area of interest for the proposed study. Some of the more detailed specifics regarding the facts provided in these tables are discussed here. The facts in the table pertain to Stopword identification in some of the Indian languages. Sahu et al. [3] investigate the effect of Stopwords on the information retrieval (IR) systems in various Indian languages in their study titled 'Effect of Stopwords in Indian Language IR'. According to the research, the deletion of Stopwords tends to increase the MAP value significantly when compared to including Stopwords. The authors also discuss the influence of Stopwords on retrieval performance based on document length and observe that in short texts, Stopwords had a smaller impact compared to longer ones for the four Indian languages that were tested. Overall, the results show that Stopword reduction can be a good approach to enhance the performance of information retrieval systems in Indian languages. Alshanik et al. [7] proposed a hyperplane-based approach in their paper 'Accelerating Text Mining Using Domain-Specific Stopword Lists'. This approach relies on encoding words in a low-dimensional vector space and computing their distance from a hyperplane. Using this approach, redundant features can be eliminated and lead to a significant reduction in text dimension. Four different methods for Punjabi Stopwords removal are presented by Kaur et al. [8] in their paper 'Stopwords Removal and Its Algorithms Based on Different Methods'. The four different methods presented include the traditional method that depends on an existing list of Stopwords, a frequency-based method, a method that removes singletons, and a method that uses a corpus of Punjabi words. After implementing each of the four approaches, their efficacies were tested. The results showed that the removal of the identified Punjabi Stopwords reduced the size of the document by 20–30%. The aggregation model for the removal of Stopwords was given by Rani et al. [9]. The model has an information model and a term weight model of two different types. A mechanical technique for extracting Stopwords from plain texts was presented by Ferilli et al. [10]. They used a geometric strategy and a frequency-based method known as TF-IDF to establish the cut-off point in the ranking of possible Stopwords. The approach was designed to be language-independent and to work well with relatively small corpora.

Based on the concept of Deterministic Finite Automata (DFA), Jha et al. [11] presents a Stopword removal procedure for Hindi. As compared to previous techniques with regard to pattern matching algorithms that were used in deleting Stopwords, the time consumed is less, so this procedure is more effective based on time and space requirements. Jayashree et al. [12] conducted a study on the efficiency of Naive Bayesian techniques for classifying

Kannada texts with the removal of Stopwords. They proposed a stop-word removal algorithm that extracts Stopwords from a given document by using pattern-matching techniques and Unicode representation of the input text data.

Ladani et al. [13] present an overview of the most important methods that have been used by researchers in their analysis of Indian and non-Indian languages during the last couple of decades to identify Stopwords. They also analysed a large number of datasets, classification strategy techniques, and models to check if the removal of Stopword is actually necessary or significant in the settings of applications related to text categorisation and information retrieval. A dictionary-based approach to Stopword elimination from the Sanskrit language was proposed by Raulji et al. [14]. Results of this Stopword removal in Sanskrit have proven to be better using several corpora of Sanskrit.

Ferilli et al. [15] tested the effectiveness of several previously proposed methods when applied to extremely small corpora, such as individual papers. The experiments demonstrated that a simple term-frequency approach is a very robust indicator that outperforms more complex approaches. To achieve faster and easier data processing for the Bengali language, Haque et al. [16] suggested a corpus-based approach to identify Stopwords in a given context and removing them.

The results of the experiment shown here indicate that the method proposed accelerates Bengali language processing in many ways. Fayaza et al. [17] introduced a method based on TF-IDF to compile a list of Stopwords for Tamil, one of the languages with limited resources. Stopword lists for general and domain-specific themes, such as local, international, sports, and entertainment subjects, were developed due to the research. Shaukat et al. [18] presented an algorithm enhancing the Stopword removal procedure from Urdu text using Deterministic Finite Automata.

They designed a set of Stopwords and a finite state machine (FSM) for those. They then implemented this FSM as a state table with 38 columns and an arbitrary number of rows (states) by utilising the random-access memory of the computer. Madatova et al. [19] designed a set of Stopwords developed from Uzbek texts by combining bigram and unigram algorithms. Their method particularly works well in agglutinative languages, where the Stopwords would be 'hidden' within the text, and therefore difficult to find.

The objective of Rani et al. [20] was to combine both statistical and knowledge-based approaches for creating corpus-specific Stopword lists for Hindi text documents. The recommended method of generating the Stopword list involves ranking the words using various techniques, followed by normalisation of results using a vote ranking system based on social choice theory. Two simple approaches to generate corpus-dependent Stopword lists were proposed by Baradad et al. [21]: keyword adjacency Stopword lists and Poisson distribution-based Stopword lists. With the aid of author-provided keywords for a set of abstracts, these techniques can be automated to produce Stopword lists that are easy to update when new texts are added to the training corpus.

The main focus of Mathews et al. [22] was the pre-processing of user-generated reviews from social networking sites in Malayalam. To reduce the word count, a range of pre-processing techniques was applied to the text data. Their experiments showed that such strategies reduced the word count by more than 20%. Multinomial Naive Bayes is a new technique in machine learning proposed by Tijani et al. [23], through which the system can automatically generate a Stopword lists and update it as new words appear. After the removal of Stopwords from the text content, this method outperformed current systems and produced 255 Stopwords, leading to a 63% text compression rate.

Methodology

A simple and often employed method is to manually scan for frequently occurring terms in a text corpus to produce a list of Stopwords. This process of manually finding Stopwords is tedious and labour-intensive, but there are many other ways to produce an effective list of Stopwords for a particular corpus. These alternate methods include statistical methods such as TF-IDF that identify and remove the most common and irrelevant phrases in a corpus, as well as machine learning systems that use classifiers to automatically detect and remove Stopwords. Other researchers have suggested a number of methods for automatically compiling a list of Stopwords to overcome the current restriction.

They are as follows:-

Statistic-based methods

- TF-IDF model: A popular statistical technique in natural language processing for determining the importance of words within a document or among a group of texts is called frequency-inverse document frequency (TF-IDF). This approach compares the frequency of a word's occurrence in a given document to its frequency across the entire dataset [24]. The fundamental tenet of TF-IDF is that phrases that are commonly used in a single document but infrequently used throughout the corpus are given a greater priority since they are more distinctive and pertinent to that particular document.

$$TF-IDF_{i,j} = tf_{i,j} \times log(N)\frac{N}{df_i}) \qquad (1)$$

where:

- TF–IDF$_{i,j}$ is the TF–IDF score of term i in document j.
- Tf$_{i,j}$ is the term frequency of term i in document j.
- N is the total number of documents in the corpus.
- df$_i$ is the number of documents that contain term i.
- *log* is the natural log function.
- BM25 model: The Bag-of-words technique is applied using the BM25 ranking algorithm to determine how relevant a document could be to a particular query, regardless of its position relative to the query being searched. It is predominantly used by search engines to index the input data and rank documents appropriately. The formula to simplify the function for information retrieval, we can ignore any repeated terms in the query [25] is as follows:

$$w(t,d) = tf_d \left(\frac{log\left(\frac{N-n+0.5}{n+0.5}\right)}{k_1((1-b))+b\frac{dl}{avdl}+tf_d} \right) \qquad (2)$$

Information-based methods

- Aggregation model. Stopwords can be eliminated from text using the aggregation model [23]. Both statistical and information-based methods are used to create a comprehensive and accurate list of Stopwords. To identify the most often used terms that are probably Stopwords, statistical methods take into account the frequency of words within a certain text corpus. On the other hand, information-based methods identify terms that are irrelevant to text analysis by using the word's context and semantic meaning. By using these two techniques, the aggregation model enhances accuracy and efficiency in tasks involving natural language processing by producing a more thorough and accurate list of Stopwords than one that is manually selected.

Implementation

This paper's implementation portion has mainly focused on developing a domain-specific Stopword list for the Assamese language. In this work, through the analysis of a corpus of writings from various areas of difference such as politics, sports, health, education, and more, a domain-specific Stopword list for the Assamese language is produced.

Types of Assamese Stopwords

A unique list of Stopwords has been developed for the Assamese language to include the fact that different languages use different terms as stop terms. The entire parts of speech in Assamese have been used while developing this list of Stopwords. This list of Stopword has been divided into seven broad domains though, to streamline it. This has been done to preserve the semantic meaning of

the language and to ensure that the Stopwords used in each domain are relevant to the situation [16]. This extensive list of Stopwords has created a system which is working with a condensed version of data while maintaining its original meaning. Because there is less search space when processing short texts, computers can process them more quickly than when processing longer words. Regardless of any domain-specific list of Stopwords, the following are examples of common Stopwords in the Assamese language that are grouped according to their part of speech(পদ):

- Adjective:ভাল(good),বহুত/বৰ(many/much), আটাইতৈক (only one) etc.
- Pronoun: সি (he), মই(me), তাই (she) etc.
- Adverb:আগেত (before),ধযযুৰে (Patiently), লেহেক (slowly), বেগাই (quickly), etc.
- Conjunction: আৰু (and), এতেক/গিতেক/অথৱাৎ (therefore) etc.
- (Preposition):ছফালে(behind),পৰা(from), উপৰত (top), তলত (under) etc.
- (Interjection) : ওহ (Oh !), ৱাঃ(WOW), চেঃ(shit) etc.

Assamese Stopword generation

Although some words are generally considered Stopwords, there are certain instances where they may play an important role or complement the entire semantic meaning of the sentence. Some examples of words that are main words for each of the parts of speech but are generally referred to as Stopwords are given in Table 12.2 [26]. We are developing a domain-specific strategy for the removal of Stopwords, using the TF-IDF technique, in an effort to address the problem at hand. The first step is finding the domain associated with some given text corpus. Most

Table 12.1 Stopwords with example.

Parts of speech	Word	Example
Adjective	Much/বহুত	চুমি বহুত ভাল ছোৱালী (Sumi is a very nice girl)
Pronoun	She/ তাই	তাই স্কুললৈ যাব (she will go to school)
Adverb	Suddenly/ হঠাতে	সি হঠাতে গাড়িখন ৰখালে (He suddenly stopped the car)
Conjunction	And/ আৰু	সি আৰু মই ভাত খাম (Me and him will eat rice)
Proposition	Top/ওপৰত	তুমি চকীখনৰ ওপৰত বহা (it in the chair)
Interjection	Yuck!!/ছিঃ !!	ছিঃ ইমান গেলা গোন্ধ (Yuck it is smelling awefull)

Source: Author

common terms in the corpus are examined and compared to a list of Stopwords that might be specific to the topic. If it matches an entry on the Stopword list, the term is removed from the text. To obtain meaning from terms in a text corpus, pre-processing has to be done. This involves breaking the text into smaller parts, eliminating punctuations, and creating the dataset. The customised Stopword list also comes in handy while stripping common meaningless terms out from the dataset.

• Corpus creation

Since Assamese is an inflectional language, writing it poses a number of problems. Spelling variations arise due to the complexity of the grammatical structure of the language. The problem is compounded by the fact that standard Assamese text is not readily available. While user-generated content on social networking sites can be a good source of information, it is often marred by grammatical mistakes and coded data [26]. We need to collect clear, grammatically correct Assamese material so that the tests are accurate and reliable. Since the content on newspaper websites is generally formal with little parallel text, collecting data from that source was feasible.

• Punctuation removal and tokenisation

Following the collection of error-free data for the experiment, we cleaned the data, removing all potential punctuation. Some of the punctuation signals used in Assamese are not addressed by the NLTK punctuation removal tool since they are specific to language; for instance, ' | ' or '. ' in English. In a similar way, we prepared an Assamese tokeniser, which splits an input text document into discreet words, phrases, or sentences, whenever necessary for analysis.

• Creation of domain-specific Stopwords

We have practically applied domain-specific Stopword removal techniques in classifying the data to seven

categories: defence, medicine, education, sports, economics, politics, and so on. For the creation of Stopwords regarding each category based on their term frequency-inverse document frequency score, we looked at how frequently a word came up in the given corpus to see if it is a Stopword. If the frequency count was high for a word, then we looked into its relevance in the document. It was a relevance check in the human sense since the meaning of some words changes depending on the way they are used [26].

If a word was discovered as irrelevant or lacked the appropriate contextual meaning, then it gets listed in the domain's list of stop terms. In such a case, if the term is very important and specific contextual value, it must be removed from the Stopword list. Hence, for each domain-specific Stopword list, we may possibly delete these words from the corpus provided above. Here, we implemented a domain-specific Stopword removing algorithm that automatically removes such operations in the process. This function is initially identifying the particular business or domain from which those corresponding Stopwords must be deleted. It enables precision, accuracy, and faster comparison analysis of the texts for use within different NLP applications.

Proposed approach

• About data: Every text collection is done manually and recorded into a CSV file in the UTF-16 LE format as a result of the first data preparation procedures. Two columns make up this data: one for the text data and one for the domain. The corpus of texts is positioned in the text data column, while the domain of the corpus is positioned in the domain column.

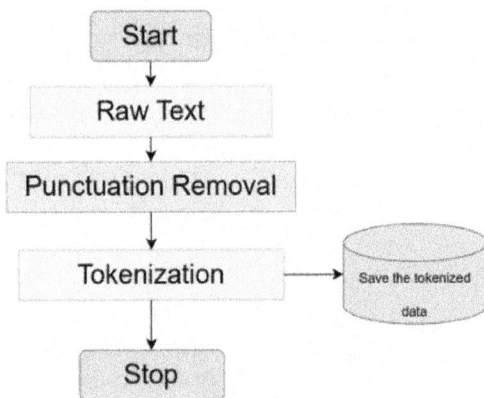

Figure 12.1 Demonstration of punctuation removal and tokenisation
Source: Author

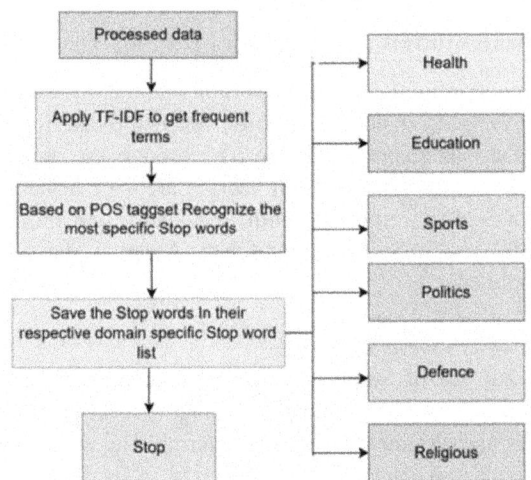

Figure 12.2 Demonstration of domain specific Stopword creation
Source: Author

- Characteristics of the dataset: The main objective of the study was to gather information to identify offensive Assamese remarks on social media platforms. The information is painstakingly assembled from a wide range of sources, from news stories to Facebook posts, comments on YouTube videos, and the like. We had to collect the data personally so that we could ensure that it was accurate and relevant because the particular kind of digital resource required for the experiment was not easily accessible.

  We considered various articles, discussions, and even movies to collect feedback on a range of controversial topics, including politics, celebrities, talk shows, morning shows, and religions such as Buddhism, Sikhism, Hinduism, Islam, Christianity, Judaism, and atheism. It was found that most of the data was entirely written in either English or Assamese, or code blended together. Thus, we retained the Assamese comments in our corpus but excluded others from this set, especially all the ones that were exclusively written in English. We also translated some comments in English which were of Assamese.

- Working of domain-specific Stopword removal
  The function for domain-specific Stopword removal, which is applied to text data, is shown in Figure 12.3. Finding the domain of the given text is the first step of the function. The reason is that this is important because different domains can have different Stopwords. Once the domain has been found, the program looks up the list of Stopwords associated with that domain. The function determines if all words in the text data are in the list of Stopwords, provided a list exists for the selected domain. If one is found, the word is removed from the text data.

Comparison and experiment analysis

The proposed technique of domain-specific Stopword removal is being evaluated by applying the text data generated from the novel method in both Support Vector Machines (SVM) and Naive Bayes (NB) classifiers [27]. The primary goal is to check if the proposed strategy is valid and compare the results developed by these two classifiers. To evaluate the performance of the classifiers, several metrics are measured in the evaluation process, which include precision, recall, accuracy, and F1-score [28].

- SVM: SVM is used in regression analysis and classification. To maximise the gap between two classes, it builds a hyperplane in the feature space [27].
- NB: A supervised learning algorithm that uses Bayes' theorem is named Naive Bayes or NB. It determines how likely each class is and makes the assumption that any properties of the input data are independent of one another. The output is then drawn from the class with the maximum probability [27].

The recommended technique that is used for removing Stopwords for text data to find inappropriate comments is by using a domain-specific list of Stopwords. This method,

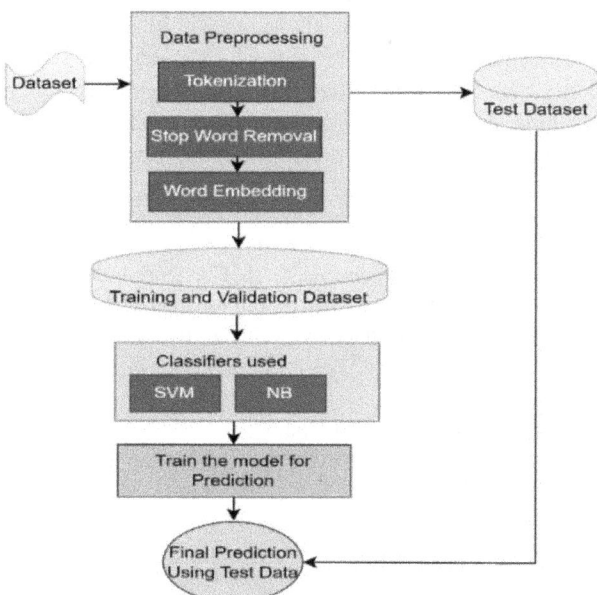

Figure 12.3 Demonstration of complete approach of Stopword removal with machine learning implementation
Source: Author

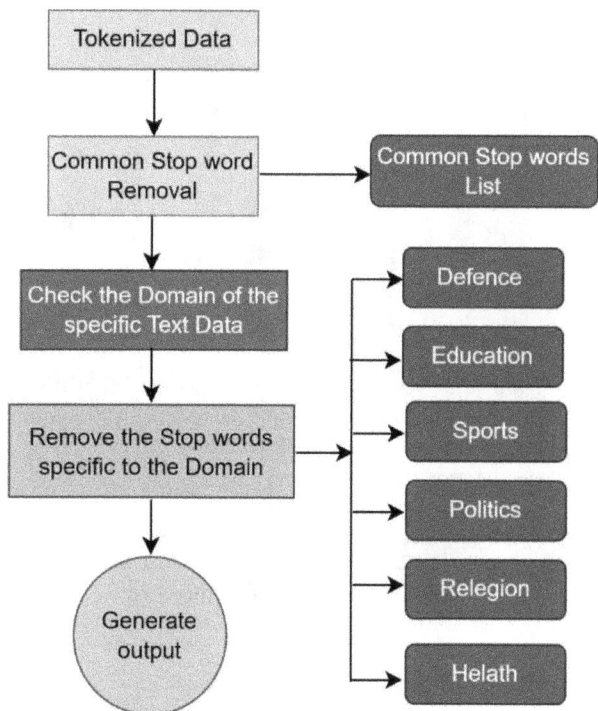

Figure 12.4 Demonstration of common and domain-specific Stopword removal
Source: Author

involving SVM and NB classifiers, better performed than previous methods only removing common Stopwords in the text. It has also reduced the experimental data by 20–30%. Tables 12.2 and 12.3 demonstrate how the proposed approach of the study brings out how important it is to use Stopwords relevant to the specific subject while pre-processing text.

This could lead to a major improvement in the accuracy of text classification models. Additionally, it implies

Table 12.2 Output of processed input data with SVM and NB Classifier

Classifier	Precision	Recall	F1
SVM	90.4	87	90.66
NB	87.33	85	89.83

Source: Author

Table 12.3 Result comparison with other languages

Language	Recall	Precision	F1
Assamese	87%	90.4%	90.66%
Bengali [16]	72.22%	100.00%	83.87%
Tamil [17]	-	-	82.0%

Source: Author

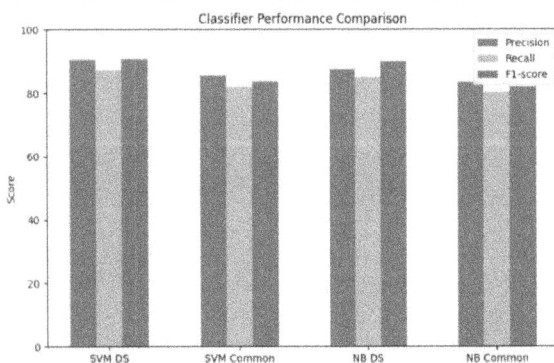

Figure 12.5 Demonstration of classifier (SVM, NB) performance comparison based on domain specific Stopwords removal and common Stopword removal
Source: Author

that harmful remarks in big datasets can be found with the help of machine learning methods like SVM and NB classifiers. Table 12.3 compares the results of our proposed approach with Bengali and Tamil.

In Figure 12.5, the bar chart depicts precision, recall, and F1-scores for two machine learning models, SVM and Naive Bayes, applied to two different Stopword-removal approaches: domain-specific (DS) and common. SVM has an insignificant benefit over Naive Bayes, but the domain-specific technique outperforms the standard approach.

Two distinct approaches are compared in Table 12.4, and the assessment criterion used to determine how effective each approach is the F1 score. Our suggested method for the Assamese language received a higher F1 score of 90.66% than the Bengali language approach, which had an F1 score of 83.87%. This suggests that, in comparison to the Bengali language approach, the suggested method for the Assamese language has done better in terms of the F1 score.

Conclusion and future work

The process of removing domain-specific Stopwords is crucial to achieve more precise results in text analysis. This is done by removing domain-specific Stopwords, which helps to emphasise key words in the text. Thus, by giving importance to terms in the text, it is possible to remove domain-specific Stopwords in the Assamese language to generate more accurate analysis. Therefore, this may lead to faster processing time and accurate results. Using the text data that we have, we pulled the domain-specific Stopwords out from our system manually. Again, we must note that these Stopwords are specific to the information collected and analysed for this study only. Future work: in the next phase, a complex algorithm will be implemented that can automatically determine which words in a specific domain are Stopwords. The approach will be done through machine learning techniques by training models on a text and recognising them as domain-specific Stopwords.

References

[1] Ghosh, K., Bhattacharya, A. (2017). Stopword removal: why bother? A case study on verbose queries. In Proceedings of the 10th Annual ACM India Compute Conference, 99–102.

Table 12.4 Comparison of the two classifiers with and without using domain-specific Stopwords corpus

Classifier	With domain-specific Stopword removal			With common Stopword removal		
	Precision	Recall	F1	Precision	Recall	F1
SVM	90.4	87	90.66	85.4	82	83.66
NB	87.33	85	89.83	83.33	80	81.83

Source: Author

[2]   Dolamic, L., Savoy, J. (2010). When stopword lists make the difference. *Journal of the American Society for Information Science and Technology*, 61(1), 200–203.

[3]   Sahu, Siba Sankar, Sukomal Pal. (2022). Effect of Stopwords in Indian language IR. *Sādhanā*, 47(1), 17.

[4]   Das, Palash, and Madhumita Barbora. (2020). Derivational morphology of Assamese lexical word categories. *Indian Journal of Language and Linguistics*, 1(2), 1–15.

[5]   Makrehchi, Masoud, Kamel, Mohamed S. (2008). Automatic extraction of domain-specific Stopwords from labeled documents. In European Conference on Information Retrieval. Berlin, Heidelberg: Springer Berlin Heidelberg, 222–233.

[6]   A. Sarma, P., Kr, S. (2019). Assamese Stop Words.

[7]   Alshanik, Farah, Amy Apon, Alexander Herzog, Ilya Safro, Justin Sybrandt. (2020). Accelerating text mining using domain-specific stop word lists. In 2020 IEEE International Conference on Big Data (Big Data). IEEE, 2639–2648.

[8]   Kaur, Jashanjot, Preetpal Kaur Buttar. (2018). Stopwords removal and its algorithms based on different methods. *International Journal of Advanced Research in Computer Science*, 9(5), 81–88.

[9]   Rani, R., Lobiyal, D. K. (2018). Automatic construction of generic Stopwords list for Hindi text. *Procedia Computer Science*, 132, 362–370.

[10] Ferilli, S. (2021). Automatic multilingual stopwords identification from very small corpora. *Electronics*, 10(17), 2169.

[11] Jha, V., Manjunath, N., Shenoy, P. D., Venugopal, K. R. (2016). Hsra: Hindi Stopword removal algorithm. In 2016 International Conference on Microelectronics, Computing and Communications (MicroCom). IEEE, 1–5.

[12] Jayashree, R., Srikanta Murthy, K., Anami, Basavaraj S. (2014). Effect of Stopword removal on the performance of Naïve Bayesian methods for text classification in the Kannada language. *International Journal of Artificial Intelligence and Soft Computing*, 4(2–3), 264–282.

[13] Ladani, D. J., Desai, N. P. (2020, March). Stopword identification and removal techniques on TC and IR applications: a survey. In 2020 6th International Conference on Advanced Computing and Communication Systems (ICACCS). IEEE, 466–472.

[14] Raulji, J. K., Saini, J. R. (2016). Stop-word removal algorithm and its implementation for Sanskrit language. *International Journal of Computer Applications*, 150(2), 15–17.

[15] Ferilli, S., Izzi, G. L., Franza, T. (2021). Automatic Stopwords identification from very small corpora. In Intelligent Systems in Industrial Applications. Springer International Publishing, 31–46.

[16] ul Haque, R., Mehera, P., Mridha, M. F., Hamid, M. A. (2019). A complete Bengali stop word detection mechanism. In 2019 Joint 8th International Conference on Informatics, Electronics Vision (ICIEV) and 2019 3rd International Conference on Imaging, Vision Pattern Recognition (ICIVPR). IEEE, 103–107.

[17] Fayaza, F., Fathima Farhath, F. (2021). Towards stop words identification in Tamil text clustering.

[18] Shaukat, K., Hassan, M. U., Masood, N., Shafat, A. B. (2017). Stop words elimination in Urdu language using finite state automaton. *International Journal of Asian Language Processing*, 27(1), 21–32.

[19] Madatov, K., Bekchanov, S., Vii, J. (2022). Dataset of stopwords extracted from Uzbek texts. *Data in Brief*, 43, 108351.

[20] Rani, R., Lobiyal, D. K. (2022). Performance evaluation of text-mining models with Hindi Stopwords lists. *Journal of King Saud University-Computer and Information Sciences*, 34(6), 2771–2786.

[21] Baradad, V. P., Mugabushaka, A. M. (2015, June). Corpus specific stop words to improve the textual analysis in scientometrics. In ISSI.

[22] Mathews, D. M., Abraham, S. (2018). Effects of pre-processing phases in sentiment analysis for Malayalam.

[23] Tijani, O. D., Onashoga, S., Akinwale, A. T. (2017). An auto-generated approach of stop words using aggregated analysis. In 13th International Conference on Information Technology Innovation for Sustainable Development (No. July).

[24] Das, B., Chakraborty, S. (2018). An improved text sentiment classification model using TF-IDF and next word negation. arXiv preprint arXiv:1806.06407.

[25] Robertson, S., Zaragoza, H., Taylor, M. (2004). Simple BM25 extension to multiple weighted fields. In Proceedings of the Thirteenth ACM International Conference on Information and Knowledge Management, 42–49.

[26] Saiyed, S., Sajja, P. (2022). Empirical analysis of static and dynamic Stopword generation approaches. In ICT Systems and Sustainability: Proceedings of ICT4SD 2021, vol. 1. Springer Singapore, 149–156.

[27] Asogwa, D. C., Chukwuneke, C. I., Ngene, C. C., Anigbogu, G. N. (2022). Hate speech classification using SVM and Naive Bayes. arXiv preprint arXiv:2204.07057.

[28] Goutte, C., Gaussier, E. (2005). A probabilistic interpretation of precision, recall and F-score, with implication for evaluation. In Advances in Information Retrieval: 27th European Conference on IR Research, ECIR 2005, Santiago de Compostela, Spain, March 21–23, 2005. Proceedings 27. Springer Berlin Heidelberg, 345–359.

13 Optimising low-resource Khasi NER through transfer learning

Ransly Hoojon[a]

IT Department, North Eastern Hill University, Shillong, India

Abstract

The goal of Named Entity Recognition (NER) is to identify the different types of entities from textual data and tag them with corresponding standard NER tags. This identification was focused on nouns that appear in the text on a generic Khasi corpus. The study pertains to Khasi language, a language characterised by a scarcity of linguistic resources. At present, there is no NER tagging tool for the language, and the development of one is challenging because there is no publicly available NER dataset in Khasi. Hence, alongside the development of the tool, the construction of the corresponding dataset is also necessary. In this work, we explore the feasibility of utilising various pre-trained transformer-based models to improve cross-lingual transfer learning (TL) for the downstream NER task. Our experiment showed that deploying pre-trained models from other resource-rich languages greatly improves the effectiveness of cross-lingual transfer learning. This can also serve as a foundational starting point for future development. Thus, we were able to work around the challenges of data scarcity and achieve promising results with the RoBERTa-base and XLM-RoBERTa-base models, with F1-scores of 91.25% and 91.17%, respectively. Our findings highlight the feasibility of using transfer learning as a means to solve problem descriptions in another language domain, especially for low-resource languages.

Keywords: Name entity recognition, khasi language, transfer learning, low resource

Introduction

Khasi is a language predominantly spoken by people residing in the state of Meghalaya, located in the north-eastern region of India, as well as in certain areas of the neighbouring country, Bangladesh. The Khasi language uses the Roman script, which consists of 23 letters. It is classified as a language within the Austro-Asiatic family and has been recognised as an associate official language of the state. Introduced in 1995)[4], NER plays a vital role in various applications such as question answering, machine translation, sentiment analysis, and text summarisation. NER helps to identify the names of people, organisations, locations, etc. mentioned in news articles or books, allowing for easier organisation and retrieval of information. Research in NER tagging in other languages has progressed by a large margin, which was not the case with Khasi. Khasi lacked many linguistic tools, and in the context of NER, no work had been reported. NER tagging is essential for supporting other NLP tasks. Hence, to develop the language in the realm of NLP, it is important that research initiatives start with the development of these tools that are preliminary to the development of others. The Khasi language features numerous distinct and interesting named entities. To highlight a few examples, the entities DAYS, like *sngi u Blei* and *sngi thohdieng*, are multi-word entities that translate to Sunday and Friday in English. The CARDINAL entity number *four*, which is written as *saw* in Khasi, one can interpret the word *saw* in multiple ways, as it can also function as an adjective. Hence, terms such as *ngut* or *tylli* are being used in conjunction with *saw* to precisely identify it as a CARDINAL entity, like *saw ngut* or *saw tylli*. These fascinating attributes of the Khasi language compelled us to contemplate the possibility of developing a system such as NER. But to do so effectively, having a substantial amount of training dataset was essential for training an NER transformer-based model. With this limitation, Khasi required alternative approaches, such as transfer learning (TL), to overcome the scarcity of labelled data. TL, which refers to the process of making use of knowledge acquired in one domain to solve problems in another domain of the targeted language, has shown promising results in various NLP tasks, including NER)[17]. By harnessing pre-trained transformer models in English and other languages, which have been trained on large amounts of data, we can benefit from the knowledge and patterns learned from these languages. The contributions we made were as follows: (i) Developed a novel dataset containing named entities in Khasi language. (ii) Created a novel spell-checking application for the Khasi language. (iii) Developed a Named Entity Recognition (NER) model customised for Khasi language. (iv) Demonstrated the benefits of using cross-lingual transfer learning in a language with limited resources.

Literature review

Research on NER has advanced in creating precise models and datasets for English, many foreign languages, and some Indian languages, but it is lacking progress for

[a]ranslyh@gmail.com

DOI: 10.1201/9781003675235-13

regional languages such as Khasi. There was a growing emphasis on low-resource natural language processing (NLP) tasks, with named entity recognition being one of them. We present some of the research conducted on Indian languages. A NER system for Hindi was created using a rule-based approach by Kaur and Kaur [6]. This required the creation of a gazetteer list for each category of named entities. The system handles three specific types of entities: monetary values, cardinal directions, and animal/bird entities, achieving a precision rate of 95.77%. Saha, Sarkar, and Mitra, [16] employed machine learning methods to develop a Hindi Named Entity Recognition system. The algorithm was trained using a dataset of 234,000 words collected from the *Dainik Jagaran* newspaper. The system was assessed using a blind test corpus that included 25,000 words, categorised into four classes. The F-measure reached 81.52%. Chopra, Joshi, and Mathur [3] implemented Hidden Markov Models (HMM) for Hindi NER. Their study demonstrated the HMM's potential to develop language-independent systems and its ease of scalability and analysis. They attained an F-Measure of 97.14% on 2,343 and 105 tokens for the training and testing data. They recognised the existence of label bias and the limitations caused by the small quantity of the training and testing data. The research carried out by Patil, Patil, and Pawar [11] on the Marathi language using an HMM achieved an F1 score of 62.70% without pre-processing and 77.79% considering pre-processing. However, there is a lack of adequate recognition of other named entities. The study suggests using pre-processing techniques like lemmatisation to improve efficiency, even if there may be associated costs. The study utilises the Conditional Random Fields (CRF) modelling technique for Hindi, incorporating 29 features such as context words, word prefixes, word suffixes, POS information, and gazetteer lists. Their results show precision, recall, and F1-score values of 72.78%, 65.82%, and 70.45%, respectively. Ajees and Idicula [1] employed CRF modelling for Malayalam. Tags like B-PER, I-PER, B-LOC, etc., were utilised, resulting in an accuracy of 92.3%. Sabane et al. [15] utilised various versions of BERT, such as baseBERT, alBERT, and RoBERTa, to create NER models for languages including Hindi and Marathi. Utilising multilingual models trained on Hindi and Marathi datasets, along with cross-lingual transfer learning approaches, can improve NER performance in situations with limited language resources. The research presented by Murthy, Khapra, and Bhattacharyya [9] demonstrated that using multilingual learning with Hindi as an assisting language, including Marathi, Bengali, Tamil, and Malayalam, yields superior performance compared to monolingual learning. The NER system was created for Manipuri)[10], an agglutinative language spoken in the northeastern part of India. The system attained an 81.12% recall, an 85.67% precision,

and an 83.33% F-score through the utilisation of CRF. Athavale V Bhardwaj [18] implemented a Hindi NER where word embeddings from an unlabelled corpus were used to train a BiLSTM model, producing an accuracy of 77.48%. **Ekbal and Bandyopadhyay [19]** suggested a hybrid Bengali NER model that combines Maximum Entropy (MaxEnt), SVM, and CRF. The model achieved an F-score of 92.55%. A Malayalam NER developed by **Ajees and Idicula [20]** uses neural networks, POS tags, and word embeddings. Training and testing were done on 20,615 sentences, which produced an accuracy of 95.3%.

Data collection

To ensure a diverse range of text, the data were collected from a variety of sources, including Khasi newspapers and books on various topics.

Pre-processing

During the data pre-processing stage, we eliminated any extraneous data, such as advertising, headers, footers, etc. In addition, we rectified any spelling mistakes and developed a spell-checker tool specifically designed for Khasi. Despite having certain shared alphabets, the current English spell checkers or OCR systems are insufficient for the Khasi language. This could be attributed to the absence of some alphabets, such as c, f, q, z, and x, in the Khasi language. Instead, Khasi incorporates unique alphabets like ñ and ï which are pronounced as eiŋ and yi:, respectively. As a final pre-processing step, a word embedding model was trained on the cleaned corpus. These word embeddings were utilised for spell-checking to further refine the generated word rankings as they took into account semantics. This is explained in detail in the Research Methodology section.

Data annotation and statistics

The annotation classes used were based on the Spacy core model detailed in Table 13.1, which comprises of 18 classes. An extra class was added, bringing the total number of classes to 19. The data gathered were annotated for named entities using a scriptable tool known as Prodigy [13]. Recipes like ner.manual and ner.correct from Prodigy were used for annotation, whereas the train curve was employed for validating the annotated data. The ner.manual recipe was used for manually annotating entities in a text, and the ner.correct recipe was employed to create gold-standard NER data. The annotators who worked on their roles were native Khasi speakers. A senior linguist regularly reviewed a random sample of annotated data to ensure quality. The learning curve analysis was performed on the annotated data using the train-curve recipe, dividing the data into subsets of 25%, 50%, 75%, and 100%, with a validation split of 20%. The graph below showed a

Table 13.1 Annotation classes.

SI No	Entity type	Description	Example
1	PERSON	People, including fictional	Joe Biden, John Smith
2	NORP	Nationalities or religious or political group	The Republican Party
3	FAC	Buildings, airports, highways etc	Logan International Airport
4	ORG	Companies, agencies, institutions, etc	Microsoft, FBI, MIT
5	GPE	Countries, cities, states	France, Chicago
6	LOC	Non-GPE locations, mountain ranges, etc	Europe, Nile River, Midwest
7	PRODUCT	vehicles, foods, etc (Not services)	iPhone, Pizza
8	EVENT	Battles, wars, sports events, etc	FIFA World Cup
9	WORK OF ART	Titles of books, songs, etc	The Mona Lisa
10	LAW	Named documents made into laws	Roe v. Wade
11	LANGUAGE	Any named language	English, Khasi
12	DATE	Absolute or relative dates or periods	06/09/23
13	TIME	Times smaller than a day	Four hours
14	PERCENT	Percentage including %	Eighty percent
15	MONEY	Monetary values including unit	Twenty Cents
16	QUANTITY	Measurements as of weight or distance	1 Kilometres, 50kg
17	ORDINAL	First, second, etc	9th, Ninth
18	CARDINAL	Numerals type	2, Three, Fifty-four
19	DAY	Day of the week	Monday, Sunday

(Source: spacy.io)

learning curve that demonstrates the progress of the model over time, starting at 0% and reaching a peak of 90% accuracy. The performance of the model on the freshly annotated training dataset was assessed using *score* and *ner* parameters, which indicated the quality of the collected and labelled data. Training on 25% of the annotated data resulted in a significant improvement in performance to roughly 85% accuracy, with a progressive rise to 90% accuracy when training on 50% of the dataset. No additional enhancement in accuracy was observed beyond 50% on the training set. Thus, collecting more data on diverse topics could enhance accuracy and broaden the coverage of the NER dataset for the language. Figure 13.1 showed the training curve result.

A pattern file containing well-known entities was put together to expedite the human annotation process and maintain consistency. The pattern file was used as a reference in the *ner.manual* procedure to streamline and reduce the human annotator's efforts during the annotation process. The hand-crafted pattern file had 8,025 entries. An example of the pattern file looked something like this: Each entry began with a label for an entity type, such as 'PERSON', followed by the lowercase pattern to be matched, for example:

"label": "PERSON", "pattern": ["LOWER": "john", "LOWER": "smith"]

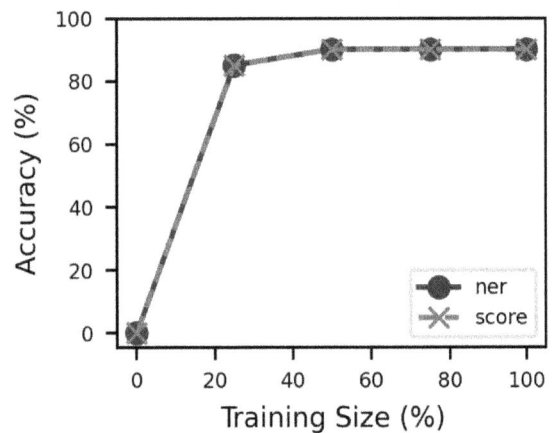

Figure 13.1 Dataset learning curve analysis
Source: Author

A total of 151,413 annotations were completed on the collected data thus far. Table 13.2 displayed the total count for each class.

Research methodology

Currently, to the best of our knowledge, there is no existing NER dataset. Also, there is no tool such as a spell checker for the Khasi language. Hence, the aim of this work was to

Table 13.2 Total count of each entity type.

Sl. No	Entity type	Total
1	GPE	45396
2	PERSON	30344
3	ORG	17687
4	NORP	16523
5	LOC	10028
6	DATE	9489
7	CARDINAL	8226
8	MONEY	2878
9	LAW	2245
10	TIME	2081
11	DAY	1498
12	PRODUCT	1263
13	EVENT	881
14	QUANTITY	849
15	PERCENT	697
16	FAC	539
17	ORDINAL	370
18	LANGUAGE	289
19	WORK OF ART	130

Source: Author

accomplish tasks not only dataset creation, but also building a Khasi spell checker and, subsequently, developing an NER tagger for Khasi through transfer learning.

Spell checker
In the process of collecting data from online sources, a common issue that arose was the occurrence of misspelled words. One potential explanation for this occurrence was the utilisation of an encoding scheme that was incompatible with the character set. In Khasi, it was worth noting that another factor contributed to the misspelling or misrepresentation of certain characters, namely the characters ñ and ï. Many users were observed to tend to replace these characters with n and i, respectively. Consequently, a multitude of written documents exhibited variations in spelling. The objective of the spell checker was not only to correct incorrectly spelled words but also to enhance textual consistency through the utilisation of spell check. The spell checker tool was designed using a hybrid approach, combining the design concept found at 'Phatpiglet/autocorrect' [12] and a word embedding model to improve its accuracy. The spell checker produced a wide range of typographical variations that handled insertion, deletion, replacement, and transposition errors of up to two characters. This limit

could be extended to three or more characters, but doing so increased the overhead of word ranking. We maintained an allowed list of 30,275 words, which also included the names of persons, locations, organisations, and other named entities. This list ensured that the spell checker did not attempt to correct these specific words. In the original implementation, a word frequency file was used to prioritise candidate words. Words with higher frequencies were given higher priority in the candidate list and word ranking. However, in our experiment, in the frequency list consisting of 23,048 words, each word was assigned the same frequency with an arbitrary integer value of 5, and we utilised word embedding to assist in selecting the correct word from the list of candidates. This technique effectively mitigated the potential bias that might have arisen from the disproportionate emphasis placed on commonly occurring words compared to less frequent ones. A few examples of online text containing spelling errors or misrepresentation and their corrections were provided.

- Uta u nongkhrong u la kylli *balci* phi la leh haduh katne? The word *balci* has been corrected to *balei*.
- Hapoh u sohphan don *shiibin* ki symboh. The word *shiibin* has been corrected to *shibun*.
- khlem ka *jingiatylliiadei* jingmut lang. The word *jingiatylliiadei* has been corrected to *jingiatylli* and *iadei* to *ïadei*.

Word embedding
The purpose of developing a word embedding model is to incorporate it into a spell checker that fixes spelling errors not only based on character differences but also on the semantic significance or resemblance of words or phrases. Word embedding maps words into dense vector representations, where words with similar meanings have similar vector representations. This allows for more accurate spell-checking by identifying words that are similar to the misspelled word. We chose to use fastText [2] as our word embedding model due to its ability to handle out-of-vocabulary problems. FastText can decompose words into smaller units, or root words, perform embedding on these units, and then recompose the original word. This allows for a better representation of words that may not be present in the training data. We trained the word embedding model on a large collection of Khasi text corpus, which consists of around 72MB of news articles and books. Gensim, a free and open-source Python library, was used to train the fastText word embedding model. The model was trained using a vector of 100 dimensions, a window size of 10, a minimum word frequency of 5, and a total of 100 training epochs. According to Mikolov, Chen, Corrado, and Dean [8], who describe the advantages and disadvantages of using Skip-gram and CBOW in word embedding, we tested both approaches, and

Table 13.3 Similar words to *thiah* which translate to *sleep*.

Khasi word	Score	English meaning
iohthiah	0.79	asleep
ïohthiah	0.783	asleep
pynthiah	0.771	put to sleep
shohsamthiah	0.728	drowse

Source: Author

our finding was that both yielded similar results. Some examples of CBOW word embedding are demonstrated in Table 13.3.

Experimental Setup

For our experiment, we adopted Spacy framework, one of the leading open-source libraries in advanced NLP. Its robust features and efficient processing made it an ideal choice for performing tasks such as named entity recognition. Spacy's configurable NLP pipe-lining framework allowed for easy integration of different pre-trained transformer-based models along with Spacy NER component. In our experiments, we started with the Spacy NER component, which was responsible for identifying named entities from text data. Then, we configured the Spacy pipeline to also include different pre-trained transformer-base models (BERT) which included both monolingual and multilingual. We selected the base version due to limited computing power available, such as bert-base-cased, ai4bharat/ indic-bert, roberta-base, xlm-roberta-base and bert-base-multilingual-cased. The main objective of integrating a transformer model was to improve the NER component's understanding of words/tokens in a sentence by making use of the embedded contextual and semantic information produced by the transformer model. During the training process with the novel Khasi NER dataset, each model's parameters were carefully calibrated to obtain maximum performance. We partitioned the dataset into 80% for training and 20% for evaluation. The evaluation process involved measuring key metrics such as precision, recall, and F1 score, which could be calculated by the following formula:

$$Precision = TP/TP + FP \tag{1}$$

$$Recall = TP/TP + FN \tag{2}$$

$$F1\text{-}Score = 2*(Precision*Recall/Precision + Recall) \tag{3}$$

where TP, TN, FN, and FP represent the terms true positive, true negative, false negative, and false positive, respectively.

Parameters optimisation and training

During the training of several transformer-based models for the Named Entity Recognition (NER) task, parameters are optimised for maximum efficiency. For bert-base-cased and base-multilingual-cased, the batch size was set to 128, while for the other models, it was set to 512. The selected optimiser for training is Adam)[7]. The warm-up learning rate started at an initial value of 0.00005 and increased linearly to a specified value over a warm-up period of 250 steps. After the warm-up period, the learning rate remained constant for the remaining 20,000 training steps. Gradient clipping was implemented with a threshold value of 1.0. Gradient clipping was a technique used to limit the values of the gradient during training to prevent the issue of exploding gradients. The application of L2 regularisation involved the use of a coefficient of 0.01 (L2 = 0.01). L2 regularisation mitigates overfitting by imposing a penalty on the magnitudes of the weights in the model. The Adam optimizer was configured with additional parameters, namely beta1 (0.9), beta2 (0.999), and epsilon (0.00000001). These factors governed the momentum, adaptive learning rates, and numerical stability of the optimisation process. The technique of gradient accumulation was employed with a factor of 3. This method aided in simulating bigger batch sizes, which enhanced the stability and convergence of training, particularly in situations when there was a constraint on GPU memory. On the other hand, the training of the Spacy Tok2Vec-NER model was optimised using the Adam optimizer. The value of Beta1, Beta2, and Epsilon was set to 0.9, 0.999, and 0.00000001, respectively, and the learning rate was set to 0.001. Gradient clipping was employed with a threshold of 1.0, while L2 regularisation was implemented with a coefficient of 0.01. The dropout rate was set at 0.1. The training process utilised a batch size of 2,000 samples and was monitored using a patience of 1,600 evaluations, with a maximum of 20,000 training steps. Evaluation was carried out at intervals of 200 steps during the training process. The batching strategy employed a 'batch by words' method with a compounding size schedule. The Tok2Vec component utilised architectures such as MultiHashEmbed and MaxoutWindowEncoder to embed and encode tokens into fixed-length vectors. The vector representations captured contextual information from the document, thus improving the performance of Named Entity Recognition (NER) and other tasks.

Results and discussion

Table 13.4 provides a concise overview of the performance of various models on the Khasi NER task, measured using three evaluation metrics: precision, recall, and F1-score. When considering the precision measure, it is evident that the roberta-base+DSNER model had the greatest precision

Table 13.4 Summary of Khasi NER model performance.

Model name	Dataset	Precision	Recall	F1-Score
default Spacy NER (DSNER)	Khasi NER	91.09	89.69	90.38
bert-base-cased + DSNER	Khasi NER	89.12	88.38	88.75
ai4bharat/indic-bert + DSNER	Khasi NER	88.04	85.19	86.59
roberta-base+ DSNER	Khasi NER	91.12	91.39	91.25
xlm-roberta-base+DSNER	Khasi NER	90.55	91.79	91.17
bert-base-multilingual -cased + DSNER	Khasi NER	90.24	90.54	90.39

Source: Author

Shillong **GPE** : Ka Dr H Gordon Robert Hospital **ORG** , Jaiaw **GPE** , ka la plie a ka bynta ai jingsumar kaba dang

shu dep pynthymmai a kaba la tip kum Emergency & Trauma Unit kaba don hapoh compound jong kane hi ka hospital.

Ka jingplie paidbah a kane ka Emergency & Trauma Unit ka la long ha ka sngi nyngkong **DAY** jong ka ta ew, kata ka 25

tarik u Lber 2019 **DATE** , ha kaba i kong Hiarmon Pyrtuh **PERSON** , Chairperson jong ka Jing aseng Kynthei KJP

Figure 13.2 Model sample output
Source: Author

score of 91.12%. The default Spacy NER (DSNER) model closely follows with a precision score of 91.09%. The precision scores of the other models range from 88.04% to 90.55%. The xlm-roberta-base+DSNER model achieved the greatest recall score of 91.79%, while the roberta-base+DSNER model followed closely with a recall score of 91.39%. The recall scores of the other models range from 85.19 % to 90.69%. The roberta-base+DSNER model achieved the greatest F1-score of 91.25%, while the xlm-roberta-base+DSNER model closely followed with an F1-score of 91.17%. The F1-scores of the other models range from 86.59% to 90.39%. From the obtained results, it could be inferred that the roberta-base+DSNER and xlm-roberta-base+DSNER models exhibited superior performance on the Khasi NER task, with excellent precision, recall, and F1-scores. The default Spacy NER (DSNER) model also demonstrated excellent performance, exhibiting a comparatively high F1-score. Conversely, the ai4bharat/indic-bert+DSNER model exhibited the lowest F1-score, suggesting that it might not have been the best choice for this specific dataset. From the resulting performance outputs, it could be observed that models trained on multilingual data did not always yield superior results, particularly when comparing a monolingual roberta-base model, Spacy NER (DSNER) model with multilingual xml-roberta-base model and others, despite the latter being specifically designed for cross-lingual tasks. This could have been attributed to factors like the fine-tuning of parameters or the nature of the dataset that was utilised during the training process. Another notable finding was

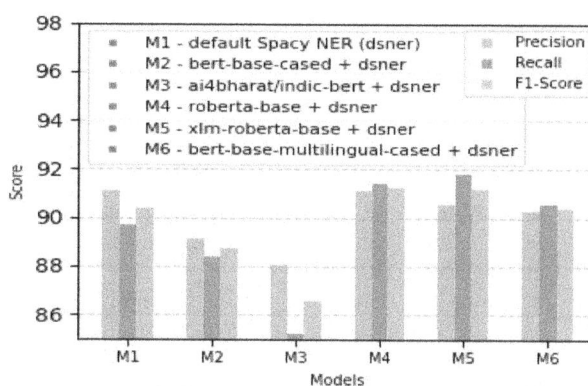

Figure 13.3 Comparison of models
Source: Author

that the ai4bharat/indicbert pre-trained model, which was trained on multiple Indian languages, performed badly on the Khasi NER task compared to other models. Besides other factors, one potential cause for this underwhelming performance could have been attributed to the linguistic structure on which the model had been initially trained. Most Indian languages typically adhere to the Subject-Object-Verb (SOV) word order)[5] and exhibit a high degree of morphological complexity. In contrast, Khasi follows the Subject-Verb-Object (SVO) word order) [14]. This could have also explained why the roberta-base model outperformed others, as it had been exclusively trained on English text that adhered to the same word order as the Khasi language. Figure 13.2 provides some output visualisations of the best performing model, while Figure 13.3 shows a graph comparing each model's performance.

Conclusion and future works

Our research primarily focused on developing a Khasi NER system, but at the same time creating a Khasi spell checker, and annotating new Khasi NER datasets was essential, as these resources were not readily available. Experimental results showed that the cross-lingual transfer learning model performed particularly well between English and Khasi, achieving better outcomes than other language pairs. While the results are promising, there is room for improvement, especially in reducing false detections caused by overlapping entities. Moving forward, we aim to expand the dataset and explore multiple approaches for NER processing. Additionally, our future work will focus on nested entity recognition as a key research area.

Acknowledgment

The annotation tool Prodigy, provided by Explosion AI Germany through Operations Specialist Anna Wieser, operates under a research license free of charge for our project. Your support was crucial in successfully completing our task. The tool has been essential in optimising our data annotation process and validation operations.

References

[1] Ajees, A. P., Idicula, S. M. (2018). A named entity recognition system for Malayalam using conditional random fields. In 2018 International Conference on Data Science and Engineering (ICDSE), 1–5.

[2] Bojanowski, P., Grave, E., Joulin, A., Mikolov, T. (2017). Enriching word vectors with subword information. In Lee, L., Johnson, M., Toutanova, K., (eds.) Transactions of the Association for Computational Linguistics, vol. 5. Cambridge, MA: MIT Press, 135–146.

[3] Chopra, D., Joshi, N., Mathur, I. (2016). Named entity recognition in Hindi using hidden Markov model. In 2016 Second International Conference on Computational Intelligence & Communication Technology (CICT), 581–586.

[4] Grishman, R., Sundheim, B. (1996). Message understanding conference- 6: a brief history. In COLING 1996 Volume 1: The 16th International Conference on Computational Linguistics.

[5] Kakwani, D., Kunchukuttan, A., Golla, S., N.C., G., Bhattacharyya, A., Khapra, M. M., Kumar, P. (2020). IndicNLPSuite: monolingual corpora, evaluation benchmarks and pre-trained multilingual language models for Indian languages. In Findings of the Association for Computational Linguistics: EMNLP 2020, 4948–4961.

[6] Kaur, Y., Kaur, E. R. (2015). Named entity recognition (NER) system for Hindi language using combination of rule based approach and list look up approach. *International Journal of Scientific Research and Management (IJSRM)*, 3(3).

[7] Kingma, D. P., Ba, J. (2017, January 29). Adam: A Method for Stochastic Optimization.

[8] Mikolov, T., Chen, K., Corrado, G., Dean, J. (2013, September 6). Efficient Estimation of Word Representations in Vector Space.

[9] Murthy, R., Khapra, M. M., Bhattacharyya, P. (2018). Improving NER tagging performance in low-resource languages via multilingual learning. *ACM Transactions on Asian and Low-Resource Language Information Processing*, 18(2), 9:1–9:20.

[10] Nongmeikapam, K., Shangkhunem, T., Chanu, N. M., Singh, L. N., Salam, B., Bandyopadhyay, S. (2011). CRF based name entity recognition (NER) in Manipuri: a highly agglutinative Indian language. In 2011 2nd National Conference on Emerging Trends and Applications in Computer Science, 1–6.

[11] Patil, N. V., Patil, A. S., Pawar, B. V. (2017). HMM based named entity recognition for inflectional language. In 2017 International Conference on Computer, Communications and Electronics (Comptelix), 565–572.

[12] Phatpiglet/autocorrect. (2024, January 12). . Python, PHAT Piglet. Available at https://github.com/phatpiglet/autocorrect

[13] Project: Prodigy Explosion Developer Tools and Consulting for AI, Machine Learning and NLP. (2024, November 25). Available at https://explosion.ai/_/project/explosion.ai

[14] Rynjah, R. K., Lyngdoh, S. A. (2022). Word order in standard Khasi and its varieties: a comparative study of change and variation, 10(12), 279–287.

[15] Sabane, M., Ranade, A., Litake, O., Patil, P., Joshi, R., Kadam, D. (2023). Enhancing low resource NER using assisting language and transfer learning. In 2023 2nd International Conference on Applied Artificial Intelligence and Computing (ICAAIC), 1666–1671.

[16] Saha, S. K., Sarkar, S., Mitra, P. (2008). A hybrid feature set based maximum entropy Hindi named entity recognition. In Proceedings of the Third International Joint Conference on Natural Language Processing: Volume-I.

[17] Smith, J. S., Nebgen, B. T., Zubatyuk, R., Lubbers, N., Devereux, C., Barros, K., Tretiak, S., et al. (2019). Approaching coupled cluster accuracy with a general-purpose neural network potential through transfer learning. *Nature Communications*, 10(1), 2903. Nature Publishing Group.

[18] Athavale, V., Bhardwaj, S., Pamecha, M., Prabhu, A. Shrivastava, S. (2016). Towards deep learning in Hindi NER: an approach to tackle the labelled data scarcity," In NLPAI, 2016, 154–160.

[19] **Ekbal, A., Bandyopadhyay, S. (2010).** Named entity recognition using appropriate unlabeled data, post-processing and voting. *Informatica (Slovenia), 34,* 55–76.

[20] **Ajees, A P., Idicula, S. M. (2018).** A named entity recognition system for Malayalam using neural networks. In *2018 8th International Conference on Advances in Computing & Communications (ICACC-2018)*, 962–969.

14 Localised detection of rice disease using YOLO 11: an advanced approach for precision agriculture

Gitanjali Patowary[a] and Dr Ridip Dev Choudhury[b]

Krishna Kanta Handiqui State Open University, India

Abstract

Food security and crop diseases remain critical challenges in many countries, significantly affecting the financial stability and livelihoods of farmers. Conventional deep learning techniques based on image processing have significant delays and high computational resources for identifying diseases. However, new technologies, such as YOLO 11, are introducing faster detection technologies, and specify multiple diseases in a single process. In this study, YOLO 11 is applied for detecting rice leaf diseases. This model was tested on a rice leaf image dataset and performed remarkably well in terms of precision and recall for many diseases. Importantly, the model achieved an average recall of 0.85 and a mean Average Precision (mAP) at IoU 0.50, improving from 0.1 to 0.5.

Keywords: YOLO, precision agriculture, rice disease

Introduction

Agriculture plays an important role in a country's economic development. It relies on the common factor of crop harvests (plants), their quantity, and quality [1]. All these factors have an impact on agriculture, including climate change, seasonal diseases, seed quality. Therefore, farmers' livelihoods improve or worsen with any change in agricultural conditions, which in turn has a direct proportional impact on a country's economic state [2].

For the purposes of food security, increased crop production, and the sustainability of farmers' livelihoods, it is essential to accurately detect rice diseases in their early stages so that substantial damage can be prevented. YOLO is an essential tool for this process because it can detect multiple rice diseases in one image in real-time, efficiently, and with little computational resources. This makes it ideal for use in agricultural settings where resources are limited.

YOLO is a real-time object recognition technique that is widely used today in all real-time applications, from vehicle number detection to disease detection. In contrast to traditional methods, where an image is scanned multiple times, this algorithm divides an image into a grid and predicts multiple bounding boxes and class probabilities for each grid cell at once. This approach makes it significantly faster. Over time, it has evolved through several iterations, each improving on its predecessor to deliver even better accuracy and performance.

YOLO, a highly effective object detection algorithm, is ideal for detecting multiple plant diseases within a single image. Its ability to identify and localise multiple diseases simultaneously makes it invaluable for early detection and intervention. By training the YOLO model on a diverse dataset of healthy and diseased plant images, it can accurately recognise various disease symptoms, such as spots, blotches, and discoloration. Once trained, the model processes images in real-time, delivering precise and timely diagnoses. This enables farmers to act quickly, applying suitable treatments or isolating infected plants, thereby reducing crop losses and enhancing overall yield.

Literature review

Xiong et al. [3] comprises a modified upgraded version of YOLOv-5model, YOLOv5-ct for better outcome of weed identification in the sector of agriculture. The YOLOv5-ct model employs a Transformer encoder emphasizing attention mechanism and localization into YOLOv-5 and a Convolutional Block Attention Module (CBAM) emphasizing weed in a noisy backdrop. By obtaining 92.4% accuracy, 89.2% recall, and 94.6% mAP, YOLOv5-ct outperform YOLOv-3 and YOLOv-5.

Alharbi et al. [9] research works on a few-shot learning model based on EfficientNet with an attention mechanism achieves 93.19% accuracy in classifying 18 wheat diseases using only 40 images. This data-efficient and computationally effective approach enhances wheat disease detection for practical agricultural applications.

Hasan et al. [10] works a hybrid model combining a deep CNN with an SVM classifier, leveraging transfer learning to re-train on 1080 images of nine rice disease types, achieving a classification accuracy of 97.5%.

Abirami et al. [11] works on an image processing and neural network-based approach for grading rice granules, utilizing median filtering, adaptive thresholding, and Canny edge detection to extract features like perimeter and area, achieving 98.7% accuracy for non-overlapping

[a]gitanjalip730@gmail.com, [b]ridipgu2010@gmail.com

DOI: 10.1201/9781003675235-14

granules. The Scaled Conjugate Gradient Training neural network model offers a robust, automated solution for rice quality assessment, adaptable to other foodstuffs.

Susa et al. [4] research facilitates the detection and classification of cotton plant leaf diseases using the YOLOv3 model. This research achieves 96.79% training accuracy, 92.26% validation accuracy, and 95.09% mean average precision. It provides detection accuracy for video frames ranging from 98% to 99%, and for live stream frames, the accuracy ranges between 74% and 99%.

Xinru et al. [5] introduced a method using enhanced YOLOv5 for detecting apple leaf diseases. This YOLOv5 incorporates the Wise-IoU (WIoU) Loss Function, which improves the detection of low-quality samples, and the CBAM (Convolutional Block Attention Module), which enhances feature extraction by focusing on critical image regions using both channel and spatial attention. The model uses a dataset of 9,145 images from Kaggle and achieves a mAP@0.5 of 85.2%.

Tasneem et al. [6] implemented YOLOv8 for plant leaf disease detection. This research utilises a dataset of 2,569 images from the PlantDoc dataset, which includes 13 plant species and 30 disease classes. The model attains a mean Average Precision (mAP) of 0.79, Precision of 0.61, and Recall of 0.71 at an Intersection over Union (IoU) of 0.5.

Srivastava et al. [7] uses YOLOv7 for detecting and classifying potato leaf disease and achieves an accuracy of 98.1%, surpassing classifiers such as ANN, SVM, and RF.

R et al. [8] utilised YOLOv3 for rapid and precise plant disease identification. It uses the Plant Village dataset, including 54,448 photos of both healthy and sick leaves across 38 categories. The model achieves accuracy between 80% and 90%.

Methodology and model specifications

The rice disease dataset (640 × 640) has been collected and labelled properly, as shown in Table 14.1 and utilised in our YOLO 11 model.

Table 14.1 Label counts for rice disease dataset.

Label	Class Name	Count
0	Bacterial_Leaf_Blight	646
1	Brown_Spot	1403
2	Healthy Leaf	821
3	Leaf_Blast	763
4	Leaf_Scald	564
5	Narrow_Brown_Leaf_Spot	737
6	Neck_Blast	951
7	Rice_Hispa	933

Source: Author

The above dataset is trained with YOLO 11. It is a model that has a total of 319 layers and 2.6 million parameters. This assists to learn difficult patterns. Well, why YOLOv11n instead of previous versions like YOLOv10, it gives the most accurate objects detection. It also processes images faster, which is more useful for real-time applications. Additionally, YOLOv11n is more efficient in using computational resources, which allows it to perform well while consuming less power.

YOLO works for rice disease detection by first collecting a dataset of rice leaf images that show various diseases, such as Leaf Blast or Brown Spot. These images are labelled by drawing bounding boxes around the affected areas and assigning each box a disease label. After preparing the data, the model is trained to recognise these diseases by feeding the labelled images into the model and adjusting the model's parameters through training.

The model was trained for 30 epochs, and the images in the training dataset were resized to 640 × 640 pixels. A batch size of 16 was used, meaning 16 images were processed at a time during training. The learning rate started at 0.001 and gradually decreased to 0.1, controlling how much the model's weights were adjusted with each step. A weight decay of 0.0005 was applied to prevent overfitting. The model uses techniques like data augmentation to make it more robust and avoid overfitting. Once trained, it can predict the disease in new rice leaf images by identifying the location of the affected areas and classifying the disease.

We opted for YOLOv11n instead of YOLOv11m and YOLOv11x because it offers an excellent balance between speed and accuracy. YOLOv11n provides expedited inference times and necessitates less processing resources relative to bigger models, making it optimal for real-time applications such as rice disease detection. While YOLOv11m and YOLOv11x provide higher accuracy, YOLOv11n delivers sufficient precision with a smaller model size, making it easier to deploy on resource-constrained environments or devices with limited GPU memory. Additionally, YOLOv11n generalises well on smaller datasets, ensuring reliable performance without overfitting, and its lightweight nature makes it suitable for large-scale use with minimal latency.

Results

The model's robust performance in detecting rice maladies such as Bacterial Leaf Blight, Neck Blast, and Rice Hispa is emphasised by the F1-Confidence Curve, which demonstrates reliable detection with high F1 scores. The model's promising capability is demonstrated by the moderate accuracy of other diseases, including Leaf Scald, Healthy Leaf, Narrow Brown Leaf Spot, and Brown Spot, while Leaf Blast indicates that there is room for improvement. The YOLOv11 model achieved an overall precision (P) of 0.667, meaning 66.7% of the detections is correct.

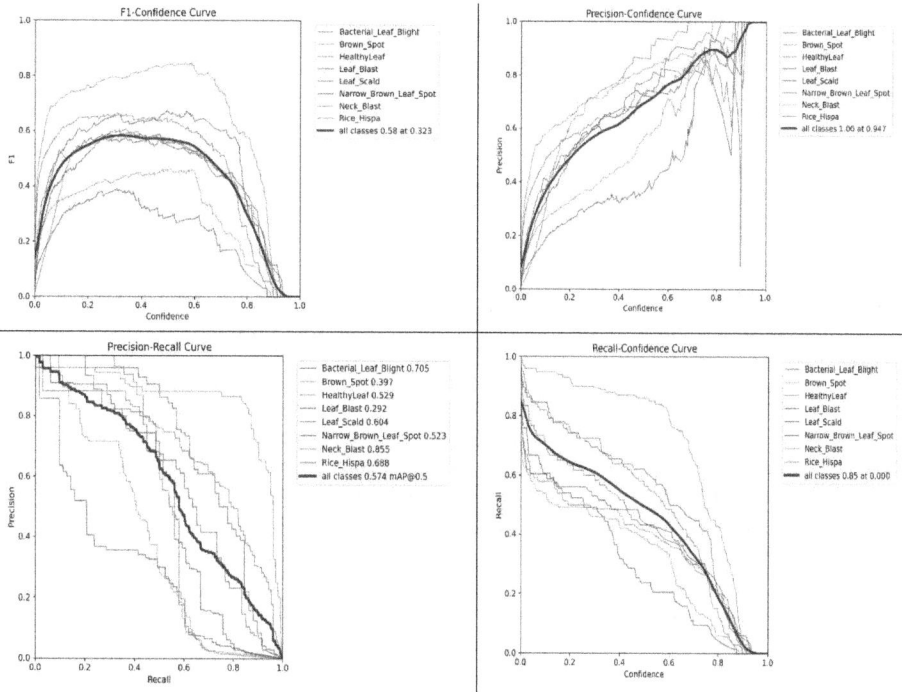

Figure 14.1 F1, precision-confidence, recall-confidence and precision-recall curve
Source: Author

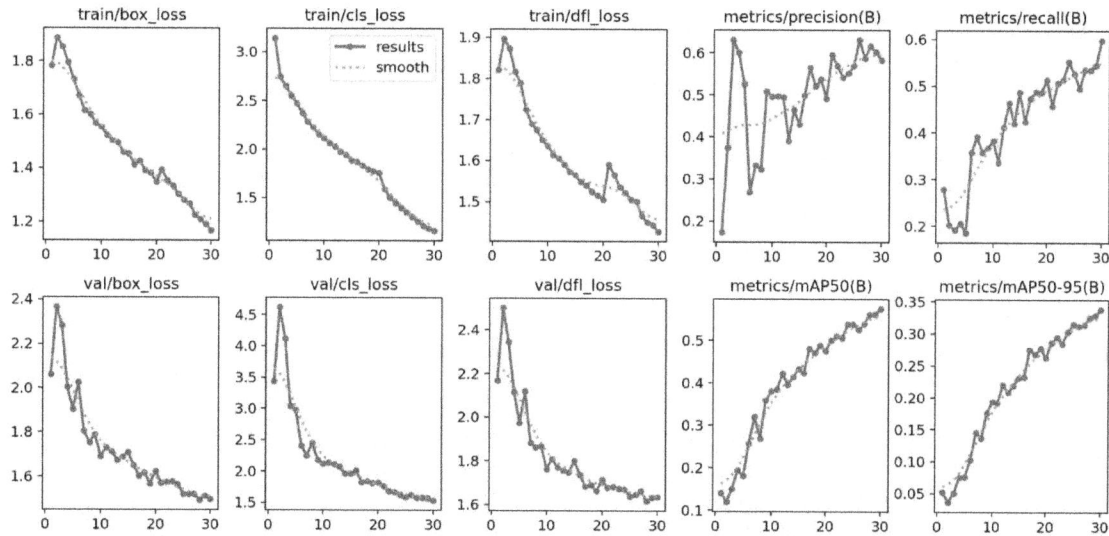

Figure 14.2 Results on training & validation loss
Source: Author

The recall (R) is 0.561, indicating that the model correctly identified 56.1% of the actual diseased leaves. The mAP50 (mean average precision at 50% IoU) was 0.573, showing a moderate ability to detect and classify rice diseases accurately.

The **Precision-Confidence Curve** graph provides insight into the model's performance, showing how its precision varies with different confidence levels across various rice disease classes. The model demonstrates strong performance at higher confidence levels, particularly for Neck Blast and Bacterial Leaf Blight. However, for diseases like Leaf Blast, the model's precision indicates a need for further refinement to enhance its accuracy and reliability.

The **Precision-Recall Curve** for the rice disease classification model highlights varying performance across different rice disease classes. The model demonstrates exceptional accuracy in identifying Neck Blast with a precision of 0.855, followed by Bacterial Leaf Blight at 0.705 and Rice_Hispa at 0.688. It also performs well for Leaf_Scald and HealthyLeaf, with precisions of 0.604 and 0.529, respectively. However, the model shows moderate performance for Narrow_Brown_Leaf_Spot (0.523) and Brown Spot (0.397), and struggles the most with Leaf_Blast, which has the lowest precision at 0.292. These results show that the model is highly effective for certain diseases, there is room for improvement in detecting others.

As seen in the Figure 14.1, the model shows outstanding accuracy in detecting Neck Blast, Bacterial Leaf Blight, and Rice Hispa, maintaining high recall rates even when operating at lower confidence thresholds. Nonetheless, it faces challenges in identifying Leaf Blast, as indicated by its lowest recall value. Even with this constraint, the model exhibits robust overall performance in the majority of diseases, achieving an average recall of 0.85. This suggests its reliability in accurately assessing the health of rice leaves.

The graph furnished illustrates the model's training and validation performance over a period of 30 epochs, emphasising a variety of metrics and losses. During training, the box loss decreases from 1.8 to 1.2, the classification loss reduces from 3.0 to 1.5, and the distribution focal loss drops from 1.9 to 1.3, reflecting improvements in localisation and classification of rice diseases. Precision and recall both increase from 0.2 to 0.6, indicating more accurate and reliable predictions. In validation, the box loss declines from 2.4 to 1.6, classification loss reduces from 4.5 to 1.5, and distribution focal loss decreases from 2.4 to 1.6, showing consistent generalisation to unseen data. The mean Average Precision (mAP) at IoU 0.50 rises from 0.1 to 0.5, while mAP at IoU 0.50 to 0.95 improves from 0.05 to 0.35, highlighting the model's growing accuracy and effectiveness in detecting objects across various thresholds.

Conclusion

The work demonstrates the effectiveness of the YOLOv11n model in detecting rice diseases with high precision and recall, particularly for diseases like Neck Blast, Bacterial Leaf Blight, and Rice Hispa. The model's ability to provide real-time, accurate detection makes it a valuable tool for precision agriculture, enabling timely interventions to mitigate crop damage. Despite its strong performance, the model shows room for improvement in detecting certain diseases such as Leaf Blast. Future work could focus on refining the model to enhance its accuracy across all disease classes. Overall, the YOLOv11n model offers a promising solution for resource-constrained agricultural settings, supporting farmers in maintaining healthy crops and ensuring food security.

References

[1] Shakeel, W., Ahmad, M., Mahmood, N. (2020). Early detection of cercospora cotton plant disease by using machine learning technique. In 30th International Conference on Computer Theory and Applications (ICCTA), 2020. https://doi.org/10.1109/iccta52020.2020.9477693.

[2] Arsenovic, M., Karanovic, M., Sladojevic, S., Anderla, A., Stefanovic, D. (2019). Solving current limitations of deep learning-based approaches for plant disease detection. *Symmetry*, 11(7), 939. https://doi.org/10.3390/sym11070939.

[3] Xiong, K., Li, Q., Meng, Y., Li, Q. (2023). A study on weed detection based on improved Yolo V5. In 4th International Conference on Information Science and Education (ICISE-IE), Zhanjiang, China, 2023, 1–4. https://doi.org/10.1109/icise-ie60962.2023.10456396.

[4] Susa, J. A. B., Nombrefia, W. C., Abustan, A. S., Macalisang, J., Maaliw, R. R. (2022). Deep learning technique detection for cotton and leaf classification using the YOLO algorithm. In 2022 International Conference on Smart Information Systems and Technologies (SIST), 1–6. https://doi.org/10.1109/sist54437.2022.9945757.

[5] Xinru, H., Limin, C., Qing, X., Chongyuan, L., Yinchai, W., Peijun, F. (2023). Plant disease detection based on improved YOLOv5. In 2nd International Conference on Robotics, Artificial Intelligence and Intelligent Control (RAIIC), Mianyang, China, 162–165. https://doi.org/10.1109/raiic59453.2023.10280962.

[6] Tasneem, R., Shahanaz, S., Vasantha, S. V., Mageswari, R. U., Priyadarsini, P. I., Reddy, M. V. (2024). Plant leaf disease detection system using single shot detector YOLOV8. In IEEE International Conference on Contemporary Computing and Communications (InC4), Bangalore, India, 2024, vol. 33, 1–6. https://doi.org/10.1109/inc460750.2024.10649160.

[7] Srivastava, A., Rawat, B. S., Bajpai, P., Dhondiyal, S. A. (2024). Potato leaf disease detection method based on the YOLO model. In 4th International Conference on Data Engineering and Communication Systems (ICDECS), 1–5. https://doi.org/10.1109/icdecs59733.2023.10502511.

[8] R, C. L. K., B, P., G, S., J, N. J., T, G., Hashim, M. (2023). Yolo for detecting plant diseases. In Third International Conference on Artificial Intelligence and Smart Energy (ICAIS), Coimbatore, India, 1029–1034. https://doi.org/10.1109/icais56108.2023.10073875.

[9] Alharbi, A., Khan, M. U. G., Tayyaba, B. (2023). Wheat disease classification using continual learning. *IEEE Access*, 11, 90016–90026. https://doi.org/10.1109/access.2023.3304358.

[10] Hasan, M. J., Mahbub, S., Alom, M. S., Nasim, M. A. (2019). Rice disease identification and classification by integrating support vector machine with deep convolutional neural network. In 2019 1st International Conference on Advances in Science, Engineering and Robotics Technology (ICASERT), 1–6. https://doi.org/10.1109/icasert.2019.8934568.

[11] S, A., P, N., H, K. (2014). Analysis of rice granules using image processing and neural network pattern recognition tool. *International Journal of Computer Applications*, 96(7), 20–24. https://doi.org/10.5120/16806-6530.

[12] Prakash, N., Rajakumar, R., Madhuri, N. L., Jyothi, M., Bai, A. P., Manjunath, M., Gowthami, K. (2022). Image classification for rice varieties using deep learning models. *YMER Digital*, 21(06), 261–275. https://doi.org/10.37896/ymer21.06/25.

15 Assamese handwritten script segmentation using deep learning approach

Saurabh Sutradhar[a] and Dr. Ridip Dev Choudhury[b]

Krishna Kanta Handiqui State University, India

Abstract

Assamese is an Indo-Aryan language spoken by approximately 15 million people, predominantly in the northeastern Indian state of Assam. Its script shares numerous similarities with Bengali and other scripts from the Indian subcontinent. Assamese comprises 11 vowels and 41 consonants, with several vowel modifiers that can combine with consonants [1]. Accurate letter and word segmentation improves OCR recognition of Assamese text, enhancing digitisation efforts. However, achieving proper segmentation of Assamese writing is challenging due to inherent variability in writing styles, scan noise, and the script's complex nature. Accurately reading and translating written records into machine-readable text requires meticulous segmentation. To address these challenges, this paper outlines a plan to explore a deep-learning framework for segmenting Assamese handwritten scripts. The development process is structured into two primary phases: image collection and segmentation. The initial phase focuses on gathering a diverse set of images from various online repositories and offline sources. In the next phase, the collected images are fed into deep learning models such as SegNet and U-Net. The segmented output is then refined using Convolutional Conditional Random Fields (CCRF). Finally, a detailed comparative analysis of SegNet and U-Net is conducted, evaluating each model's ability to accurately segment Assamese handwritten scripts.

Keywords: Assamese handwritten scripts, image segmentation, SegNet, U-Net, convolution conditional random fields; comparative analysis

Introduction

The Assamese script is a variant of the Eastern Nagari script and is used to write the Assamese language, predominantly spoken in the Indian state of Assam [2]. Due to the variety of phonetic characters and the prevalence of compound characters, separating written Assamese script requires advanced algorithms to reliably identify and isolate individual letters and words from continuous text [3]. Segmentation of Assamese handwritten script involves dividing the text into its basic components, such as lines, phrases, and symbols, for further analysis, processing, or comprehension. This step is crucial in developing OCR systems, which aim to convert handwritten or printed content into machine-readable format [4,8]. As an abugida, where consonant-vowel pairs are written as units, the Assamese script presents unique segmentation challenges due to its intricate combinations of letters and diacritic marks [5].

The fundamental purpose of segmenting handwritten Assamese text is to accurately identify and extract components from a continuous flow of text. This process serves as the foundation for subsequent applications, such as text recognition, digital transcription, and language analysis [6]. Variations in writing styles among individuals are reflected in differences in size, form, and style, as well as the spacing between words and lines. Similar to the Bengali script, Assamese characters can change form when combined with others or when certain consonant marks are added [7]. The presence of conjunct vowels and diacritical symbols, used to represent vowel sounds, further complicates the segmentation process. In handwritten texts, especially in cursive styles, characters and words may overlap or touch, posing significant challenges for algorithms attempting to identify and distinguish individual characters. Overcoming these challenges requires innovative approaches that are robust enough to handle the diverse and intricate nature of written Assamese characters [8].

The application of deep learning for segmenting Assamese handwriting scripts represents a significant advancement in document analysis and OCR technology. Deep learning models can leverage extensive datasets of handwritten Assamese texts to accurately identify and separate the script into its components, such as letters, words, and lines [9]. This process involves several fundamental strategies and techniques, each tailored to address the unique challenges of segmenting Assamese script. Convolution Neural Networks (CNNs) are at the forefront of deep learning applications for image recognition. They excel at identifying distinct patterns and features in images, making them ideal for handwriting segmentation. Recurrent Neural Networks (RNNs), along with their advanced variant, Long Short-Term Memory Networks (LSTMs), are designed to process sequential data. These networks are particularly effective for Assamese script

[a]saurabh.sutradhar@kkhsou.in, [b]RIDIPGU2010@gmail.com

DOI: 10.1201/9781003675235-15

segmentation because they can recognise the ordered structure of text and handle characters that change shape depending on their position within a word [10]. The integration of deep learning in Assamese handwritten script segmentation opens up new research and application possibilities. These include the development of more sophisticated OCR systems, the digital preservation of historical documents, and increased access to handwritten texts in the Assamese language [11]

The contributions of the implemented model are as follows:

- To create a database for Assamese handwritten scripts.
- To study and analyse the effectiveness of deep learning framework for segmenting Assamese handwritten script.

The rest of the paper comprises the following components: Section 2 involves a comprehensive review of research literature on segmentation models for Assamese handwritten scripts. Section 3 describes the proposed methodology. Lastly, Sections 4 and 5 cover the presentation of results and conclusions, respectively.

Literature review

In 2015, Borah S. et al. [12] introduced a character segmentation method utilising horizontal and vertical projections for an Artificial Neural Network (ANN)-based recognition system. The approach involves pre-processing scanned handwritten documents, segmenting characters using projection profiles, and extracting geometric features to train the ANN for improved recognition accuracy. Projection profiles may work well with neatly written text but struggle with cursive and overlapping characters, and the method is also mistakenly segmented by connected strokes and ligatures.

In 2018, R. Bania et al. [13] presented a computational model encompassing pre-processing, text segmentation, and feature extraction from individual characters. The segmentation process utilises a global projection profile approach to identify the upper, middle, and lower zones of words. The study combines diagonal features with texture features derived from the Gray Level Co-occurrence Matrix (GLCM) and employs an Artificial Neural Network for classification, achieving a recognition accuracy of 94.35%. In this work, researchers use Ostu's global projection for segmentation.

In 2022, Husam et al. [14] proposed a method that utilises the horizontal linear intensity of the image to define letter boundaries and differentiate between characters. The proposed approach involves three pre-processing stages: filling closed and open gaps (where circles are missing), removing commas to delineate ligature regions and prevent character overlap, and cropping the word image to eliminate excess whitespace. Filling open and closed gaps aims to enhance the pixel intensity of characters, followed by analysing vertical densities.

In 2024, Prarthana Dutta et al. [15] proposed a top-down segmentation technique that isolates lines from handwritten documents to segment words and characters. The method employs a raster scanning object detection technique to enhance character segmentation. The approach achieved an average segmentation accuracy of 93.61% for lines, 85.96% for words, and 88.74% for characters in both Assamese and Telugu scripts. The raster scan methodology is computationally expensive and may result in low accuracy in large documents with noise.

In 2024, Sukha Deep et al. [16] implemented a method developed for dividing documents based on projection profiles, statistical analysis, and script architecture. This method is capable of segmenting text lines that are closed, curved, skewed, or closely spaced, even when they exhibit wide variations in pattern and size. For word separation, an endpoint identification method has been devised to accurately segment phrases, including those with intra-word gaps. Training ANN for segmentation is resource intensive and that's why researchers have applied this method only to limited datasets, which may lead to overfitting.

Proposed methodology

Problem statement

Segmenting Assamese handwriting scripts is a complex task that plays a crucial role in various applications, from digitising historical manuscripts to developing robust OCR systems tailored to the Assamese language. This process involves breaking down handwritten text into its basic components, such as characters, words, and lines, to facilitate further analysis and recognition. The uniqueness of the Assamese script, with its distinct characters, modifiers, and conjuncts, presents both opportunities and challenges for the development of effective segmentation techniques. The research gap in traditional deep learning-based segmenting Assamese handwriting scripts model is given below.

- Traditional segmentation models often struggle with the variability in handwriting styles, sizes, and forms, as well as inconsistencies in character spacing and alignment. These challenges can hinder the development of segmentation models that perform reliably across diverse handwriting samples. However, the methodology introduced in this study effectively addresses these issues, leading to notable improvements in model accuracy.
- Traditional methods frequently suffer from overfitting, where a model excels on training data but struggles with new information, particularly when the

dataset lacks diversity. In contrast, the proposed approach utilises an innovative strategy that effectively mitigates and reduces the effects of overfitting, thereby enhancing the robustness and generalisability of the segmentation model.

Methodology

The given flowchart represents the procedure for our study. The process begins with the acquisition of handwritten images, which have been obtained from various sources through scanning, photographing, or direct digital input. These raw images often contain noise, distortions, and variations in intensity, making pre-processing a crucial step. During pre-processing, techniques such as grayscale conversion, binarisation, normalisation, noise reduction, and resizing are applied to enhance the quality of the images, ensuring they are suitable for segmentation. Once pre-processed, the images are fed into two deep learning-based segmentation models: SegNet and U-Net. These models independently process the images and generate segmented outputs, which are then stored for further analysis. To enhance the accuracy of segmentation, the outputs from both SegNet and U-Net are further refined using Conditional Random Field with Convolutional Neural Networks (CCRF). This step helps in reducing segmentation errors, improving character boundary detection, and refining pixel classification. Once the segmented images are obtained from CCRF, a comparative analysis is conducted. In the next section, we will describe these steps in detail (Figure 15.1).

Handwritten image acquisition

For this study, we have collected handwritten images from various sources, both online and offline. Most of the online images were taken from a link of the "https://www.scribd.com/document/192654741/UPSC-Assamese-Literature-Notes-Part-2" on 24-08-2024. This dataset contains almost 83 documents, which have an average of 300 to 500 words each. Another 83 documents are collected from the Department of Assamese, Krisnakanta Handiqui State Open University in offline mode. A sample of the collected dataset is shown in Figure 15.2.

Pre-processing

The collected raw images are often in RGB (Red, Green, Blue) format, which contains unnecessary colour information for segmentation. Therefore, the images are converted into grayscales, which reduces computational complexity while preserving text information. Handwritten documents generally contain unwanted noise. To address this issue, noise reduction techniques, such as Gaussian filtering and median filtering, are applied as both are the common filters used for textual data. After noise reduction, the images are resized to 512 × 512 pixels, and then Otsu's Thresholding method is used for binarisation.

Deep learning model

In our work, deep learning architectures, SegNet and U-Net have been chosen for the segmentation of Assamese handwritten scripts. These models are chosen for their proven effectiveness in image segmentation tasks to enhance accuracy specifically for Assamese script segmentation.

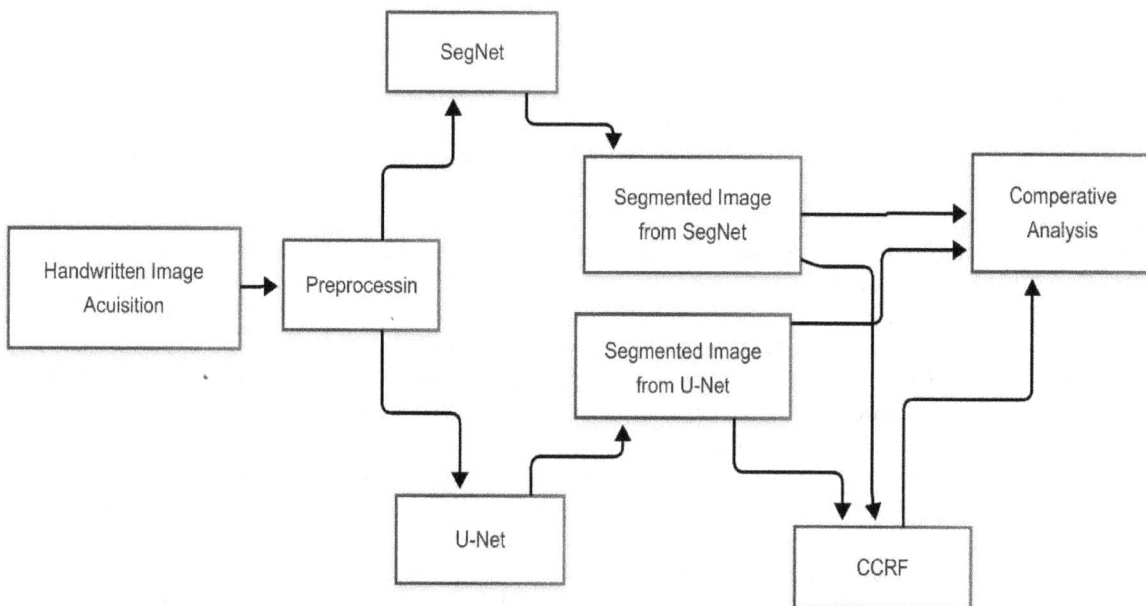

Figure 15.1 Propose methodology for this study
(Source: Author's own compilation)

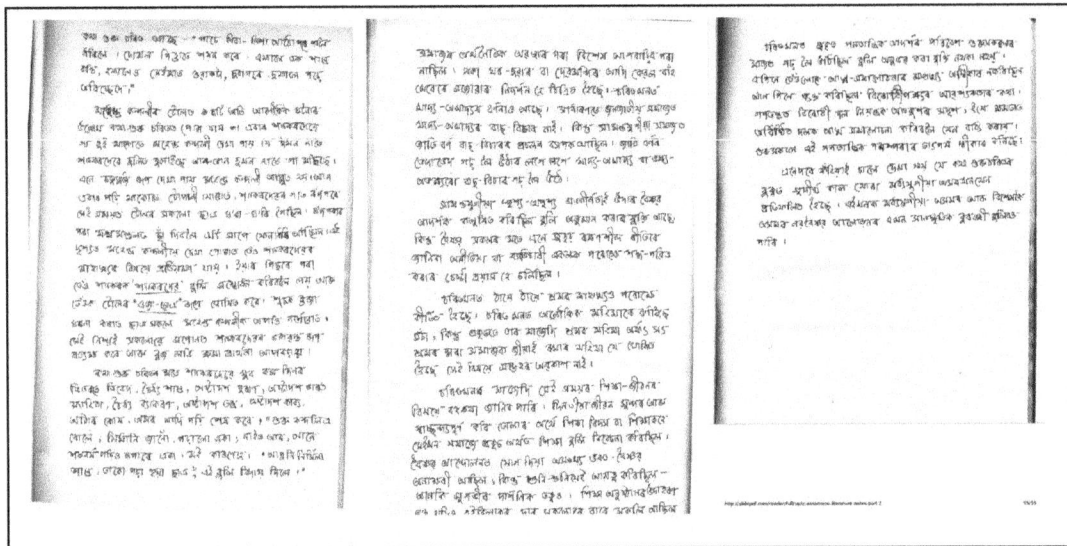

Figure 15.2 Sample Images of collected dataset
Source: Author

SegNet [17] is a deep learning architecture designed for semantic segmentation tasks. It is an encoder–decoder network that assigns a class label to each pixel in an image, making it highly effective for various segmentation tasks, including handwriting script segmentation. Here, the collected image is the input to this phase, SegNet's design includes an encoder network identical to the VGG16 network and a corresponding decoder network, allowing it to capture high-level features and provide accurate pixel-wise predictions. Unlike the U-Net network, which sends encoder feature maps to up-sampling decoding devices, resulting in significant memory needs, SegNet only recombines indices, requiring less memory.

The SegNet architecture consists of two main paths: the down-sampling (encoding) path and the up-sampling (decoding) path, followed by a pixel-wise classification layer. In the encoding path, there are 13 convolutional layers, mirroring the initial 13 layers of the VGG-16 network. Each encoder level corresponds to a decoder layer. The decoding system also comprises 13 convolutional layers. The final output of the decoder path feeds into a multi-class SoftMax classification algorithm, which assigns class probabilities to each pixel. Specifically, the encoding path starts with five convolution blocks, each beginning with a 2×2 max-pooling operation with a stride of 2 for downsampling. These convolution blocks consist of multiple layers of 3×3 convolutions, followed by batch normalisation and ReLU activation. The first two convolution blocks have one layer each, while the latter three blocks have three layers each. The decoding path mirrors the encoder path but uses upsampling instead of max-pooling. Upsampling involves taking the output of the preceding layer and combining it with the max-pooling indices from the corresponding encoder layer. The final layer of the decoder produces high-dimensional feature maps, which are then fed into a softmax classification layer to independently classify each pixel. This structured approach allows SegNet to efficiently and accurately perform image segmentation, making it a robust solution for applications requiring detailed and precise image analysis. Finally, the SegNet-based segmented images are produced from the Decoder phase.

U-Net [18] is a deep learning architecture widely used for image segmentation tasks, characterised by its symmetric design that processes images through two main segments. The first segment is the contracting path, which follows the typical CNN structure. In this approach, each block consists of two 3×3 convolution layers, followed by a max-pooling layer and ReLU activation function. This configuration is repeated multiple times to progressively reduce spatial dimensions while capturing essential features from the input image. The second component is the expanding path, which utilises 2×2 up-convolutions to upsample feature maps. During this process, feature maps from each layer in the contracting path is resized and concatenated with its unsampled counterparts. This is followed by two consecutive 3x3 convolutions with ReLU activation. Finally, a 1x1 convolution layer is employed to reduce feature map sizes and generate the segmented final image. Cropping is necessary to align the feature maps properly and remove pixels with minimal contextual information, ensuring that the resulting feature maps maintain relevant spatial details. The U-Net architecture forms a characteristic U-shape, enabling the network to effectively capture and utilise contextual information from overlapping areas in the image. This design allows U-Net to perform precise image segmentation by leveraging both

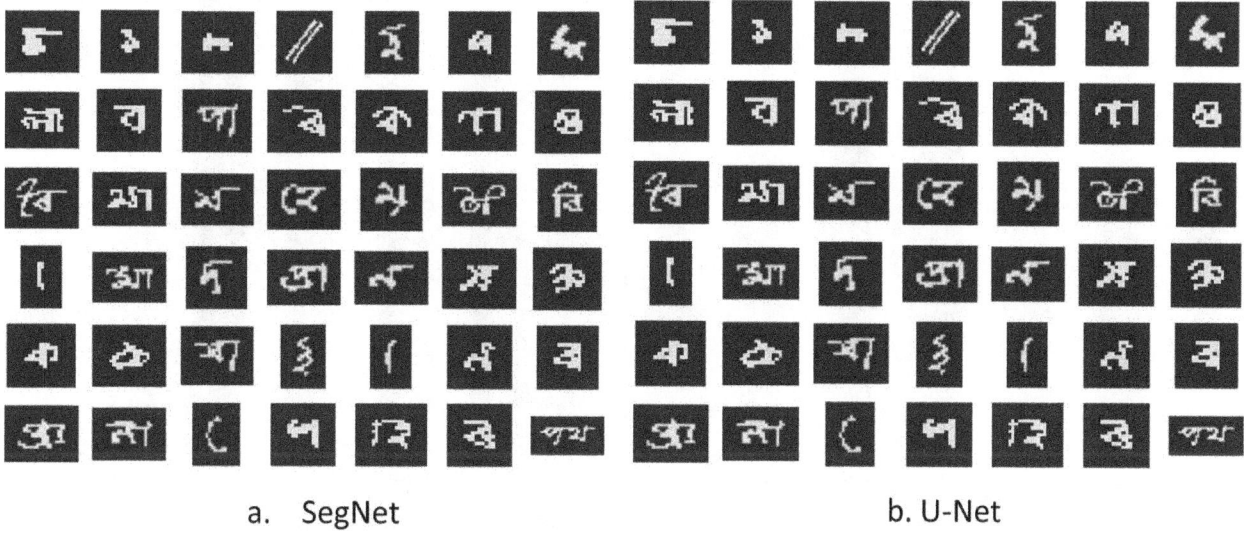

a. SegNet   b. U-Net

Figure 15.3 Samples of segmented images a. SegNet segmentation & b. U-Net segmentation
Source: Author

the detailed features captured in the contracting path and the contextual information integrated through the expanding path (Figure 15.3).

The energy value of the U-Net is represented in Eq. (15.1).

$$EN = \sum m(k) \log\left(ty_p(k)\right) \tag{15.1}$$

Here, the pixel-wise SoftMax feature used in the final feature map is denoted. ty_p and represented in Eq. (15.2).

$$ty_p = \frac{\exp E\left(q_p(k)\right)}{\sum_{p=1}^{p} \exp(E_p(k)')} \tag{15.2}$$

The process of activation in a channel p is illustrated as q_p. Throughout this procedure, the U-Net displays results as high values ranging between 0 and 1. Finally, a feature based on U-Net has been obtained.

A key difference between U-Net and SegNet is their memory usage and processing technique. U-Net utilises skip connections, where encoder feature maps are directly concatenated with their corresponding decoder layers, allowing fine-grained spatial details to be preserved at every stage. This increases memory consumption but enhances feature propagation, making it well-suited for high-resolution segmentation. On the other hand, SegNet employs index recombination, where only the max-pooling indices are transferred to the decoder, significantly reducing memory overhead while maintaining segmentation accuracy.

Conditional Random Fields (CCRF) [19] are widely used for post-processing in image segmentation tasks to refine pixel-wise classification results. Unlike SegNet

and U-Net, which perform primary segmentation, CCRF enhances the segmented output by incorporating contextual information, ensuring spatial consistency in predictions' operations by modelling dependencies between neighbouring pixels, making it highly effective for overcoming segmentation artefacts such as noise, boundary inconsistencies, and misclassifications. In this phase, segmented images from SegNet and U-Net serve as inputs, and CCRF is applied to refine these outputs by enforcing spatial smoothness constraints while preserving important structural details.

The CCRF model consists of unary and pairwise potential functions. The unary potential is derived from the initial segmentation probability maps obtained from SegNet and U-Net, while the pairwise potential ensures label coherence between adjacent pixels. This approach effectively smooths segmentation boundaries and corrects misclassified pixels by considering contextual dependencies.

To achieve this, CCRF employs Gaussian pairwise potentials to penalise abrupt pixel label transitions unless strong evidence exists. The energy function of the CCRF model is formulated as:

$$E(x) = \sum_{\{i\}} \psi_u(x_i) + \sum_{\{i,j\}} \psi_p(x_i, x_j)$$

where x represents the pixel labels, $\psi_u(x_i)$ corresponds to the probability of a pixel x_i belonging to a particular class $\psi_p(x_i, x_j)$, and enforces consistency between neighbouring pixel labels. CCRF employs Gaussian pairwise potentials to penalise abrupt pixel label transitions unless strong evidence exists. This ensures that segmentation

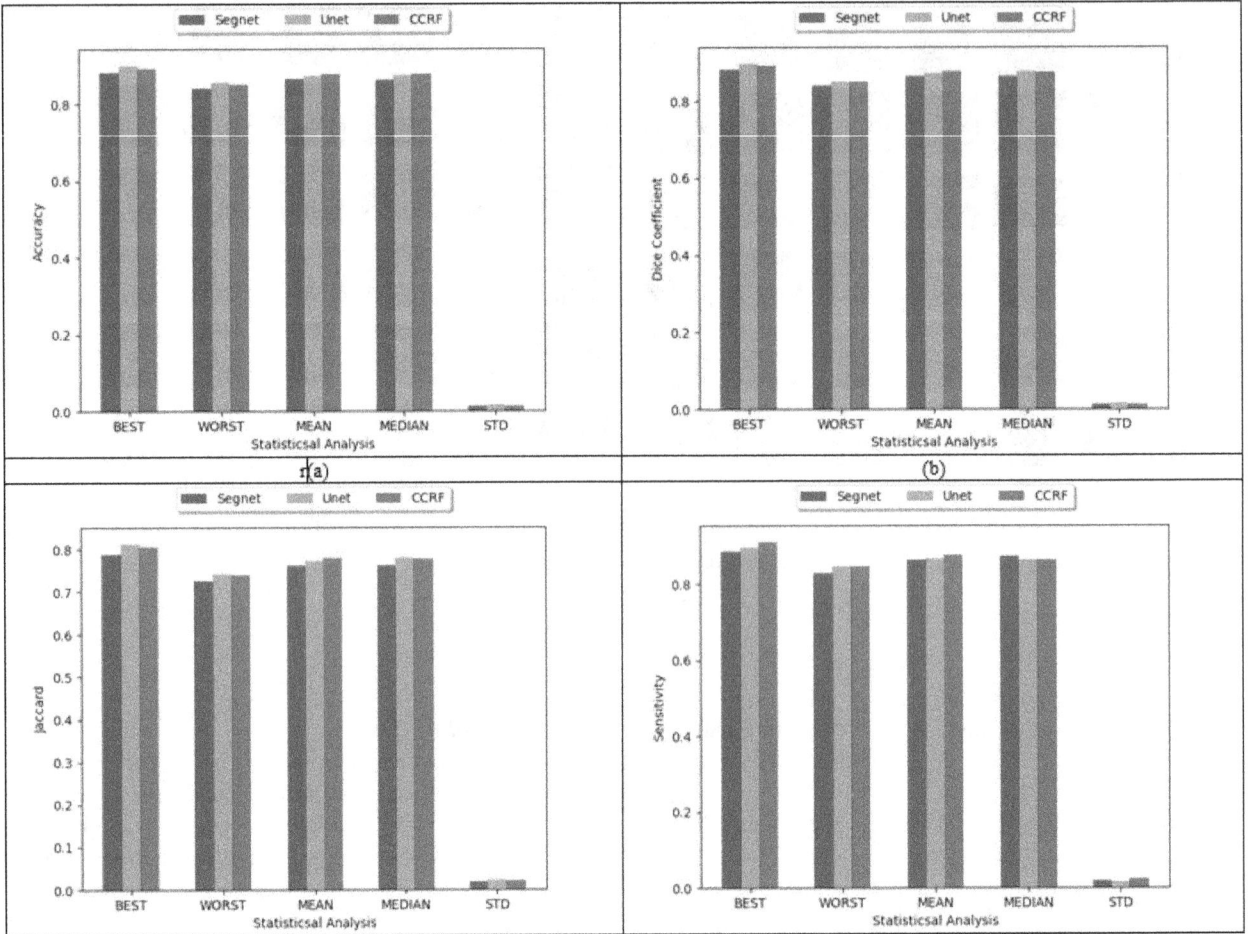

Figure 15.4 Statistical analysis of our study (a) Accuracy, (b) Dice coefficient, (c) Jaccard, and (d) Sensitivity
Source: Author

results are more spatially coherent while maintaining important structure details.

Comparative analysis
Comparative analysis is essential to evaluate the effectiveness of segmentation models. SegNet, U-Net, and CCRF each contribute unique strengths to the segmentation pipeline. CCRF-refined segmentation results are undergoing comparative analysis alongside raw segmentation outputs from SegNet and U-Net. Performance is evaluated using standard metrics such as Intersection over Union (IoU), Dice Score, and pixel accuracy, ensuring a robust evaluation of the different approaches.

Result and analysis

Experimental set-up
In our study, we worked with a dataset comprising 163 scanned handwritten documents. Each document was manually annotated using LabelMe to generate ground truth segmentation masks. The primary challenges included dealing with overlapping characters, accurately segmenting modifiers such as *Akar* and *Ukar*, and handling the variability in handwriting styles, which made consistent segmentation difficult. We utilised Google Collab with a Tesla T4 GPU for our work. We have used Python 3.8 for scripting and model development, TensorFlow 2.6 and PyTorch 1.10 for deep learning implementations, and OpenCV and NumPy for pre-processing and data augmentation.

Analysis
Our evaluation metrics considered Accuracy, Dice Coefficient, Jaccard Index (Intersection over Union – IoU), and Sensitivity (Recall) for the comparative analysis of our results, as they are the common parameters to evaluate such work [20].

Accuracy measures the overall correctness of pixel classification by comparing the number of correctly classified pixels (both foreground and background) to the total number of pixels in the image. It is computed as the ratio of the sum of true positives and true negatives to the total number of pixels, including false positives and false negatives. While accuracy is a straightforward and widely used metric, it can be misleading in segmentation tasks, particularly when the background pixels vastly outnumber the foreground pixels. In such cases, a high accuracy value may not necessarily indicate a well-performing segmentation model, as the model could achieve high accuracy simply by classifying most pixels as background.

Accuracy evaluates overall pixel classification correctness, but it can be misleading in segmentation tasks where background pixels dominate. While a high accuracy might suggest good performance, it may simply indicate that the model is classifying most pixels as background rather than truly distinguishing characters. The *Dice Coefficient*, also known as *the F1 Score* for segmentation, is a better metric for evaluating overlap between predicted and actual segmentation masks. It is more sensitive to class imbalances, making it effective for handwritten text segmentation, where character pixels constitute a small portion of the image. Similarly, the *Jaccard Index (IoU)* measures segmentation accuracy by considering the ratio of correctly classified character pixels to the total predicted and ground-truth pixels. IoU is a stricter metric than Dice, as it penalises false positives and false negatives more aggressively, ensuring precise boundary detection. *Sensitivity* (Recall) quantifies the model's ability to correctly identify character pixels while minimising false negatives, which is crucial for maintaining the integrity of complex Assamese scripts, including modifiers and Juktakshar.

The results indicate that all three models perform well across different statistical measures, including best, worst, mean, median, and standard deviation. In terms of accuracy, all models achieve consistently high values, with U-Net and CCRF slightly outperforming SegNet, suggesting that these architectures handle pixel-wise classification with greater precision. The Dice Coefficient, which evaluates the overlap between predicted and ground-truth segmentation, remains high across all models, with CCRF achieving the best performance, indicating its superior ability to preserve fine character details as it is a refinement process.

Similarly, the Jaccard Index, which is a stricter measure of segmentation accuracy, follows a similar trend, with CCRF demonstrating the highest overlap with ground truth, followed closely by U-Net, while SegNet lags slightly behind. The sensitivity metric, which measures how effectively the models detect character pixels without missing relevant regions, shows a comparable trend, with CCRF and U-Net outperforming SegNet, suggesting that SegNet may miss certain character strokes or modifiers, leading to higher false negatives. The standard deviation (STD) across all metrics is relatively low, indicating stable

and consistent performance across different test cases. Overall, CCRF achieves the best results as it refines the output of SegNet and U-Net and takes only the value of the highest evaluation matrix among those. U-Net and SegNet perform slightly lower but good solutions in resource-constrained environments.

Conclusion

This study explored the segmentation of Assamese handwritten scripts using deep learning models, specifically SegNet and U-Net, with further refinement through Convolutional Conditional Random Fields (CCRF). The small dataset size remains a significant limitation in our study. Deep learning models require large-scale annotated data for optimal performance; yet, manual annotation of handwritten documents is highly labour-intensive and time-consuming. As a result, the segmentation results remain somewhat inconclusive due to the dataset's limited scope. Expanding the dataset with automated or semi-automated annotation methods could improve model generalisation and robustness.

For future work, SegNet's superior performance in word segmentation can be leveraged as an intermediate step. First, words can be segmented using SegNet, and then modifiers (Matras) can be removed to simplify character extraction. Finally, SegNet can be reapplied to segment individual connected components, improving character-level segmentation accuracy. This multi-step approach could enhance segmentation performance while reducing computational complexity, making Assamese script digitisation more efficient and scalable.

References

[1] Bharath, A., Madhvanath, S. (April 2012). HMM-based lexicon-driven and lexicon-free word recognition for on-line handwritten Indic scripts. *IEEE Transactions on Pattern Analysis and Machine Intelligence*, 34(4), 670–682.

[2] Al Hamad, H. A. , Abu Zitar, R. (Aug. 2010). Development of an efficient neural-based segmentation technique for Arabic handwriting recognition. *Pattern Recognition*, 43(8).

[3] Ghosh, S., Chatterjee, A., Sen, S., Kumar, N., Sarkar, R. (2021). CTRL – CapTuRedLight: a novel feature descriptor for online Assamese numeral recognition, 80, 30033–30056.

[4] Bhattacharya, U., Chaudhuri, B. B. (Mar. 2009). Handwritten numeral databases of Indian scripts and multistage recognition of mixed numerals. *IEEE Transactions on Pattern Analysis and Machine Intelligence*, 31(3), 444–457.

[5] Bharath, A., Madhvanath, S. (Apr. 2012). HMM-based lexicon-driven and lexicon-free word recognition for on-line handwritten Indic scripts. *IEEE Transactions on Pattern Analysis and Machine Intelligence*, 34(4), 670–682.

[6] Keysers, D., Deselaers, T., Rowley, H. A., Wang, L.-L., Carbune, V. (Jun. 2017). Multi-language online handwriting

recognition. *IEEE Transactions on Pattern Analysis and Machine Intelligence*, 39(6), 1180–1194.

[7]   Malhotra, R., Addis, M. T. (2023). End-to-end historical handwritten Ethiopic text recognition using deep learning. *IEEE Access*, 11, 99535–99545.

[8]   Chherawala, Y., Roy, P. P., Cheriet, M. (Dec. 2016). Feature set evaluation for offline handwriting recognition systems: application to the recurrent neural network model. *IEEE Transactions on Cybernetics*, 46(12), 2825–2836.

[9]   Barriere, C., Plamondon, R. (Feb. 1998). Human identification of letters in mixed-script handwriting: an upper bound on recognition rates. *IEEE Transactions on Systems, Man, and Cybernetics, Part B (Cybernetics)*, 28(1), 78–81.

[10]   Lorigo, L. M., Govindaraju, V. (May 2006). Offline Arabic handwriting recognition: a survey. *IEEE Transactions on Pattern Analysis and Machine Intelligence*, 28(5), 712–724.

[11]   Namboodiri, A. M., Jain, A. K. (Jan. 2004). Online handwritten script recognition. *IEEE Transactions on Pattern Analysis and Machine Intelligence*, 26(1), 124–130.

[12]   Sin, B.-K., Ha, J.-Y., Oh, S.-C., Kim, J. H. (Apr. 1999). Network-based approach to online cursive script recognition. *IEEE Transactions on Systems, Man, and Cybernetics, Part B (Cybernetics)*, 29(2), 321–328.

[13]   Buoy, R., Iwamura, M., Srun, S., Kise, K. (2023). Toward a low-resource non-Latin-complete baseline: an exploration of Khmer optical character recognition. *IEEE Access*, 11, 128044–128060.

[14]   Borah, S., Konwar, S. (2015). Segmentation of Assamese handwritten characters based on projection profiles. *International Journal of Computer Applications*, 130(17), 12–16.

[15]   Bania, R. K., Khan, R. U. (2018). Handwritten Assamese character recognition using texture and diagonal orientation features with artificial neural network. *International Journal of Computer Applications*, 180(17), 1–6.

[16]   Al Hamad, H. A., Abualigah, L., Shehab, M., Al-Shqeerat, K. H. A., Otair, M. (2022). Improved linear density technique for segmentation in Arabic handwritten text recognition. *Multimedia Tools and Applications*, 81, 28531–28558.

[17]   Dutta, P., Muppalaneni, N. B. (2024). A top-down character segmentation approach for Assamese and Telugu handwritten documents. *Journal of Ambient Intelligence and Humanized Computing*, 15, 3275–3287.

[18]   Kaur, S., Bawa, S., Kumar, R. (2024). Heuristic-based text segmentation of bilingual handwritten documents for Gurumukhi-Latin scripts, 83, 18667–18697.

[19]   Badrinarayanan, V., Kendall, A., Cipolla, R. (Dec. 2017). SegNet: a deep convolutional encoder-decoder architecture for image segmentation. *IEEE Transactions on Pattern Analysis and Machine Intelligence*, 39(12), 2481–2495.

[20]   Zeng, Z., Xie, W., Zhang, Y., Lu, Y. (2019). RIC-Unet: an improved neural network based on Unet for nuclei segmentation in histology images. *IEEE Access*, 7, 21420–21428.

[21]   Golpardaz, M., Helfroush, M. S., Danyali, H. (Sep. 2020). Nonsubsampled contourlet transform-based conditional random field for SAR images segmentation. *Signal Processing*, 174.

[22]   Müller, D., Soto-Rey, I., Kramer, F. (2022). Towards a guideline for evaluation metrics in medical image segmentation. arXiv preprint arXiv:2202.05273. https://doi.org/10.48550/arXiv.2202.05273

16 Clustering of foods based on nutritional profile using principal component analysis and k-nearest neighbor classification

Laishram Maria Devi[1,a], Chingriyo Raihing[1,b] and Satishchandra Salam[2,c]

[1]Department of Home Science, Dhanamanjuri University, Imphal, 795001, Manipur, India

[2]Department of Mechanical Engineering, Manipur Institute of Technology, Manipur University, Imphal, 795001, India

Abstract

With the growing demand for personalised nutrition and the prevention of lifestyle-related disorders, understanding food selection based on nutritional content has become increasingly important. This study applies machine learning techniques to a comprehensive dataset of food nutritional profiles, aiming to identify clusters of food items based on their nutrient composition. Principal component analysis is used to reduce the dimensionality of the dataset and provide a two-dimensional visualisation of the data. Subsequently, k-nearest neighbor (k-NN) clustering is applied to group the foods. Initial analysis suggests that around 9 to 10 clusters offer reasonable groupings, as indicated by low Davies-Bouldin index value of around 1.5. The study presents clusters that potentially offer insights into dietary planning by identifying groups of nutritionally similar foods. This approach can also aid in designing personalised diets, suggesting alternative foods with similar nutrient profiles. In conclusion, such studies demonstrate the potential of machine learning for food clustering, nutritional analysis, and diet planning.

Keywords: Cluster, nutrition, k-NN, principal component analysis, diet planning

Introduction

The role of nutrition in maintaining health and preventing lifestyle-related disorders has become increasingly well-recognised in recent years. As individuals seek to choose their diets to meet personalised health goals, the need for advanced, data-driven approaches to food categorisation has grown [2,9]. Foods vary widely in their nutrient content, and the ability to systematically categorise them based on nutritional similarities can have broad applications, from personalised nutrition recommendations to health-focused food labelling.

Machine learning (ML) methods offer a powerful toolkit for analysing large and complex datasets [7] and have been increasingly applied in fields such as engineering, bioinformatics, genomics, and health sciences [6,12]. In the context of food analysis, clustering techniques in particular can reveal hidden structures within data, allowing for the grouping of similar food items based on their nutritional composition [1]. These groupings can provide valuable insights for individuals seeking to plan balanced diets, or for healthcare professionals who design nutrition-based interventions.

In this study, we apply two ML techniques—principal component analysis (PCA) and k-nearest neighbors (k-NN)—to a dataset of 549 food items, with the goal of identifying clusters of foods that share similar nutrient profiles. PCA serves as a dimensionality reduction technique, allowing us to visualise the data in two dimensions while preserving the most significant variance in the dataset. k-NN clustering is then used to classify the food items into groups based on their proximity in the reduced feature space. By identifying natural clusters of foods, our method offers a data-driven approach to food categorisation that could be used for customised dietary suggestions, or to guide health-conscious food ingredient labelling.

This work aims to explore the potential of these ML methods to categorise foods in a way that is not solely dependent on traditional classifications, such as calorie content or macronutrient balance, but instead takes into account the entire nutritional profile. Ultimately, the findings from this work can contribute to ongoing efforts in personalised nutrition and diet optimisation, laying the groundwork for future enhancements using more advanced ML techniques such as domain adaptation and transfer learning.

Literature review

In recent years, nutritional epidemiology has increasingly embraced advanced multivariate techniques to analyse complex dietary datasets, providing new insights into

[a]marialaishram3@gmail.com, [b]chingriyo@gmail.com, [c]satisji@gmail.com

DOI: 10.1201/9781003675235-16

dietary patterns and their health implications. Traditional approaches, which focus on individual nutrients or foods, are being supplemented by more holistic methods that capture the synergies in how foods are consumed. Statistical techniques such as principal component analysis (PCA) cluster analysis have been adopted to reveal hidden dietary patterns.

To cite a few examples, Fransen et al. [3] emphasised the importance of validation in PCA-based dietary studies, demonstrating that appropriate component selection ensures meaningful food groupings. Maugeri et al. [4] further highlighted the value of integrating PCA with clustering techniques, showing that combining hierarchical and k-means clustering enhances the robustness of dietary pattern analysis. Also, Schulz et al. [8] classified analysis methods into hypothesis-based, exploratory, and hybrid approaches, noting that exploratory techniques like PCA and clustering are particularly effective for large datasets.

Solans et al. [10] introduced compositional PCA as an alternative method for deriving dietary patterns, offering a more balanced perspective on food intake behaviours. Zhang et al. [12] compared PCA with principal balances analysis (PBA), demonstrating that PBA improves interpretability and links dietary patterns to health outcomes, such as a reduced risk of hypertension. Papastratis et al. [5] extended this work by applying AI-driven generative models for personalised nutrition planning, marking a shift towards machine learning-based dietary recommendations.

Our study builds on this body of work by applying PCA for dimensionality reduction and k-nearest neighbors (k-NN) for clustering foods based on their nutritional profiles. PCA, by reducing the high-dimensional data into two principal components, allows us to visualise the clusters in a lower dimensions while preserving the essential variance of the dataset. Unlike some of the more sophisticated methods discussed in the literature, we focused on these well-established techniques to maintain clarity and interpretability, given the exploratory nature of our analysis.

Aims and objectives

The study aims to apply ML techniques to identify clusters of foods based on their nutritional profiles. Specifically, the objectives are to:

- apply PCA to reduce the dimensionality of the food item dataset, and visualise patterns in two dimensions.
- use k-NN clustering to group foods by nutritional similarity.
- analyse correlations among nutrients to uncover key relationships.
- visualise a similarity matrix to validate clusters and propose potential food substitutions for diet planning.

Novelty of the study

This study uniquely applies PCA and k-NN to a large nutritional dataset to explore natural food clusters based on comprehensive nutrient profiles. Unlike traditional approaches that focus on isolated nutrients or predefined groups, we adopted a data-driven clustering method, supported by a food similarity matrix to highlight nutritional equivalencies, offering fresh insights for personalised diet recommendations.

Methodology

Dataset description

The dataset comprises nutritional profiles for 549 food items, detailing both macronutrient (carbohydrates, proteins, fats, fibre) and micronutrient (vitamins, minerals) composition. It was sourced from a public domain dataset hosted at Kaggle [11]. Out of 34 variables, calorific value and nutritional density were excluded, as response variables and were not used as features for clustering. The list of nutrients used as features for the classification training were as follows: 1. Fat, 2. Saturated fats, 3. Monosaturated fats, 4. Polyunsaturated fats, 5. Carbohydrates, 6. Sugars, 7. Protein, 8. Dietary fibre, 9. Cholesterol, 10. Sodium, 11. Moisture content, 12. Vitamin A, 13. Vitamin B1, 14. Vitamin B11, 15. Vitamin B12, 16. Vitamin B2, 17. Vitamin B3, 18. Vitamin B5, 19. Vitamin B6, 20. Vitamin C, 21. Vitamin D, 22. Vitamin E, 23. Vitamin K, 24. Calcium, 25. Copper, 26. Iron, 27. Magnesium, 28. Manganese, 29. Phosphorus, 30. Potassium, 31. Selenium, 32. Zinc.

Data pre-processing

There were a few entries with faulty entries of zeros. They were removed from the dataset. Also, probably due to data entry errors, a few extreme outliers (very large z values) were observed and removed, resulting in 549 usable food items from the original dataset. So, less than 5% of the total dataset were removed during pre-processing of the dataset. Further, to improve computational integrity, the features were scaled through column-wise normalisation to prevent larger-ranged features from dominating the analysis.

Correlation analysis

Firstly, the correlations between different nutritional components were conducted to understand their relationships before applying ML techniques. Understanding these correlations is essential for proper feature selection and dimensionality reduction, as highly correlated features can affect the performance of clustering algorithms. In this study, a correlation matrix was computed by using the Pearson correlation coefficient for the 32 nutritional attributes of food items, where values closer to 1 indicate a strong positive correlation, and values closer to -1

indicate a strong negative correlation. It was computed using the function '*corrcoef*' in MATLAB 2018a.

Principal component analysis (PCA)

The dataset, consisting of 549 food types with 32 different nutritional attributes, was transformed into a lower-dimensional space while retaining the most significant variation in the data. It was projected onto two principal components (PC1 and PC2), retaining the most significant variance. This facilitated visualisation of food groupings based on their nutritional content.

k-nearest neighbor (k-NN) clustering

k-NN clustering was applied to the PCA-reduced data. Different values of k were tested to group foods by nutritional similarity. Euclidean distance was used to measure the proximity of food items in the reduced feature space. Further, Davies-Boulding index was computed to assess the quality of clustering.

Results

Correlation matrix

Fats: Saturated fats, monounsaturated fats, and polyunsaturated fats show a relatively high correlation with total fat content (0.8 to 0.9 range), indicating that as the total fat content increases, these types of fats tend to increase as well (Figure 16.1). This is expected, as these are subcategories of total fat.

Sugars and carbohydrates: Carbohydrates are moderately correlated with sugars (around 0.6–0.7). This makes sense as sugars are a significant component of total carbohydrates. However, dietary fibre is less correlated with sugars, suggesting that foods high in carbohydrates may not necessarily be high in sugars but could be rich in other carbohydrate forms, such as fibre.

Protein: Protein shows weaker correlations with most other attributes, except for some minerals (like phosphorus and magnesium). This indicates that protein content varies independently from many other macronutrients like fats or carbohydrates.

Vitamins: Vitamins, particularly the B vitamins (Vitamin B1, B2, B6), show some internal correlations, likely because foods rich in one B vitamin tend to contain others. Interestingly, Vitamins A and C do not show strong correlations with these, reflecting that their sources in food can vary significantly.

Minerals: Calcium, magnesium, and phosphorus show moderate to strong correlations (around 0.5–0.7), indicating that foods rich in one of these minerals may likely contain others. Potassium and sodium, despite both being electrolytes, show a weak correlation, which is a reflection of dietary sources differing for these minerals.

Sodium and cholesterol: Interestingly, sodium does not show strong correlations with fat or cholesterol levels, suggesting that foods high in sodium are not necessarily high in fats or cholesterol. This has dietary implications as it shows sodium-heavy processed foods may differ from those rich in fats and cholesterol.

In addition to these nutritional insights, it also implied that there are at least some dependent relationships among the nutrients. Based on this, PCA was applied to eliminate such redundant features for further attempt to cluster the foods.

PCA analysis

Figure 16.2 shows the variance explained by the principal components. It can be observed that the first two components explain for approximately 38% of the variance, which further consolidates that there are

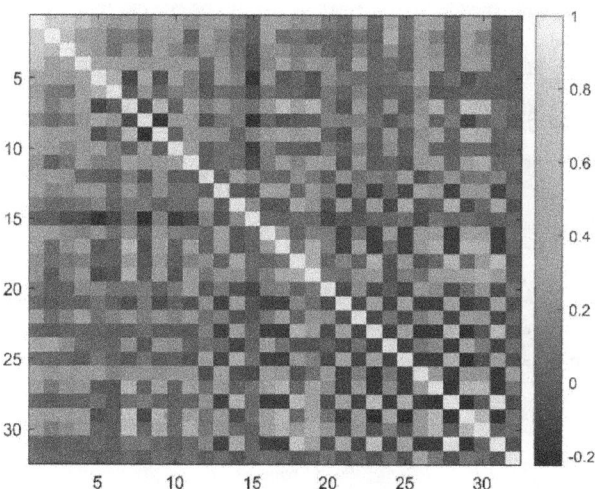

Figure 16.1 Correlation matrix of the nutrient attributes
Source: Author

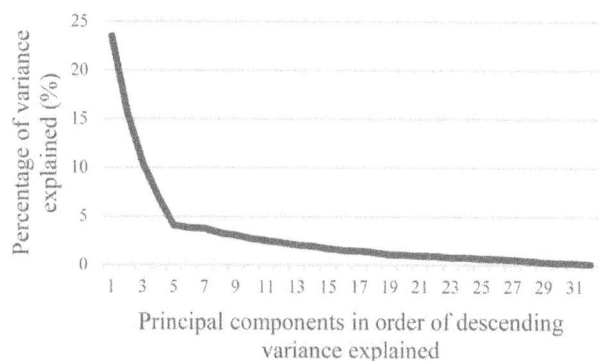

Figure 16.2 Variance explained by the principal components
Source: Author

significantly linear dependencies between the observed set of nutrient attributes.

The dimensionality reduction through application of PCA allowed us to visualise the distribution of food items in a two-dimensional space (Figure 16.3), where nutritionally similar foods clustered closely together. Separate clusters formed for different food categories, indicating underlying patterns in the nutrient profiles.

However, several food items were positioned far from the dense clusters, indicating unique nutritional profiles. These could be specialty food items that have unusually high or low concentrations of specific nutrients, such as oils, which are high in fats, or fortified foods, which are high in specific vitamins or minerals.

k-NN clustering
After applying PCA, the reduced-dimensional dataset was used for k-NN clustering. The goal of this clustering was to group food items based on similarity in their nutritional content. Different values of 'k' were tested to optimise the clustering, and Euclidean distance was used as the distance metric. The clusters were then mapped back to the PCA space for visualisation, as shown in Figure 16.3. Here, it was observed that 9–10 clusters yielded the lowest Davies-Bouldin index of around 1.5, and therefore the clustering was presented for 10 clusters.

It can be observed that classification algorithm successfully identified several well-defined clusters in the PCA-reduced space. These clusters correspond to food types with similar nutritional compositions. For example, one cluster may represent protein-rich foods, such as meat and legumes, while another may represent carbohydrate-dense foods, like grains and cereals.

Food similarity matrix
The above findings are indicative of similarity among the food items at least in the principal component space, if not in nutritional space. To consolidate on this, a food similarity index was computed between all pairs of the food entries by calculating the Euclidean distance between their coordinates in the PC1 and PC2 space. This provides a measure of how close or similar are the food items to each other.

Figure 16.4 shows the food similarity matrix thus calculated. The diagonal entries are the self-computed distances, and therefore are of zero values (similarity to self). It can be observed that while there are a few exceptionally large Euclidean distances between the food item pairs, there are a few food item pairs which show really low values. The groups thus created by the collection of these lowly distanced food items i.e., highly similar food items should be replaceable with each other when observed from their overall nutritional profiles.

On the other hand, those food items which exhibit very low similarity could be those food groups which have specific nutritional characteristics. For example, entries 305 and 306, which exhibited overall low similarities, were cheeseburgers from Burger King and McDonald's, respectively. It is interesting to observe that while they show low similarity with other food items, they share high similarity between themselves. Such observations consolidate the concept of food similarity index thus calculated here.

Implications for food science
First, the ability to group food items based on their nutritional composition can offer new insights into food categorisation beyond traditional methods. By relying on nutritional profiles rather than predefined food groups, we can uncover unexpected relationships between foods

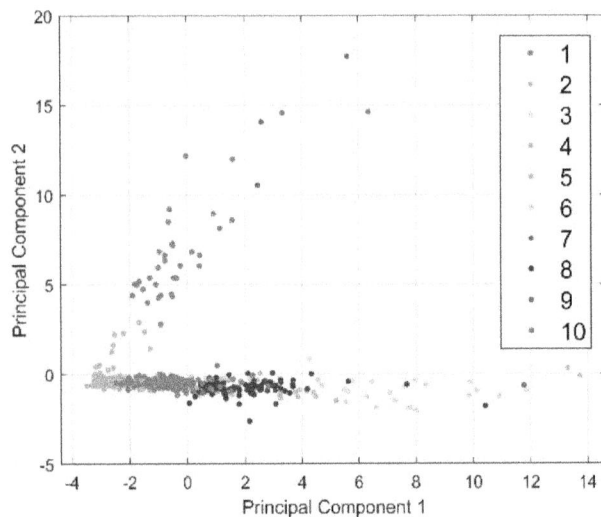

Figure 16.3 PCA cluster plotted against the first two principal components
Source: Author

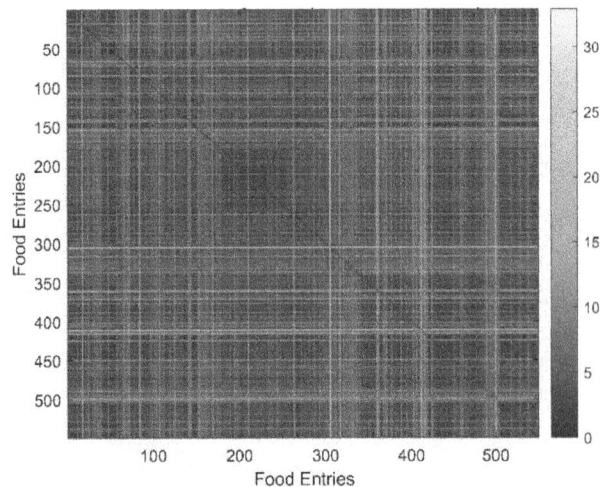

Figure 16.4 Food similarity matrix
Source: Author

that might not be apparent in conventional categories. For example, the clustering results suggest that certain foods typically considered dissimilar, such as nuts and certain fish, might share similar nutritional profiles, potentially offering interchangeable roles in diets based on their nutrient content.

The use of PCA to reduce dimensionality also allows us to focus on the most significant nutritional variables, filtering out less impactful factors. This approach can help researchers and dietitians better understand which nutrients are driving the differences between various food groups. In particular, the separation along the first two principal components indicates that macronutrients like fats, proteins, and carbohydrates are the most significant factors in distinguishing food groups, while micronutrients like vitamins and minerals also play a supporting role.

From a practical perspective, these findings can be used to develop a food similarity index (FSI) for personalised nutrition, where people may need to swap certain food items due to allergies, dietary restrictions, or personal preferences. By clustering foods with similar nutritional profiles, consumers can maintain a balanced diet even when making substitutions.

However, care should be taken during interpretation of these results since there is a lack of standardisation in food preparation methods that inevitably introduces uncertainty into the nutritional data, which may impact the accuracy of clustering results. Different preparation techniques (e.g., raw, cooked, or processed) can significantly alter the nutrient composition of food items, potentially distorting the true nutritional similarities between them. Future studies could address this issue by incorporating probabilistic approaches, such as Bayesian inference, to model the uncertainty in the reported nutrient values. A Bayesian framework could estimate the true nutrient distribution for each food item, allowing for more robust clustering results that account for variation in preparation methods.

Implications for machine learning

From a ML perspective, this study showcases the effectiveness of dimensionality reduction and clustering in analysing complex, high-dimensional datasets. By using PCA, we were able to condense a 32-dimensional nutritional dataset into two principal components while still retaining a significant amount of the original variance. This demonstrates PCA's utility as a pre-processing step, especially when working with datasets where features are highly correlated or redundant.

The application of k-NN clustering on the PCA-reduced data highlights the importance of feature selection and reduction for improving the performance of unsupervised learning algorithms. Additionally, the approach used here can be transferred to local food types. This would provide valuable practical values in the development of alternative dietary patterns for local food recipes even when limited by location and seasonal availability.

Conclusion

In this study, PCA and k-NN clustering to a dataset containing the nutritional profiles of 549 food items. By using PCA, we reduced the dimensionality of the data from 32 nutritional features to a few key components, preserving most of the variance. This simplification allowed us to apply k-NN clustering more effectively, identifying distinct groupings based on nutrient similarities. The clustering results suggest the potential for developing a food similarity index, which could help guide dietary choices by offering nutritionally similar alternatives.

References

[1] Aguilar O., Ojeda D. L., Aguirre A., Gonzalez P.A. (2022). A decision tree-based classifier to provide nutritional plans recommendations. In The 17th Iberian Conference on Informations Systems and Technologies (CISTI), Madrid, Spain, 1–6.

[2] Bailey, R. L., Mitchell, D. C., Miller, C., Smiciklas-Wright, H. (2007). Assessing the effect of underreporting energy intake on dietary patterns and weight status. Journal of the American Dietetic Association, 107(1), 64–71.

[3] Fransen, H. P., May, A. M., Stricker, M. D., Boer, J. M., Hennig, C., Rosseel, Y., and Beulens, J. W. (2014). A posteriori dietary patterns: how many patterns to retain? The Journal of Nutrition, 144(8), 1274–1282.

[4] Maugeri, A., Barchitta, M., Favara, G., La Mastra, C., La Rosa, M. C., Magnano San Lio, R., Agodi, A. (2022). The application of clustering on principal components for nutritional epidemiology: a workflow to derive dietary patterns. Nutrients, 15(1), 195.

[5] Papastratis, I., Konstantinidis, D., Daras, P., Dimitropoulos, K. (2024). AI nutrition recommendation using a deep generative model and ChatGPT. Scientific Reports, 14(1), 14620.

[6] Salam, S., Verma, T. N. (2020). Identifying empirically important variables in IC engine operation through redundancy analysis. In Advances in Applied Mechanical Engineering: Select Proceedings of ICAMER 2019. Springer Singapore, 45–51.

[7] Salam, S., Verma, T. N. (2021). An empirical investigation on the influence of operating conditions on internal combustion engine behavior and their graded significance. Energy Sources, Part A: Recovery, Utilization, and Environmental Effects, 1–19.

[8] Schulz, C. A., Oluwagbemigun, K., Nöthlings, U. (2021). Advances in dietary pattern analysis in nutritional epidemiology. European Journal of Nutrition, 1–16.

[9] Sogari G., Argumedo V., Gomez M., Mora C. (2018). A study using an ecological model for healthy behaviour. Nutrients, 10(12), 1823.

[10] Solans, M., Coenders, G., Marcos-Gragera, R., Castelló, A., Gràcia-Lavedan, E., Benavente, Y., Moreno,

V., Pérez-Gómez, B., Amiano, P., Fernández-Villa, T., Guevara, M. (2019). Compositional analysis of dietary patterns. Statistical Methods in Medical Research, 28(9), 2834–2847.

[11] Utsav Dey. (2024). Food Nutrition Dataset [Data set]. Kaggle. https://doi.org/10.34740/KAGGLE/DSV/8820139.

[12] Zhang, F., Tapera, T. M., Gou, J. (2018). Application of a new dietary pattern analysis method in nutritional epidemiology. BMC Medical Research Methodology, 18, 1–10.

17 Estimating crop yields using machine learning

Shrungashri Chaudhary[a] and N Kishorjit[b]

PhD Scholer, Assistant Prof. IIIT Manipur, India

Abstract

The goal of this research is to have crop prediction for distinct crops in Manipur and, with the help of this research and advancements in technology, we can bring a positive drastic change in the production of the crops annually. Therefore, the aim of this research is to identify top five crops that have the highest yield production in Manipur and, using our research knowledge, we can increase their production considerably. The objective is to identify the right time and right season during which the crops can be sowed and harvested. Timely sowing and harvesting can be a game-changer for farmers, which, in turn, will save our farmers from unpredictable losses and which will assure for successful crop yield prediction. A plethora of research has been conducted on crop yield prediction. The most commonly used prediction models are ML models, such as Linear regression model, Decision tree, SVM, and Random Forest, which have performed well in terms of accuracy. There are also Deep learning models are available, such as LSTM model. Various evaluation metrics, such as RSME, R-squared, and MAE, are used to quantify the accuracy of these prediction models.

Keywords: Random forest (RF), SVM, decision tree (DT), R2, RMSE, MAE

Introduction

Agriculture is the basic source of income for most states in India. Nearly 70% of the population is dependent on agricultural income for their survival. Neglect in the agricultural research domain, farmers' suicides due to untimely rainfall, and alarming climatic conditions have made the agricultural domain critically important. There are also factors affecting the crop production, including rainfall, temperature, soil health, season, and socio-economic factors. Therefore, it becomes crucial for farmers to timely sow and harvest their crops to save them from huge losses. With proper research knowledge, predictions can be done for timely sowing and harvesting of crops, which will save farmers from untimely climatic condition consequences.

There are many prediction models available for crop prediction which will help farmers to know when to timely sow and harvest their crops. Prediction models, such as machine learning models like Linear Regression, ANN, Decision Tree, Random Forest, Support Vector Machines, are used for crop yield estimation. According to many researches in machine learning models with specific amount of data, Random Forest has outperformed other models. Deep learning models, such as LSTM, can also use used. When there is more complex data available, the LSTM model has the highest accuracy in predicting the events compared to machine learning models. Deep learning models are designed to handle where the data is huge and complex. The aim of LSTM is to predict the event based on the learned patterns the LSTM has been trained on. LSTM can retain memory for large amounts of data due to LSTM's ability for long-term dependencies. CNN is also one of the Deep Learning models which take images or videos data as the input. Feature selection and feature enhancement techniques are applied after which images are trained for pattern recognition and algorithm is trained to learn about the cropping pattern which add to the more accuracy of the results. Based on the available data (such as temperature, soil type, rainfall, socio-economic factors), machine learning models are the best for this research for estimating crop production.

Literature review

The paper proposed by Dr. Namita Kale et al. [7] introduced the idea of estimation for crop yield using deep learning and satellite imagery techniques. Deep learning algorithms like CNN and RNN are used to extract data and analyse spatial and temporal information. The use of the Tweak Chick swarm Optimization method (TCSO) and VGG models outperformed various models in terms of probable accuracy. It closely observes the health of crop growth and soil with the help of satellite imaginary obtained from Landsat, Sentinel, and MODIS in real time. Crop yield data is gathered from filed surveys or remote sensing data or imaginary data, then preprocessing is done to normalise the data. Remote sensing data were translated into histograms, which in turn were translated into pixel intensities, which served as a model input. CONV3d, CNN, and LSTM were employed to learn about the temporal information, with the ReLU activation function. The TensorFlow library was used to implement the deep learning network. After this, feature selection from the pre-processed data is done using TCSO method. Using discrete hybrid deep belief network with combination of VGG NET classifies the rank and categorise the

[a]shrungashri@iiitmanipur.ac.in, [b]kishorjit@iiitmanipur.ac.in

DOI: 10.1201/9781003675235-17

crop based on yield and this information is used to train the algorithm on a labelled dataset. To increase prediction, more elements like meteorological information, soil information, and previous crop yield data are added while training the model. This model helps in increasing the labelled training data.

The paper proposed by Kavita Jhajhariaa et al. [1] discusses prediction of crop yield using ML techniques for five identified crops. Algorithms used were RF, Support Vector Model, Gradient descent, LSTM, and Lasso regression technique. Among all, the RF model performed better and the results were cross-validated using RSME and MAE methods. This paper intended using these algorithms helped farmers to have better understanding about crop selection. Most commonly used approaches are multivariant regression, DT decision tree, association rule mining, and ANN. Data retrieval was done from the government official website, acquiring a total of 3447 rows and 7 columns of data. Yield, soil type, and rainfall were taken as independent variables. The unique soil characteristics of each district were identified. After which, soil type was taken as categorical data, and soil data created as a dummy variables. Later, rainfall data was incorporated to the dataset, by identifying the amount of precipitation for particular crop, and by calculating the mean rainfall value for the months corresponding to that season. Precipitation during July, August, September, and October was considered for crops in the kharif season. Data (yield, soil type, rainfall) and variables (state, district, area, production, months, annual total) were pre-processed to transform raw data into a flawless dataset. Data encoding: Dataset have two types of variables: continuous and categorical variables. The categorical data been encoded using methods like label encoder or one-hot encoder, creating dummy variables. Next, standardisation of dataset using StandarScaler from sciKit Library is applied to the dataset to get desired output.

End steps that is large number of training sets are taken using the SciKit-learn library, and import of train-test-split module (it splits the data) with test set of 2% and random state set to 71%, thereby giving most accurate possible result. The analysis showed the water intensive crops, i.e., depending on rainfall, has a dip in production, while other crops with the help of irrigation showed better results in the production with respect to rainfall and soil type. Each crop was examined separately. For Bajra, this crop has been consistent in its production since 1997–2019. This can be factored due to better seasonal rainfall, modern irrigation methods, better fertiliser, and better production techniques. For wheat, there was a drop from 1997 to 2003, and the area gradually started increasing its production in 2019. The conclusion tells us that the area under production has increased, but the area under the cultivation is consistent or not increased due to limited land resources. Random forest model outperformed deep learning Gradient Descent and LSTM model in terms of accuracy. As Random Forest can perform better on specific parameters, Gradient Descent and Lasso regression required large amount of data.

Research work proposed by P.Priya*1, U.Muthaiah in this paper authors have proposed the idea of crop yield prediction using supervised machine learning algorithm that is Random Forest. In this proposed idea, the model is trained using Random Forest algorithm until the desired level of accuracy is achieved on the training data. The dataset used includes precipitation, production, temperature, and rainfall, which is a collection of decision trees. These decision trees are applied on remaining data users for accurate classification.

Dataset used in research are the records from Indian government. Parameters taken are rainfall, kharif and rabi season, maximum temperature, crop production in tons. RF is used to estimate results, each tree is trained by using a subset of training data. Trees are been trained on the subset which are being selected at random. The algorithm converts the data into csv file and load those datasets and splits the loaded data into training data and test data as 67 and 33 percentage. Separate the training data by class. Then calculate mean standard deviation for required tuple and summarize the datasets. Now compare the summarize data list and the original data set and calculate the probability. The largest probability is taken as prediction. Therefore, RF algorithm was used to estimate the accurate crop production yield with most accurate result as an output.

Saeed Khakil*, Lizhi Wang in their published paper they have proposed crop yield prediction for soyabean and corn yield in the USA using historical data of 2016, 2017 and 2018 and achieved RMSE: 9% and 8% respectively. This paper presented a deep learning framework model using CNN and RNN for crop yield prediction along with Random Forest, Deep network fully connected neural network and LASSO. CNN - RNN model were used to capture time dependencies and genetic improvement of seeds. For accuracy in prediction accuracy, weather prediction, soil condition and management practices were used. This paper incorporated the used of hybrid CNN-RNN. CNN process data (ie. Signals and sequences, images and videos) having multiple pooling and convolutional layers with few fully connected dense layer. Number of filters, padding and strides framework is designed. RNN is used to capture time dependencies of sequential data as RNN store the history of past element in sequential order which proved to be very beneficial while traning the model. To remove the vanishing gradient problem LSTM RNN was used.

W-CC and S-CNN used to capture the linear and non-linear effect of weather and soil data respectively.

After the pooling the output of S-CNN is given to fully connected layer that has 40 neurons. RNN is given 5 years yield dependencies. LSTM RNN has 64 hidden units. All weights were provided with Xavier method with stochastic gradient descent. Adam optimizer were used with learning rate 0.03%. ReLU actication function was used for CNN and FC layer. The output layer used linear activation function. The project was implemented in python using tensor flow library. Hybrid CNN and RNN outperformed other models like Random Forest algorithm.

Pre-processing

Dataset

The First thing that we will need is to collect data for our research from different sources. Therefore, data was collected from different government offices like Agricultural dept. of Manipur, ICAR dept, Horticulture dept. of Manipur, also many governmental websites were used to get different data like rainfall, soil type and temperature. There are two forms of data they are independent data and dependent data. Dependent data is that data which we want to get predicted like Production. Production field will be used as the dependent variable and independent variables like area, season, whole year, rainfall, temperature are the variables on which the Dependent variable relies.

Data encoding

The next step is to create a flawless data. Now, replace the missing values with the mean values. Then, convert the categorical data (Season column) into the numerical form. This is called as Data encoding. In this, we have used Label encoding. For example, Karif stands for 1, Rabi stands for 2, and so on. Once we have encoded season column with numbers, there is no need of month column, so we just delete it from our dataset. We also do not need district and state columns, so we will also delete both these columns from our dataset.

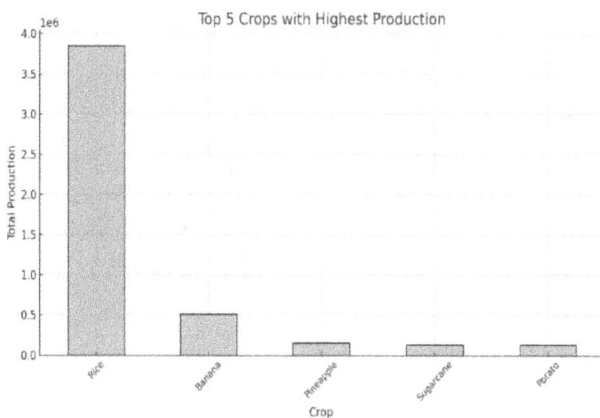

Figure 17.1 Decision tree

Source: Author

After data encoding, we will identify top five crops produced in Manipur, irrespective of their district. With the help of statistical analysis, we found that Rice, Banana, Pineapple, Sugarcane, and potato are the most produced crops, respectively.

Now, we have area, rainfall, and temperature as independent variables. With the help of Year and Rainfall columns, we identify the average rainfall for each season, i.e., June to September, October to December, and January to March. In this way, there is rainfall data available for that particular year and season, indicating how much rainfall has taken place. Using statistical analysis, we find that the rainfall and temperature parameters were almost similar, and there was no drastic difference in the Rainfall parameter.

Standardisation of features

In this process, all the features are brought to same scale range, after which the data is ready to be given as an input to machine learning models. It is also called as data normalisation. Feature scaling is applied to the independent variables. Now, the feature scaling is fit to both training and testing data.

Now import libraries like pandas, numpy and scikitlearn for basic functioning. Saranya et. al [4].

Reading the .csv file (Manipurdataset.csv), use *rainfall* as the independent variable, removing District and State columns from the data frame. Select the top five most produced crops from the lists over the past 10 years. Finally, split the data into testing and training data frames.

Models

Decision tree

Decision trees are used for predicting the class of a given dataset.

The decision tree regression model was applied to predict the 'Production' column. Here are the evaluation metrics:

- MSE: 5.22%
- R^2: 88.94% (accuracy)

This indicates that the model explains approximately 88.94% of the variance in the data.

Random forest

The RF regression model was applied to predict the 'Production' column. Here are the evaluation metrics:

- MSE: 3.78%
- R^2: 91.99%

The Mean Squared Error (MSE) for the Random Forest model corresponds to approximately 3.78% of the squared mean of the actual production values.

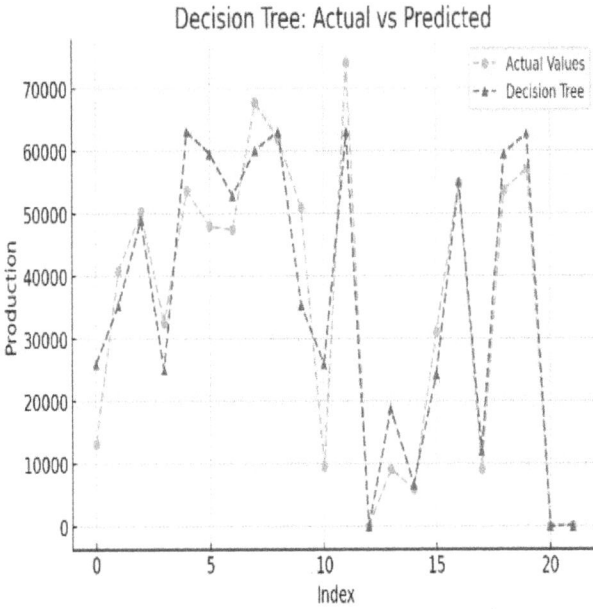

Figure 17.2 Decision tree

Source: Author

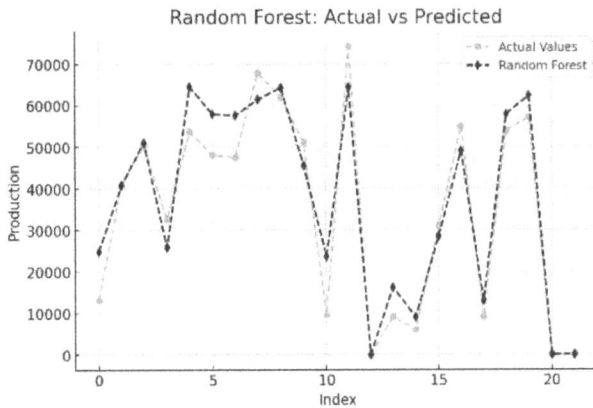

Figure 17.3 Random forest

Source: Author

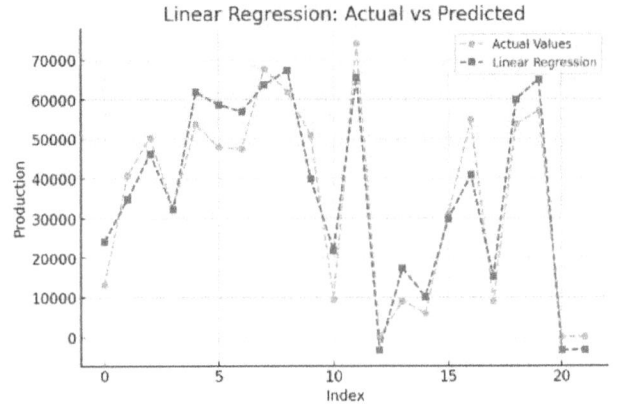

Figure 17.4 Linear regression

Source: Author

Table 17.1 Optimal parameter resulted from grid search.

Models	R-squared	MSE
Decision tree	88.94%	5.22%
Random forest	91.99%	3.78%
Linear regression	89.9%	21.88%
Linear regression	89.9%	21.88%

Source: Author

This further confirms the model's improved performance compared to the decision tree model.

Linear regression model

The linear regression model was applied with 'Production' as the dependent variable. The evaluation metrics are as follows:

- MSE: 21.88% (the model's predictions deviate by about 21.88% from the actual production values)
- R^2: 89.9% (which indicates that about 89.9% of variation)

This suggests that the model performs well in predicting production based on crop year, area, and season.

*R*2, MSE, and MAE

These are evaluation matrices used to quantify the accuracy of the predictive models.

*R*2 method

The *R*-squared method is used to see how much close our predicted results are to the actual values. This method tells us how different models are performing and indicates the accuracy of the models.

MSE: Mean squared error tells us the error percentage. This method tells us how our predicted values are far from the actual values.

Conclusion

Based on the results from models, we came to the following conclusions. Linear regression shows much difference between actual and predicted values. It may due to its inability to capture complex patterns well. Linear regression is not ideal for datasets where there are nonlinear dependencies present. Therefore, the prediction deviated significantly in some cases.

In matters of computational efficiency, Decision tree can be used. Decision tree is better than Linear regression but less stable than Random forest due to overfitting tendencies. Decision tree captures non-linearity better than Linear regression. But it may overfit the training data which may cause sharp fluctuations in prediction. However, it is more flexible compared to Linear regression in terms of accuracy.

For RF, it provides the best balance, with smoother predictions that closely align with actual values. RF has outperformed both other two models in terms of prediction accuracy. It is the average of multiple decision trees, which improves prediction accuracy and reduces overfitting. With the least error and greater accuracy, RF model is more reliable than the other two models. Therefore, Random forest is the most accurate model for estimating production, as it captures complex patterns better than the other two models. The Linear regression model is recommended only when the relationship between the data is linear, as it does not capture variations effectively.

Therefore, in terms of increase produce, the study has found that rice is grown here in all seasons like Rabi, Kharif, and Autum. However, when data was segregated seasonally, it was found that Kharif season is best for rice cultivation, as it has the highest rice cultivation. The observation states that factors like rainfall, temperature, and soil type were constant, with no drastic change in these factors. The production was mostly affected due to socio-economic factors, such as changes in methods of cultivation, more use of fertilisers, agricultural awareness programme by government, timely research, various governmental schemes for agriculture, subsidise support, and so on. These factors were the reason for major positive impact on agricultural output.

There are also deep learning models which can alleviate the accuracy of the models. For example, if Remote Sensing Data is used, it can be used to study how the climate change is affecting the cropping patterns and what measures can be taken so that the farmers can be prepared in future for dire consequences of upcoming climate change challenges. The accuracy of the models majorly depends on how large and flawless the dataset can be. Therefore, estimating prediction can directly help the stakeholders to take calculative decision about their investments, and save them from huge losses and debt.

References

[1] Kavita, Ms, Pratistha Mathur. (2020) Crop yield estimation in India using machine learning. In 2020 IEEE 5th International Conference on Computing Communication and Automation (ICCCA), 220–224. doi:10.1109/ICCCA49541.2020.9250915.

[2] Vaishali Kadway, Research Scholar, Barkatullah University, Bhopal, India. Sanjeev Gour, Associate Professor, Career College, Bhopal, India. Swati Namdev Assistant Professor, Career College, Bhopal, India. "Analysis and prediction of crop production using machine learning techniques"

[3] Saeed Khaki1, Lizhi Wang, Sotirios V. Archontoulis. A CNN-RNN framework for crop yield prediction. Original Research published: 24 January 2020. doi: 10.3389/fpls.2019.01750.

[4] Saranya, C. P., Guru Murthy, B., Karuppasamy, M., Sunmathi, M., Shree Sakthi Keerthna, S. (March 2020). Coimbatore Institute of Engineering and Technology, Coimbatore, A survey on crop yield prediction using machine learning algorithms" 2020 IJRAR.

[5] Priya, P., Muthaiah, U., Balamurugan, M. (April 2018). Predicting yield of the crop using machine learning algorithm. *International Journal of Engineering Sciences & Research Technology*". ISSN: 2277-9655.

[6] SML Venkata Narasimhamurthy, AVS Pavan Kumar. Rice crop yield forecasting using random forest algorithm. *International Journal for Research in Applied Science & Engineering Technology (IJRASET)*. ISSN: 2321-9653; IC Value: 45.98; SJ.

[7] Dr. Namita Kale, Dr. S. N Gunjal, Dr. Manoj Bhalerao, Dr. H. E. Khoxdke, Santosh Gore, Dr. B. J. Dange. (25/07/2023). Crop yield estimation using deep learning and satellite imagery. *International Journal of Intelligent Systems and Applications in Engineering*.

18 Detection of sensitive information using CRF in unstructured text

Longjam Velentina Devi[1,a] and Navanath Saharia[2,b]

[1]Research Scholar, CSE Department, Indian Institute Information and Technology, Manipur, India

[2]Assistant Professor, CSE Department, Indian Institute Information and Technology, Manipur, India

Abstract

This work presents a method for detecting sensitive information in text data using a CRF model. This strategy detects sensitive information word by word, as well as the links between them, as opposed to the traditional method, which detects sensitive information at the sentence or text level. This approach can be used for privacy purposes, where security needs to be implemented selectively to the sensitive data rather than the entire dataset. By applying this method, data can be made available to researchers while limiting access to sensitive information. The proposed model achieves a precision score of 0.98 and an accuracy of 98%, while identifying sensitive information setting a benchmark in the area of privacy preservation.

Keywords: Privacy, security, CRF, sensitive information, machine learning

Introduction

Sensitive information detection is one of the most important concern area where there is full of personal sensitive information content such as healthcare, finance, and personal privacy, so as to safeguard those information from the unauthorised access and breaches, ensuring compliance with regulations such as GDPR Regulation [1] and HIPAA Gostin et al [2], while maintaining trust in digital systems. Protected health information (PHI), financial records, and personally identifiable information (PII) Kulkarni and Cauvery [3] can now be more accurately identified across a variety of formats and platforms thanks to recent developments in artificial intelligence (AI) and machine learning processing. This approach will be enabling to use in an organisation for risk reduction and incident response in addition to increasing the effectiveness of data security measures. Integrating this methodology into the current security frameworks is essential for responding to changing threats and preserving strong defences against data breaches, as businesses depend more and more on digital solutions. This integration facilitates real-time monitoring and analysis, allowing organisations to swiftly respond to potential vulnerabilities and safeguard their sensitive information more effectively. A large amount of textual data is generated daily across social media, emails, blogs, and other communication platforms that offer opportunities and challenges. One of the most prominent issues in case of privacy preservation is to identify the correct sensitive information and further protecting sensitive data, such as private information, private company information, and other types of private material. In addition to improving security, putting thorough data classification and protection procedures into practice guarantees regulatory compliance, which builds trust with both stakeholders and clients.

This work introduces a new method for leveraging the CRF model to identify sensitive (S) and non-sensitive (N) words. Sequential dependencies were well captured by the CRF model, confirming its applicability for sequence labelling applications. The contextual features were essential because they made it possible for the model to efficiently use the information contained in the surrounding tokens. As the model is able to access the surrounding tokens, it is able to investigate the effect of each other. For example, in the context of 'Customer care number', 'Customer' may not be sensitive, but in the case of 'Customer number', it will be considered as a sensitive one. The contributions can be categorised in the following ways:

1. Word sensitive information detection: In contrast to the traditional approach, this method seeks to detect sensitivity at the word level as well, that concentrate solely on classifying complete texts or sentences. This method can be implemented in the area where there is more concern for securing only in some part of the text, instead of implementing security in the whole text, which is more of a tedious one.
2. Sensitive information detection in unstructured text: This method can detect sensitivity not only in the structured text but also from the unstructured text also.
3. Improved content security: The technique makes sure that non-sensitive content is still available while

[a]velentina@iiitmanipur.ac.in, [b]nsaharia@iiitmanipur.ac.in

DOI: 10.1201/9781003675235-18

preserving security by separating sensitive words, rather than identifying full sentences or documents.

This paper is structured as follows: Section 2 provides the review about the existing method, outlining their pros and cons. Section 3 explains the proposed method and presents the experimental results. Section 4 provides the conclusion of the paper.

Literature review

With the advancement in the technologies, privacy is becoming a big question as whenever a person tries to access any service, their personal private information is provided in return of those services they requested, so it is the responsibility of the application server to secure those information. Therefore, many organisations and institutions began giving their preferences in the protection of sensitive information from cyberattacks [4]. Traditional rule-based approaches, such as regular expressions, have commonly been used to identify personally identifiable information (PII) [3]. However, these methods often fail to detect the sensitivity due to its complexity and unpredictable nature of natural languages, leading to a gap in their research. Various machine learning techniques are being used to capture the unique characteristics of the sensitive information, providing and improvement in detecting sensitive information. To detect sensitive information in Electronic Health Records, for example, Zhang and Jiang [5] used machine learning. By utilising metadata engineering over 30 features, they achieved 99% accuracy, much outperforming conventional techniques. Similarly, Zalte and Shah [6] showed the promise of deep learning in medical text analysis by classifying sensitive clinical records using a BiLSTM model in conjunction with BERT, obtaining an accuracy of almost 93%.

Another sensitive information detection model, suggested by Neerbek (2020), Huo and Jiang [7], and Anand et al. (2023), performed better in capturing sentence level features and attention mechanisms.

Despite having specified some of the methods that maximise accuracy, the methods are only applicable to situations where structured data collection or other operating contexts or sentences are built. This eventually reduces their practical scope and makes it difficult to work in a situation where there is need to publish dataset for open access. For instance, passive datasets used for collaborative research or self-initiated projects, such as charismatic megafauna and coral reefs datasets, can be over-thinned out without compromising their usability by entire blocks of data extracts. In contrast, the method described here provides a wider approach by examining data at the word level, which allows the identification and remoteness of sensitive information more easy. As a result, only the sensitive words can be masked and the set can then

be transferred in its original form without being biased. Instead, such action will improve the safety as well as non-sensitive data accessibility and inclusion, which makes this approach suitable for use in occasion when a part of the data needs to be protected.

Proposed technique

In this experiment, we propose a Conditional Random Fields (CRF)-based approach to identify sensitive information from text data. In this example, CRF is used because it is specifically defined for sequence modelling that could scan on the text data and come up with patterns. This approach is different from traditional methods, which flag sensitive data only at the sentence or document level, rather addressing a bigger picture by breaking down every word of text in it and classifying whether it holds sensitive information. The aim is to serve as a privacy-preserving, context-aware, fine-grained sensitivity classification service for various use-cases (e.g. Data Privacy, Regulatory Compliance, Secure communication, and others) across different domains such as eHealthcare, Fintech, and Critical Infrastructure. Our approach is based on powerful CRF, which bring sequence level relationship and it works very well on NLP tasks. This section discusses the core components of the approach, including dataset preprocessing, model architecture, training procedures, and evaluation metrics, followed by an analysis of the results obtained from the experiment.

Data pre-processing and preparation

We have collected tweets from Twitter(X) using the Twitter API from 01/01/2001 to 31/12/2022, and are able to collect 102,429 tweets, out of which only 1,014 tweets contains sensitive data. We conducted a range of cleaning steps to get the data ready for analysis. Stop words and two letter word were removed, except for those found in email addresses to avoid extra noise. On the other hand, several parts of the text, which are punctuation marks, but have no relation with email address, were deleted, and these pre-processing tasks are done in different phases, such as: punctuation marks are replaced with space characters, which is a very similar technique used across the Language Processing Toolkit (Nltk).

Tokenisation: In this study, each tweet was reached using the Python's .split() method. This step resulted to the generation of tokenised sequences for each sentence's tokens accompanying its encoded labels. Similarly, the encoded labels corresponding to each tweet were tokenised for the purpose alignment.

Padding and truncation: All the tweets were padded or truncated based on the requirement to maintain uniformity in the input length.

Handling imbalanced data: Class imbalance was also a big concern during the validation dataset, and to balance the classes during validation, Random OverSampling was done. This ensured that sensitive labels were present and adequate in number during evaluation leading to reliable metrics. However, there was no trouble on how the training dataset was handled as it was left as it was so as to maintain the original fair distribution of both sensitive and non-sensitive data. The validation set was oversampled using RandomOverSampler from imbalanced-learn to achieve a better proportion of sensitive and non-sensitive labels. This is in the attempt to boost the model as much as possible from the evaluation point of view concerning the valid complaint about the minority class, and the result should be better and more broadly trustworthy.

Features extraction

In this experiment, a comprehensive set of features was engineered to capture both word-specific attributes and contextual information, ensuring that the model could effectively differentiate between sensitive and non-sensitive tokens. For each token in a sentence, several lexical features were extracted to represent its intrinsic properties. These features included the lowercase form of the token to normalise variations in capitalisation and capture case-insensitive patterns. Additionally, binary features were used to indicate whether a token was entirely in uppercase, title-cased, or numeric, as these attributes are often indicative of sensitive information, such as names, titles, or numerical identifiers. These word-specific features provided the foundation for understanding the basic characteristics of each token. Contextual information was another critical aspect of feature extraction in this experiment. The model incorporated information about the tokens immediately preceding and succeeding the current token. For example, the lowercase, uppercase, title-case, and numeric features of the preceding (-1:word.lower()) and succeeding ($+1$:word.lower()) tokens were included to provide a broader context for each word. This allowed the model to understand the relationships between neighbouring tokens, which is particularly important for detecting multi-word entities such as addresses or phone numbers. To account for positional significance within a sentence, special features were added to indicate the beginning of a sentence (BOS) and the end of a sentence (EOS). These markers helped the model distinguish tokens that appeared in unique sentence positions, which may carry specific semantic implications, such as the start of an address or the conclusion of a formal statement. By combining word-level attributes with contextual and positional features, the feature extraction process created a rich representation of the dataset for the CRF model. This enabled the model to learn complex patterns and dependencies within the text, ultimately improving its ability to accurately identify sensitive information at the word level. This comprehensive approach to feature extraction ensured that the CRF model

was equipped to handle the nuanced and varied nature of real-world textual data.

Model architecture

The model architecture for sensitive information detection consists of several interconnected components, designed to process raw tweets, extract meaningful features, and train a Conditional Random Fields (CRF) model for identifying sensitive data at the word level. In this setup, the CRF model is modified to give the output in the form of binary class, that is 'S' and 'N'.

As depicted in Figure 18.1, the process begins with the dataset consisting of tweets, which may contain sensitive information such as names, addresses, phone numbers, or other personal identifiers. Further, pre-processing steps, mentioned in *Data pre-processing and preparation* section, are implemented. After that, feature extraction is performed based on many features such as Word-level features, where attributes like lowercase transformation, capitalisation, numeric detection, and title case detection are extracted for each word. Contextual features, where information from the previous and next words in the sentence is included to capture dependencies between tokens. Positional features, where special markers, such as 'beginning of sentence' (BOS) and 'end of sentence' (EOS), are added to help the model understand the position of tokens within the text. The training and the validation dataset contain an imbalance dataset, as the number of sensitive words in a sentence is less as compared to the non-sensitive data. Therefore, we apply oversampling on the sensitive class of the validation set using RandomOverSampler. This provides appropriate representation of all classes and prevents the model from getting biased over dominate class (majority), so that it can predict under represented label as well. The CRF model [7] of the contextual features is trained based on raw (unannotated) token sequences, and predicts a binary label for each node in this sequence that indicates whether its corresponding original token was

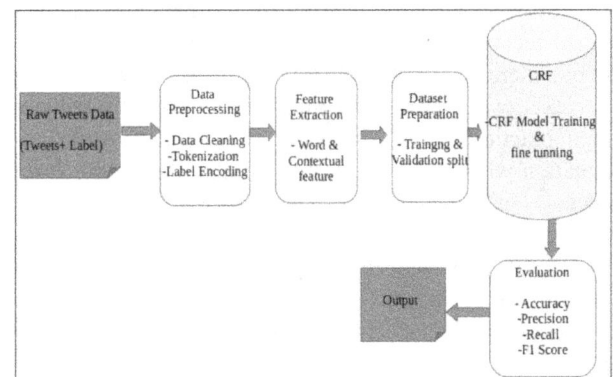

Figure 18.1 Overall data flow of the proposed method
Source: Author

sensitive or non-sensitive. The CRF model takes relation among tokens and features into account to capture this dependency in the data. It returns the result that could be sensitive and non-sensitive class.

Training and evaluation

The CRF model is trained using L-BFGS optimizer (Limited Memory Broyden Fletcher Goldfarb Shanno algorithm). This method is used to reduce memory usage and computational cost. Two regularisation parameters, c1 (L1 regularisation) and c2 (L2 regularisation), are tuned to prevent overfitting or making too dull/insensitive can be generalised. The above steps help create a trade-off between model complexity and accuracy, to avoid overfitting the training data. In the training, the model learns to classify each token as either 'S' (sensitive) or 'N' (non-sensitive) by leveraging dependencies between adjacent tokens. The dataset was performed using grid search method to find the best parameters, and the optimal parameters are shown in Table 18.1.

Performance evaluations for imbalanced class and after implementing Random oversampling: Model performance on the imbalanced dataset and results after random oversampling are provided in Tables 18.2 and 18.3. These results demonstrate the problems when working with imbalanced data, and how oversampling was able to counteract it in terms of detection accuracy, for better representation of minority class 'S'. In a first place, the model showed issues with class 'S', getting a precision of 0.77, recall of 0.70, and F1-score of 0.73, with an overall accuracy of 96%. Once oversampling is done for the

minority class, the model performance has increased in case of 'S'. Precision went up to 0.96, showing more false positives in the process, and recall also increased instead of going down: from 0.77 (in sensitive tokens) to 0.80, which means the ability for detecting true positive has improved too. As a consequence, a better balance between precision and recall was achieved making the F1-score to 0.88 Furthermore, the accuracy of the model increased to 98%, which indicates that dealing with imbalanced class distributions greatly contribute for better performance. This improvement is further reflected in the ROC curves (Figures 18.3 and Figure 18.4) for class 'S' and class 'N', respectively. Figure 18.3 demonstrates that after oversampling, the ROC-AUC for class 'S' reached 0.991, indicating a strong distinction between positive and negative samples. Similarly, Figure 18.4 presents the ROC-AUC of 0.991 for class 'N', confirming that the model effectively distinguishes this class as well.

Performance based on different regularisation values: Performance was evaluated based on three different values 0.01, 1.0, 10.0 of c1 and c2. Table 18.2 shows the performance result at different c1 and c2, while Figure 18.2

Table 18.2 Models performance on imbalanced class.

Parameter name	Value
Accuracy of the model	96%
Precision for 'S' Class	0.77
Recall for 'S' Class	0.70
F1-Score for 'S' Class	0.73

Source: Author

Table 18.1 Optimal parameter resulted from grid search.

Parameter name	Value
c1	0.1
c2	0.1
max-iterations	100
No of K-Folds	3

Source: Author

Table 18.3 Models performance after Randomsampling.

Parameter name	Value
Accuracy of the model	98%
Precision for 'S' Class	0.96
Recall for 'S' Class	0.80
F1-Score for 'S' Class	0.88

Source: Author

Table 18.4 Performance based on different regularisation values.

	c1 = 0.1, c2 = 0.1		c1 = 1.0, c2 = 1.0		c1 = 10.0, c2 = 10.0	
	S	N	S	N	S	N
Precision	0.968	0.990	0.961	0.981	0.982	0.973
Recall	0.807	0.999	0.650	0.999	0.489	1.000
F1-Score	0.881	0.994	0.775	0.990	0.653	0.986

Source: Author

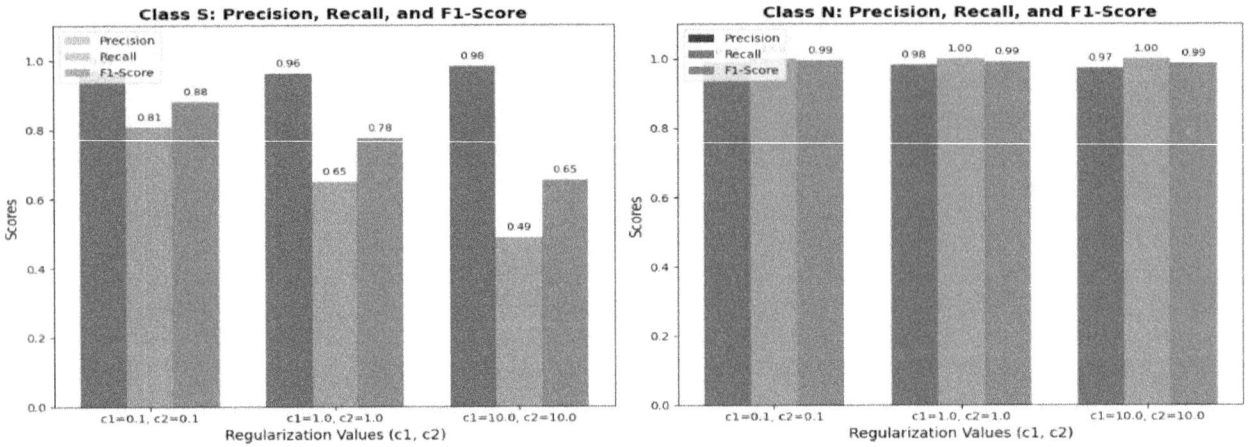

Figure 18.2 Visualization of performance based on different regularisation values
Source: Author

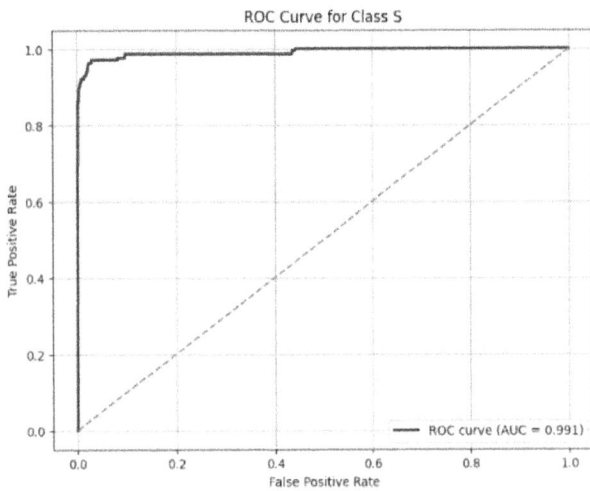

Figure 18.3 ROC curve For S class
Source: Author

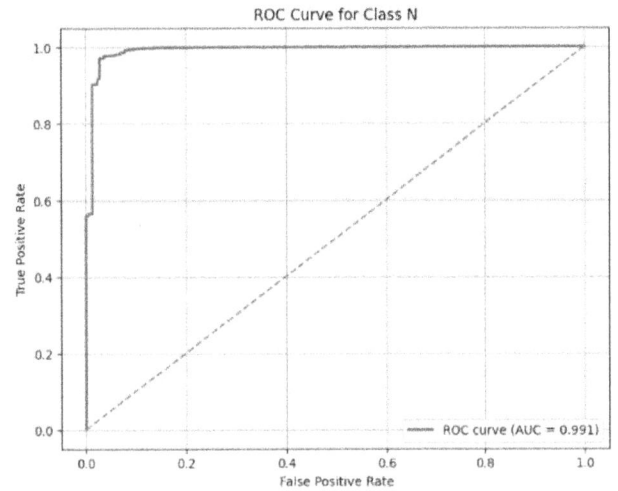

Figure 18.4 ROC curve For N class
Source: Author

shows the comparison of the performance matrices such as Precision, Recall, and F1-Score at different regularisation values. The result indicate that Class S, precision was consistently high across all configurations (0.961–0.982), revealing the model's performance in reducing false positives. But, recall declines sharply from 0.807 under low regularisation ($c1 = 0.1$, $c2 = 0.1$) to 0.489 under high regularisation ($c1 = 10.0$, $c2 = 10.0$), indicating that, under high regularisation, the model struggles to identify positive sensitive tokens. In contrast, Class N maintains consistently high precision (0.973 to 0.990) and near-perfect recall (0.999 to 1.000) across all settings, with F1-scores ranging from 0.986 to 0.994. These findings indicate that while the model is robust for non-sensitive tokens regardless of regularisation, low regularisation values provide the best balance for sensitive token detection by achieving higher recall and F1-scores.

Conclusions

This method of sensitive information detection using CRF showed high performance in detecting sensitive information. It is unique due to its granularity, which leads to the possibility of recognising sensitive information at the token level to implement selective masking, preserving the correctness of all non-sensitive content. This method of implement robust feature engineering, and random sampling is used to overcome imbalance class, the model achieves remarkable performance, with an accuracy of up to 98%, a precision of 0.96, and an F1-score of 0.88 for the minority class 'S'. These results indicate the method's ability to overcome the challenges of imbalanced datasets and its applicability for requiring secure data. Overall, the proposed approach establishes a strong benchmark for sensitive information detection and provides a scalable solution for enhancing data privacy and security.

References

[1] Regulation P. (2016). Regulation (EU) 2016/679 of the European Parliament and of the Council. Regulation (EU), 679.

[2] Gostin, L. O., Levit, L. A., Nass, S. J. (eds.). (2009). Beyond the HIPAA Privacy Rule: Enhancing Privacy, Improving Health Through Research.

[3] Kulkarni, P., Cauvery, N. K. (2021). Personally identifiable information (PII) detection in the unstructured large text corpus using natural language processing and unsupervised learning technique. International Journal of Advanced Computer Science and Applications, 12(9).

[4] Kitsios F, Chatzidimitriou E, Kamariotou M. (2022). Developing a risk analysis strategy framework for impact assessment in information security management systems: A case study in it consulting industry. Sustainability, 14(3), 1269.

[5] Zhang, K., Jiang, X. (2023). Sensitive data detection with high-throughput machine learning models in electrical health records. In AMIA Annual Symposium Proceedings. American Medical Informatics Association, vol. 2023, 814.

[6] Zalte, J., Shah, H. (2024). Contextual classification of clinical records with bidirectional long short-term memory (Bi-LSTM) and bidirectional encoder representations from transformers (BERT) model. Computational Intelligence, 40(4), e12692.

[7] Huo, L., Jiang, J. (2023). Research on intelligent perception algorithm for sensitive information. Applied Sciences, 13(6), 3383.

19 A home-based skill improvement tool utilising smartphone sensors

Sudipta Saha[a], Saikat Basu[b] and Koushik Majumder[c]

Department of Computer Science and Engineering Maulana Abul Kalam Azad University of Technology, West Bengal, India

Abstract

Motor skills play a vital role at all stages of human life, while visuospatial working memory (VSWM) is a crucial determinant of human behaviour. Children with special needs and elderly individuals often experience impairments in both of these areas. However, there needs to be a comprehensive home-based tool that can help to enhance both motor skills and working memory. To address this gap, we have developed an Android app called 'Play with Fun,' which features multiple games. One specific game, 'Roll and Drop Ball,' aims to ameliorate fine motor skills and working memory. This app includes additional games that embed standard tasks to effectively evaluate fine motor skills and working memory. The game 'Roll and Drop Ball' incorporates wrist exercises, such as flexion, extension, ulnar deviation, and radial deviation, to support upper limb rehabilitation. The study has two main objectives: to assess the concordance between scores from 'Roll and Drop Ball' and those from other standardised tasks using machine learning, and to evaluate the effectiveness of the game 'Roll and Drop Ball' as a tool for improving fine motor skills and working memory.

Keywords: Fine motor, working memory, game, machine learning

Introduction

Motor skills involve coordinating and moving body muscles to perform specific tasks. Gross motor skills rely on the movement of larger muscles, such as those in the arms, legs, and torso, while fine motor skills involve the movement of the smaller muscles in the hands, wrists, fingers, and feet. We use fine motor skills in everyday activities, such as bathing, shaving, fastening clothes, combing hair, holding a spoon, writing, and operating devices [18]. Manual dexterity is required for fine motor tasks because good hand, eye, and brain coordination are necessary [11]. Visuospatial Working Memory (VSWM) enables individuals to retain and manipulate images in their minds. The visuospatial working memory span refers to the ability of temporarily holding and manipulating visual and spatial information. A decline in this ability can lead to ineffective learning and work performance, and impairments may hinder individuals from leading normal lives. In adults, VSWM capabilities are linked to professional development, while in children, VSWM is associated with academic success [15].

Poor fine motor skills and working memory (WM) abilities are common in children with disorders such as motor dysgraphia [13], developmental coordination disorder [3], and attention deficit hyperactivity disorder [8]. Elderly individuals also often experience difficulties with WM and fine motor skills [5]. Proper training and exercises can help children and adults to enhance their fine motor and WM skills.

In a study by Yu et al. (2018) in [9], motor skill interventions successfully improved motor competence among children with Development Coordination Disorder (DCD). Similarly, another study [2] observed positive changes in the motor skill of adults with autism spectrum disorder and intellectual disability following a 12-week motor control intervention. Training can enhance performance in various WM activities (Gathercole et al., 2019). Most fine motor training methods rely on physical apparatus [2], gaming consoles, and independent sensors [16]. This equipment may be too expensive for low- and middle-income Indian households. Additionally, practicing WM and fine motor exercises may seem time-consuming in today's hectic world.

A new Android app called 'Play With Fun' has been created in this study. This app consists of four games. The first game, 'Roll and Drop Ball' (RADB), incorporates tasks that focus on fine motor skills and VSWM. This game aims to enhance users' visuospatial working memory and fine motor skills through regular practice. Since RADB is a newly developed game, its effectiveness as an exercise tool must be validated. Two elderly individuals were tested with this game to assess their improvement through consistent use. Their progress was measured against several standard, well-accepted tests. In addition to RADB, the app 'Play With Fun' includes three other games, each designed to implement established fine motor and WM. Lastly, the app features 'Digit Span' (DS), which utilises the Digit Span test, the most commonly used assessment for verbal WM [6]. In this study, 25 adults were evaluated

[a]sudipta.30.saha@gmail.com, [b]saikatbasu@gmail.com, [c]koushikzone@yahoo.com

DOI: 10.1201/9781003675235-19

using all four games. Correlation and machine learning models were employed to determine the agreement between the scores of the standard tests and those from RADB.

The remainder of the paper is organised as follows. Section 2 reviews the existing literature. Section 3 presents the novelty of the present work. Section 4 describes the details of the game RADB. Section 5 explains the research methodology. Section 6 discusses the results of the present work. Section 7 discusses the facts discovered from the results. Section 8 concludes the paper.

Literature review

Numerous studies have shown that training programs delivered through computers or smartphones can enhance fine motor and WM skills. One study [17] examined the effectiveness of such training on schoolchildren's WM and mathematical abilities. The program lasted 13 weeks, consisting of two weekly sessions, each lasting 30 minutes. According to the findings, children in the training group scored better than in the control group in several areas, including arithmetic fluency, math grades, reading skills, inhibition, and non-verbal IQ. Another study [10] utilised gesture-based games to train children (aged 5–8) in WM and basic mathematics concepts. This program spanned over eight months and included four weekly sessions, each lasting 45 minutes. The children's progress was assessed using the forward and backward Corsi blocking-tapping tests and the forward and backward Digit Span tests. The study reported significant improvements in the children's WM and basic mathematics skills. In a separate study [20], 34 younger and 27 older adults participated in four weeks of intensive computer-based WM training. Both age groups showed enhanced WM performance in the tasks they trained on and in structurally similar but untrained tasks. In another study [21], involving 156 healthy older adults, participants were randomised to either the piano training or music listening group. They were encouraged to attend at least 20 sessions over six months. The post-test results indicated that practicing the piano led to more significant improvements in fine motor skills than merely listening to music. Two interactive online computer games were developed to enhance hand-eye coordination and manual dexterity [12]. Unlike the control group, children in the intervention group were given a choice between these two online games to play at home over four weeks. At the end of the intervention, children in the intervention group significantly outperformed those in the control group on the manual dexterity sub-tasks of the Movement Assessment Battery for Children-2.

In the current work, a game has been developed to help individuals improve their fine motor skills and WM at a minimal cost. The Android platform was chosen to make this training tool accessible to middle-class and lower-class households by eliminating the need for expensive trainers and equipment. Users need to practice the game regularly. The game is designed to be a training tool that can be played at home, allowing people the flexibility to engage with it whenever it is convenient for them without disrupting their schedules.

Novelty of the work

This work develops a gaming exercise tool, 'RADB,' designed to enhance fine motor skills and VSWM using a standard smartphone. In this exercise, users must tilt the device in the correct direction to play the game rather than relying on the touchscreen features of the smartphone. As a result, it is expected that using a smartphone will not negatively impact their fine motor skills. This study not only aims to assess the tool's effectiveness for improving WM and fine motor skills but also evaluate the concordance between scores from the game RADB and other standardised tasks using machine learning.

Developed game

Game description

The game 'Roll and Drop Balls' (RADB) consists of 10 levels. Stimuli are shown for 10–20 seconds before each level's game begins (increases with level difficulties). The game's stimuli are randomly 'holes' throughout the device screen, with an animal/object assigned to each hole. There are 8 holes up to level 4 and 10 holes beyond level 4. The player must carefully observe and remember the animal/object corresponding to each hole's position. A start button will appear as soon as the stimuli disappear, inviting the player to start the game. The gaming time is fixed at 120 seconds for the first four levels and 180 seconds for the last six. During playtime, the holes remain in their original locations as stimuli, but associated objects/animals vanish from the device screen. The main goal of each level is to tilt the device as often as possible to drop a red ball into good holes and avoid bad ones within the specified time limit. The sorts of animals/objects associated with holes increase as the degree of challenge climbs (the highest level allows for a maximum of nine distinct types of animals/objects). Good holes (add scores) and bad ones (subtract scores) are the two main categories of holes. Textual and pictorial instructions help identify which holes are good or bad. When the player successfully puts the red ball into a hole, the corresponding animal/object will temporarily emerge, accompanied by a short matching sound. The game's screen prominently features a countdown of the remaining time and the updated scores. The idea of 'old' versus 'new' holes is introduced, and it applies only to the good holes, to encourage players finding and dropping the ball into the yet-uncovered good holes. Dropping the red ball into new good holes earns more points than using old ones. A new good hole turns into an old good hole after the first time it is used.

Embedded fine motor tasks
Fine motor skills include the capacity to control the wrists and hands with the help of the small hand muscles. The main goal of the game RADB is to tilt the mobile device using the wrists to drop a red ball into certain holes. This game includes fine motor tasks, as described by Cratty (1962). This game includes wrist workouts, including flexion, extension, radial deviation, and ulnar deviation. This game may be played with one hand or both. When a player attempts to bring the red ball lower, the player's wrists flex, and when he tries to move it higher, they extend. When the player uses only one hand to move the red ball left or right, ulnar and radial deviations are more apparent; however, they are not when he uses both hands.

Embedded working memory tasks
WM is the capacity to retain and process goal-oriented information simultaneously. To achieve a high score in the game RADB, a player must figure out the higher-scoring holes to drop the ball. To drop the red ball into the intended holes, players need to tilt the gadget in the right direction simultaneously. A player now has to remember triple things to locate the majority of scoring holes: animals or objects that are connected to each hole's location (as suggested by stimuli), which animals or objects increase or decrease the score and by how much (as directed), and the previously covered good holes (old good holes). Finding the best holes is a WM processing activity, while recalling the previously mentioned details is a storage task.

Methodologies

Participants and experimental procedures
This study includes 25 healthy individuals from West Bengal, India (12 males and 13 females). Convenience sampling was used to choose the participants. To be eligible, participants must not have been diagnosed with a neurological illness and must be between the ages of 18 and 60. Those who agreed to participate provided their informed consent prior to the examination. The participants' ages varied from 18.16 to 52.83 years (average = 27.41 years, standard deviation = 10.14). Participants were asked to play the first four levels of the game RADB. They were also tested using three standard tasks to find the concordance of the RADB games' scores with the standard tasks' scores. These three tasks are embedded in the TM, BT, and DS games. All participants are tested with the above four games using the same Samsung tablet in a laboratory setup. Two case studies are also performed on older

individuals. One is male (age = 76.67 years), and another is female (age = 68.58 years). These case studies aim to determine whether playing the game RADB improves the fine motor and WM skills of two elderly individuals. It is suggested that they spend two weeks practicing RADB. They are advised to practice the game's first four levels during the first week and the last six during the second week, on weekdays only. The TM, BT, and DS games are used to assess these two participants both before and after. Our representative visited these two persons' homes during practice and assessment times.

Testing games
Trail Making game (TM game) is based on the Trail Making Test (TMT), which evaluates executive functions such as inhibition, perceptual speed, and fine motor ability [14]. The original TMT uses paper and pencil for the test and consists of Part A and Part B. Part A involves drawing lines in ascending order between consecutive numbers from 1 to 25. Part B requires drawing lines in ascending sequence between numbers and letters (1 to 13 and A to L). Park and Schott [14] found that the digital version of TMT outperforms the paper-pencil-based version in terms of reliability, sensitivity, and clinical significance. The TM game used in this work implements the digital version of TMT Parts A and B in Levels 1 and 2, respectively. The game's critical quantifiable parameter is the time required to accomplish each level in seconds.

Block Tapping game (BT game): The Corsi block-tapping test is considered the 'gold standard' in neuropsychology for assessing working memory and short-term visuospatial memory [1]. Forward and backward block tapping are two parts of this task. In this task, the participant must touch the blocks in the same order as the experimenter or in the opposite order. A board with nine pseudorandom cubic blocks is used for this task. A digital Corsi Block Tapping task is implemented in the BT game. Automated assessments of span, improved presentation timing accuracy, and simplicity of setup are some benefits of the digital Corsi block tapping over the manual version [4]. There are two levels to the BT game that was utilised in this experiment. The first level implements forward block-tapping tasks, while the second level implements backward tasks. For each of the two game levels, there are 16 trials altogether. The trial number indicates how the block length grows gradually from two to nine, with two trials conducted for each length. The level concludes after all 16 trials have been completed, or the player cannot touch the blocks in the proper order for both trials with a specific length. The length of the last correctly remembered sequence is taken as a forward or backward block span, depending on the required recall order. Thus, the number of blocks recalled is the unit of measurement of this game.

Digit Span game (DS game): One of the oldest and most used verbal short-term and WM tests is the forward and backward digit span test [6]. The calculation of 'digit span' uses previously spoken digit sequences that need to be remembered either forward or backward. Two levels of the DS game were used in this experiment. The forward-digit span task is used in level one, while the backward-digit span job is used in level two. The player hears a series of one-digit numbers spoken aloud in this game. The next step is for players to write down and submit the numbers they heard, either in the same order (level one) or in the opposite order (level two). Digit sequences have two trials per length, beginning with a list length of two digits and increasing by one. There are 16 trials in total for each of the two-game levels. The level ends when the player completes all 16 trials or cannot remember the numbers in the correct sequence for both trials of a certain length. Depending on the necessary recall order, the length of the previously successfully remembered sequence is either taken as a forward or backward digit span. Thus, the number of digits recalled is the unit of measurement of this game.

Data collection and analysis

All gaming data were automatically stored in the built-in SQLite database on Android devices. Scores from the first four levels of the RADB game were collected from 25 participants aged between 18 and 60 years. Total scores (RADBTScore) were calculated from the individual level score. In addition to the total raw scores, four parameters were derived from the game: 'total number of ball drops' (RADBTMovement), 'total positive drops' (RADBTPMovement), 'total number of new good holes covered' (RADBTNewHoleCov), and 'total scores based on the speed at which new holes were covered' (RADBTScoreOnHFC). In the rest of the sections, the TM game scores for Levels 1 and 2 are represented as TMA and TMB, respectively. In the BT game, the scores for Levels 1 and 2 are labelled BTF (Block Tapping – Forward) and

BTB (Block Tapping – Backward). Similarly, for the DS game, the scores for Levels 1 and 2 are referred to as DSF (Digit Span – Forward) and DSB (Digit Span – Backward). During the first week, two elderly individuals were advised to practice the first four levels of the game RADB on five weekdays. In the second week, they were encouraged to practice the last six levels of the game on weekdays. We took into account their individual scores from each day. Additionally, we compared their pre-test and post-test scores on the standard games.

Results

Descriptive statistics of the scores obtained by the 25 participants in the games RADB, TM, BT, and DS are listed in Table 19.1. Pearson's correlation coefficients are determined between all the RADB game's parameters and the scores of other games. These are shown in Table 19.2.

Independent samples *t*-tests were conducted between the male and female groups' scores. Neither the RADB parameters, nor the TMA, TMB, BTF, BTB, DSF, or DSB

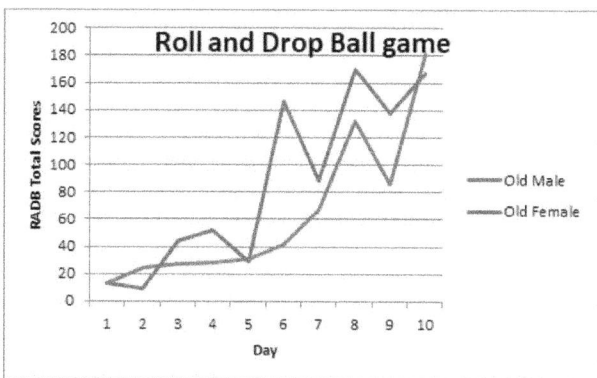

Figure 19.2 Pre-test and post-test scores of two elderly persons in TM game
Source: Author

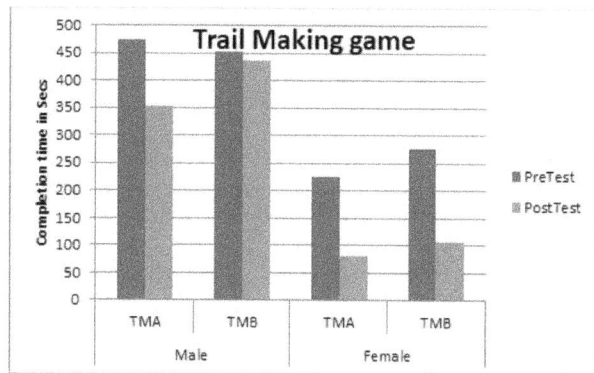

Figure 19.1 RADB practice scores of two elderly persons
Source: Author

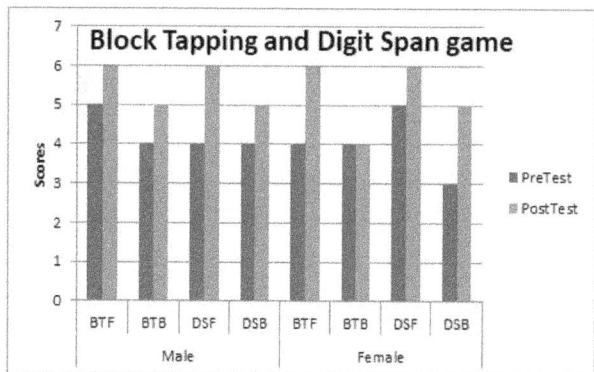

Figure 19.3 Pre-test and post-test scores of two elderly persons in BT and DS games
Source: Author

Table 19.1 Descriptive statistics of games' scores.

Games' parameter	Max	Min	Average	Standard deviation
RADBTScore	120	32	78.12	20.28
RADBTMovement	67 31	31	52.32	8.90
RADBTPMovement	63	25	47.32	8.84
RADBTNewHoleCov	19	12	18.08	1.58
RADBTScoreOnHFC	190	128	172.72	14.29
TMA	151	51	104.92	25.38
TMB	260	72	162.96	52.23
BTF	9	4	6.16	1.28
BTB	8	2	5.32	1.46
DSF	8	4	6.28	1.06
DSB	9	2	5.56	1.68

Source: Author

Table 19.2 Correlation coefficients of games' parameters.

Games' parameter	RADBT score	RADBT movement	RADBTP movement	RADBT NewHoleCov	RADBT ScoreOnHFC
TMA	−0.103	−0.259	−0.168	−0.262	−0.348
TMB	−0.272	−0.118	−0.158	−0.460*	−0.501*
BTF	0.224	0.090	0.128	0.323	0.422*
BTB	0.146	0.212	0.156	0.547**	0.498*
DSF	0.223	0.153	0.176	0.583**	0.596**
DSB	−0.112	−0.171	−0.192	0.076	0.135

'*' and '**' indicate correlation coefficients are significant at a 95% and 99% confidence level confidence level respectively

Source: Author

Table 19.3 Regression results.

Games score	Linear regression result		Decision tree regression result		
	Highest adjusted R^2 value	Predictors	R^2 value	Tree depth	Predictors' importance
TMA	0.168	RADBTScore, RADBTMovement, RADBTPMovement, RADBTNewHoleCov, RADBTScoreOnHFC	0.279	1	RADBTPMovement: 1
TMB	0.218	RADBTScoreOnHFC	0.482	2	RADBTPMovemen:0.56, RADBTScoreOnHFC:0.37, RADBTScore: 0.07
BTF	0.142	RADBTScoreOnHFC	0.213	1	RADBTPMovement:1.0
BTB	0.269	RADBTNewHoleCov	0.666	3	RADBTNewHoleCov: 0.58, RADBTScore: 0.17, RADBTPMovement: 0.16, RADBTScoreOnHFC:0.09
DSF	0.33	RADBTScoreOnHFC	0.328	1	RADBTNewHoleCov: 1.0
DSB	−0.005	RADBTPMovement	0.633	2	RADBTScoreOnHFC: 0.37, RADBTScore: 0.31, RADBTPMovement: 0.31

Source: Author

Table 19.4 Elderly persons' practice score on Roll and Drop Ball.

	Day 1	Day 2	Day 3	Day 4	Day 5	Day 6	Day 7	Day 8	Day 9	Day 10
Elderly male	13	24	27	28	31	41	67	132	86	181
Elderly female	13	9	44	52	29	146	89	170	138	167

Source: Author

Table 19.5 Elderly persons' pre-test and post-test scores on standard games.

	Play time	TMA	TMB	BTF	BTB	DSF	DSB
Elderly male	Pre-test	473	453	5	4	4	4
	Post-test	354	437	6	5	6	5
Elderly female	Pre-test	226	276	4	4	5	3
	Post-test	80	106	6	4	6	5

Source: Author

scores, showed significant differences between the male and female groups.

Regression analysis is performed to examine the relationship between scores from the game RADB and the performance on the other standard tasks. The parameters from the RADB game served as predictors, while the scores from other games are treated as dependent variables in the regression analysis.

Both Multiple Linear Regression and Decision Tree Regression are utilised in this process. For the Decision Tree Regressor, the GridSearchCV tool is employed with five-fold cross-validation to determine the optimal tree depth. Table 19.3 presents the results of both Linear and Tree regression. For Linear Regression, the best R^2 values, along with their corresponding predictors, are listed. For the Decision Tree Regressor, R^2 values, optimal depths, and feature importance are provided.

The 10-day practice scores for two elderly participants using the RADB are shown in Table 19.4 and are graphically represented in Figure 19.1. Their pre-test and post-test scores on the standardised game are displayed in Table 19.5, with corresponding pre-test and post-test scores illustrated in a bar chart in Figures 19.2 and 19.3.

Discussion

The RADB game's parameters—RADBTScore, RADBTMovement, and RADBTPMovement—do not correlate significantly with other games' scores. Both TMA and TMB have negative correlations with all RADB game's parameters. TM games measured parameter TMA, and TMB measures the time required to complete the tasks. The people who lack the required skills need more time to complete the tasks. TMA shows insignificant negative correlations with the RADB game's parameters. However,

TMB shows significant negative correlations with RADB game's RADBTNewHoleCov and RADBTScoreOnHFC. In the TM game Part B, the player needs to remember the next alphabet while the current target is the number, or vice versa. It is a working memory task absent in Part A of the TM game. Both BTF and BTB show significant positive correlations with RADBTScoreOnHFC, but only BTB shows a significant positive correlation with RADBTNewHoleCov. BT game's backward is more challenging than its forward. The backward task requires a working memory update. Given that RADB games involve visuospatial working memory tasks, BTB is more strongly correlated with RADB parameters RADBTNewHoleCov. The DS game's forward score (DSF) strongly correlates with RADBTNewHoleCov and RADBTScoreOnHFC, but surprisingly, the backward score of DSB is not correlated with any Roll and Drop Ball game's parameters.

The insignificant results obtained by independent samples *t*-tests indicate that neither the skill needed to play RADB games nor any other standard games differ significantly between males and females.

Linear regression results indicate that the linear model does not fit well while predicting other game scores from RADB game parameters. The maximum adjusted R^2 value is obtained in the DSF prediction case; in most cases, RADBTScoreOnHFC predicts other games' scores most accurately. Decision tree regression provides better results than the linear model, and the highest R^2 value was obtained for BTB (0.666).

The 10-day practice scores for two elderly participants using the RADB game show that the male participant's scores mainly increase day by day, while female participants' scores show ups and downs. In the case of the 'TM' game, after 10-day practice scores, the time required to complete Parts A and B was reduced for both the male and female participants. This result indicates that their fine motor skill improves. Block Tapping and Digit Span games scores of post-test for both male and female participants increased in all the cases except one, suggesting an improvement in their working memory span.

Conclusion

In this study, we developed a smartphone-based tool for home use to improve fine motor skills and visuospatial working memory. The results from two case studies indicate that two weeks of practice with this tool improved these areas for two elderly participants. We also sought to find a correlation between scores from the newly developed game and some well-established tests, revealing a high correlation in certain instances. By using a decision tree regression model, we could also predict some standard task scores from the game with moderate accuracy. In the future, we plan to test the effectiveness of this newly developed game as a skill-improvement tool for both adults and children with neurological diseases.

References

[1] Arce, T., McMullen, K. (2021). The Corsi block-tapping test: evaluating methodological practices with an eye towards modern digital frameworks. *Computers in Human Behavior Reports*, 4, 100099.

[2] Azar, N. R., McKeen, P., Carr, K., Sutherland, C. A., Horton, S. (2016). Impact of motor skills training in adults with autism spectrum disorder and an intellectual disability. *Journal on Developmental Disabilities*, 22(1), 28.

[3] Bernardi, M., Leonard, H. C., Hill, E. L., Botting, N., Henry, L. A. (2018). Executive functions in children with developmental coordination disorder: a 2-year follow-up study. *Developmental Medicine & Child Neurology*, 60(3), 306–313.

[4] Brunetti, R., Del Gatto, C., Delogu, F. (2014). eCorsi: implementation and testing of the Corsi block-tapping task for digital tablets. *Frontiers in Psychology*, 5, 939.

[5] Curreri, C., Trevisan, C., Carrer, P., Facchini, S., Giantin, V., Maggi, S., ..., Sergi, G. (2018). Difficulties with fine motor skills and cognitive impairment in an elderly population: the progetto veneto anziani. *Journal of the American Geriatrics Society*, 66(2), 350–356.

[6] Dassanayake, T. L., Hewawasam, C., Baminiwatta, A., Ariyasinghe, D. I. (2021). Regression-based, demographically adjusted norms for victoria stroop test, digit span, and verbal fluency for Sri Lankan adults. *The Clinical Neuropsychologist*, 35(suppl. 1), S32–S49.

[7] Hyde, C., Fuelscher, I., Sciberras, E., Efron, D., Anderson, V. A., Silk, T. (2021). Understanding motor difficulties in children with ADHD: a fixel-based analysis of the corticospinal tract. *Progress in Neuro-Psychopharmacology and Biological Psychiatry*, 105, 110125.

[8] Jane, J. Y., Burnett, A. F., Sit, C. H. (2018). Motor skill interventions in children with developmental coordination disorder: a systematic review and meta-analysis. *Archives of Physical Medicine and Rehabilitation*, 99(10), 2076–2099.

[9] Lakshmi, D., Ponnusamy, R. (2021). Working memory enhancement during early childhood based on the utilization of interactive gesture game-based learning. Indonesian *Journal of Electrical Engineering and Computer Science*, 21(2), 768–775.

[10] Matheis, M., Estabillo, J. A. (2018). Assessment of fine and gross motor skills in children. In Handbook of Childhood Psychopathology and Developmental Disabilities Assessment, 467–484.

[11] McGlashan, H. L., Blanchard, C. C., Sycamore, N. J., Lee, R., French, B., Holmes, N. P. (2017). Improvement in children's fine motor skills following a computerized typing intervention. *Human Movement Science*, 56, 29–36.

[12] Momeni, M. S., Hajebi, M. Z., Monirpour, N. (2022). Predicting working memory (visual-spatial) based on motor skills and self-regulation from the perspective of Barkly theory in dysgraphia children 8-12 years old.

[13] Park, S. Y., Schott, N. (2022). The trail-making-test: comparison between paper-and-pencil and computerized versions in young and healthy older adults. *Applied Neuropsychology: Adult*, 29(5), 1208–1220.

[14] Ramos, A. A., Hamdan, A. C., Machado, L. (2020). A meta-analysis on verbal working memory in children and adolescents with ADHD. *The Clinical Neuropsychologist*, 34(5), 873–898.

[15] Ruiz-Rodriguez, A., Martinez-Garcia, A. I., Caro, K. (June 2019). Gesture-based video games to support fine-motor coordination skills of children with autism. In Proceedings of the 18th ACM International Conference on Interaction Design and Children, 610–615.

[16] Sánchez-Pérez, N., Castillo, A., López-López, J. A., Pina, V., Puga, J. L., Campoy, G., ..., Fuentes, L. J. (2018). Computer-based training in math and working memory improves cognitive skills and academic achievement in primary school children: behavioral results. *Frontiers in Psychology*, 8, 2327.

[17] Saha, S., Basu, S., Majumder, K., Das, S. (2024). Integrating smartphone sensor technology to enhance fine motor and working memory skills in pediatric obesity: a gamified approach. *International Journal of Next-Generation Computing*, 15(1).

[18] Von Bastian, C. C., Langer, N., Jäncke, L., Oberauer, K. (2013). Effects of working memory training in young and old adults. *Memory & Cognition*, 41, 611–624.

[19] Worschech, F., James, C. E., Jünemann, K., Sinke, C., Krüger, T. H., Scholz, D. S., ..., Altenmüller, E. (2023). Fine motor control improves in older adults after 1 year of piano lessons: analysis of individual development and its coupling with cognition and brain structure. *European Journal of Neuroscience*, 57(12), 2040–2061.

20 Offline handwritten character recognition of Manipuri Script using convolutional neural network (CNN) and vision transformer (ViT)

Elangbam Binoy Singh[a] and Thokchom Tangkeshwar Singh[b]

Department of Computer Science, Manipur University, India

Abstract

This paper explores the applications of deep learning models for offline handwritten character recognition of Manipuri Script. We implement two distinct models: a Convolutional Neural Network (CNN) and a pre-trained Vision Transformer (ViT-B/16). The performance of these models in accurately recognising the handwritten characters of Manipuri Script is presented. The experiments are conducted on a dataset of total 18,020 images of handwritten characters of Manipuri Script. A recognition rate of 96.70% is achieved with the CNN model, while the pre-trained ViT model obtains a recognition rate of 95.48%.

Keywords: Vision transformer, convolutional neural network, manipuri script, handwritten character recognition

Introduction

In computer vision and pattern recognition, handwritten character recognition (HCR) is a vital field of research. It aims to enable machines to accurately interpret and digitise human handwriting. The ability to convert handwritten text in the form of machine-readable formats covers versatile applications, including document digitisation, postal address reading, automated form processing, and signature verification. Despite of significant advancements, HCR remains a challenging task because of the variation in writing styles of individuals, inconsistencies in stroke patterns, and distortions caused by different writing instruments or surfaces. Recent developments in CNNs and transformer-based models, like Vision Transformers (ViTs), have pushed the boundaries of accuracy and efficiency in this domain, offering new solutions to previously unsolved problems in recognising handwritten characters.

Literature survey

A brief survey of Manipuri Script
Manipuri, also called Meeteilon [1,9], Meiteiron, and Meithei [5] in linguistic literature, is the official language of the State of Manipur. It is one of the 22 official Indian languages. The Manipuri Script, or Meitei Mayek Script, is used to write Manipuri language. It is the mother tongue of the ethnic group Meitei/Meetei. However, apart from the Meiteis/Meetei, Meitei Pangals (Manipuri Muslims) also speak Manipuri as their mother tongue.

Manipuri is a tonal language of the Tibeto-Burman language family [1,9]. This script consists of Iyek Ipee/ Mapung Iyek, Lonsum Iyek, Cheitap Iyek, Cheishing Iyek, and Khudam Iyek as depicted in Figure 20.1.

A brief survey of HCR research in Manipuri Script
Handwritten character recognition (HCR) involves several steps, namely (a) image acquisition, (b) pre-processing, (c) feature extraction, (d) classification, and (e) post-processing. Out of all the stages listed, feature extraction and classification are the primary ones that have been shown to have the biggest effects on a system's character recognition system. Different techniques were found to be proposed for character recognition of Manipuri Script. Research works reported in HCR of Manipuri Script are summarised in Table 20.1.

For recognition of numeral figures of Manipuri Script, Laishram et al. [7] achieved 85% accuracy using a Neural Network (NN) as the classifier and pixel density of binary pattern as the feature. Similarly, Maring and Dhir [11] achieved 85.58% accuracy using an SVM classifier and a Gabor filter-based technique.

Singh [14] used probabilistic and fuzzy features with a 2-layer feed forward backpropagation NN as classifier, achieved 90.3% accuracy for the recognition of Manipuri Script. In another approach, a recognition rate of 97.5% and 98% was achieved using hybrid feature and Gabor wavelet feature, respectively, for handwritten discrete word recognition in the upper, middle, and lower zones, which were recognised using five trained Multi-layer Perceptrons (MLPs) and decoded into editable text.

Nongmeikapam et al. [12] achieved 94.29% accuracy using k-Nearest Neighbor (k-NN) as the classifier and

[a]abnoy.elang@gmail.com, [b]drtangkeshwar@gmail.com, tangke.th@manipuruniv.ac.in

DOI: 10.1201/9781003675235-20

Histogram of Oriented Gradient (HOG) as the feature extractor on a dataset of 56 classes.

Hijam and Saharia [4] achieved a 96.24% accuracy in handwritten characters recognition of Manipuri Script using a CNN architecture. It was experimented on a dataset of 60,285 handwritten characters of Manipuri Script. Inunganbi et al. [6] achieved 98.70% accuracy in handwritten characters recognition of Manipuri Script using a three-channel CNN of gradients and grayscale images.

Pre-processing and segmentation

The goal of pre-processing and segmentation is to produce images that are easy for the recognition process to operate on more accurately and then present segmentation of lines, segmentation of digits or characters, and ensure uniformity.

We consider *Iyek Ipee, Lonshum Iyek, Cheitap Iyek*, and *Cheising Iyek* for recognition. The images are collected using 50 characters per page of an A4 sheet, written by students of different age groups. The pages are scanned using HP LaserJet M1005 MFP scanner with a 300 dpi resolution and are saved as jpg files. Figure 20.2(a) represents a sample of the scanned image.

We edit all scanned images by making some changes to the exposure, contrast, sharpness, etc. to make the characters in the image clear and free from noise. Then using Python script, we load each image using OpenCV and convert it to grayscale, simplifying it to a single channel. Then, we convert the image to binary using a threshold, setting pixels above 128 to black and below 128 to white. Dilation is applied to thicken characters, bridging gaps for easier detection. We extract contours, isolate characters based on bounding rectangles, and save them, resulting in a dataset of 13,250 images (100 × 168) as depicted in Figure 20.2(b).

In addition to the self-collected dataset, a pre-processed dataset of 2,700 images handwritten Manipuri alphabet characters from seniors and faculty members have been provided by my project supervisor and a pre-processed dataset of 5,580 images of handwritten Manipuri script (128 × 128) was downloaded from Kaggle [13]. First, we convert all the 5,580 images downloaded from Kaggle to its negative. Next, we resized both the dataset into a dimension of 100 × 168. Then compiling it, we obtain a dataset of 8,280 images for all the 57 characters of Manipuri Script: 200 images for each of the 27 classes, and 96 images for each of the remaining 30.

As we excluded *Khudam Iyek* and to fulfil the need of a balanced dataset for recognition, we remove some specific images from dataset of 13,250 images and dataset of 8,280

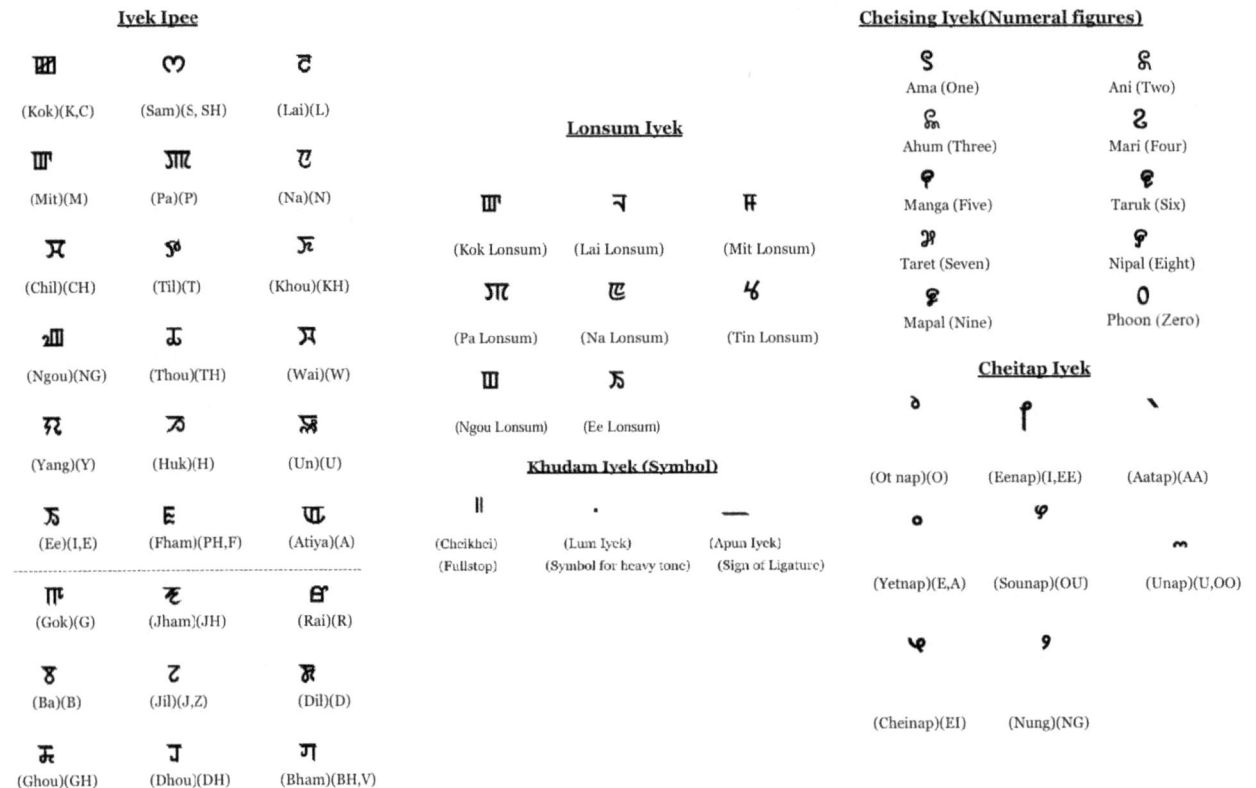

Figure 20.1 Manipuri Script

Source: Author

images. Finally, we compile the two datasets resulting in a balanced dataset of total 18,020 images of dimension 100x168, containing 340 images for each class (53 classes excluding *Khudam Iyek)* of Manipuri Script.

CNN and pre-trained ViT model

CNNs and ViTs are both widely used for image classification tasks, but they approach the problem differently. CNNs use convolutional layers to learn local spatial features, progressively building up hierarchical representations of an image through filters that detect edges, textures, and patterns [3]. This is complemented by pooling layers to reduce dimensionality, and fully connected layers for classification. In contrast, ViT-B/16 model leverages the Transformer architecture, where images are split into patches that are treated as input tokens. These patches are processed by multi-head self-attention layers, capturing both local and global dependencies across the image [2]. Unlike CNNs, which excel

in learning local features, ViT's global attention allows it to handle complex patterns with greater contextual awareness, particularly when pre-trained on large datasets. Both architectures have their strengths, with CNNs being more efficient on smaller datasets and ViTs showing superior performance when large-scale pre-training is involved.

Proposed CNN model

The proposed CNN model has four convolutional layers, four MaxPooling layers, and a fully connected layer having three dense layers with an output layer of 53 neurons with a softmax activation function, as depicted in Figure 20.3. For convolutional layers, the filter size is 3×3, having a stride rate of 1 and 'valid' padding. For the first convolutional layer, 16 filters were employed, 32 filters for second, 64 filters for third, and 128 for the fourth convolutional layer. In the case of four MaxPooling layers, the filter size and stride rate are of 2×2.

Table 20.1 Manipuri script's works in HCR.

Work	Size of the dataset	No. of classes	Classifier(s)	Test accuracies
Laishram et al. [7]	1,000	10	NN	85%
Laishram et al. [8]	1,000	43	NN	80%
Maring and Dhir [11]	7,200	10	SVM	89.58%
Singh [13]	594	27	NN	90.3%
	1,000	10	MLP	96%
	3,040	36	MLP	90.14%
	1,890	27	MLP	92.96%
	1,000	10	MLP	97.5%
	800	10	MLP	98%
Nongmeikapam et al. [12]	5,600	56	k-NN	94.29%
Hijam et al. [4]	60,285	37	CNN	96.24%
Inunganbi et al. [6]	14,700	35	CNN	98.70%

Source: Author

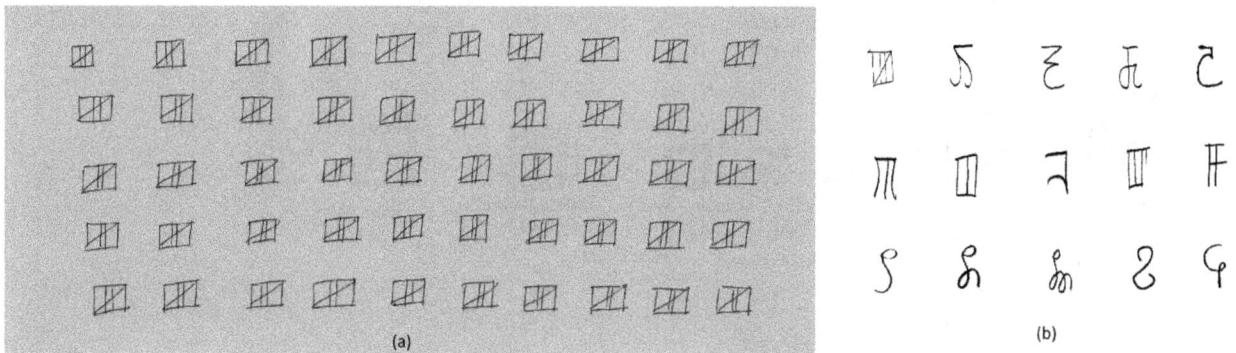

Figure 20.2 (a) A scanned image and (b) Some segmented handwritten characters images of Manipuri Script
Source: Author

Pre-trained ViT-B/16 model

The ViT-B/16 model which is pre-trained on ImageNet21k and further fine-tuned on ImageNet2012 is a Vision Transformer model designed for image classification tasks [10]. This model, known as ViT-B/16, has 85,798,656 parameters.

After the output of the transformer block, a Flattened layer reshapes the data into a single dimension of 768 units, which does not add any parameters. It is then accompanied by Batch Normalisation layer with 3,072 parameters, helping to normalise the output and stabilise training. The first dense layer reduces the dimensionality from 768 to 128 units and includes 98,432 parameters, computed as $(768 \times 128) + 128$ for the weights and biases. Another Batch Normalisation layer with 512 parameters is applied next. The second dense layer further reduces the dimensionality to 64 units, involving 8,256 parameters. Following this, a third dense layer with 2,080 parameters reduces the dimensionality to 32 units. Finally, the last dense layer, responsible for the classification output, maps the 32 units to 53 classes and has 1,749 parameters. Each dense layer's parameters are calculated as (input units × output units) + output units for the weights and biases, ensuring the model can effectively learn and make predictions based on the input features.

Experimental results

The models are experimented on a dataset that encompasses 53 classes (excluding Khudam Iyek), each class consisting of 340 images with dimensions of 100 × 168 (pixels), resulting in a total of 18,020 images of handwritten characters of Manipuri Script (as summarised in Table 20.2). The dataset is partitioned as 10,812, 3,604, and 3,604 images for training, validation, and testing, respectively.

During runtime, images of 100 × 168 dimensions are resized to 100 × 100 dimension for the proposed CNN model and 224 × 224 dimension for the proposed pre-trained ViT-B/16 model.

The CNN architecture is built and compiled. The training dataset is loaded, processed, and fitted into the model. We train different models. Out of them, we saved the best one. After training, we get 98.73% training accuracy, 96.81% validation accuracy, and 96.70% testing accuracy. On the other hand, the pre-trained ViT model "ViT-B/16" gave 98.91% training accuracy, 95.56% validation accuracy, and test accuracy of 95.48% on the same dataset in just 10 epochs. A comparative performance of the two models is summarised in Table 20.3.

Table 20.2 Description of dataset

Attributes	Values
Number of classes	53 (excluding Kudam Iyek)
Number of sample images in each class	340
Total number of image samples	18,020
Size of images	100 × 168
Format	JPG

Source: Author

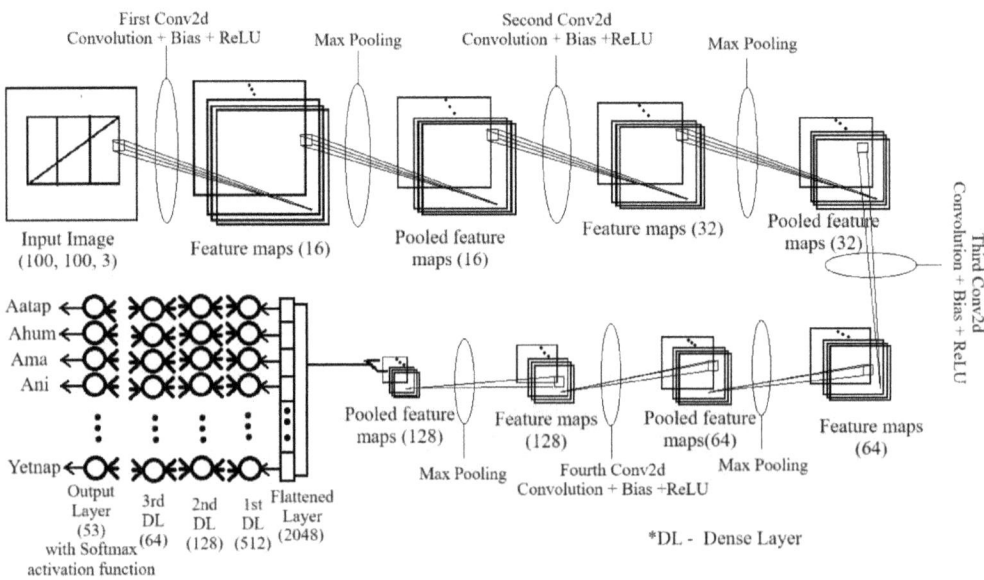

Figure 20.3 Proposed CNN model

Source: Author

Table 20.3 Comparative results of CNN and ViT-B/16.

Parameter	CNN	ViT-B/16
1. Train Loss	0.0387	1.3498
2. Test Loss	0.1996	1.4361
3. Training Accuracy	98.73%	98.91%
4. Test Accuracy	96.70%	95.48%
5. Number of epochs	80	10

Source: Author

Conclusion

Our experiments with the models: the proposed CNN model along with the pre-trained ViT-B/16, demonstrated efficiency of deep learning approaches in recognising and classifying handwritten characters of Manipuri Script. The accuracy obtained in the experiments is encouraging, as it facilitates recognition of the characters of the script among the different styles of handwriting in the dataset. Also, it seems that it can overcome the challenges posed by the significant similarity between certain characters.

In future work, word recognition of Manipuri Script may be investigated using Transformer architecture. Emphasis can be on increasing the size and diversity of dataset, including the symbols (Khudam Iyek). Optimising the model, like fine-tuning the hyperparameters of the existing models, can help in increasing the recognition rate of handwritten characters.

References

[1] Chingtamlen, W. (2007). A Short history of Kangleipak (Manipur) Part II. Kangleipak Historical & Cultural Research Centre, Sagolband Thangjam Leirak, Imphal.

[2] Dosovitskiy, A., Beyer, L., Kolesnikov, A., Weissenborn, D., Zhai, X., Unterthiner, T., Dehghani, M., Minderer, M., Heigold, G., Gelly, S., Houlsby, N. (2020). An image is worth 16 × 16 words: Transformers for image recognition at scale. arXiv preprint arXiv:2010.11929.

[3] Gonzalez, R. C., Woods, R. E. (2018). Digital Image Processing (4th ed.). Pearson.

[4] Hijam, D., Saharia, S. (2018). Convolutional neural network based Meitei Mayek handwritten character recognition. In 10th International Conference, IHCI 2018, Allahabad, India, December 7–9, 2018, Proceedings (pp. 223-234). Springer. https://doi.org/10.1007/978-3-030-04021-5_19

[5] Hodson, T. C. (1908). The Meitheis. Low Price Publications, Delhi.

[6] Inunganbi, S., Choudhary, P., Manglem, K. (2020). Handwritten Meitei Mayek recognition using three-channel convolution neural network of gradients and gray. Computational Intelligence, 1–17. https://doi.org/10.1111/coin.12392

[7] Laishram, R., Singh, A. U., Singh, N. C., Singh, A. S., James, H. (2012). Simulation and modeling of handwritten Meitei Mayek digits using neural network approach. In Proceedings of the International Conference on Advances in Electronics, Electrical and Computer Science Engineering-EEC, 355–358.

[8] Laishram, R., Singh, P. B., Singh, T. S. D., Anilkumar, S., Singh, A. U. (2014). A neural network based handwritten Meitei Mayek alphabet optical character recognition system. In: 2014 IEEE International Conference on Computational Intelligence and Computing Research (ICCIC). IEEE, 1–5.

[9] Mangang, N. K. (2003). Revival of a Closed Account: A Brief History of Kanglei Script and the Birth of Phoon (zero) in the World of Arithmetic and Astrology. Sanmahi Laining Amasung Punshiron Khupham (Salai Punshipham), Lamshang, Imphal.

[10] Morales, F. (n.d.). ViT-B_16_imagenet21k+imagenet2012.npz. GitHub. Retrieved from https://github.com/faustomorales/vit-keras/releases/download/dl/ViT-B_16_imagenet21k+imagenet2012.npz

[11] Maring, K. A., Dhir, R. (2014). Recognition of Cheising Iyek/Eeyek-Manipuri digits using support vector machines. IJCSIT, 1(2).

[12] Nongmeikapam, K., Manipur, I., Kumar, I.W.K., Singh, M.P. (December 2017). Exploring an efficient handwritten Manipuri Meetei-Mayek character recognition using gradient feature extractor and cosine distance based multiclass k-nearest neighbor classifier. In Proceedings of ICON-2017, Kolkata, India. NLPAI, 328–337.

[13] Singh, P. (2019). A benchmark dataset for Manipuri MeeteiMayek handwritten character recognition. IEEE Dataport. https://doi.org/10.5061/dryad.r4xgxd27w

[14] Singh, T. T. (2017). Off-line Handwritten Character Recognition of Manipuri Script. PhD Thesis, Thapar University, Patiala.

21 Denoising in Terahertz images

Phibansabeth Nongkseh[1,a] and Debdatta Kandar[2,b]

[1]Department of Information Technology, North Eastern Hill University (NEHU) Shillong, Meghalaya, India

[2]Professor, Department of Information Technology, North Eastern Hill University (NEHU) Shillong, Meghalaya, India

Abstract

This study enhances concealed object detection in Terahertz (THz) imaging by addressing noise challenges through advanced denoising techniques. THz imaging, known for penetrating non-metallic and non-polar materials, faces low signal-to-noise ratio (SNR) and resolution issues that limit detection accuracy. It evaluates methods such as wavelet-based denoising, Gaussian filtering, mean filtering, and NL-means filtering, using a dataset of 3,157 annotated images. The best performance was obtained with wavelet-based denoising, and the peak signal-to-noise ratio, structural similarity index measure, and MSE were 42.96, 0.98, and 4.87, respectively. Using the denoised images as inputs for the YOLOv8 object detection model increased performance and enhanced accuracy, with a mean average precision (mAP@0.5) of 0.584 and fewer classification errors and losses during training. This will illustrate the need for pre-processing to improve the quality of object detection. Future studies should consider improving the existing datasets and optimising the model for real-world applications.

Keywords: Terahertz imaging, concealed object detection, denoising technique, wavelet filter, Gaussian filter

Introduction

THz radiation, with frequencies ranging from microwaves to infrared light (0.1 to 10 THz), has attracted a lot of attention for its excellent performance in imaging applications. Unlike visible light, THz waves can penetrate most non-metallic and non-polar materials, except water and metals [2]. This unique feature has made THz imaging worthwhile in several fields, including security screening, manufacturing quality control, biomedical imaging, and art restoration. Moreover, THz radiation is non-ionising, thus safe for biological tissues, and its sensitivity to material differences allows for the detection of concealed objects, structural flaws, or chemical compositions hidden beneath surfaces. However, it is very challenging to detect concealed objects using THz imaging due to the low signal-to-noise ratio that most of these systems experience [1]. The SNR can obscure the crucial information; hence, identifying the object becomes a problem, especially when the material density is high, thereby preventing THz waves from penetrating [13]. Researchers are working to address these challenges, but progress has been inconsistent, and more efficient detection methods are needed.

In recent years, machine learning architectures have shown potential for solving complex imaging challenges, including object detection [10,11]. However, modifying these models for THz images necessitates meticulous fine-tuning and optimisation to suit their distinct features. A major challenge is the lack of publicly accessible datasets, which limits the creation and training of effective models.

Based on these findings, our work is aimed towards exploring and identifying techniques used to reduce noise in THz images for better accuracy in detection. We take a comprehensive review of filtering techniques and systematically consider noise reduction in THz imagery. Our methodology includes an extended comparative analysis of various filtering algorithms, assessing their performance against a publicly available dataset. This enables us to identify and select the optimum technique based on quantitative metrics of performance, while preserving important features from image and minimising harmful noise artefacts. The filtered outputs are then integrated into the detection framework. This approach, while fundamentally rooted in essential image processing principles, highlights the significant impact that targeted pre-processing can have on downstream detection tasks in challenging THz imaging scenarios.

The rest of the paper is structured as follows. Section 2 reviews the extant literature. Section 3 describes the description of dataset. Section 4 explains the research methodology. Section 5 discusses the results and discussion. Section 6 summarises the paper.

Literature review

Recent studies highlight the significant influence of object placement on detection accuracy, particularly in concealed item detection, where complexities arise from factors such as occlusion and variability in positioning. A YOLOv3-based approach has shown promising results in this domain; however, it faces critical challenges, including

[a]awahlangphibansabeth@gmail.com, [b]kdebdatta@gmail.com

DOI: 10.1201/9781003675235-21

imbalanced training data that skews model performance towards more prevalent classes and difficulties in accurately identifying small target classes due to limitations in resolution and scale [7]. These issues must be addressed to increase the performance of YOLOv3 for hidden object detection Liang et al. improved the detection process by integrating the hard example mining technique with RetinaNet, but these have limitations, such as small datasets and difficulties in small object detection [8].

With the progress on the hurdle of technology in this field, researchers are making significant strides to overcome technological limitations. Danso et al. modified the YOLOv5 model by incorporating the BiFPN network in the detection of concealed harmful weapons [5]. Architectures such as YOLOv3-53 [9] and SSD-ResNet-50 [3] have promising applications in real-time. Zhang et al. developed a whole other paradigm using a combination of thresholding and enhanced Faster R-CNN through transfer learning to deal with one of the continuing problems in the field [15]. There still exists a limitation which is the lack of spatial resolution and exhibits classification errors. Therefore, there is still a need to improve the model to classify objects accurately.

Noise reduction in images needs to be optimised to compromise between the suppression of signal interference, image detail preservation, and SNR improvement [14]. Simple techniques like median and Gaussian smoothing only help suppress interference and retain very crucial image components. Shen et al. employed a combination of Non-Local (NL) means and anisotropic diffusion algorithms for processing passive THz images and showed strong results when evaluated in terms of Mean Squared Error (MSE). However, the effectiveness of their approach was constrained by the limited size of the dataset used in their study [9]. Z. Chen et al. addressed interference issues, such as Newton's rings and white noise, in THz transmission images by using an improved mean filtering technique alongside DBSCAN clustering, effectively identifying and eliminating residual noise artefacts [4]. Danso et al. explored non-linear filtering techniques for enhancing THz images, systematically evaluating various noise reduction methods under different types of interference. The researchers demonstrated that the combined application of mean and median filters was particularly effective in mitigating Gaussian noise, especially at higher noise intensities [6]. All these different approaches represent ongoing work to improve image clarity and reduce unwanted artefacts across different imaging applications.

Dataset description

This study uses the only publicly available active THz imaging dataset for concealed object detection [7], serving as a vital resource for advancing research in this challenging field. The dataset contains 3,157 image samples with 1,347 concealed objects, as mentioned in Table 21.1, labelled in JPEG format for images and in XML format for object labels. The dataset has a split of 70% for training, 20% for validation, and the last 10% for testing, all of which creates a highly suitable setting for evaluation. With the imbalanced classes and a lot of noise in the images, achieving highly accurate detection is quite challenging. Hence, there is a need for more advanced methods in THz imaging research.

The images in this dataset were captured utilising an active THz imaging system endowed by the China Academy of Engineering Physics at an operational frequency of 140 GHz and using an array scanning mode at a resolution of 5 mm × 5 mm. Ten human subjects were involved in the acquisition: four males and six females. These humans were kept with something concealed in their clothing from different positions. Each image is either a front or back shot of the subject and has from 0 to a maximum of three hidden objects. One can see that it adds uniqueness and variety to the dataset. The imaging process processes these scenarios as they exist in practice, and this makes the dataset very useful in research into sophisticated techniques in detection of concealed objects.

Methodology and model specifications

The methodology for detecting concealed objects using active THz imaging is carried out in two phases:

Let the set of input images be {I1, I2,...,IN}, and the corresponding annotation be {A1,A2,...,AN, where I_i Î $R^{H \times W}$ (height H and width W) is the *i*-th image and A*i* is its annotation.

Table 21.1 Description of the concealed items and quantity of each object of the dataset.

Class	Items	Quantity
GA	Gun	116
KK	Kitchen knife	100
SS	Scissors	96
MD	Metal dagger	64
CK	Ceramic knife	129
WB	Water bottle	107
KC	Key chain	78
CP	Cell phone	129
CL	Cigarette lighter	163
LW	Leather wallet	78
UN	Unknown	289

Source: Author

Phase 1: Denoising
Each image Ii will undergo the denoising process using a denoising filter F, such as wavelet denoising, Gaussian filter, or NL-means filter. The denoised image $I_i^{denoise}$ is given by:

$$I_i^{denoise} = F(I_i)$$

where F is the denoising function that improves image quality by reducing noise while preserving key image features.

Phase 2: Object detection
After denoising, the enhanced images $I_1^{denoise}, I_2^{denoise} \dots I_N^{denoise}$ are fed into a YOLO model for object detection. Let the detection result for the *i*-th image be denoted by $\hat{A}_i$, which contains the predicted bounding boxes and class labels. The object detection can be expressed as:

$$\hat{A}_i = YOLO(I_i^{denoise})$$

where $\hat{A}$ represent the set of predicted annotation for the denoised image $I_i^{denoise}$.

The integration of traditional signal processing techniques with machine learning enhances the model's capability to handle noisy and complex inputs, ensuring reliable performance even under challenging conditions. This makes the proposed system well-suited for applications requiring high precision and reliability in active THz imaging.

Data pre-processing
This study explores various noise reduction techniques, focusing on filtering methods commonly employed in research. Advanced approaches are being investigated to enhance signal clarity and reduce interference effectively. Several techniques tested on our dataset are described below.

1. Mean filter: The mean filter reduces noise by replacing each pixel with the average value of its neighbouring pixels within a kernel. Mathematically, for a given pixel at position (i, j) in an image (I), the mean filter output (I'(i, i)) is calculated as:

$$I'(i, j) = \frac{1}{k*k}\sum_{m=-\frac{k}{2}}^{\frac{k}{2}}\sum_{n=-\frac{k}{2}}^{\frac{k}{2}} I(i + m, j + n)$$

where $k*k$ represents the dimensions of the kernel, and $I(i + m, j + n)$ denotes the intensity values of the neighbouring pixels.

2. Gaussian filter: The filter reduces noise and smooths an image by applying a weighted average where weights are defined by the Gaussian function:

$$G(x, y) = \frac{1}{2\pi\sigma^2} e^{-\frac{x^2+y^2}{2\pi\sigma^2}}$$

here, σ is the standard deviation determining the filter's spread. For a pixel at (I,j), the filtered intensity is:

$$I'(i, j) = \sum_m \sum_n I(i + m, j + n).G(m, n)$$

This approach smooths the image while preserving edges, as weights decrease with distance from the kernel centre.

3. NL Means filter: NL Means calculates a weighted average of all pixels, giving higher weights to similar patches. Mathematically, the NL Means filter estimates the denoised intensity $(\hat{Z}(i))$ of a pixel *i* as:

$$\hat{Z}(i) = \frac{\sum_{j \in \Omega} w(i,j)I(j)}{\sum_{j \in \Omega} w(i,j)}$$

where the weight $w(i,j)$ is defined as:

$$w(i, j) = e^{-\frac{\|P_i-P_j\|^2}{h^2}}$$

here, $\|P_i - P_j\|^2$ is the squared distance between patches, and *h* controls weight decay, ensuring noise reduction while preserving details.

4. Wavelet-based denoising: This reduces noise by decomposing image $I(x,y)$ into wavelet coefficients $W(u,v)$ via the discrete wavelet transform (DWT). A threshold *T* is applied to suppress noise:

$$\hat{W}(u, v) = \begin{cases} W(u, v) - T, & |W(u, v)| > T \\ 0, & |W(u, v)| \leq T \end{cases}$$

The inverse DWT then reconstructs the denoised image, retaining features while removing noise.

Object detection model
Most of the existing work in concealed object detection makes use of YOLO (You Only Look Once) model. In addition to this, the study presents work based on the application of YOLOv8 along with a denoising approach to greater performance in detecting concealed objects. The speed, accuracy, and customisation features of YOLOv8 make it the most suitable option for real-time applications such as surveillance and autonomous systems.

This model has an architecture that is anchor-free, thus simplifying the training process as well as performance across different datasets [12]. With great mAP

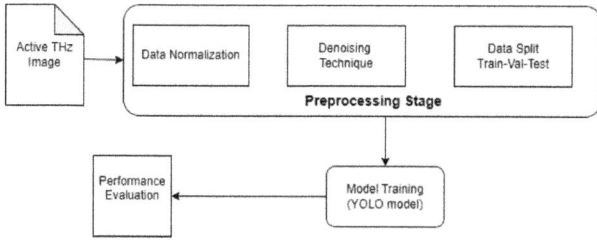

Figure 21.1 Workflow diagram of the proposed methodology
Source: Author

Table 21.2 Performance evaluation of denoising techniques.

Denoising technique	Avg PSNR	Avg SSIM	Avg MSE
Mean filter	34.99	0.99	21.03
Gaussian filter	36.61	0.99	14.48
NL Means filter	40.69	0.98	5.96
Wavelet denoising	42.99	0.98	4.87

Source: Author

performance and speedier inference times, YOLOv8 really does keep object detection reliable but not speedy processing. Additionally, there's some robust advanced data augmentation like MixUp and Mosaic, which makes this further promising for real-world application scenarios and establishes its foundation as one of the most reliable options for object detection tasks.

Results and discussion

Multiple denoising techniques were applied and subsequently evaluated against standard image quality metrics such as Peak Signal-to-Noise Ratio (PSNR), Structural Similarity Index Measurement (SSIM), and Mean Squared Error (MSE). Among the various methods implemented, wavelet denoising appears to give the best result, with a maximum PSNR of 42.96, an SSIM peak of 0.98, and a minimum MSE of 4.87, with respect to others as shown in Table 21.2 that is fed as input to the YOLOv8 model.

To assess the effect of denoising on detection of objects, the YOLOv8 model was tested on original images as well as wavelet-denoised images. Also, the confusion Matrix for YOLOv8 of every detection object with and without wavelet denoising is shown in Figures 21.2 and 21.3, in which classification accuracy and detection performance improvements were associated with the image processing effects of wavelet denoising. Additionally, the overall performance of the YOLOv8 model with wavelet denoising, including mAP@0.5 is 0.584, as shown in Figure 21.4.

The loss graph in Figure 21.5 illustrates the training progress, showing reduced loss during training when

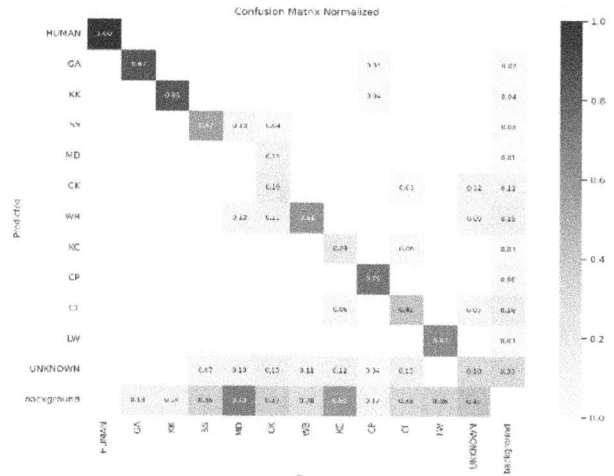

Figure 21.2 Confusion matrix for YOLOv8 without wavelet denoising, showing the classification performance across different object classes
Source: Author

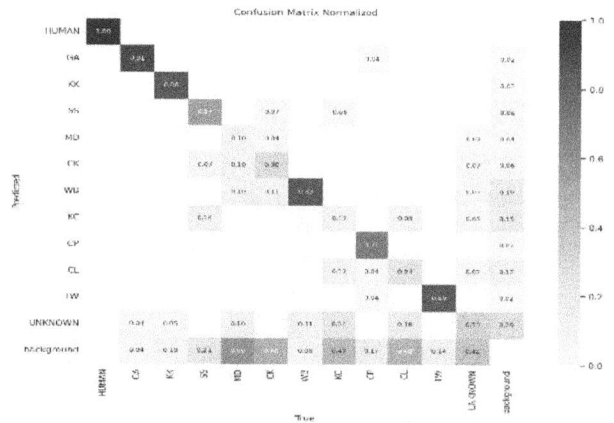

Figure 21.3 Confusion matrix for YOLOv8 with wavelet denoising, showing classification performance across classes
Source: Author

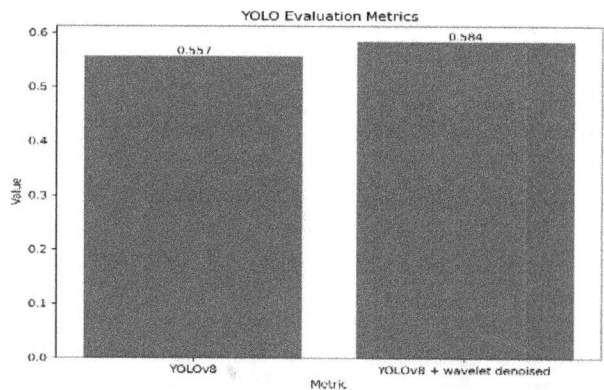

Figure 21.4 mAP@05 score of the detection method
Source: Author

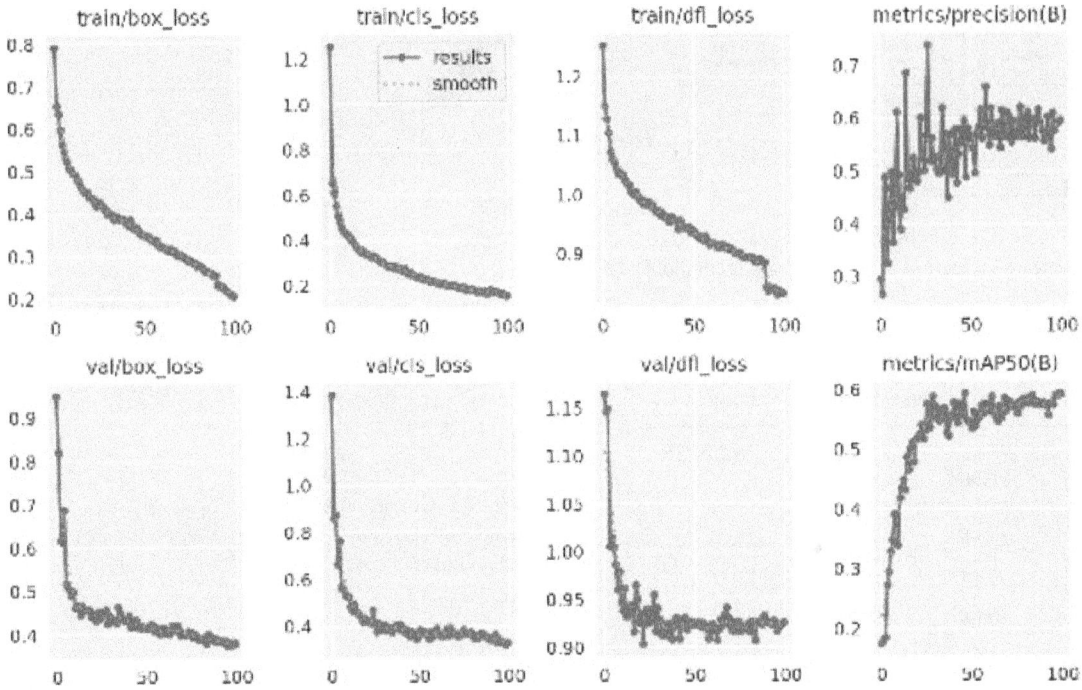

Figure 21.5 Training loss curve of YOLOv8 with wavelet denoising
Source: Author

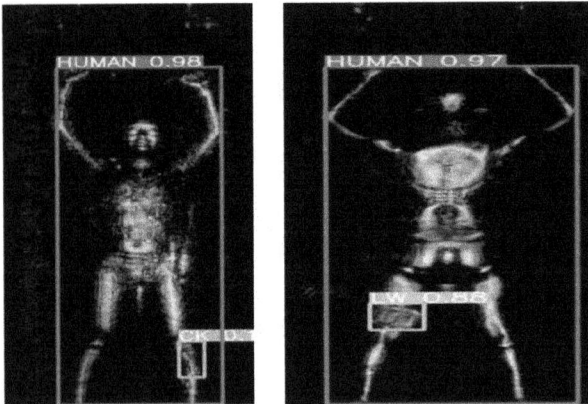

Figure 21.6 Detected object using YOLOv8 with wavelet denoising
Source: Author

wavelet denoising is applied, which highlights its positive influence on object detection performance. Figure 21.6 presents the qualitative detection outcomes of YOLOv8 following integration with wavelet denoising, particularly highlighting better object localization. While wavelet denoising improved detection performance, further accuracy gains can be achieved by exploring advanced deep-learning techniques and model optimisation strategies. However, data imbalance and noise levels in datasets remain critical challenges. Imbalance, where certain classes (e.g., CK, CL, CP) dominate over others (e.g., MD), leads to biased models with poor minority class performance, as seen in Figure 21.2, where the MD class shows lower accuracy. Noise, such as mislabelled data or outliers, exacerbates the issue by causing overfitting, where models learn noise instead of meaningful patterns, reducing generalisation, and increasing training time. Addressing these challenges is essential for enhancing model reliability and performance.

Conclusion

In this study, we explored the use of various denoising techniques to enhance the quality of THz images and improve object detection accuracy. Among the methods, wavelet denoising demonstrated the best performance, achieving the highest PSNR, lowest MSE, and comparable SSIM, as evaluated on standard metrics. When integrated with the YOLOv8 model, wavelet-denoised images significantly improved detection performance, with high mAP scores and reduced training loss compared to non-denoised images. These findings highlight the critical role of targeted preprocessing in enhancing THz imaging systems. Future work will focus on addressing challenges such as dataset scarcity and optimising deep learning models to further enhance detection accuracy.

References

[1] Bandyopadhyay, A., Sengupta, A. (2022). A review of the concept, applications and implementation issues of terahertz spectral imaging technique. *IETE Technical Review*, 39(2), 471–489.

[2] Bründermann, E., Hübers, H. W., Kimmitt, M. F., Bründermann, E., Hübers, H. W., Kimmitt, M. F. (2012). Terahertz imaging. *Terahertz Techniques*, 301–340.

[3] Cheng, L., Ji, Y., Li, C., Liu, X., Fang, G. (2022). Improved SSD network for fast concealed object detection and recognition in passive terahertz security images. *Scientific Reports*, 12(1), 12082.

[4] Chen, Z., Zou, Z., Liu, H., Zhang, N., Yan, H., Feng, J., Wang, C. (2022, August). A compound noise reduction algorithm for terahertz imaging. In 2022 47th International Conference on Infrared, Millimeter and Terahertz Waves (IRMMW-THz). IEEE, 1–2.

[5] Danso, S. A., Shang, L., Hu, D., Odoom, J., Liu, Q., Nana Esi Nyarko, B. (2022). Hidden dangerous object recognition in terahertz images using deep learning methods. *Applied Sciences*, 12(15), 7354.

[6] Danso, S., Liping, S., Deng, H., Odoom, J., Appiah, E., Etse, B., Liu, Q. (2021). Denoising terahertz image using non-linear filters. *Comput. Eng. Intell. Syst*, 12.

[7] Liang, D., Xue, F., Li, L. (2021). Active terahertz imaging dataset for concealed object detection. *arXiv preprint arXiv:2105.03677*.

[8] Li, L., Xue, F., Liang, D., Chen, X. (2021). A hard example mining approach for concealed multi-object detection of active terahertz image. *Applied Sciences*, 11(23), 11241.

[9] Pang, L., Liu, H., Chen, Y., Miao, J. (2020). Real-time concealed object detection from passive millimeter wave images based on the YOLOv3 algorithm. *Sensors*, 20(6), 1678.

[10] Park, H., Son, J. H. (2021). Machine learning techniques for THz imaging and time-domain spectroscopy. *Sensors*, 21(4), 1186.

[11] Shen, X., Dietlein, C. R., Grossman, E., Popovic, Z., Meyer, F. G. (2008). Detection and segmentation of concealed objects in terahertz images. *IEEE Transactions on Image Processing*, 17(12), 2465–2475.

[12] Terven, J., Córdova-Esparza, D. M., Romero-González, J. A. (2023). A comprehensive review of YOLO architectures in computer vision: From YOLOV1 to YOLOV8 and YOLO-NAS. *Machine Learning and Knowledge Extraction*, 5(4), 1680–1716.

[13] Tian, N., Wang, X., Xu, Y., Wang, R., Lian, G., Zhang, L. (November 2022). Deep learning enabled hidden target detection in terahertz images. In 2022 International Conference on Sensing, Measurement & Data Analytics in the era of Artificial Intelligence (ICSMD). IEEE, 1–6.

[14] Trofimov, V. A., Trofimov, V. V., Shestakov, I. L., Blednov, R. G. (May 2016). Concealed object detection using the passive THz image without its viewing. In Passive and Active Millimeter-Wave Imaging XIX. SPIE, vol. 9830, 88–98.

[15] Yuan, M., Zhang, Q., Li, Y., Yan, Y., Zhu, Y. (2021). A suspicious multi-object detection and recognition method for millimeter wave SAR security inspection images based on multi-path extraction network. *Remote Sensing*, 13(24), 4978.

22 A novel ensemble rank aggregation feature selection algorithm for classification

Nur Alom[a] and Rubul Kumar Bania[b]

Department of Computer Science, Birangana Sati Sadhani Rajyik Vishwavidyalaya, Golaghat, Assam, India

Abstract

In contemporary data analysis time, managing the high dimensionality or a large number of features in a dataset is essential, making feature selection a critical pre-processing step for classification tasks. Identifying the most relevant features not only streamlines the model but also reduces unnecessary complexity, which can otherwise degrade classification performance. This paper introduces and evaluates a novel ensemble rank aggregation feature selection algorithm designed for classification. The method integrates multiple base feature selectors—Information Gain, Gain Ratio, One-Rule, ReliefF, and Symmetrical Uncertainty—to effectively identify relevant features from high-dimensional datasets. The aggregation process uses mean and min-majority ranking methods and, subsequently, their intersection, leveraging the power of each single selector to generate a robust feature subset. Experiments were conducted on five different medical domain datasets. Experimental results show that the proposed ensemble method rationalises the model by decreasing the feature set while maintaining or even improving classification performance.

Keywords: Rank aggregation, feature selection, medical data, classification

Introduction

Feature Selection (FS) is a vital pre-processing technique in data mining and machine learning domains, with over three decades of extensive research and development. The increasing size of datasets, both in the number of instances and features, particularly in fields such as medical research, network intrusion detection, and customer behaviour analysis, poses significant scalability and performance challenges for learning algorithms [6]. FS tackles these issues by enhancing the efficiency and accuracy of classification models, eliminating redundant, irrelevant, or noisy features from high-dimensional datasets. The FS process involves selecting a subset of relevant features based on specific evaluation criteria. Broadly, FS methods can be grouped into feature subset selection and feature ranking. Feature ranking evaluates each feature individually, assigning a score and sorting them accordingly, while feature subset selection identifies a subset of features that collectively enhance model performance [9].

Recently, ensemble FS has gained attention in research communities [3]. Aiming to improve the classification performance, this approach integrates various ranking techniques to form a unique ranking list. The ensemble technique typically involves two steps:

i. Producing various ranking lists: These lists are obtained from various ranking techniques.
ii. Aggregating rank lists: An aggregation function is used to get an amalgamated list.

The second step is especially critical, as the options of aggregation methods can affect to the effectiveness of the ensemble [7].

Initial studies on the ensemble of feature ranking techniques were done by Rokach and Chizi [2]. This experiment was conducted to check whether an ensemble of feature subsets improves classification accuracy over separate rankers. The dataset utilised for this experiment was taken from UCI machine learning repository. Another study on ensemble of FS techniques by Saeys [1] involved two datasets and two wrapper approaches, using symmetrical uncertainty, relief, random forests and linear support vector machines as FS techniques.

Our work primarily focuses on the ensemble of filter-based model of feature ranking techniques. The key advantage of using a filter-based model is its independence from the learning model, ensuring an unbiased selection process. Additionally, filter models enable algorithms to maintain a simple and straightforward structure.

By observing the requirements of FS model to study the data of medical and healthcare domain, the main objectives of this study are:

i. To review the existing base FS model that can be applied to prepare an EFS model for analysis of medical domain datasets.
ii. To prepare an ensemble mechanism for robust selection of features from a pool feature rank lists.

---

[a]trynuralom@gmail.com, [b]rubul.bania@gmail.com

DOI: 10.1201/9781003675235-22

The structure of the rest of the article is outlined below. In Section 2, we investigate into a review of previous studies related to ensemble FS problem. Section 3 demonstrates the proposed methodology. In Section 4, we have presented the experimental results. Section 5 covers the conclusions and future work for this research.

Proposed method

The pipeline of the proposed methodology is shown in Figure 22.1, which will help in better understanding the proposed method. In the proposed method, there are two main phases: the base selector and the aggregator.

In base selector, there are five different ranking methods, namely Information Gain (IG), Gain Ratio (GR), One-Rule (One-R), ReliefF, and Symmetrical Uncertainty (SU) [4,5,7]. These base feature selectors produce distinct ranking lists of features present in a dataset. They generate the ranking list by computing the relevant score using the method-specific criterion, and then rank all the features in descending or ascending order of their scores. After that, finally generate a ranked list of features, with the top-ranked features being the most relevant according to the method. The different ranking methods of base selectors are explained below:

1. Information Gain (IG): One of the most popular methods for evaluating features is IG, a commonly used univariate model. It is based on information theory and helps rank features of a dataset according to how well they meet a specific evaluation criterion. After ranking, the top features are selected using suitable thresholds. IG reduces uncertainty when the value of a feature is unknown, thereby helping to identify the correct class. Before ranking the features, the entropy of the distribution is calculated to measure uncertainty. This uncertainty can be determined by the entropy of the distribution, sample entropy, or model entropy of the dataset. IG is defined as the amount by which the entropy of X decreases to reflect additional information about X provided by Y and it is expressed as

$$IG(X|Y) = H(X) - H(X|Y)$$

where,
$H(Y)$: entropy of the class variable and
$H(Y|X)$: conditional entropy of the class given the feature.

2. Gain Ratio (GR): GR is a widely used univariate model based on information theory, designed to refine the concept of IG. One of its key purposes is to correct IG's bias towards features with many distinct values. GR assigns a high value when the data in a dataset D is evenly distributed and a low value when

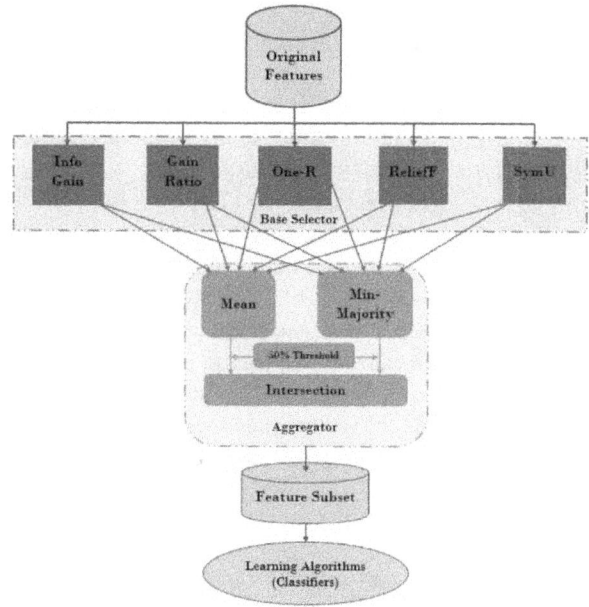

Figure 22.1 Block diagram of the proposed ensemble FS mechanism
Source: Author

most of the data is concentrated in a single branch of a feature. By taking into account the number and size of branches, GR adjusts IG using a factor known as split information (SplitInfo). This SplitInfo is calculated using the entropy distribution of the feature's instance values. GR of a feature Y for a feature value v can be calculated as follows:

$$GR(X) = \frac{IG(X)}{SplitInfo(X)}$$

where

$$SplitInfo = -\sum_{i=1}^{v} \frac{|Di|}{|D|} * \log_2 \frac{|Di|}{|D|}$$

The SplitInfo value measures the amount of information gained by splitting the dataset D into v parts based on the findings from the assessment of feature X. Each partition, D_i, contains the samples in D where the value of X is v.

3. One-Rule (One-R): This technique is similar as utilising single-level decision trees. It calculates the predictive correctness of all features by making rules relying on each single feature. As both training and test datasets are applied, classification accuracy can be calculated for each rule and, hence, for each feature. Based on these classification results, a ranked list of features is generated.

4. ReliefF: The ReliefF filter works by selecting a subset of instance randomly from dataset and detecting its nearest neighbour, in which one from like class and another from unlike class. After that, it updates

the relevance weight for each feature by comparing the feature values of neighbours and randomly selected instance. The idea for this technique is straightforward: a good feature should differentiate instances from unlike classes while having harmonious values for instances within like class.

5. Symmetrical Uncertainty (SU): SU is built on the concepts of information theory. It corrects the bias of IG towards features with higher values by regulating its range. An SU value of 0 means that features X and Y are not dependent, and a value of 1 implies that knowing the value of one feature is certain to predict the value of the other. Additionally, SU treats both features in a pair symmetrically.

$$SU(X,Y) = 2 * \frac{IG(X|Y)}{H(X) + H(Y)}$$

For aggregation, there are three main steps namely mean, min-majority, and intersection. First, mean rank lists and min-majority rank lists are found out by using all the rank lists generated by base selectors. After that, the intersection of those two lists is found out by applying a 50% threshold value to both lists. The steps are explained in detail below:

1. Mean: The main motive is to find the mean value for each feature's ranks produced by different ranking methods of base selectors, and give the high importance to the lower mean value. That means for a particular feature, lower the mean value higher the rank value. Suppose, there are m number of ranking lists, then the mean value M_i of ith feature (F_i) will be

$$M_i = \frac{\sum (ranks\ of\ F_i\ across\ m)}{m}$$

where,
M represents the mean value of each feature.
Now, to find the rank list, sort the mean value of each feature in ascending order so that lower mean value feature will get higher rank. The feature indices of the sorted list are the final rank list produced by mean method.

2. Min-majority: The main idea behind the min-majority is that for a particular feature, if a rank value appears majority times across all ranking methods, then that rank value will be the min-majority rank value for that feature. Otherwise, the minimum rank value of that particular feature across the methods will be the min-majority rank for that feature. Suppose, there are m number of ranking lists and r is a rank value which appears majority time (i.e., $m/2$) across m for a particular feature F_i, then r will be the min-majority rank of F_i. Otherwise, the minimum r across m will be the min-majority for F_i.

Now, rank the list in ascending order by min-majority rank value so that lower rank will be assigned higher importance. The feature indices of the sorted list of min-majority rank form the last rank list generated by min-majority method.
Upon completion of these two steps, a 50% cut-off is applied to both mean and min-majority rank list i.e., top 50% features from mean rank list and to 50% features from min-majority rank list are chosen for third step.

3. Intersection: In the third and final step, determine the common elements which occur in both mean and min-majority rank list after imposing 50% threshold on both the rank lists.
After the successful completion of the third step, whatever elements found are the selected features by the proposed method.

Classification
Once the pertinent features have been selected from the entire feature set available in the dataset, the performance of the suggested FS method should be tested with the classifier. So, to perform this task, classification is done using k-fold cross validation by taking $k = 10$ on five popular classifiers, namely k-Nearest neighbour (k-NN), Decision Tree (DT), Naïve Bayes (NB), Random Forest (RF), and Support Vector Machine (SVM) [8].

Dataset and experiment setup
To conduct this experiment, five different datasets of the medical domain is used. These datasets are collected from the UCI machine learning repository [https://archive.ics.uci.edu/datasets], namely Hepatitis, Lung Cancer, Dermatology, Breast Cancer Wisconsin Diagnostic (BCWD), and Arrhythmia. Table 22.1 highlights the details of the used datasets.
Procedure and experimental setup are as follows:

1. The configuration of the system used for the experiment is AMD Ryzen 5 4500U with Radeon Graphics with 2.38 GHz clock speed and 8 gigabytes of random access memory.

Table 22.1 Dataset details.

Dataset	Instances	Features	Classes
Hepatitis	155	19	2
Dermatology	366	34	2
BCWD	569	30	2
Lung Cancer	32	56	3
Arrhythmia	452	279	16

Source: Author

2. The Waikato Environment for Knowledge Analysis (WEKA) [https://ml.cms.waikato.ac.nz/weka] software is used for the base selector and the datasets are converted to .arff file format so that WEKA can read the file.
3. For the aggregation, Python is used as a programming environment with Google Colab. Data frame data structure is used for processing different rank lists.

Experimental results and analysis

This section discusses the experimental results obtained from different classifiers namely, KNN, DT, NB, RF, and SVM with different datasets for the proposed ensemble FS technique. The classification accuracies of different classifiers with the full, original set of features are shown in Table 22.2. Out of five dataset, three shows better results in original feature set with SVM classifier.

In Tables 22.3 to 22.7, results of the classifiers with reduced feature sets on five different datasets are shown. Threshold value for selecting the ranked features for the various FS methods namely GR, IG, One-R, ReliefF, and SU is set as 50%.

In the hepatitis dataset, out of five classifiers, KNN, DT, NB have shown the exactly same percentage value

of correctly classified instances with selected features of the proposed method and the original dataset, which is clearly visible in Tables 22.2 and 22.3. And in RF and SVM, the percentage value is slightly degraded with the proposed method than the original dataset. Table 22.3 shows the detailed results of the hepatitis dataset. But the proposed method achieved 84.51% in correctly classifying instances, which is comparatively a better result than the other ranker methods.

In the case of dermatology dataset with the KNN classifier, the percentage of accurately classified instances is declined by 0.27%. But from Table 22.4, it can be seen that with the rest of the classifiers, the proposed method shows exactly the same value as the original dataset. And interestingly, NB, RF, and SVM show the same values with the original dataset as well as the proposed method, i.e., 99.72%. With GR ranker, it shows 100% of correctly classified instances, and with the proposed methods NB, RF, and SVM, it shows 99.72% of correctly classified instances.

Table 22.5 shows experimental results for the BCWD dataset. From the table, it is clearly visible that RF shows domination results for maximum feature section methods.

Table 22.2 Classification result of original feature set of different datasets.

Dataset	Classifiers				
	KNN	**DT**	**NB**	**RF**	**SVM**
Hepatitis	80.64	83.87	84.51	**85.16**	**85.16**
Dermatology	99.18	98.36	99.72	99.72	99.72
BCWD	95.95	93.32	92.97	95.95	**97.71**
Lung Cancer	37.5	50.00	**62.50**	46.87	40.62
Arrhythmia	52.87	64.38	62.38	67.25	**70.13**

Source: Author

Table 22.3 Comparison of classification result in hepatitis dataset.

FS methods	Classifiers				
	KNN	**DT**	**NB**	**RF**	**SVM**
GR	80.64	**83.22**	**83.22**	81.93	**83.22**
IG	**85.16**	84.51	**85.16**	82.58	83.87
One-R	81.29	80.64	82.58	81.29	**83.22**
RefliefF	80.00	79.35	**82.58**	81.29	**82.58**
SU	82.58	83.22	**85.81**	85.16	83.87
Proposed method	80.64	83.87	**84.51**	83.23	83.87

Source: Author

Table 22.4 Comparison of classification result in dermatology dataset.

FS methods	Classifiers				
	KNN	**DT**	**NB**	**RF**	**SVM**
GR	99.45	98.36	99.45	99.45	**100**
IG	98.63	98.36	**99.72**	99.45	**99.72**
One-R	98.63	98.36	99.45	99.45	**99.72**
RefliefF	**99.72**	98.36	**99.72**	99.18	**99.72**
SU	99.18	98.36	**99.72**	**99.72**	**99.72**
Proposed method	98.90	98.36	**99.72**	**99.72**	**99.72**

Source: Author

Table 22.5 Comparison of classification result in BCWD dataset.

FS methods	Classifiers				
	KNN	**DT**	**NB**	**RF**	**SVM**
GR	92.97	92.44	92.44	**94.72**	94.02
IG	92.97	92.26	92.44	**94.72**	94.02
One-R	92.97	92.26	92.44	**94.37**	94.02
RefliefF	96.13	94.55	94.55	96.66	**97.01**
SU	92.97	92.26	92.44	**94.72**	94.02
Proposed method	93.49	92.79	92.97	**94.20**	93.67

Source: Author

The proposed method achieved 94.20% of correctly classified instances with RF, which is nearly equal to other FS methods with RF.

The NB classifier has fully dominated other classifiers on the lung cancer dataset, with comparatively high values of correctly classified results. From Table 22.6, it can be seen that GR has the highest percentage, i.e., 78.12%, while the proposed method and IG also performed well, with 71.87% of correctly classifying instances in the dataset.

In the arrhythmia dataset, the SVM classifier has dominated all the classifiers with each FS technique, except for the proposed method. Table 22.7 explains the experimental results of the arrhythmia dataset. The proposed method correctly classifies 71.23% in the RF classifier, whereas One-R has 73.00% correctly classified with the SVM classifier.

From the above experimental results using five diverse medical datasets and classifiers, including KNN, DT, NB, RF, and SVM, highlighted the efficiency of the proposed method. For the hepatitis dataset, ranker methods have the highest accuracy of 85.81% with 9 features, while the proposed method has 84.51% with 8 features, and for the dermatology dataset, maximum rankers achieved 99.72%

Table 22.6 Comparison of classification result in lung cancer dataset.

FS methods	Classifiers				
	KNN	DT	NB	RF	SVM
GR	59.37	62.50	**78.12**	59.37	68.75
IG	53.12	65.62	**71.87**	65.62	59.37
One-R	43.75	50.00	**65.62**	59.37	50.00
RefliefF	53.12	59.37	**71.87**	68.75	65.62
SU	53.12	62.5	**68.75**	65.62	62.5
Proposed method	53.12	62.5	**71.87**	59.37	53.12

Source: Author

Table 22.7 Comparison of classification result in arrhythmia dataset.

FS methods	Classifiers				
	KNN	DT	NB	RF	SVM
GR	55.08	65.26	67.92	71.23	**72.12**
IG	55.08	65.26	67.92	70.35	**71.90**
One-R	54.42	65.70	65.48	71.68	**73.00**
RefliefF	54.20	66.59	65.26	69.46	**71.01**
SU	55.08	65.26	67.92	70.57	**71.68**
Proposed method	55.97	65.04	66.59	**71.23**	70.79

Source: Author

of the highest accuracy with 17 features, which is the same as for the proposed method with only 14 features. But for the BCWD dataset, with 15 features, one ranker achieved 97.01% accuracy, which is the highest value and most of the rankers achieved 94.72% accuracy, with 13 features, while the proposed method achieved 94.20% accuracy. Again, for the lung cancer dataset, one ranker obtained 78.12% accuracy with 28 features, but in the case of the proposed method, with 25 features, it obtained 71.87% accuracy. Now, for the endmost and largest dataset, i.e., arrhythmia, containing 279 features, out of which, with 139 features, one ranker has obtained 73% accuracy, but with 116 features, the proposed method has obtained 71.23% accuracy. In most cases, it provides competitive or better classification accuracy than the original dataset and individual ranking techniques. From the above inquiry, it is clear that the proposed ensemble FS method can be a fruitful solution for the FS problem.

Conclusion and future work

In modern-day data analysis, the high dimensionality of data or huge features of data makes FS an important pre-processing step for classification. Finding out the most important features allows the classification model to run smoothly because the engagement of too many irrelevant features causes the increment of complexity for the learning model and ultimately affects the classification performance. In this research, a new ensemble rank aggregation FS algorithm is proposed and it is also tested for classification tasks. By combining multiple base feature selectors—IG, GR, One-R, ReliefF, and SU—the proposed method effectively identified relevant features from high-dimensional datasets. The aggregation process, which employed mean and min-majority ranking methods followed by intersection, demonstrated its capability to balance the strengths of individual selectors and produce a robust feature subset. The experimental results on five different diverse datasets confirm that reducing the feature set with the ensemble method can simplify the model while maintaining or enhancing performance. In the future, we aim to apply our proposed FS method to various datasets, incorporating advanced re-sampling techniques and exploring its integration with other machine learning models.

References

[1] Saeys, Y., Abeel, T, Y., Peer, Y. V. (2008). Robust feature selection using ensemble feature selection techniques, W. Daelemans et al. (eds.): In ECML PKDD. Part II, LNAI 5212, 313–325.

[2] Rokach, L., Chizi, B., Maimon, O. (2006). Feature selection by combining multiple methods. Advances in Web Intelligence and Data Mining, 295–304.

[3] Hoque, N., Singh, M., Bhattacharyya, D. K. (2018). EFS-MI: an ensemble feature selection method for classifica-

tion: an ensemble feature selection method. *Complex & Intelligent Systems*, 105–118.

[4] Abut, F., Akay, M. F., George, J. (2 019). A robust ensemble feature selector based on rank aggregation for developing new VO2max prediction models using support vector machines. *Turkish Journal of Electrical Engineering & Computer Science*, 3648–3664.

[5] Borandag, E., Ozcift, A., Kilinc, D., Yucalar, F. (2019). Majority vote feature selection algorithm in software fault prediction. *ComSIS*, 515–539.

[6] Seijo-Pardo, B., Bolón-Canedo, V., Alonso-Betanzos, A. (2019). On developing an automatic threshold applied to feature selection ensembles. *Information Fusion*, 227–245.

[7] Bania, R. K., (2022). R-GEFS: condorcet rank aggregation with graph theoretic ensemble feature selection algorithm for classification. *International Journal of Pattern Recognition and Artificial Intelligence*, 2250032.

[8] Claude, E., Leclercq, M., Thébault, P., Droit, A., Uricaru, R. (2024). Optimizing hybrid ensemble feature selection strategies for transcriptomic biomarker discovery in complex diseases. *NAR Genomics and Bioinformatics*, lqae079.

[9] Morán-Fernández, L , Bolón-Canedo, V., (2023). Finding a needle in a haystack: insights on feature selection for classification tasks. *Journal of Intelligent Information Systems*. https://doi.org/10.1007/s10844-023-00823-y.

23 Analysing the impact of Word2Vec on stance detection in low-resource languages: a case study on Manipuri editorial articles

Pebam Binodini[1,a], Sunita Sarkar[1] and Kishorjit Nongmeikapam[2]

[1]Department of Computer Science and Engineering, SOT, Assam University, Silchar, Assam, India

[2]Department of Computer Science and Engineering, IIIT Manipur, Mantripukhri, Manipur, India

Abstract

This research examines how the size of datasets, methods for resampling, and the choice of models impact stance detection using Word2Vec embeddings. The experiments are conducted on a newly compiled dataset that includes articles in Manipuri, a language with limited resources. The effectiveness of four machine learning algorithms—Support Vector Machine (SVM), k-Nearest Neighbors (k-NN), Logistic Regression (LR), and Random Forest—was tested using datasets of various sizes (300, 450, 600, 800, and 1,000 articles). We used SMOTETomek resampling techniques to address the data imbalance issue. The results further revealed that, among all the models, Random Forest, alongside SMOTETomek with an F1-score of 80%, performed significantly better than all the other models in stance detection tasks. In particular, SMOTETomek significantly enhanced the performance of other models, such as k-NN and Logistic Regression, mainly in the highest combination of articles. The results are promising, particularly considering the limited resources available for this language segment and the inherent challenges associated with Word2Vec embeddings. These findings emphasise the importance of balanced data, diverse datasets, and robust models in stance detection, demonstrating Random Forest's suitability.

Keywords: Stance detection, Word2Vec, low-resource languages, Manipuri editorials, machine learning classifiers, custom word embeddings

Introduction

Stance detection is a vital component of natural language processing (NLP) focused on identifying a text's position regarding a specific subject. It is widely used in political research, opinion analysis, and social media monitoring, particularly for spotting bias and misinformation. The Word2Vec framework [1] has gained popularity in high-resource languages for understanding the complex relationships between words through vector representation. The effectiveness of Word2Vec in low-resource languages still needs to be explored. It has been utilized successfully in many natural language processing tasks, such as sentiment analysis, text classification, and stance detection, because it can represent words as dense vectors, enhancing the classification of opinions within relevant contexts. Compared to high-resource languages where pre-trained embeddings like BERT exist, low-resource languages often need to train embeddings from scratch on domain-specific datasets, posing additional challenges in resource-constrained settings.

In this experiment, Word2Vec embeddings are explored using Manipuri editorial articles across various dataset sizes,and also focus on the class imbalance in stance recognition. The "discuss" class dominates, while the 'for' and 'against' classes are in the minority. To improve performance, SMOTETomek is applied to generate synthetic samples for the minority classes and to remove overlapping samples [12][8]. The goal is to provide insights into the best natural language processing (NLP) practices within resource-constrained contexts.

Section 2 presents the related works, Section 3 discusses the methodology followed, Section 4 narrates the results and discussions, and Section 5 narrates the conclusion and future directions.

Related Works

Word2Vec, by Mikolov et al. [1], is a well-known method for generating embeddings that capture syntactic and semantic relationships, making it suitable for NLP tasks like stance detection, sentiment analysis, and text classification. Training embeddings from scratch is particularly advantageous in low-resource and domain-specific contexts, as it enables models to capture nuanced cultural and contextual distinctions,as shown in the work ofRinni Janati et al. [2] andMourad et al. [3]. The analysis conducted by Mohammad et al. [4] utilized Word2Vec embeddings to describe training features that enhanced classification accuracy for stance categories such as "for," "against," and "discuss." Significant research has been done on stance detection by applying word embeddings trained on larger text corpora, particularly using the Word2Vec modeling framework. Косів et al. [5] showed that while TF-IDF

[a]tambipebam@gmail.com

DOI: 10.1201/9781003675235-23

performed better in certain scenarios, Word2Vec provided valuable semantic information, particularly when used with CNN and LSTM models. Koсiв et al. [5] also confirmed the effectiveness of SMOTETomek in political stance classification; however, its application to low-resource languages is still limited. Additionally, Augenstein et al. [6] demonstrated that Word2Vec performs even more effectively when combined with target-specific encodings, making it easier to identify stances on targets that had not been previously encountered. Domingues et al. [7] evaluated data imbalance approaches using machine learning and word embeddings, emphasizing their impact on minority class prediction. Tran et al. [9] confirmed the effectiveness of training Word2Vec embeddings from scratch in Vietnamese stance detection tasks. Dey et al. [10] adopted a two-phase LSTM with attention to improve the stance detection task. Ghanem et al. [11] and Leng et al. [8] illustrate that techniques like SMOTE and Tomek Links can improve the performance of classifiers for minority classes.

Methodology

This section discusses the steps conducted during the experiment, and the model architecture is shown in Figure 23.1. The process starts with data collection and annotations, followed by pre-processing to cleanse the data. Next, the embeddings generated by Word2Vec are given as input for each combination of articlesubsets containing 300, 450, 600, 800, and 1000 articles. Each combination undergoes a train-test split with an 80:20 ratio. SMOTETomek is employed to balance the training data for combinations of 600, 800, and 1000 articles. The processed data is provided to the classifiers for training, and the model's effectiveness is evaluated. The experiment is conducted on a system with 32 GBof RAM with an AMD Ryzen 7 processor.

Data collection
The dataset was gathered from Sanaleibak[1], a local newspaper, and consists of editorial articles covering various domains with a primary focus on governance-related topics.

Data annotations
Each article was annotated by three native Meitei speakers, who labelled the stance towards state governance. The final stance label was determined by majority voting, where at least two annotators had to agree. The dataset was categorised into three stance classes: **For**: Supporting governance policies, **Against:** Opposing governance policies, and **Discuss**: Neither supporting nor opposing. This process ensured high-quality annotations for accurate stance detection. Table 23.1 shows the distribution of articles across stance categories, and a sample annotations of dataset is shown in Table 23.2.

[1]https://sanaleibak.in

Pre-processing
The annotated dataset underwent several preprocessing steps to prepare it for stance detection. First, the script was converted to Unicode to ensure consistent encoding. Second, the script was normalized to remove inconsistencies, and then tokenized for further processing. After

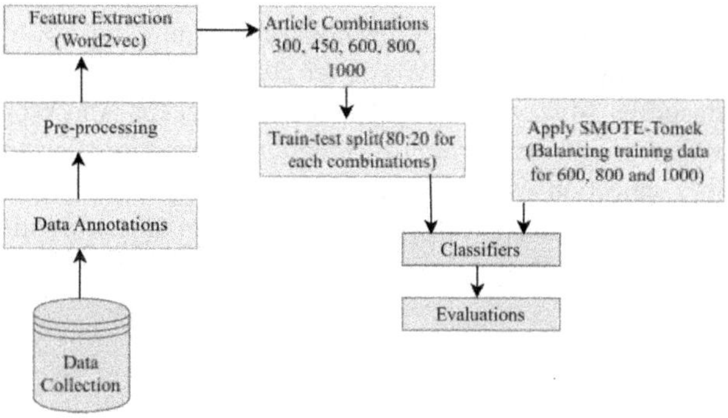

Figure 23.1 Proposed architecture model
Source: Author

Table 23.1 Dataset distribution.

For	Against	Discuss
150	440	605

Source: Author

Table 23.2 Sample dataset annotations.

Sl. No.	Article Body	Label
1	ꯒꯣꯙꯇꯤꯔꯦꯛꯑꯨꯗ꯭ꯍꯥꯞꯒꯤ ꯙꯥꯡꯍꯥꯡꯎꯝꯒꯤ ꯗꯨꯁꯥꯗ ꯇꯪꯗꯨꯄꯣ ꯍꯤꯔꯦꯀꯇ ꯑꯦꯞꯒꯥꯄꯒꯤ ꯍꯨꯁꯥꯎꯒꯤ ꯅꯦꯕꯅꯒꯤ ꯑꯩꯒꯤꯑꯣ ꯇꯃꯣꯛꯇꯤꯟ ꯍꯤꯔꯦꯀꯒꯥꯄꯤ ꯒꯣꯙꯨꯗ꯭ꯗꯇꯛ ꯐꯍꯥꯐꯝꯒꯤꯗ ꯐꯣꯟꯐꯝꯂ ꯐꯥꯑꯣꯗꯥꯅꯇꯤ…	Against
1	In a parliamentary democratic society, government leaders must remember that the people are the true leaders…	
2	ꯍꯇꯐꯒꯥꯕꯤ ꯂꯨꯇꯔꯍꯇ ꯍꯤꯄꯥꯞ ꯍꯥꯞꯅꯥꯐꯒꯨꯐ ꯇꯑꯨꯍꯔꯇꯔꯗ ꯂꯨꯞꯗꯐꯤꯦꯇꯥꯡꯒꯤ ꯏꯒꯥꯐ ꯁꯥꯞꯨꯈꯦꯐꯡꯗ ꯍꯤꯦꯍꯔꯣꯟ ꯗꯗ …	For
2	The Manipuri community must focus on improving financial stability for self-reliance..	
3	ꯑꯐ ꯇꯣ ꯑꯥꯐ ꯒꯨꯐ ꯇꯨꯍꯤꯕꯐ ꯇꯐꯑꯨꯍꯧꯦꯐ ꯍꯤꯦꯍꯇꯤ ꯇꯊꯥꯑꯣꯒꯤꯄꯤ ꯒꯥꯐꯥꯐ ꯍꯤꯇꯤꯂꯣꯙꯦꯐ ꯐꯇꯍꯐꯥꯏ ꯂꯨꯎꯍ ꯍꯇꯐꯒꯥꯕꯥꯈ ꯇꯦꯔꯐꯤꯋꯝꯒꯤ …	Neutral
3	Since Narendra Modi became Prime Minister, he has visited Manipur four times, including this visit.	

Source: Author

this, a custom list of Manipuri stop words was created and removed. These steps optimized the dataset for embedding generation and other feature extraction methods, ensuring a high-quality foundation for stance detection.

Feature extractions

Feature extraction transforms raw data into meaningful numerical representations, enhancing model performance by focusing on relevant information and minimising noise. This study employs the Word2Vec skip-gram model introduced by Mikolov et al. [1]. The core concept is to provide a probability to each surrounding word according to its connection with the center word. Mathematically, the probability of a context word w_j occurring in the surroundings of a center word w_i and is given in Equation 23.1:

$$P(w_j/w_i) = \frac{\exp(e_j^T c_i)}{\sum_{k=1}^{N} e_k^T c_i} \qquad (23.1)$$

Where: c_i represents the embedding vector of the target word w_i and e_j represent the embedding vector of the context word w_j, and N is the total size of the vocabulary.

The denominator in the equation normalises probabilities by adding all language terms in N, ensuring an appropriate probability distribution. Word2Vec is implemented using Gensim library in python. Algorithm 23.1 elaborates on the steps taken to create the custom Word2Vec model for the document vector.

Class balancing

The SMOTETomek [13] technique was applied to address class imbalance in the training set. Developed by Chawla et al. [12] and refined by Tomek [14], this technique combines SMOTE (Synthetic Minority Oversampling Technique) with Tomek Links. SMOTE generates synthetic samples for underrepresented classes, while Tomek Links removes noisy or overlapping examples from the majority class, ensuring a cleaner and more balanced dataset.

Classifiers: Here, four machine learning classifiers are used for the experiment. They are Support vector machines, logistic regression, k-NN, and Random Forest.

Evaluation metrics

Precision: Precision calculates how accurately the model identifies true positives among all predicted positives and is shown in Equation 23.2.

$$\text{Precision} = \frac{\text{Correct Positive(CP)}}{\text{Correct Positive(CP)+Wrong Positive(WP)}} \qquad (23.2)$$

Recall: Recall, sometimes referred to as sensitivity, measures how well the model identifies all actual positives and is shown in Equation 23.3.

$$\text{Recall} = \frac{\text{Correct Positive(CP)}}{\text{Correct Positive(CP)+Missed Positive(MP)}} \qquad (23.3)$$

F1 score: The F1 score blends precision and recall into a single value using their harmonic mean and is shown in Equation 23.4.

$$F1\ Score = \frac{2 \times (Precision \times Recall)}{Precision + Recall} \qquad (23.4)$$

Accuracy: Accuracy measures the proportion of correct predictions made by the model and is shown in Equation 23.5.

$$Accuracy = \frac{CP + CN}{CP + CN + WP + MP} \qquad (23.5)$$

Where, CP: Instances correctly predicted as positive, CN: Instances correctly predicted as negative, WP (Wrong Positive): Instances incorrectly predicted as positive, and MP(Missed Positive): Instances positive but predicted as negative.

Algorithm 23.1: Custom Word2Vec model

Input: Pre-processed tokenised dataset `data['filtered_tokens']`.
Output: Trained Word2Vec model `word2vec_model` and document vectors.
1. Convert `data['filtered_tokens']` into a list of tokenised documents:
`documents = data['filtered_tokens'].tolist()`.
2. Train a Word2Vec using the following parameters:
- vector_size: 100, window: 5, min_count: 1, workers: 4, epochs: 10
3. Compute Document Vectors:
a. Retrieve word vectors for words in the document that are present in `word2vec_model.wv`.
b. Compute the average of these vectors for the document.
c. Return a zero vector if no valid words are present.
4. Output the trained `word2vec_model` and computed document vectors using the `document_vector(doc)` function.

Experimental setup

In this experiment, we train an exclusive Word2Vec with a vector size of 100 dimensions, a context window of 5, and trained across 10 epochs utilising four CPU threads for computational efficiency. Following the generation of document embeddings, combinations of 300, 450, 600, 800, and 1,000 are produced. For each combination, 80% of the data is allocated for training and 20% for testing. SMOTETomek is utilised exclusively on the training data for the combinations of 600, 800, and 1,000, as the 300 and 450 combinations provide adequate data. The testing data retains its original form without any adjustments to represent the actual data appropriately.

Results and discussion

Tables 22.3–23.6 summarise the performance metrics of machine learning classifiers on the test set. Random

Table 23.3 Support vector machines.

No of articles	Train time (s)	Test time (s)	Accuracy	Precision	Recall	F1 score
300	0.01802	0.000648	63.33	69.91	63.33	60.37
450	0.024657	0.000926	55.56	58.67	55.56	51.88
600	0.093132	0.003442	60.45	60.48	60.45	60.46
800	0.238579	0.006221	60.25	58.71	60.25	58.88
1,000	0.321669	0.008846	60.54	61.80	60.54	60.84

Source: Author

Table 23.4 k-Nearest neighbor.

Number of articles	Train time (s)	Test time (s)	Accuracy	Precision	Recall	F1 score
300	0.000258	0.001402	43.33	37.23	43.33	39.98
450	0.000287	0.001878	48.89	48.98	48.89	48.88
600	0.000317	0.002353	65.78	68.12	65.78	62.86
800	0.000327	0.002385	64.86	64.63	64.86	63.02
1,000	0.000329	0.002978	65.46	68.44	65.46	63.93

Source: Author

Table 23.5 Logistic regression.

Number of articles	Train time (s)	Test time (s)	Accuracy	Precision	Recall	F1 score
300	0.03119	0.000159	51.67	57.43	51.67	49.44
450	0.01753	0.000123	48.89	46.84	48.89	45.38
600	0.03441	0.000133	60.47	59.68	60.47	59.42
800	0.02479	0.000110	61.08	61.55	61.08	61.15
1,000	0.03486	0.000135	63.85	64.33	63.86	63.36

Source: Author

Table 23.6 Random forest.

Number of articles	Train time (s)	Test time (s)	Accuracy	Precision	Recall	F1 score
300	0.16397	0.002552	55.00	61.48	55.00	51.44
450	0.227082	0.002664	55.56	53.79	55.56	53.15
600	0.438122	0.00338	75.68	75.34	75.68	75.44
800	0.627895	0.005377	79.92	79.71	79.92	79.76
1,000	0.811445	0.004524	80.40	79.94	80.40	80.01

Source: Author

forest outperformed others, achieving the highest accuracy (80.40%) and F1 score (80.01) with 1,000 articles, making it ideal for large datasets despite longer training times. SVM performed reasonably well, with ~60% accuracy on larger datasets and minimal processing overhead. k-NN improved with dataset size. Logistic regression consistently improved with larger datasets, achieving 63.85% accuracy and 63.36 F1 score with 1,000 articles, offering a computationally efficient alternative. Overall, random forest proved the most reliable for stance recognition on large datasets.

Table 23.8 demonstrates similarity clusters for the word 'Sarkar,' meaning 'government,' highlighting the embeddings' effectiveness in representing governance-related vocabulary. As the size increases the clarity of words also increases. This study addressed the lack of pre-trained embeddings for low-resource languages and showcased the benefits of custom embeddings in natural language processing tasks.

Table 23.7 F1 score comparison without and with SMOTETomek.

Variables	Combinations	F1 score without SMOTE	F1 score with SMOTE
LR	600	53.00	59.42
	800	57.00	61.15
	1,000	59.66	63.36
SVM	600	58.50	60.46
	800	56.75	58.88
	1,000	58.50	60.84
k-NN	600	56.66	63.02
	800	57.50	62.86
	1,000	55.00	63.93
RF	600	61.16	75.00
	800	63.70	79.00
	1,000	64.87	80.00

Source: Author

Table 23.8 Similar words for the word 'Sarkar'.

Combinations	Words
300	ꯆꯇꯣꯤꯂꯦ, ꯇ, ꯂꯥꯒꯔꯒꯤ, ꯃꯃꯥꯒꯤ, ꯀꯣꯝꯤꯆꯂꯦ, ꯁꯦꯥꯔꯦ, ꯇꯤꯂ, ꯆꯝꯅ, ꯀꯔꯦꯢꯤꯅꯦꯢꯝ, ꯑꯤꯕꯔꯥ
	Electionna, Ta, sarkargi, mamangi, commissionna, wangkhei, north, jammu, president, oiraba
450	ꯃꯤꯅꯤꯁꯇ꯭ꯔꯤ, ꯀꯥꯌꯕꯦꯥꯝ, ꯃꯤꯝ, ꯐꯥꯔꯥꯕꯦꯆ, ꯆꯤꯝꯅꯤꯔꯒꯣ, ꯀꯀꯆꯤ, ꯁꯤꯂꯦꯝ, ꯑꯤꯕꯔꯥ, ꯀꯔꯦꯝ, ꯗꯤꯒꯦꯤꯇꯤ
	Ministry, cabinet, mipham, phaobadasu, chungsinnaraga, kakching, supreme, oiraba, oorep, deputy
600	ꯂꯥꯒꯕꯦ, ꯆꯇꯦꯥꯔ, ꯀꯆꯤꯀꯤꯒꯦꯝ, ꯇꯤꯆꯤ, ꯀꯣꯝꯤꯆꯂꯦ, ꯆꯇꯣꯔꯢꯕꯦꯅ, ꯀꯥꯌꯕꯦꯆꯣ, ꯁꯦꯥꯔꯦ, ꯑꯤꯒꯤꯔꯤꯀꯀꯦꯥ, ꯂꯥꯒꯔꯒꯤ
	Sarkarna, handakta, canchipurda, liching, commission, electoral, cabinetta, supreme, agriculture, sarkargi
800	ꯀꯆꯃꯤꯔ, ꯃꯤꯝꯃ, ꯒꯦꯥꯔ, ꯔꯦꯡ, ꯂꯥꯒꯕꯦ, ꯁꯦꯥꯇ, ꯂꯦꯤꯒꯤꯔꯤ, ꯒꯣꯔꯢꯀꯃꯦꯢꯤꯅꯦꯝ, ꯆꯝꯁꯝ, ꯇꯤꯁꯤ
	Kashmirda, miphamda, bharat, Rahul, sarkarna, houseta, leiringeida, government, jammu, list
1,000	ꯔꯦꯡ, ꯇꯤꯔꯦꯔꯦ, ꯆꯦꯥꯇꯣꯔꯒꯤ, ꯃꯤꯝꯃ, ꯐꯥꯔꯦꯣꯝ, ꯀꯔꯦꯤꯒꯣꯔꯤꯀ, ꯇ꯭ꯔꯤꯕꯦꯀꯦꯂꯗꯦ, ꯀꯆꯃꯤꯔ, ꯇꯤꯁꯤ, ꯃꯤꯅꯤꯁꯇ꯭ꯔꯤꯒꯇꯦ
	Rahul, leadersu, centregi, miphamda, finance, pirakkhiba, tribunalna, kashmirda, list, ministryna

Source: Author

Table 23.7 clearly shows that the F1 scores improved across all classifiers with resampling, particularly for random forest and k-NN. This highlights the importance of balanced training data for making stable and accurate predictions, especially in resource-limited areas where data shortages exacerbate class imbalance problems.

Logistic regression is the most efficient classifier, taking just 0.034 seconds to train and 0.000135 seconds to test with 1,000 articles, achieving 63.85% accuracy and a 63.36 F1 score. In contrast, random forest, which has the highest accuracy at 80.40% and an F1 score of 80.01, takes longer to train at 0.811 seconds. Thus, Logistic regression is ideal for resource-constrained applications, while random forest is better for accuracy-critical scenarios.

The confusion matrix in Figure 23.2 shows that SVM and k-NN, despite moderate accuracy, struggle with 'Neutral' and 'Against' confusion. Logistic regression improves precision and reduces misclassification. Overall, random forest is the top classifier, followed by logistic regression, while SVM and k-NN face significant class confusion.

Conclusion and future directions

This study provides a novel dataset of Manipuri editorial articles, offering a significant resource for improving research in low-resource natural language processing. It evaluates the potential of Word2Vec embeddings—trained from scratch—to reflect Manipuri's linguistic subtleties

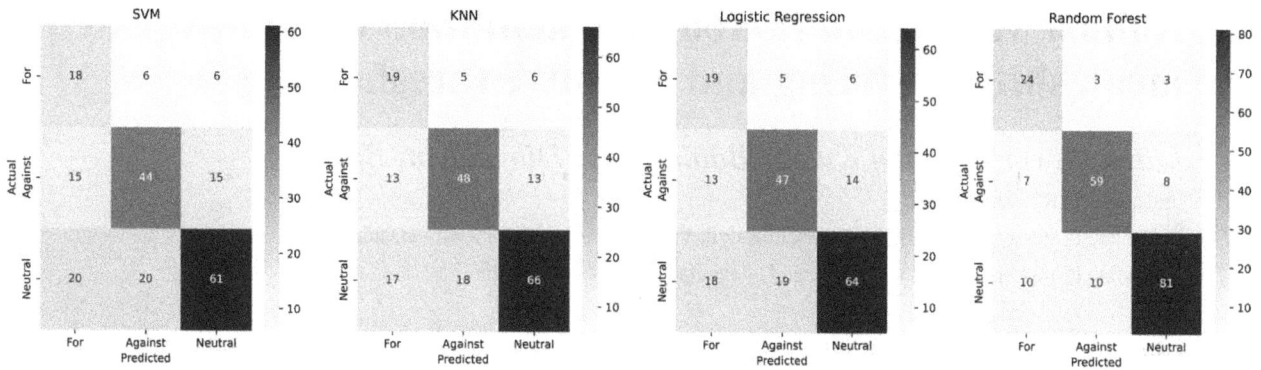

Figure 23.2 Confusion matrix for 1,000 combinations across all four classifiers
Source: Author

across various dataset options. The research effectively tackled the issue of class imbalance in stance detection by using the SMOTETomek method, markedly enhancing the classification efficacy for minority stances like 'for' and 'against.'

The findings indicate that Random Forest was the superior classifier, particularly with the most significant combinations of datasets, although logistic regression functioned as a viable option in resource-limited situations. Future researchers may incorporate transformer-based embeddings, such as BERT, optimised for low-resource languages to improve stance identification efficacy. Extending this methodology to more low-resource languages may confirm its generalisability and facilitate broader progress in multilingual NLP. These findings provide practical insights and a basis for additional investigation in stance identification and low-resource language processing.

References

[1] Mikolov, T. (2013). Efficient estimation of word representations in vector space. *arXiv preprint arXiv:1301.3781.*

[2] Jannati, R., Mahendra, R., Wardhana, C. W., & Adriani, M. (2018, November). Stance classification towards political figures on blog writing. In *2018 International Conference on Asian Language Processing (IALP)* (pp. 96-101). IEEE.

[3] Mourad, S. S., Shawky, D. M., Fayed, H. A., & Badawi, A. H. (2018). Stance detection in tweets using a majority vote classifier. In *The International Conference on Advanced Machine Learning Technologies and Applications (AML-TA2018)* (pp. 375-384). Springer International Publishing.

[4] Mohammad, S., Kiritchenko, S., Sobhani, P., Zhu, X., & Cherry, C. (2016, June). Semeval-2016 task 6: Detecting stance in tweets. In *Proceedings of the 10th International Workshop on Semantic Evaluation (SemEval-2016)* (pp. 31-41).

[5] Косів, Ю. А., &Яковина, В. С. (2022). Three language political leaning text classification using natural language processing methods. *Прикладніаспектиінформаційнихт ехнологій,* 5(4), 359-370.

[6] Augenstein, I., Rocktäschel, T., Vlachos, A., &Bontcheva, K. (2016). Stance detection with bidirectional conditional encoding. *arXiv preprint arXiv:1606.05464.*

[7] Domingues, G. C. (2023). Evaluating data imbalance approaches for classifying semantic relations using machine learning and word embeddings.

[8] Leng, Q., Guo, J., Tao, J., Meng, X., & Wang, C. (2024). OBMI: oversampling borderline minority instances by a two-stage Tomek link-finding procedure for class imbalance problem. *Complex & Intelligent Systems,* 1-18.

[9] Tran, O., Phung, A. C., & Ngo, B. X. (2022, June). Using convolution neural network with BERT for stance detection in Vietnamese. In *Proceedings of the Thirteenth Language Resources and Evaluation Conference* (pp. 7220-7225).

[10] Dey, K., Shrivastava, R., & Kaushik, S. (2018). Topical stance detection for Twitter: A two-phase LSTM model using attention. In *Advances in Information Retrieval: 40th European Conference on IR Research, ECIR 2018, Grenoble, France, March 26-29, 2018, Proceedings 40* (pp. 529-536). Springer International Publishing.

[11] Ghanem, B., Rosso, P., & Rangel, F. (2018, November). Stance detection in fake news a combined feature representation. In *Proceedings of the first workshop on fact extraction and VERification (FEVER)* (pp. 66-71).

[12] Chawla, N. V., Bowyer, K. W., Hall, L. O., &Kegelmeyer, W. P. (2002). SMOTE: synthetic minority over-sampling technique. *Journal of Artificial Intelligence Research,* 16, 321-357.

[13] Batista, G. E., Prati, R. C., & Monard, M. C. (2004). A study of the behavior of several methods for balancing machine learning training data. ACM SIGKDD explorations newsletter, 6(1), 20-29.

[14] Tomek, I. (1976). Two modifications of CNN. *IEEE Transactions on Systems, Man, and Cybernetics, SMC-6(6),* 769-772. https://doi.org/10.1109/TSMC.1976.5408784

24 Robust MRI brain tumour segmentation using integrated mean shift clustering and adaptive thresholding

Maya Pawar[1,a], Rubul Kumar Bania[2,b] and Dibya Jyoti Bora[3,c]

[1]Independent Researcher

[2]Associate Professor, Birangana Sati Sadhani Rajyik Vishwavidyalaya, Golaghat, Assam, India

[3]Associate Professor, The Assam Kaziranga University, Jorhat, Assam, India

Abstract

Precise and efficient brain tumour segmentation uin Magnetic Resonance Imaging (MRI) scans plays avital role in diagnosing neurological conditions and guiding appropriate treatment strategies. However, conventional segmentation methods face significant challenges, such as intensity inhomogeneity, noise, and overlapping tissue boundaries. This paper introduces a Hybrid Segmentation Approach Using Mean Shift Clustering and Adaptive Thresholding to address these limitations and enhance the accuracy of MRI brain tumour analysis.

The proposed method utilises the complementary strengths of two robust techniques: mean shift clustering and adaptive thresholding. Mean shift clustering, a non-parametric clustering method, effectively handles noise and intensity variations in MRI images while preserving the structural details of the brain tissues. Adaptive thresholding, specifically Otsu's method, ensures the delineation of tumour regions by dynamically segmenting the image based on pixel intensity distributions. By integrating these two approaches, the hybrid method achieves a balance between precision and computational efficiency.

This research contributes to the field of medical image analysis by providing a reliable and computationally efficient segmentation technique that can assist clinicians in the early detection and treatment of brain tumours. The experimental results demonstrate that the proposed approach achieves higher efficiency compared to traditional algorithms.

Keywords: MRI brain tumour segmentation, mean shift clustering, adaptive thresholding, medical image analysis, tumour boundary delineation

Introduction

Magnetic Resonance Imaging (MRI) is a widely adopted, non-invasive imaging technique for diagnosing brain tumoursand assisting in treatment planning. Precisely segmenting tumours from MRI scansis essential to ensure accurate diagnosis and effective therapeutic intervention. Traditional segmentation methods often struggle with challenges such as intensity inhomogeneity, noise, and overlapping tissue regions [1]. To overcome these limitations, a hybrid segmentation approach is introduced that combines mean shift clustering and adaptive thresholding techniques.

Brain tumours refer to irregular cell growths in the brain, which may be classified as either non-cancerous (benign) or cancerous (malignant.). Early detection and precise delineation of these tumours are vital for successful treatment outcomes. Manual segmentation of brain tumours is often labor-intensive and subject to individual interpretation, highlighting the necessity for automated and dependable segmentation techniques.

Literature review

Several segmentation methods have been developed for brain tumour analysis, including thresholding, region growing, clustering, and machine learning-based approaches. Each of these methods has its advantages and limitations.

Thresholding techniques
Thresholding is a basic segmentation segmentation method that distinguishes foreground elements from the backgroundby applying a predefined intensity threshold. Introduced by Otsu, in 1979 [2], this [2] global thresholding technique is well-known for automatically selecting the bestthreshold by reducing theminimising variance within each class. However, thresholding techniques often struggle with intensity inhomogeneity and noise, leading to poor segmentation results.

Region growing techniques
These techniques initiate with selected seed points and expand the region by adding adjacentu pixels that exhibit comparable characteristics. These techniques are effective for segmenting homogeneous regions but are sensitive to noise and require accurate seed point selection [3].

Clustering techniques
Methods like k-means and fuzzy c-meansare commonly used togroup pixels into clustersbased on similarity [5].

[a]maya4djb@gmail.com, [b]rubul.bania@gmail.com, [c]research4dibya@gmail.com

DOI: 10.1201/9781003675235-24

Mean shift clustering, as introduced by Comaniciu and Meer, [4,5] is a non-parametric technique known for its robustness against noise and intensity changes. One of its key advantages is that it does not necessitate prior information about the number of clusters, which is particularly beneficial for segmenting medical images.

Machine learning-based techniques

Machine learning approaches, particularly those utilizing convolutional neural networks (CNNs) and other deep learning architectures, have demonstrated exceptional effectiveness inthe segmentation of medical images. These methods require large annotated datasets for training and substantial computational resources [6].

Hybrid approaches

Hybrid methods integrate several segmentation techniques to capitalize on the unique advantages each offers. For instance, combining clustering and thresholding techniques can improve segmentation accuracy by handling noise and intensity variations more effectively [7].

Methodology

The proposed hybrid segmentation approach integrates mean shift clustering with adaptive thresholding to improve the accuracy of MRI brain tumour segmentation. A step-by-step explanation of the methodology is provided below.

Image pre-processing

Initially, the MRI brain scans undergo preprocessingto improve image contrast and suppress noise. This involves normalisation and filtering techniques processes. Histogram equalisationis applied to enhance contrast, while Gaussian filtering is used to smooth the images and eliminate unwanted noise.

Convert image to LAB colour space

The input MRI image is transformed from the BGR colour space to the LAB colour space to more effectively distinguish between intensity and colour information[8]. Due tou its perceptual uniformity, the LAB color space is particularly well-suited for clustering tasks [9]. In this color space, the L component denotes the intensity or lightness, whereas the A and B components capture the chromatic opposition between colorsu.

Mean shift clustering

Mean shift clustering is applied to the LAB imagefor segmentation. The spatial radius is set to 10, the colour radius to 20, and the maximum number of iterations to 100. Mean shift clustering effectively groups similar pixels, creating a segmented image with distinct regions [10].

The clustering process involves iteratively shifting each pixel to the mean of its neighbourhood until convergence.

Adaptive thresholding

The segmented image from the mean shift clustering is converted to grayscale. Adaptive thresholding, specifically Otsu's method, is applied to obtain a binary mask. Otsu's technique selects the most suitable threshold by minimising thevariance within individual pixel classes [11]. This step ensures that the segmented regions are accurately delineated from the background.

Combine segmentation results

The binary mask generated through adaptive thresholding is utilized to refine the results of mean shift clustering. A bitwise AND operation is performed between the original image and the binary mask to produce the final segmented image. This integration enhances the accuracy and robustness of the segmentation by combining the advantages of both methods Figure 24.1.

Flowchart of the proposed approach

The algorithm for the proposed hybrid segmentation approach is as follows:

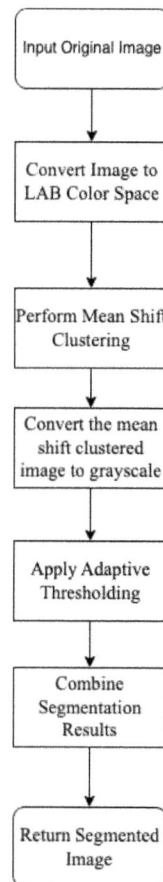

Figure 24.1 Flowchart of the proposed approach
Source: Author

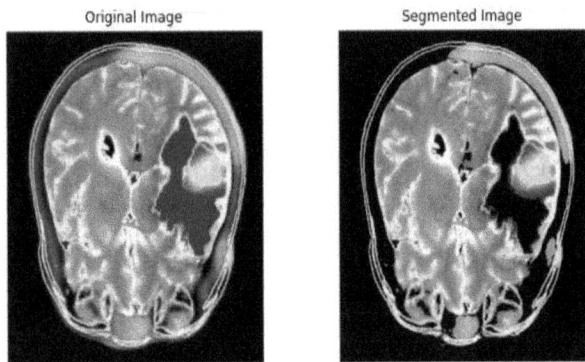

Figure 24.2 Original image and the result obtained using the proposed approach
Source: Author

Experimental results

The proposed approach was tested on a dataset of MRI brain images, which included various types of tumours. Its segmentation performance was assessed using evaluation measures including the Dice Similarity Coefficient (DSC), Jaccard Index, and processing time. The findings revealed that the hybrid technique offered superior accuracy and reliability compared to conventional approaches Figure 24.2.

Dataset
This study makes use of a set of MRI brain scanswith annotated tumour regions. The images were obtained from publicly available databases, particularly the Brain Tumour Segmentation (BRATS) challenge dataset [12]. The dataset includes several MRI modalities—T1, -weighted, T1 contrast-enhanced (T1c,), T2-weighted, and FLAIR—each paired with expert-provided tumour segmentation masks.

Evaluation metrics
The performance of the segmentation method was evaluated using the following metrics:

1. Dice similarity coefficient (DSC): Measures the overlap between the ground truth and the segmented regions.
2. Jaccard index: Measures the similarity between the ground truth and the segmented regions.
3. Computation time: Measures the time required to perform the segmentation.

The table below presents a summary ofsummarises the results, with each value representing the average obtained from evaluating 100 MRI images in Table 24.1 .

Table 24.1 Experimental results.

Method	DSC	Jaccard Index	Computation Time (s)
Mean shift clustering	0.78	0.65	12.5
Adaptive thresholding	0.82	0.70	**8.3**
Proposed hybrid method	**0.88**	**0.78**	10.2

Source: Author

Results
The proposed hybrid method achieved a higher DSC and Jaccard index compared to individual mean shift clustering and adaptive thresholding methods. This indicates better overlap and segmentation quality. The computation time was also competitive, making the method suitable for clinical applications where quick processing is essential.

Conclusion

This study proposed a hybrid method for MRI brain tumor segmentation by integrating mean shift clustering with adaptive thresholding techniques tumour. The method leverages the strengths of both techniques, resulting in improved segmentation accuracy and efficiency. Future research will aim to enhance optimising the algorithm's efficiency and assess its effectiveness across more extensive and diverser datasets. Additionally, integrating advanced machine learning techniques could further enhance the segmentation performance.

References

[1] Bauer, S., Wiest, R., Nolte, L.-P.,Reyes, M. (2013). A survey of MRI-based medical image analysis for brain tumor studies. *Physics in Medicine and Biology*, 58(13),R97–R129.

[2] Otsu, N. (1979). A threshold selection method from gray-level histograms. *IEEE Transactions on Systems, Man, and Cybernetics*, 9(1),62–66.

[3] Adams, R.,Bischof, L. (1994). Seeded region growing. *IEEE Transactions on Pattern Analysis and Machine Intelligence*, 16(6),641–647.

[4] Comaniciu, D., Meer, P. (2002). Mean shift: arobust approach toward feature space analysis. *IEEE Transactions on Pattern Analysis and Machine Intelligence*, 24(5),603–619.

[5] Bora, D. J. (2017). Performance comparison of k-means algorithm and fcm algorithm with respect to color image segmentation. *International Journal of Emerging Technology and Advanced Engineering*, 7(8), 460–470.

[6] Ronneberger, O., Fischer, P.,Brox, T. (2015). U-Net: convolutional networks for biomedical image segmentation.

In Proceedings of the International Conference on Medical Image Computing and Computer-Assisted Intervention (MICCAI), 234–241.

[7] Liu, X., Zhang, Y., Chen, M.,Zhao, G. (2013). A hybrid method for medical image segmentation using fuzzy clustering and region growing. In Proceedings of the International Conference on Bio informatics and Biomedicine (BIBM), 29–36.

[8] Bora, D. J., Gupta, A. K., Khan, F. A. (2015). Comparing the performance of L* A* B* and HSVcolor spaces with respect to color image segmentation. arXiv preprint arXiv:1506.01472.

[9] Kakarala, R. G.,Hartley, S. K. (2012). Choosing a color space for image segmentation using machine learning. In Proceedings of the International Conference on Image Processing (ICIP), 3757–3760.

[10] Tomasi, C.,Manduchi, R. (1998). Bilateral filtering for gray and color images. In Proceedings of IEEE International Conference on Computer Vision (ICCV), 839–846.

[11] Ojala, T., Pietikainen, M.,Maenpaa, T. (2002). Multiresolution gray-scale and rotation invariant texture classification with local binary patterns. *IEEE Transactions on Pattern Analysis and Machine Intelligence*, 24(7),971–987.

[12] Menze, B. H., et al. (2015). The multimodal brain tumorimage segmentation benchmark (BRATS). *IEEE Transactions on Medical Imaging*, 34(10),1993–2024.

25 Feature-driven classification of AI and human text: insights from traditional and neural models

Potshangbam Kirankumar Singh[a], Ksh Nareshkumar Singh[b] and H. Mamata Devi[c]

Department of Computer Science, Manipur University, India

Abstract

The rapid advancements in artificial intelligence (AI) have led to the generation of text that closely mimics human writing, posing challenges in distinguishing between AI-generated and human-written content. This paper introduces a feature-driven approach to address this challenge, employing six feature-engineering techniques: perplexity, burstiness, grammatical errors, repetition, coherence, and stylometry. These features are analysed using six classification algorithms: Multinomial Naïve Bayes (NB), Bernoulli NB, Stochastic Gradient Descent Classifier (SGDClassifier), Logistic Regression, Gated Recurrent Unit (GRU), and Feedforward Neural Network (FNN). Among these, Logistic Regression, SGDClassifier, and GRU deliver consistently high performance across diverse datasets. This research provides valuable insights into feature-driven classification methods, supporting applications such as content moderation, academic integrity, and misinformation detection.

Keywords: AI-generated text, human-written text, classification, feature engineering, machine learning algorithms

Introduction

The development of Artificial Intelligence (AI) and Natural Language Processing (NLP) has led to significant advancements in the generation of human-like text by machines. AI models, such as GPT and BERT, can now produce text that closely resembles human writing, often indistinguishable from content created by people. While this progress offers exciting possibilities for automation, creativity, and content generation, it also presents a set of complex challenges. The ability of AI to generate text that mimics human language raises critical issues in various domains, including legal concerns, societal impacts, and the reliability of digital information [17]. These challenges are particularly pronounced in fields where the authenticity and origin of content are essential, such as journalism, academic integrity, and the prevention of misinformation [9,13].

In response to this challenge, this paper presents a feature-driven approach for distinguishing AI-generated text from human-written content. We use the advanced feature-engineering techniques to analyse and identify distinctive characteristics of both types of text. Specifically, we focus on six key features: perplexity, burstiness, grammatical errors, repetition of words, coherence, and stylometry. These features represent different linguistic and structural patterns that can be used to differentiate human writing from AI-generated text. These extracted features are then fed into six machine learning algorithms: Multinomial Naïve Bayes (NB) [11,14], Bernoulli Naïve Bayes (NB) [11], Stochastic Gradient Descent Classifier (SGDClassifier) [3], Logistic Regression [5], Gated Recurrent Unit (GRU) [4], and Feedforward Neural Network (FNN) [16]. Each of these algorithms brings unique strengths to the classification task. For instance, Logistic Regression and SGDClassifier perform exceptionally well on large, feature-rich datasets, while GRU excels in tasks requiring an understanding of sequential context and flow, making it well-suited for longer or more complex pieces of text.

This study not only advances the state-of-the-art in AI text detection but also highlights the importance of feature-driven classification methods in tackling the growing challenge of AI-authored content. By providing a scalable and reliable framework for distinguishing between human and AI text, this research contributes valuable insights for a range of applications, including content moderation, academic plagiarism detection, and the broader goal of ensuring the authenticity and trustworthiness of digital information in an increasingly AI-driven world.

The remainder of this paper is structured as follows: First, a review of the literature is presented, followed by an explanation of the methodology employed. Subsequently, the experimental results are discussed, and finally, the study is concluded.

Literature review

The challenge of distinguishing between human-written and AI-generated text has garnered significant attention in recent years due to the increasing sophistication of AI systems. Numerous studies have explored various

[a]potshangbamkirankumar34@gmail.com, [b]nareshksh2711@gmail.com, [c]mamata_dh@rediffmail.com

DOI: 10.1201/9781003675235-25

approaches to tackle this issue, focusing on machine learning algorithms, feature engineering techniques, and dataset diversity. In the early research efforts, Beresneva [2] conducted one of the foundational studies, systematically reviewing machine learning-based methods for identifying computer-generated text. This work highlighted critical challenges, including scalability, adaptability, and accuracy, especially as text-generation technologies advanced. Subsequent research introduced more sophisticated techniques, such as the use of stylistic features like lexical diversity, sentence structure, and syllable counts, as explored by Shah et al. [17], whose models achieved 93% accuracy with the help of explainable AI tools like LIME and SHAP. Nguyen et al. [13] emphasised the role of feature engineering, employing methods like TF-IDF and n-gram modelling to train machine learning classifiers like XGBoost and SVM. However, they encountered limitations in capturing sequential dependencies and adapting to complex datasets. Akram [1] evaluated six detection tools across diverse datasets and highlighted significant performance variations, with accuracy ranging from 55% to 97%. Meanwhile, Mindner et al. [12] created a comprehensive corpus of human-written, AI-generated, and AI-rephrased texts to evaluate detection tools. Their system achieved up to 98% accuracy for AI-generated content but struggled with rephrased text, illustrating the evolving complexity of AI outputs and the need for adaptable detection methods.

More recent studies have addressed specific challenges in detecting AI-generated text. Maktab Dar Oghaz et al. [10], investigated the use of deep transformer models for detecting and classifying ChatGPT-generated content, demonstrating the potential of advanced neural architectures in this domain. Prova [15] investigated the use of modern NLP and machine learning models, including XGB, SVM, and BERT, for detecting AI-generated text. BERT emerged as the most effective, achieving 93% accuracy. Georgios [7] examined linguis- tic features, revealing measurable differences in syntax, pronoun usage, and lexical patterns between human and AI writing. Meanwhile, Huang et al. [9] proposed robust detection systems that resist adversarial attacks, demon- strating improved accuracy under adversarial conditions. Genre-specific complexities were explored by Hayawi et al. [8], who analysed diverse content types, including essays and stories, highlighting difficulties in identifying creative AI-generated texts. These findings collectively underscore the importance of combining feature engi- neering with adaptable machine learning frameworks, while pointing to unresolved challenges, such as detect- ing rephrased content, handling diverse datasets, and ensuring robustness against adversarial manipulations. Another studied conducted by Fraser et al. [6] reviewed statistical methods, including perplexity andentropy, for identifying AI-generated text.

They also examinedthe effects of decoding strategies like top-k sampling on detectability. Their comprehensive review underscored theneedtoupdatedetectionmethod- stoaccommodate newer systems, such as GPT-4, which exhibit advanced capabilities.

Collectively, these studies highlight the importance of feature engineering, dataset diversity, and model adaptability in distinguishing human-written and AI-generated texts. While early approaches focused on simpler statistical and stylistic features, recent research emphasises robust machine learning models and adversarial robustness. However, challenges remain in handling complex datasets, sequential dependencies, and creative or rephrased content, necessitating continued research in this evolving field.

Study of methodology

The proposed framework integrates feature engineering and machine learning algorithms to distinguish AI-generated text from human-written content. The methodology consists of several key steps, including data collection, text pre-processing, feature engineering, and algorithm implementation.

Text Pre-processing

Text pre-processing is an essential step to standardise and prepare raw text for analysis. This process involved lowercasing, punctuation removal, tokenisation, stopword removal, lemmatisation, and normalisation. These steps ensured the data was consistent and suitable for feature extraction and classification.

Feature engineering

It is the process of selecting, transforming, and creating features (attributes) from the raw data that can be used by machine learning models for classification or prediction tasks. In this work, the following terms *perplexity, burstiness, grammatical errors, repetition of words, coherence, and stylometry* refer to features used in feature engineering for identifying AI-generated versus human-generated text.

Perplexity: it measures how well a language model predicts a sample of text. It assesses how 'surprising' or 'natural' a sequence of words is.

Burstiness: it refers to how certain words or phrases appear in clusters or bursts within a text, rather than being evenly distributed. It captures the irregularity in the way humans write.

Grammatical Errors: this feature identifies mistakes in grammar, such as subject–verb disagreement, sentence fragments, or incorrect tense usage.

Repetition of Words: this feature analyses how frequently specific words or phrases are repeated across the text.

Coherence: it evaluates the logical flow and connectivity between sentences and ideas throughout the text.

Stylometry: it involves analysing writing style using statistical metrics like word length, sentence length, and vocabulary diversity.

Classifying algorithms

These extracted features were fed into classifying algorithms to perform binary classification (i.e. AI text or human generated text). Six algorithms are used for this binary classification: Multinomial Naïve Bayes, Bernoulli Naïve Bayes, Stochastic Gradient Descent Classifier, Logistic Regression, GRU (Gated Recurrent Unit), and FNN (FeedForwarding Neural Network). Each algorithm captures distinct linguistic and stylistic features to enhance classification accuracy. These algorithms, applied to diverse feature sets, complement each other by capturing both shallow and deep linguistic patterns, enabling accurate differentiation between AI-generated and human-written text.

Experimental results

Data collection overview

This study utilised four labelled datasets obtained from platforms such as Kaggle. These datasets were designed to capture different text-generation scenarios:

i.   Dataset 1: Comprising of 1,103 rows and 2 columns. This dataset included AI-generated and student-written text. The labels categorised the content as "ai" or "student" (for human-generated text). This dataset was uploaded by Prajwal Dongre.
ii. Dataset 2: Featuring 29,145 rows, this dataset contained student essays and essays generated by large language models (LLMs). Texts were labelled as '0' (human-written) or '1' (AI-generated). It was uploaded by Sunil Thite.
iii. Dataset 3: A sentence-level dataset derived from 1,018 articles (509 AI-generated and 509 human-generated). These articles were divided into 7,344 sentences (4,008 AI-generated and 3,336 human-generated) to assess classification at a granular level. It was uploaded by Mahdimaktabdar.
iv. Dataset 4: A large dataset containing approximately 28,500 essays written by humans or generated by advanced LLMs. This dataset was uploaded by Murugesann.

Results and discussion

The accuracy results for the six classification algorithms are summarised in Tables 25.1–25.4, corresponding to each dataset. Detailed observations and discussions are provided below:

Discussion on Dataset 1: Logistic regression and SGDClassifier exhibited near-identical top performance, with accuracies of 94.57% and 95.02%, respectively. Their success

Table 25.1 Algorithms performance result on Dataset 1.

Algorithms	Accuracy
Bernoulli NB	74.66
Multinomial NB	77.83
SGDClassifier	95.02
Logistic regression	94.57
GRU	90.50
FNN	74.66

Source: Author

Table 25.2 Algorithms performance result on Dataset 2.

Algorithms	Accuracy
Bernoulli NB	79.62
Multinomial NB	60.11
SGDClassifier	94.80
Logistic regression	95.11
GRU	90.75
FNN	79.62

Source: Author

Table 25.3 Algorithms performance result on Dataset 3.

Algorithms	Accuracy
Bernoulli NB	58.07
Multinomial NB	55.00
SGDClassifier	45.20
Logistic regression	66.78
GRU	68.96
FNN	58.07

Source: Author

Table 25.4 Algorithms performance result on Dataset 4.

Algorithms	Accuracy
Bernoulli NB	53.39
Multinomial NB	59.81
SGDClassifier	78.58
Logistic regression	86.60
GRU	81.40
FNN	53.39

Source: Author

can be attributed to their ability to effectively process feature-rich datasets. GRU, with an accuracy of 90.50%, also performed well due to its sequential modelling capability. However, simpler models like Naïve Bayes and FNN struggled to handle the feature complexity, achieving accuracies below 78%.

Discussion on Dataset 2: Logistic regression slightly outperformed SGDClassifier, achieving the highest accuracy of 95.11%. The ability of these algorithms to process high-dimensional, large-scale data effectively underscores their robustness. GRU's performance (90.75%) highlights its sequential strengths but also reveals its limitations in high-dimensional feature spaces. Meanwhile, Naïve Bayes models were more suited to binary or categorical data, explaining their limited effectiveness with accuracies of 79.62% (Bernoulli NB) and 60.11% (Multinomial NB)

Discussion on Dataset 3: Sentence-level data presented unique challenges due to reduced context, making it difficult for algorithms like SGDClassifier (45.20%) to discern patterns effectively. GRU (68.96%) excelled in this scenario, leveraging its ability to capture sequential relationships within shorter text spans. Logistic regression achieved moderate success (66.78%), but its reliance on global feature patterns limited its performance in this granular setting. Simpler algorithms, such as Naïve Bayes, struggled due to the absence of strong word-frequency distributions at the sentence level.

Discussion on Dataset 4: Logistic regression outperformed other models (86.60%), demonstrating its robustness across diverse datasets. GRU followed closely (81.40%), particularly excelling in capturing sequential dependencies in long-form content. SGDClassifier (78.58%) performed reasonably well but lagged behind GRU and Logistic regression. Simpler models like Naïve Bayes and FNN failed to adapt to the complexity of this large dataset, with accuracies below 60%.

Cross-dataset insights

i.   Strengths of advanced models: Logistic regression and SGDClassifier consistently performed well across all datasets, confirming their ability to handle high-dimensional, feature-rich data. GRU excelled in scenarios requiring sequential modelling, particularly with sentence-level data and long-form content.

ii. Limitations of simpler models: Multinomial NB, Bernoulli NB, and FNN consistently underperformed in complex datasets. These models were unable to fully utilise the nuanced features extracted during pre-processing, limiting their utility for distinguishing AI-generated content.

iii. Role of features: The engineered features—such as perplexity, burstiness, and coherence—played a crucial role in improving classification accuracy. Sequential features, like coherence and stylometry, were particularly beneficial for GRU, while statistical features, like perplexity, favoured Logistic regression and SGDClassifier.

iv. Parameters: The algorithms, Bernoulli NB, Multinomial NB, SGDClassifier and Logistic regression were used with default functionality. FNN and GRU were used with 100 number of epochs, optimizer. zero_grad() to avoid the previous epoch gradient and loss.backward() to compute the gradient of the loss.

Implications

The findings highlight the importance of selecting classification models based on dataset characteristics:

- For large-scale datasets with complex structures, Logistic regression and SGDClassifier offer robust performance.
- For shorter text spans or content with strong sequential dependencies, GRU is the most effective.
- Simpler models like Naïve Bayes may still be suitable for smaller or less complex datasets where interpretability is key.

The results emphasise the necessity of tailoring algorithms to specific dataset challenges. This study contributes a scalable framework for distinguishing AI-generated and human-written content, supporting applications in content moderation, academic integrity, and combating misinformation.

Conclusion

This study presents a robust, feature-driven approach for distinguishing AI-generated text from human-written content, leveraging advanced feature engineering and machine learning algorithms. The study highlights that while traditional linear classifiers like Logistic regression and SGDClassifier perform consistently well across various datasets, sequential models like GRU are crucial for handling short-form or context-dependent text. In the future, we can further improve classification by integrating advanced neural architectures like transformers, or combining linear and sequential models in an ensemble framework could be explored. Additionally, expanding feature engineering techniques and testing on more diverse datasets could yield deeper insights into the strengths and limitations of these algorithms. This work contributes to the growing field of AI-generated content detection, offering scalable and reliable solutions for real-world applications.

References

[1] Akram, A. (2023). An Empirical Study of AI-generated Text Detection Tools.

[2] Beresneva, D. (2016). Computer-generated text detection using machine learning: a systematic review.

[3] Bottou, L. (2010). Large-scale machine learning with stochastic gradient descent. In Lechevallier, Y., Saporta, G. (eds.) Proceedings of COMPSTAT'2010. Physica-Verlag HD. https://doi.org/10.1007/978-3-7908-2604-3_16

[4] Chung, J., Gulcehre, C., Cho, K., Bengio, Y. (2014). Empirical evaluation of gated recurrent neural networks on sequence modeling. https://arxiv.org/abs/1412.3555

[5] Cox, D. R. (1958). The regression analysis of binary sequences. *Journal of the Royal Statistical Society: Series B (Methodological), 20*(2), 215–232. https://www.jstor.org/stable/2983890

[6] Fraser, K. C., Dawkins, H., Kiritchenko, S. (2024). Detecting AI-generated text: factors influencing detectability with current methods. https://arxiv.org/abs/2406.15583

[7] Georgios P. (2024). Differentiating between human-written and AI-generated texts using linguistic features automatically extracted from an online computational tool. https://arxiv.org/abs/2407.03646

[8] Hayawi, K., Shahriar, S., Mathew, S. S. (2024). The imitation game: detecting human and AI-generated texts in the era of ChatGPT and BARD. *Journal of Information Science*,
https://doi.org/10.1177/01655515241227531
https://www.kaggle.com/datasets/mahdimaktabdar/chatgpt-classification-dataset, https://www.kaggle.com/datasets/murugesann/ai-essay-detection-daigt-v2-dataset-with-typos https://www.kaggle.com/datasets/prajwaldongre/llm-detect-ai-generated-vs-student-generated-text https://www.kaggle.com/datasets/sunilthite/llm-detect-ai-generated-text-dataset[17] .

[9] Huang, G., Zhang, Y., Li, Z., You, Y., Wang, M., Yang, Z. (2024). Are AI-generated text detectors robust to adversarial perturbations? *Proceedings of the 62nd Annual Meeting of the Association for Computational Linguistics (Volume 1: Long Papers).* Association for Computational Linguistics, 6005–6024.

[10] Maktab Dar Oghaz, M., Dhame, K., Singaram, G., Babu Saheer, L. (2023). Detection and classification of ChatGPT generated contents using deep transformer models. *Frontiers in Artificial Intelligence.*

[11] McCallum, A., Nigam, K. (1998). A comparison of event models for Naïve Bayes text classification. http://www.cs.cmu.edu/~mccallum/papers/bow.ps

[12] Mindner, L., Schlippe, T., Schaaff, K. (2023). Classification of human- and AI-generated texts: investigating features for ChatGPT. In Schlippe, T., Cheng, E.C.K., Wang, T. (eds.) Artificial Intelligence in Education Technologies: New Development and Innovative Practices. AIET 2023. Lecture Notes on Data Engineering and Communications Technologies. Springer, Singapore, vol. 190. https://doi.org/10.1007/978-981-99-7947-9_12

[13] Nguyen, T. T., Hatua, A., Sung, A. H. (2023). How to detect AI-generated texts?

[14] Nigam, K., McCallum, A. K., Thrun, S., Mitchell, T. (2000). Text classification from labeled and unlabeled documents using EM. https://doi.org/10.1023/A:1007692713085

[15] Prova, N. N. I. (2024). Detecting AI-generated text based on NLP and machine learning approaches. https://arxiv.org/pdf/2404.10032

[16] Rumelhart, D. E., Hinton, G. E., Williams, R. J. (1986). Learning representations by back-propagating errors. *Nature, 323*(6088), 533–536. https://doi.org/10.1038/323533a0

[17] Shah, A., Ranka, P., Dedhia, U., Parsad, S., Muni, S., Bhowmick, K. (2023). Detecting and unmasking AI-generated texts through explainable artificial intelligence using stylistic features. https://thesai.org/Publications/ViewPaper?Volume=14&Issue=10&Code=IJACSA&SerialNo=110

26 A novel stacking ensemble machine learning approach for landslide susceptibility mapping

Moziihrii Ado[1,a], Nongmaithem Kane[2,b] and Khwairakpam Amitab[3,c]

[1]Research Scholar, Department of Information Technology, North-Eastern Hill University, Shillong, India

[2]Junior Research Fellow, Department of Information Technology, North-Eastern Hill University, Shillong, India

[3]Assistant Professor, Department of Information Technology, North-Eastern Hill University, Shillong, India

Abstract

This paper presents a novel stacking ensemble technique for generating a landslide susceptibility map (LSM). The primary objective is to improve the accuracy of machine learning (ML) techniques-based LSM generation. The proposed stacking model has a base layer and a meta layer. The base layer generates the meta-features, which the meta-learner uses to predict the landslide probability. The base layer includes six learners: Logistic Regression, XGBoost, SVM, LightGBM, Random Forest, and k-NN classifiers. The proposed model uses the Extremely Randomised Tree classifier as the meta-learner. Meghalaya is considered as the case study area. We used 2,070 landslide and non-landslide points with 16 landslide causative factors. The proposed model is evaluated using accuracy, F1-score, Matthews correlation coefficient (MCC), and area under the receiver operating characteristic curve (AUC). The proposed model has an accuracy of 0.88, an F1-score of 0.89, an MCC of 0.79, and an AUC of 0.95, suggesting an improved overall performance. A reliable LSM for Meghalaya is generated using the proposed model.

Keywords: Machine learning, stacking, landslides, susceptibility mapping

Introduction

Landslides are natural disasters that damage vegetation and property, lead to loss of life, impact economic activities, and delay medical emergency response due to the closure of essential highways. A landslide susceptibility map (LSM) can be used to mitigate landslides and their aftereffects. Several factors influence the occurrence of landslides. Natural factors, including intense rainfalls and earthquakes, can trigger landslides. Other factors include geographical, hydrological, and topographical features. Human activities, including unplanned urbanisation, road construction, and deforestation, also contribute to landslides.

Existing studies [14,20] have employed different machine learning (ML) techniques for LSM generation. The ML models determine the relationship between different landslide causative factors (LCFs) and the occurrence of landslides to generate LSMs. Different LCFs have been used in existing studies. There has been a significant interest in landslide susceptibility and disaster-related studies using ML approaches. Existing studies suggest that hybrid, ensemble, and deep learning models outperform standalone techniques [3,16].

Our study proposes a novel stacking ensemble learning technique to generate an LSM for Meghalaya. The proposed stacking architecture consists of the base and meta layers. The base layer consists of six learners, including Logistic regression (LGR), Extra gradient boosting (XGBoost), k-nearest neighbour (k-NN), Support vector machine (SVM), Random forest (RF), and Light gradient boosting machine (LightGBM). The Extremely randomised tree (ExtraTree) is used as the meta-learner. The area under the receiver operating characteristic curve (AUC), accuracy, F1-score, and Matthews correlation coefficient (MCC) are used to evaluate the model's performance. The proposed model and composite raster image of the LCFs are used to generate LSM for Meghalaya.

Literature review

Only a few studies on landslide susceptibility mapping for Meghalaya exist. The study by Agrawal and Dixit [1] used Shannon entropy, Analytic hierarchy process (AHP), Fuzzy-AHP, and Frequency ratio. It used 15 LCFs and found that AHP performed better, with an F1-score of 0.884 and an AUC of 0.913. Another study by Agrawal and Dixit [2] applied advanced ML models like Artificial Neural Network (ANN), XGBoost, RF, k-NN, and SVM for generating LSM for the Shillong Plateau region. The study results showed that XGBoost and RF had the highest performance, with an AUC of 0.971. Badavath et al. [4] employed the AHP technique for generating LSM for the West-Jaintia Hills district. The study subdivides 10 LCFs and weighted them accordingly to generate the final susceptibility map. The AHP had an AUC value of 0.82.

[a]moziihrii@nehu.ac.in, [b]nongmaithemkane@gmail.com, [c]kamitab@nehu.ac.in

DOI: 10.1201/9781003675235-26

Several studies have applied stacking ensemble learning techniques. Hu et al. [11] developed a novel stacking architecture with ANN and C4.5 to generate an LSM. The stacking model had the best robustness compared to other ensemble techniques. Fang et al. [7] explored stacking, along with other ensemble techniques. The study used different Neural Networks, SVM, and LGR as the base learners, with LGR as the meta-learner. Huan et al. [13] compared different stacking ensemble approach combinations, including RF + LGR, Gradient Boosted Decision Tree (GBDT) + LGR, and XGBoost + LGR. The GBDT + LGR performed best with an AUC value of 0.8168. Other ensemble techniques employed by existing studies include bagging and boosting. These methods generally have homogenous weak base learners. Existing study findings suggest that the stacking ensemble technique combines heterogeneous ML models to take advantage of diverse algorithms to produce a more robust ensemble model. Therefore, this study explores stacking ensemble learning techniques to generate a reliable LSM.

Dataset

Study area
The study area, Meghalaya, belongs to the northeastern region of India. This region is known to experience many landslides due to its hilly terrain and the heavy rainfall it receives during the monsoon season. However, there are fewer studies of the region. Meghalaya, like other northeastern states, experiences many landslides. According to Dikshit et al. [6], the National Highway (NH)-40, which starts from Jorabat and terminates at Dawki, and NH-44, which starts from Civil Hospital Shillong and terminates at Malidhar, are prominent unstable zones. The state's geographical area is 22,429 km2, and it shares its border with Assam and Bangladesh. It receives heavy rainfall, about 1234 mm to 7467 mm annually [2].

Landslide and non-landslide data
Various techniques are employed to collect existing landslide data, such as field visits and historical databases from governmental or educational institutions. Point data is the preferred option for landslide representation since it can provide a uniform representation of landslides of diverse sizes [12]. The historical landslide data for the study were acquired using the Geological Survey of India (GSI) Bhukosh portal [9]. Recent landslide points were generated using coordinates of landslides obtained from Northeast Eastern Space Applications Centre's landslide monitoring web portal North Eastern Regional Node for Disaster Risk Reduction [21]. A balanced non-landslide dataset is also required to generate an LSM. The tool developed by Huang et al. [15] was used to generate the non-landslide points. The proposed model was trained and evaluated using 2,070 data points.

Landslide causative factors
Landslide causative factors are the features used by the ML model to classify a region into landslide and non-landslide areas. Different numbers of causative factors have been used in existing studies, with studies using up to 60 factors [18]. There is no fixed number of required causative factors, and determining which factors to use can be a challenging decision [17] and the use of relevant features can greatly improve the predictive capability of ML models. Different researchers have used different approaches to select the optimum causative factors. Some of these approaches include Pearson correlation [11], multi-collinearity using variance inflation factors and tolerance [19]. The 16 optimal LCFs used in our study were selected using the recursive feature elimination with cross-validation technique.

Our study uses 16 landslide causative factors. The ASTER Global Digital elevation model V3 data from USGS Earth Explorer [24] was used to generate the topographical features like aspect, elevation, slope degree, topographic wetness index (TWI), and sediment transport index (STI). Other factors used in the study include average annual rainfall data, which were generated using the India Meteorological Department's (IMD's) 0.25 × 0.25 annual gridded rainfall data [22]. A number of factors were obtained from GSI's Bhukosh portal [9], including lithology, geomorphology, fault, lineament, waterways, and roads. The Normalised difference vegetative index (NDVI) was generated using remote sensing data from Harmonised Sentinel-2 MultiSpectral Instrument, Level-2A. The soil sub-surface moisture (SuSM) data was collected from NASA-USDA SMAP [5], and the data for soil texture from FAO, & IIASA [8]. Table 26.1 provides a brief information on the causative factors.

Methodology and proposed model specifications

The study uses 2,070 landslide and non-landslide points to extract the LCFs values. The extracted landslide causative factors are pre-processed using SKlearn MinMaxScaler to provide bounded normalised numerical data and OneHotEncoder to ensure the encoded data are nominal. The pre-processed data is split in the ratio of 80:20 for training and testing, respectively. The novel stacking model is trained on the training data. The stacking ensemble approach was selected due to its improved generalisation capability by leveraging the strengths of heterogeneous learners. The model is evaluated using accuracy, F1-score, MCC, and AUC metrics on the testing data. The proposed model and the composite raster image of all the LCFs were used to generate the LSM for Meghalaya.

Proposed model specifications
The proposed model, shown in Figure 26.1, is a stacking ensemble technique that combines heterogeneous

models to improve generalisation capability. The stacking approach trains a meta-learner on the meta-features generated by the base learners, allowing the meta-learner to learn from the meta-features to improve prediction accuracy. The model has a base layer and a meta layer. The base layer uses six ML algorithms to generate an intermediate dataset called the meta-features. The meta-layer consists of a single ML algorithm trained on the meta-features data. The layered approach of stacking ensemble learning enables the improved generalisation capability of the final stacked model. For our study, the base learners include LGR, RF, LGBM, XGBoost, SVM, and k-NN. The hyperparameters for each base-layer ML model were tuned using Bayesian optimisation (BO).

The ExtraTree classifier is employed as the meta-classifier. Like RF, it creates a forest of decision trees and splits nodes based on random subsets of features. However, unlike RF, the ExtraTree uses the whole training dataset to grow the trees, and the splitting of node is done randomly [10]. The stacked ensemble model had better accuracy, F1-score, MCC, and AUC values with ExtraTree as the meta-classifier. The hyperparameter for the ExtraTree was tuned using the BO technique. The final parameters after tuning were max_depth = 20, n_estimators = 60, min_samples_split = 2, and min_samples_leaf = 1.

Results and discussion

The performance of the base-learners and the proposed model was comparatively evaluated using accuracy, F1-score, MCC, and AUC metrics. The performance of all the models is provided in Table 26.2. Our proposed model has an improved accuracy value of 0.88, F1-score of 0.89, MCC of 0.79, and an AUC value of 0.95 compared to the base-learners and other ensemble approaches like RF (bagging), XGBoost (boosting), and LightGBM (boosting) techniques.

The evaluation metrics results indicate the improved performance of the proposed model over the individual base-learners. This indicates that the novel stacking ensemble technique, with LGR, XGB, SVM, k-NN, RF, and LGBM as the base learners and ExtraTree as the meta-learner can be considered a reliable ensemble technique to improve the accuracy of an LSM. Our study findings imply that powerful ML models like LGBM, XGB, and RF, which have performed well in existing landslide susceptibility studies, can be combined using the stacking ensemble technique to get a more robust ML model. Using SHapley Additive exPlanations [23], feature importance was explored, where the distance from roads, slope degree, NDVI, road density, and soil texture are shown to have the most influence on LSM generated using the proposed model.

Figure 26.2 presents the final LSM for Meghalaya, and the state's susceptibility to landslide is provided as a probability ranging between 0 and 1. Regions with a susceptibility value nearer to 1 are highly susceptible to landslides. Based on the LSM, regions near roads are more susceptible to landslides. The susceptibility map can aid

Table 26.1 Landslide causative factors.

Causative factors	Importance
Aspect	Influences soil moisture content, vegetation cover and thickness of the soil
Elevation	Influences vegetation cover
Slope degree	Steep slope are more prone to landslides
TWI	Soil moisture and water accumulation potential
STI	Potential of sediment transport in the region
Avg. rainfall	Considered a trigger of landslide as continues rain can destabilise the soil
Lithology	Influences the geomorphological process
Geomorphology	Landforms can influence landslides
Dist. fault	Fault movement can weaken rocks
Dist. lineament	Proximity increases weathering and fracturing of rocks
Dist. roads	Destabilisation of the earth surface due to construction of roads
Road density	Increase in road development increases landslides due to destabilisation of the earth surface
Dist. water ways	Influences soil moisture and slope stability
NDVI	Vegetation cover reduces landslides
SuSM	Increase in soil moisture content reduces the shear strength of a slope
Soil texture	Certain soil are more prone to landslides

Source: Author

Table 26.2 Landslide causative factors.

	Accuracy	F1-score	MCC	AUC
LGR	0.79	0.79	0.58	0.88
XGB	0.85	0.87	0.73	0.93
SVM	0.83	0.83	0.67	0.91
k-NN	0.85	0.86	0.71	0.92
RF	0.86	0.86	0.73	0.94
LGBM	0.86	0.86	0.72	0.94
Proposed	**0.88**	**0.89**	**0.79**	**0.95**

Source: Author

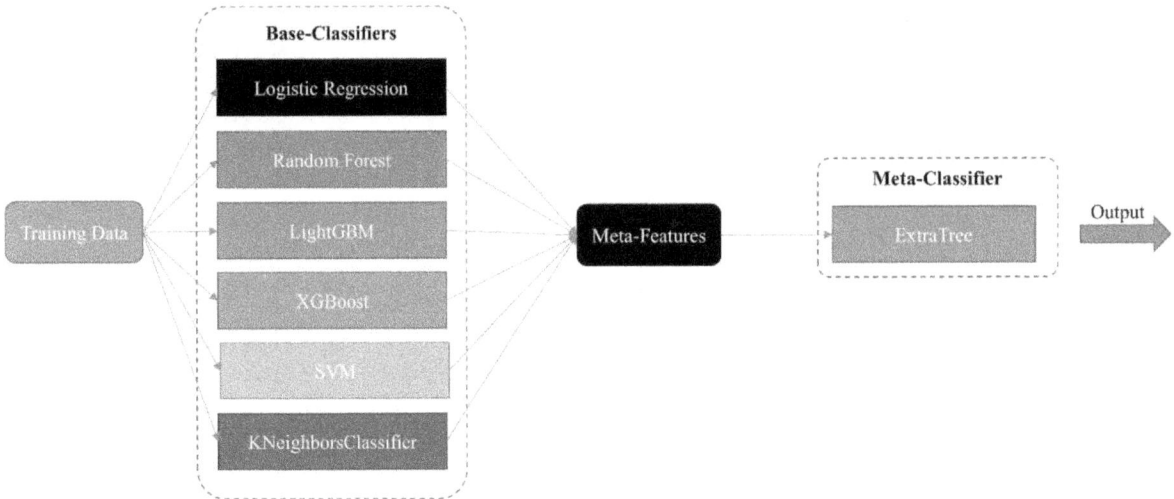

Figure 26.1 Proposed novel stacking model
Source: Author

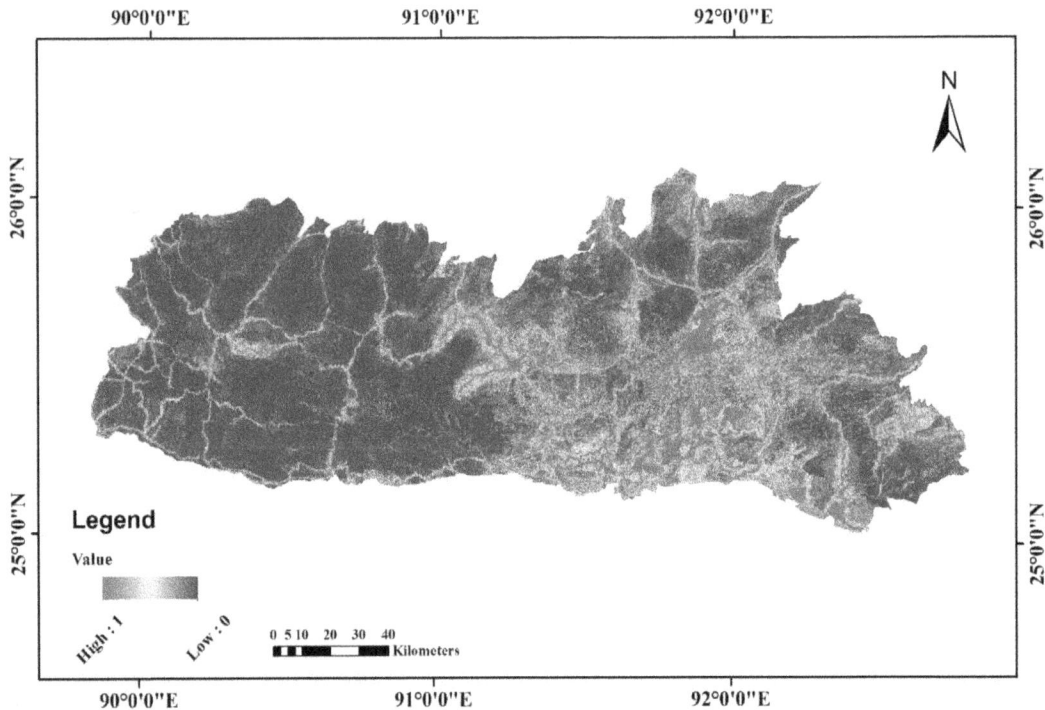

Figure 26.2 Landslide susceptibility map for Meghalaya generated using the novel stacking model
Source: Author

in landslide mitigation and disaster management efforts for the state. Regarding computational requirements, the proposed model can run on basic hardware configuration; however, CPU and wall time increase as training data increases, implying that higher configuration may be required for larger datasets. With the required computational resources, the proposed stacking approach can scale well to handle large-scale applications. Adjustments to the model can optimise it to run on lower computation configurations when required.

Conclusion

The study introduced a novel stacking ensemble technique to generate an LSM for Meghalaya. The study uses 2,070 landslide and non-landslide data to extract the point values of 16 landslide causative factors, which are used to train and test the proposed model. The proposed model has six base learners with ExtraTree as the meta-learner. The proposed model achieved an accuracy of 0.88, F1-score value of 0.89, MCC value of 0.79, and an AUC value of 0.95, showing improvement over the standalone base learners. The results imply that the stacking approach can improve upon existing models by heterogeneously combining them to provide a more reliable approach for generating landslide susceptibility maps. A reliable LSM was generated for Meghalaya using the proposed model, which can assist in landslide mitigation and management efforts. Future studies can explore an automated selection of optimal base-learners and meta-learners, as our study currently randomly selects base and meta-learners based on their performance and application in existing studies. Determining the balanced trade-off between computational cost and the prediction performance of the stacking technique in LSM applications can be explored in future work.

Acknowledgement

Financial support for this study was provided by Science and Engineering Research Board, Department of Science and Technology, Government of India; Grant No. EEQ/2023/000124 for the implementation of "Landslide Susceptibility Mapping and Monitoring using Machine Learning and IoT: A case study of Meghalaya, India".

References

[1] Agrawal, N., Dixit, J. (2022). Assessment of landslide susceptibility for Meghalaya (India) using bivariate (frequency ratio and Shannon entropy) and multi-criteria decision analysis (AHP and fuzzy-AHP) models. *All Earth*, 34(1), 179–201.

[2] Agrawal, N., Dixit, J. (2023). GIS-based landslide susceptibility mapping of the Meghalaya-Shillong Plateau region using machine learning algorithms. *Bulletin of Engineering Geology and the Environment*, 82(5), 170.

[3] Azarafza, M., Azarafza, M., Akgün, H., Atkinson, P. M., Derakhshani, R. (2021). Deep learning-based landslide susceptibility mapping. *Scientific Reports*, 11(1), 24112.

[4] Badavath, N., Sahoo, S., Samal, R. (2024). Landslide susceptibility mapping for West-Jaintia Hills district, Meghalaya. *Sādhanā*, 49(1), 52.

[5] Bolten, J., Crow, W. (2012). Improved prediction of quasi-global vegetation conditions using remotely-sensed surface soil moisture. *Geophysical Research Letters*, 39(19).

[6] Dikshit, A., Sarkar, R., Pradhan, B., Segoni, S., Alamri, A. (2020). Rainfall induced landslide studies in Indian Himalayan region: a critical review. *Applied Sciences*, 10(7), 2466.

[7] Fang, Z., Wang, Y., Peng, L., Hong, H. (2021). A comparative study of heterogeneous ensemble-learning techniques for landslide susceptibility mapping. *International Journal of Geographical Information Science*, 35(2), 321–347.

[8] FAO, IIASA. (2023). Harmonized World Soil Database Version 2.0. Rome and Laxenburg. FAO & International Institute for Applied Systems Analysis (IIASA). https://doi.org/10.4060/cc3823en

[9] GSI. (2024). Bhukosh | Gateway to all Geoscientific Information of GSI. Retrieved 2 September 2024, from https://bhukosh.gsi.gov.in/Bhukosh/MapViewer.aspx

[10] Geurts, P., Ernst, D., Wehenkel, L. (2006). Extremely randomized trees. *Machine Learning*, 63, 3–42.

[11] Hu, X., Zhang, H., Mei, H., Xiao, D., Li, Y., Li, M. (2020). Landslide susceptibility mapping using the stacking ensemble machine learning method in Lushui, Southwest China. *Applied Sciences*, 10(11), 4016.

[12] Hu, X., Mei, H., Zhang, H., Li, Y., Li, M. (2021). Performance evaluation of ensemble learning techniques for landslide susceptibility mapping at the Jinping County, Southwest China. *Natural Hazards*, 105, 1663–1689.

[13] Huan, Y., Song, L., Khan, U., Zhang, B. (2023). Stacking ensemble of machine learning methods for landslide susceptibility mapping in Zhangjiajie City, Hunan Province, China. *Environmental Earth Sciences*, 82(1), 35.

[14] Huang, Y., Zhao, L. (2018). Review on landslide susceptibility mapping using support vector machines. *Catena*, 165, 520–529.

[15] Huang, W., Ding, M., Li, Z., Zhuang, J., Yang, J., Li, X., Meng, L., Zhang, H., Dong, Y. (2022). An efficient user-friendly integration tool for landslide susceptibility mapping based on support vector machines: SVM-LSM toolbox. *Remote Sensing*, 14(14), 3408.

[16] Kadavi, P. R., Lee, C. W., Lee, S. (2018). Application of ensemble-based machine learning models to landslide susceptibility mapping. *Remote Sensing*, 10(8), 1252.

[17] Kavzoglu, T., Colkesen, I., Sahin, E. K. (2019). Machine learning techniques in landslide susceptibility mapping: a survey and a case study. *Landslides: Theory, Practice and Modelling*, 283–301.

[18] Koukis, G., Ziourkas, C. (1991). Slope instability phenomena in Greece: a statistical analysis. *Bulletin of Engineering Geology & the Environment*, 43(1).

[19] Li, R., Wang, N. (2019). Landslide susceptibility mapping for the Muchuan County (China): a comparison between bivariate statistical models (woe, ebf, and ioe) and their ensembles with logistic regression. *Symmetry*, 11(6), 762.

[20] Merghadi, A., Yunus, A. P., Dou, J., Whiteley, J., ThaiPham, B., Bui, D. T., ..., Abderrahmane, B. (2020). Machine learning methods for landslide susceptibility studies: a comparative overview of algorithm performance. *Earth-Science Reviews*, 207, 103225.

[21] NESAC. (2024). Seasonal Landslide Inventory for NER-2023. North Eastern Space Applications Centre. https://nerdrr.gov.in/LandslideDSS/

[22] Pai, D., Rajeevan, M., Sreejith, O., Mukhopadhyay, B., Satbha, N. (2014). Development of a new high spatial resolution (0.25 × 0.25) long period (1901–2010) daily gridded rainfall data set over India and its comparison with existing data sets over the region. *Mausam*, 65(1), 1–18.

[23] Scott, M., Su-In, L. (2017). A unified approach to interpreting model predictions. *Advances in Neural Information Processing Systems*, 30, 4765–4774.

[24] USGS – U.S. Geological. (2024). EarthExplorer. https://earthexplorer.usgs.gov/

27 Advancements in network security threat detection systems for wireless sensor networks: a review of machine learning and deep learning techniques

Bikash Kalita[a] and Satyajit Sarmah[b]

Department of Information Technology, Gauhati University, Guwahati-14, India

Abstract

The rapid growth of technology and wireless sensor networks (WSNs) has increased the need for secure communication, leading to a rise in attacks on these networks. Intrusion detection and prevention have become crucial in ensuring the safety of these systems. To address this, this research presents a performance analysis and comparison study of different machine learning and deep learning methods for security threat detection and prevention WSNs. The open nature of wireless communication in these networks poses security risks, making intrusion detection systems (IDS) essential. The article provides a comprehensive overview of IDS designed for WSNs, analysing and comparing their advantages and disadvantages. Additionally, the paper offers guidelines on potential IDS for WSNs and highlights open research issues in this field. It also covers the methodology, dataset, and performance of the models used in the study, emphasising the importance of detecting intrusions before they can cause harm to the network or compromise sensitive information.

Keywords: Wireless sensor networks, machine learning, deep Learning, intrusion detection system

Introduction

'A wireless sensor network (WSN) is a self-organising, infrastructure-less network composed of small, autonomous sensor nodes' [14]. These nodes are equipped with sensing, processing, and communication capabilities, and they operate on low power and low-cost principles. They function as battery-operated devices that collaborate to form a network, enabling efficient data gathering, processing, and communication without the need for a centralised infrastructure [3]. Figure 27.1 represents a general architecture of a WSN.

WSNs are used in a variety of scientific and technological domains because of their easy and affordable deployment capabilities. Information gathering on human activities and behaviour, such as healthcare, military surveillance and reconnaissance, and traffic, are some of these (Chandre et al., 2020). WSNs are also used to track environmental and physical events, such as wildfires, pollution, earthquakes, ocean and wildlife conditions, and water quality. They are also used to monitor industrial locations for jobs like guaranteeing building safety and keeping an eye on machinery performance [1].

Network security measures include intrusion prevention, intrusion detection, and intrusion tolerance. Intrusion prevention measures stop an adversary at the network edge and include group keys, authentication schemes, and survey tools. Security in WSNs is a major issue, especially when they are used for mission-critical operations [12]. For instance, it is crucial to stop the unauthorised distribution of private patient health details to outside parties in healthcare applications. In military applications, security measures are of paramount importance because any security flaws in the network could result in casualties [10].

For some attacks, intrusion detection is inefficient, such as attacks involving insiders/authenticated nodes. To deal with these attacks, strong systems, like parallel or k-of-n designs, need to take intrusion resistance into account, or dynamic techniques such as load balancing mechanisms.

In addition to discussing threat detection in wireless contexts, this survey article outlines the parameters that are utilised to classify IDS and offers an arrangement for categorising current IDS techniques. It also discusses lessons learned and suggests future research directions.

Related work

In this section, we examine existing research in the field to identify common methodologies and analyse any gaps that may exist. The following literature review provides a comprehensive discussion of relevant studies that have already been conducted.

Ioannou et al. [17] introduces a robust anomaly-based IDS called mIDS, which leverages Binary Logistic Regression (BLR) for classifying local sensor activity as either benign or malicious. The proposed system is evaluated through tests conducted on routing layer attacks, demonstrating mIDS's ability to accurately identify malicious behaviour with a success rate ranging from 88% to 100%.

[a]bikax99@gmail.com, [b]satyajitnov2@gmail.com

DOI: 10.1201/9781003675235-27

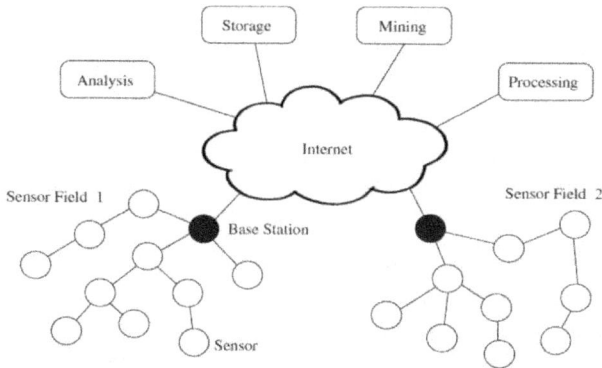

Figure 27.1 A general architecture of WSN [13]
Source: Author

In a paper, Sharma et al. [34] addressed intrusion detection using a specialised attack model containing blackhole, vampire, jelly-fish, and grey-hole attacks. The performance of normal networks, simple IDS schemes, and the hybrid IDS scheme is analysed. The proposed hybrid 3-tier IDS scheme, which integrates clustering, RC4, and digital signatures, shows consistent delay, making it suitable for larger networks. The proposed IDS is designed to work efficiently and intelligently in the presence of multiple attacks.

According to the study by Bace and Mell [8], a detection system is either software, hardware, or both that can automatically process and keep track of events taking place within a system or a network. It is a method of analysing things in light of security issues. The frequency of network attacks has increased during the last few days. To secure and maintain the security infrastructure of most organisations, intrusion detection must be added. The major purpose of the research was to improve user comprehension of security objectives in detection techniques by focusing on intrusion detection. It also shows the need for and decision-making process when choosing a certain system and network environment.

Yan et al. [39] suggested using a hybrid IDS built within the cluster head. The model included anomaly and abuse detection modules. By utilising the benefits of abuse detection and anomaly detection, the detection rate will be increased while the false positive rate will be decreased. The findings of the detect module are integrated, and the types of attacks are reported using a decision-making module.

In a vast sensor network, Roy et al. [31] developed a secure data aggregation approach. By fusing multipath routing protocols with duplicate-insensitive algorithms, a robust aggregation framework known as synopsis diffusion was proposed to several academic communities. It is mostly employed to forecast the count and sum rather than message losses brought on by node and transmission problems. However, it might not solve the issue of erroneous sub-aggregate values, which are a result of compromised nodes. To safeguard the data against attacks, a summary diffusion strategy is used, in which some compromised nodes are responsible for the fake sub-aggregate values. A revolutionary lightweight verification algorithm was the main emphasis of this approach, along with the detection of any fake contributions.

Based on the genetic algorithm, artificial immune system, and artificial neural network (ANN)-based IDS methodologies utilised in WSN, Alrajeh and Lloret [5] conducted a critical study. The work focuses on the use of artificial immune systems that are based on IDS in WSNs. The artificial immune technique detects security threats against WSNs based on anomalies. They created a complex architecture that is fully integrated with the human immune system and the security system. It discusses several traffic monitoring methods and defines a few different attack types using the provided database. The creation of tables with internal caches is the mechanism.

In sensor networks, Mamun et al. [23] suggested an anomaly detection approach. It compared the hybrid approaches used to detect the anomaly process with the standard genetic algorithm. Many sensor shortcomings and incorrect measurements lead to external disturbances in sensor networks. These could make a range of applications, including healthcare, more susceptible to attack and hamper fast and effective responses. It is completely measured in some instances using inaccurate measures of medical equipment.

Gavric and Simic [16] looked into the overview of DoS attacks and also put out a practical method for interference attacks. In this article, the authors present a variety of attacks that have been examined in terms of performance metrics like throughput, delay, transmission time, etc. Because it pays less attention to the multi-layer level, the method used to discover interference attacks is not very effective.

Mahalakshmi and Subathra (2018) highlighted the security upkeep problems in wireless sensor networks. The authors have focused on these difficulties and suggested a sleep attack detection method to lessen the power consumption of the sensor nodes. They also took into account the Denial of Sleep Attacks. However, one of the crucial elements to be examined for calculating the performance, which is less descriptive, is the energy model of the sensor network.

Y. Li et al. [22] have presented the issues with ratio-based DoS attacks and isolated state assessments of cyber-physical systems. A sensor node uses a non-wired network that is breached by an intruder to send its local approximation to a remote estimator. The study is conducted using a two-player game with possible numerous power levels. Using the present state and data gathered from earlier time steps, the authors then constructed a

Markov Game Framework to describe the interactive decision-making process. If the proposed method and the equations are not compatible, this framework may become more difficult.

A distributed and adaptive approach for a selective jamming attack in relation to time division multiple access has been put forth by Tiloca et al. [37] and is based on wireless sensor networks. The sensor nodes receive the time slots in advance, which are then utilised for a few consecutive frames. Only a single-channel TDMA network is used to calculate the network's overall performance. This restriction prevents the other layer attacks that are timed to occur within those time periods from being recognised in the assigned transmission.

Patil and Chaudhari [29] developed methods for detecting denial of service in WSNs. Numerous protocols have been suggested for detection, and a fuzzy logic-based immune system has also been proposed for the detection of denial-of-service attacks. The writers did not concentrate on the experimental findings, simply presenting the theoretical solutions.

Xie et al. [38] focused on the inclusion of an approximated sample covariance matrix in the identification of anomalies in WSN. Here, a prediction variance detector that can identify a wide range of abnormalities is studied using a collection of nearby data segments. However, because this is the main focus, and only happens when there is an excess of data flow, it is extremely dependent on the assumption that the information is spatially connected.

Security threats

There are two basic types of security attacks against WSNs: active attacks and passive attacks [24]. Passive attacks involve covert attackers who either eavesdrop on communications or damage networks by destroying essential components. Additional kinds of passive attacks include intercepting, node malfunctioning, node tampering/destruction, and traffic analysis [11].

Active attacks

In active attacks, an adversary actively interferes with the operations of the targeted network. These attackers employ various techniques such as routing attacks, eavesdropping, and the fabrication of false data streams. The actions carried out by active attackers include the insertion of erroneous data into the WSN, impersonation of network entities, and alteration of packets. These attacks aim to affect the network's operations and can be detected. For instance, active attacks may cause networking services to be discontinued or degraded [11]. These attacks involve unauthorised individuals who not only observe and intercept packet exchanges within the network but also actively manipulate the data stream. Here are some examples of active attacks [9]:

Selective forwarding: Malicious nodes selectively forward certain messages while discarding others, leading to the loss or manipulation of critical data.

Spoofed, altered, or replayed routing information: Attackers manipulate routing information within the network, either by changing, spoofing, or replaying routing information, resulting in disruption of traffic flow.

Sybil attacks: Sybil attacks specifically target fault-tolerant technologies such as distributed storage, multipath routing, and topology maintenance within the network. This attack involves an attacker creating multiple identities to deceive the network.

Wormhole attacks: Attackers create tunnels or 'wormholes' within the network, capturing data packets from one location and relaying them to another, causing disruption or unauthorised access.

HELLO packet attacks: Malicious nodes employ abnormally high transmission power to send or replay HELLO packets between nodes, exploiting vulnerabilities in routing protocols.

Sinkhole attack: This attack involves diverting traffic from a specific area of the network. It can be executed by either resourceful outsiders or malevolent insiders. The attacker misleads neighbouring nodes by advertising a false optimal route or by forging a route advertisement.

Black-hole attack: In this attack, a malicious node advertises false paths as optimal routes during the path-finding process, leading other nodes to route their traffic through the attacker. This can result in decreased network performance and disrupted communication.

Passive information gathering: If sensor nodes are not adequately encrypted, adversaries with sufficient resources can passively gather sensitive information from the network.

False Node: Attackers add false nodes to the network, injecting harmful data or disrupting the network's functionality.

Passive attacks

Passive attacks refer to unauthorised individuals who engage in the observation and interception of packet exchanges within a wireless sensor network (WSN). These passive attackers, functioning as typical nodes, carry out various activities such as data collection from the WSN, monitoring communication channels, and eavesdropping. Common examples of passive attacks include [9]:

Attacks on privacy: Passive attacks are also known as attacks on privacy, as they compromise the confidentiality of the network by intercepting sensitive information.

Monitoring and eavesdropping: The attackers can easily gain access to the content of communications by covertly listening in on data transmissions.

Traffic analysis: By analysing the traffic patterns and behaviour of the network, adversaries can potentially extract valuable information, even if the messages are encrypted.

Node camouflage: Adversaries may insert their own nodes into the network or compromise existing nodes to blend in and mimic normal node behaviour. This allows them to intercept and manipulate packets without arousing suspicion.

Security requirements

The WSN communication method uses a multi-hop system administration and allows for one-to-many or many-to-one communication patterns as well as unicast or broadcast data transfer among the other nodes [36].

Making WSN a tamper-proof network and ensuring secure data transfer are the main goals of WSN communication. The communication in WSNs strives to accomplish a number of security objectives to maintain the network's availability, integrity, confidentiality, authentication, access control, and data freshness. The following definitions apply to these security goals:

Confidentiality: Preventing unauthorised disclosure by protecting the transmitted data from unauthorised users.

Integrity: Ensuring that the messages' content is unchanged and undamaged to guard against unauthorised alterations or fabrications.

Authentication: Ensuring the identity of the end user or node and confirming the communication's source in order to stop impersonation or unauthorised access.

Access control: Limiting access to the network and its resources to authorised parties exclusively while thwarting attempts by unauthorised parties to get access.

Availability: Making sure that the necessary services and network resources are continually accessible to authorised parties while minimising downtime or service denial.

Data freshness: Preventing the usage of stale or compromised data by ensuring that the data being communicated is current, accurate, and not revealed to unauthorised users.

Scalability: Creating a network that can support many nodes and effectively handle the network's expansion to meet changing demands.

Security goals

To preserve the confidentiality, integrity, and availability of the data transmitted within the network while also authenticating the participating entities and guaranteeing

authorised access control, WSNs can achieve these security objectives.

Addressing security attacks on networks involves three primary components [15].

Prevention: The primary goal of this component is to actively stop attacks before they happen. The objective is to put strategies and countermeasures in place that can successfully defend against targeted attacks.

Detection: If an attacker is able to bypass the prevention measures, it indicates a failure in defending against the attack. In these situations, the security solution must immediately enter the detecting phase. This involves identifying the current attack and the compromised nodes.

Mitigation: The third and last phase is to mitigate the effects of the attack after it has already occurred. This involves taking steps to safeguard the network against additional compromise, as well as removing the affected nodes from the routing tables of the network.

By incorporating these three components—prevention, detection, and mitigation—network security solutions can effectively defend against attacks, identify ongoing attacks, and respond to mitigate the impact of attacks.

Intrusion detection systems

In this section, we outline current security threat detection methods and discuss how well they function with different wireless networks.

Figure 27.2 presents a hierarchical taxonomy of Network Security Threat Detection Systems for WSNs, categorising detection approaches into Traditional, Machine Learning-based, and Deep Learning-based methods. Traditional methods include signature-based, anomaly-based, rule-based, and statistical techniques, which rely on predefined patterns or statistical analysis to detect threats. ML-based detection is further divided into supervised, unsupervised, and reinforcement learning (RL) approaches, highlighting techniques such as Decision Trees (DT), Support Vector Machines (SVM), k-Nearest Neighbor (k-NN), and Deep Q-Networks (DQN). The DL-based category encompasses Convolutional Neural Networks (CNN), Long Short-Term Memory (LSTM), Recurrent Neural Networks (RNN), and hybrid models like CNN-LSTM. Additionally, the figure outlines key challenges in ML/DL-based threat detection, including data imbalance, high false positive rates, and computational complexity, as well as future research directions, such as explainable AI, federated learning, and adaptive learning methods to enhance threat detection capabilities in resource-constrained WSN environments. security threat detection systems for WSNs, categorising detection methods, along with key challenges and future research directions.

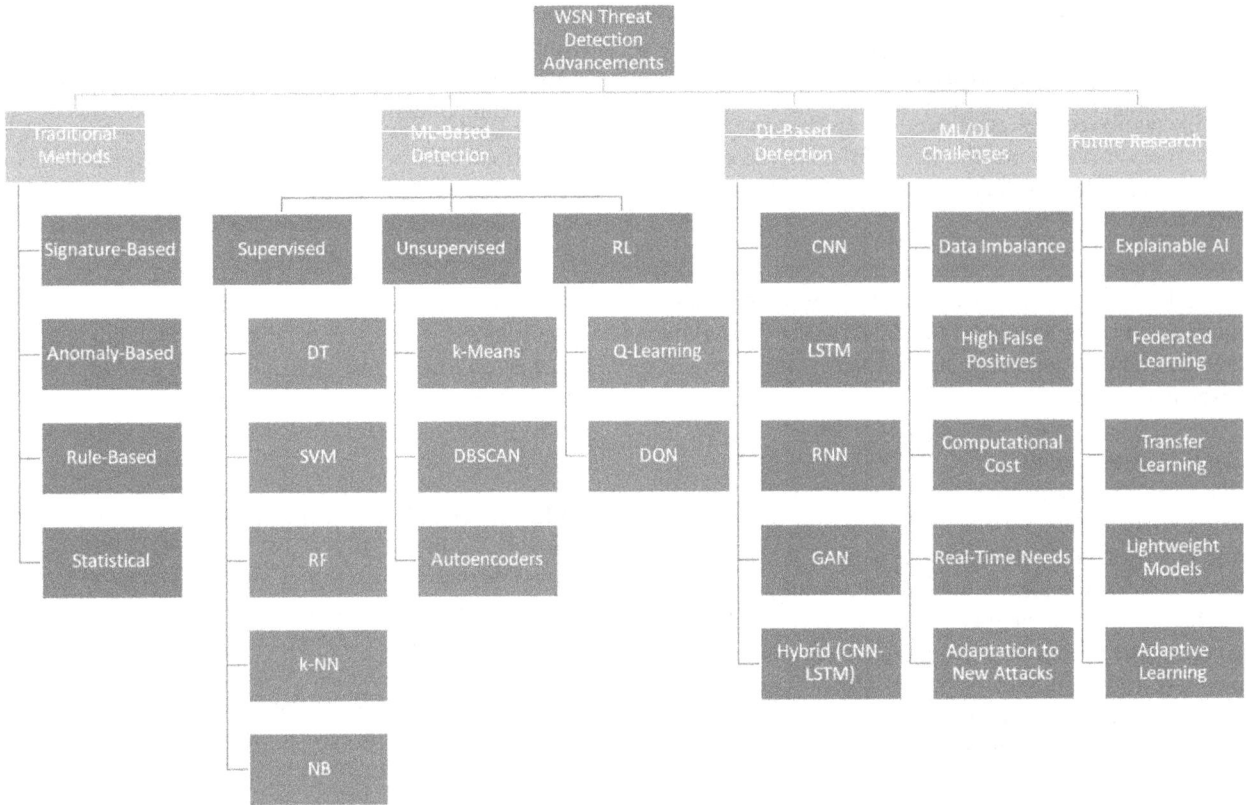

Figure 27.2 A hierarchical taxonomy of network
Source: Author

Anomaly-based intrusion detection

Anomaly-based security detection techniques search for unusual runtime characteristics. These methods, which include neural networks, Bayesian classifiers, can be used with discrete, continuous, and multidimensional datasets.

Anomaly-based approaches eliminate the requirement to maintain an attack dictionary up-to-date and to completely label all known threat characteristics, but are susceptible to false positives and the training/profiling phase is vulnerable.

Anomaly-based detection relies on statistical behaviour modelling to identify abnormalities. It involves creating profiles of the typical behaviour exhibited by network members and flagging any significant deviations that exceed a predefined threshold. However, the drawback of this method is that the regular profiles must be updated frequently due to the potential rapid changes in network behaviour. This can increase the workload on sensor nodes with limited resources [6]. According to Kachirski and Guha [20], this model achieves precise and reliable intrusion detection with minimal false positives and false negatives when the observed network follows static behavioural patterns. One notable advantage of this detection method is its effectiveness in identifying novel or previously unknown attacks.

Insider attacks are difficult to execute because they focus on system disruption, but outside attackers cannot generate transitions or transformations.

Specification-based IDS uses no user, group or data profiling and instead measures deviation from legitimate behaviours.

Signature-based intrusion detection

This technique involves creating signatures or profiles of known attacks to identify potential threats. For example, a signature for a brute force password attack could be 'three failed login attempts within five minutes.' According to Sobh [35], this detection method offers the advantage of having a low rate of false positives since it can accurately and effectively identify known threats. Signature-based methods are efficient against outsider attacks because they only react to known bad behaviour. However, a limitation of misuse detection is its inability to detect novel types of attacks that have not been previously profiled.

Reputation management

A reputation administrator needs to stay on alert to look for fraudulent nodes that want to improve their reputation in addition to detecting selfish behaviour in nodes. Approaches to reputation management work exceptionally well in large networks where it is impossible to build relationships based on initial trust.

Research gap

A comprehensive overview of the existing literature has been presented in Table 27.1, encompassing a summary of the works along with their respective advantages and limitations, as well as details regarding the dataset and techniques employed. This table serves as a valuable resource for understanding the current state of research in the field, enabling researchers and readers to gain insights into the various approaches and their corresponding strengths and weaknesses.

By presenting a concise summary of the literature, Table 27.1 highlights the diverse range of methodologies and data sources utilised in prior studies. This compilation not

Table 27.1 Comprehensive overview of WSNs, covering technological advancements and limitation.

SI. No.	Authors	Dataset used	Algorithms/techniques used	Advantages	Limitation
1	Nancy et al. [27]	KDD'99 cup	Fuzzy temporal decision tree algorithm	Reduction in false positive rate, delay in network and energy	Intelligent agents for communication in a distributed environment for enhancing the performance can be applied.
2	Pan et al. [28]	NSL KDD and UNSW NB15	NB, NN, SVM, DT, and RF classifiers.	Lightweight and intelligent, requires consideration of energy efficiency, safety, and real-time	Only supervised ML techniques have been applied.
3	Alsulaiman and Al-Ahmadi [7]	WSN DS	NB, NN, SVM, DT, and RF classifier	RF model has shown a good accuracy.	Limited to detect particular types of attack
4	W. Li et al. [21]	Not Applicable	k-NN classification algorithm	The system has achieved efficient, rapid intrusion detection	Detects only flooding attacks
5	Murugesh and Murugan [25]	WSN DS	'Archimedes Optimization with Variational Auto encoder based Attack Detection and Classification (AOVAE ADC)'	Increases accuracy	Deep learning related fusion approaches can be applied
6	Selvakumar et al. [33]	Network trace datasets	'Fuzzy Rough sets for attribute selection and Allen's interval algebra, fuzzy and rough set based nearest neighborhood algorithm (FRNN)'	Reduced the false alarm rate and increased detection accuracy.	Genetic-based feature refinement can be applied.
7	Salmi and Oughdir [32]	Computer generated WSN dataset	CNN-LSTM Based Approach	Reliable and effective	Limited to detect particular types of attack
8	Murugesh and Murugan [26]	WSN DS	'Optimal Quantum Neural Network, Spotted Hyena Optimizer with Quantum Neural Network for DDoS Attack Classification (SHOQNN-AC) Technique'	Efficient in detection of DDoS Attack	Limited to DDoS Attack Classification

Table 27.1 Comprehensive overview of WSNs, covering technological advancements and limitation—cont'd

SI. No.	Authors	Dataset used	Algorithms/techniques used	Advantages	Limitation
9	Almomani and Alenezi [4]	WSN DS	8 Data mining techniques (DMTs)	Efficient in detection of attacks	Model complexity
10	Rezvi et al. [30]	WSN DS	'k-NN, Naive Bayes, Logistic Regression, support vector machine (SVM) and ANN algorithms'	High detection rates	Deep learning model can be used
11	Aiswarya K and Vijitha S [2]	Self-created dataset	Random Forest, SVM, and NN,	High accuracy and detection probability, and low false alarm	Useful for jamming detection only
12	Jiang et al. [19]	WSN-DS	SLGBM	Better in F-measure	Hybridisation techniques can be applied for better accuracy.
13	Alsahli et al. [6]	KDD99 and WSN datasets	Naïve Bayes (NB), IBK, Random Forest	High accuracy and less time consuming	Can address different types of attacks
14	Ismail et al. [18]	WSN DS	Naive Bayes, LightGBM algorithm	Provides multilayer detection	robust and lightweight system can be considered

Source: Author

only aids in identifying the prevailing trends and patterns within the field but also facilitates the identification of gaps or areas requiring further investigation. Researchers and practitioners can leverage the information in this table to inform their own work, building upon the existing knowledge base, and potentially addressing the limitations highlighted in the literature.

The analysis of existing literature highlights two important aspects for Intrusion Detection Systems (IDS). Firstly, it emphasises the necessity for IDS to possess both speed and adaptability with minimal data inputs, owing to the rapid influx of data. While a significant portion of current research concentrates on classification methods, limited attention has been given to feature selection techniques. Furthermore, the literature reveals certain limitations associated with classification methods, particularly the lack of diversity among employed classifiers. This presents an opportunity for further investigation into the impact of parameter optimisation on the classification process.

Conclusion and suggestion

To analyse and detect network security threats in WSN, a thorough assessment of the available research has been conducted in the field of machine learning and deep learning approaches. Various methods are employed to identify threats to network security and to enhance system functionality and safety. Solutions to the many types of cyberthreats are reviewed and compiled, with a focus on recent ML technologies and methodologies.

Future research in the field of IDS in WSN can explore several areas. This includes repurposing existing work, investigating new IDS techniques to enhance the performance of current systems, and incorporating multi-trust mechanisms for intrusion detection. There is a need for tools or methodologies to transform complex systems into formal models, along with languages or schemas for expressing these models. IDS can be improved by adjusting the result according to the type and capability of the opponent, identifying compromised segments, and taking appropriate actions such as stopping operations, rekeying/resetting passwords, and resuming operations progressively.

Additionally, future research can focus on application layer data auditing to minimise the period between attack resistance and identification. Combining security threat detection methods with loss threshold monitoring can enhance effectiveness against environmental noises and smart attackers. To analyse IDS outcomes, determine the best protocol settings, encourage creativity in IDS design, and evaluate the effect of system performance, model-based analytic approaches ought to be created and evaluated. Deepening the understanding of attacker models and devising responses to maximise system lifetime should be pursued. Lastly, a dynamic control function that adjusts detection strength based on runtime detection of attacker strength should be designed.

References

[1] Abdullah, M. A., Alsolami, B. M., Alyahya, H. M., Alotibi, M. H. (2018). Intrusion detection of DoS attacks in WSNs using classification techniques. *Journal of Fundamental and Applied Sciences*, 10(4S), 298–303. https://doi.org/10.4314/jfas.v10i4s.94

[2] Aiswarya K, Vijitha S. (2022). Jamming attack detection using machine learning algorithms in wireless network. *International Journal of Advanced Research in Science, Communication and Technology*, 36–43. https://doi.org/10.48175/ijarsct-5896

[3] Akyildiz, I. F., Su, W., Sankarasubramaniam, Y., Cayirci, E. (2002). Wireless sensor networks: a survey. *Computer Networks*, 38(4), 393–422. https://doi.org/10.1016/S1389-1286(01)00302-4

[4] Almomani, I., Alenezi, M. (2018). Efficient denial of service attacks detection in wireless sensor networks. *Journal of Information Science and Engineering*, 34(4), 977–1000. https://doi.org/10.6688/JISE.201807_34(4).0011

[5] Alrajeh, N. A., Lloret, J. (2013). Intrusion detection systems based on artificial intelligence techniques in wireless sensor networks. *International Journal of Distributed Sensor Networks*, 2013. https://doi.org/10.1155/2013/351047

[6] Alsahli, M. S., Almasri, M. M., Al-Akhras, M., Al-Issa, A. I., Alawairdhi, M. (2021). Evaluation of machine learning algorithms for intrusion detection system in WSN. *International Journal of Advanced Computer Science and Applications*, 12(5), 617–626. https://doi.org/10.14569/IJACSA.2021.0120574

[7] Alsulaiman, L., Al-Ahmadi, S. (2021). Performance evaluation of machine learning techniques for DOS detection in wireless sensor network. *International Journal of Network Security & Its Applications*, 13(2), 21–29. https://doi.org/10.5121/ijnsa.2021.13202

[8] Bace, R., Mell, P. (2001). NIST Special Publication on Intrusion Detection Systems.

[9] Biswas, S., Adhikari, S. (2015). A survey of security attacks, defenses and security mechanisms in wireless sensor network. *International Journal of Computer Applications*, 131.

[10] Butun, I., Morgera, S. D., Sankar, R. (2014). A survey of intrusion detection systems in wireless sensor networks. *IEEE Communications Surveys and Tutorials*, 16(1), 266–282. https://doi.org/10.1109/SURV.2013.050113.00191

[11] Çayirci, E. (Erdal), Rong, Chunming. (2009). Security in Wireless Ad Hoc and Sensor Networks. Wiley.

[12] Chen, X., Makki, K., Yen, K., Pissinou, N. (2009). Sensor network security: a survey. *IEEE Communications Surveys and Tutorials*, 11(2), 52–73. https://doi.org/10.1109/SURV.2009.090205

[13] Dargie, W., Poellabauer, C. (2011). Fundamentals of wireless sensor networks: theory and practice. *Fundamentals of Wireless Sensor Networks: Theory and Practice*. John Wiley and Sons Ltd. https://doi.org/10.1002/9780470666388

[14] El, I. M. M., Ramakrishnan, E. S. (2013). Wireless Sensor Networks : From Theory to Applications. https://doi.org/10.1201/B15425

[15] Fuchsberger, A. (2005). Intrusion detection systems and intrusion prevention systems. *Information Security Technical Report*, 10(3), 134–139. https://doi.org/10.1016/j.istr.2005.08.001

[16] Gavrić, Ž., Simić, D. (2018). Overview of dos attacks on wireless sensor networks and experimental results for simulation of interference attacks. *Ingenieria e Investigacion*, 38(1), 130–138. https://doi.org/10.15446/ing.investig.v38n1.65453

[17] Ioannou, C., Vassiliou, V., Sergiou, C. (2017). An intrusion detection system for wireless sensor networks. In Proceedings of the 24th International Conference on Telecommunications: Intelligence in Every Form, ICT 2017. https://doi.org/10.1109/ICT.2017.7998271

[18] Ismail, S., Dawoud, D., Reza, H. (2022). A lightweight multilayer machine learning detection system for cyber-attacks in WSN. In 2022 IEEE 12th Annual Computing and Communication Workshop and Conference, CCWC 2022, 481–486.

[19] Jiang, S., Zhao, J., Xu, X. (2020). SLGBM: An intrusion detection mechanism for wireless sensor networks in smart environments. *IEEE Access*, 8, 169548–169558.

[20] Kachirski, O., Guha, R. (2003). Effective Intrusion detection using multiple sensors in wireless ad hoc networks. In Proceedings of the 36th Hawaii International Conference on System Sciences (HICSS'03).

[21] Li, W., Yi, P., Wu, Y., Pan, L., Li, J. (2014). A new intrusion detection system based on KNN classification algorithm in wireless sensor network. *Journal of Electrical and Computer Engineering*, 2014. https://doi.org/10.1155/2014/240217

[22] Li, Y., Quevedo, D. E., Dey, S., Shi, L. (2017). SINR-based DoS attack on remote state estimation: a game-theoretic approach. *IEEE Transactions on Control of Network Systems*, 4(3), 632–642.

[23] Mamun, Q., Islam, R., Kaosar, M. (2014). Anomaly detection in wireless sensor network. *Journal of Networks*, 9(11), 2914–2924. https://doi.org/10.4304/jnw.9.11.2914-2924

[24] Mohammadi, S., Jadidoleslamy, H. (2011). A comparison of link layer attacks on wireless sensor networks. *International Journal on Applications of Graph Theory In Wireless Ad Hoc Networks And Sensor Networks*, 3(1), 35–56. https://doi.org/10.5121/jgraphhoc.2011.3103

[25] Murugesh, C., Murugan, S. (2022). Evolutionary optimization with variational auto encoder based denial of service attack detection and classification in wireless sensor networks. In Proceedings – International Conference on Augmented Intelligence and Sustainable Systems, ICAISS 2022, 994–1000.

[26] Murugesh, C., Murugan, S. (2023). Modelling of optimal quantum neural network for DDoS attack classification in wireless sensor networks. In Proceedings of the 3rd International Conference on Artificial Intelligence and Smart Energy, ICAIS 2023, 1054–1059. https://doi.org/10.1109/ICAIS56108.2023.10073673

[27] Nancy, P., Muthurajkumar, S., Ganapathy, S., Santhosh Kumar, S. V. N., Selvi, M., Arputharaj, K. (2020). Intrusion detection using dynamic feature selection and fuzzy temporal decision tree classification for wireless sensor networks. *IET Communications*, 14(5), 888–895. https://doi.org/10.1049/iet-com.2019.0172

[28] Pan, J. S., Fan, F., Chu, S. C., Zhao, H. Q., Liu, G. Y. (2021). A lightweight intelligent intrusion detection model for wireless sensor networks. *Security and Communication Networks*, 2021. https://doi.org/10.1155/2021/5540895

[29] Patil, S., Chaudhari, S. (2016). DoS attack prevention technique in wireless sensor networks. *Procedia Com-*

puter Science, 79, 715–721. https://doi.org/10.1016/j.procs.2016.03.094

[30] Rezvi, A., Moontaha, S., Trisha, K. A., Cynthia, S. T., Ripon, S. (2021). Data mining approach to analyzing intrusion detection of wireless sensor network. *Indonesian Journal of Electrical Engineering and Computer Science*, 21(1), 516–523. https://doi.org/10.11591/ijeecs.v21.i1

[31] Roy, S., Conti, M., Setia, S., Jajodia, S. (2012). Secure data aggregation in wireless sensor networks. *IEEE Transactions on Information Forensics and Security*, 7(3), 1040–1052. doi.org/10.1109/TIFS.2012.2189568

[32] Salmi, S., & Oughdir, L. (2022). CNN-LSTM based approach for Dos attacks detection in wireless sensor networks. *IJACSA International Journal of Advanced Computer Science and Applications*, 13(4). www.ijacsa.thesai.org

[33] Selvakumar, K., Karuppiah, M., SaiRamesh, L., Islam, S. H., Hassan, M. M., Fortino, G., Choo, K. K. R. (2019). Intelligent temporal classification and fuzzy rough set-based feature selection algorithm for intrusion detection system in WSNs. *Information Sciences*, 497, 77–90. https://doi.org/10.1016/j.ins.2019.05.040

[34] Sharma, M. R., Agarwal, V. K., Kumar, N., Kumar, S. (2020). Integrated Intrusion Detection System (IDS) for Security Enhancement in Wireless Sensor Networks. https://doi.org/10.4018/978-1-7998-5068-7.ch009

[35] Sobh, T. S. (2006). Wired and wireless intrusion detection system: classifications, good characteristics and state-of-the-art. *Computer Standards and Interfaces*, 28(6), 670–694. https://doi.org/10.1016/j.csi.2005.07.002

[36] Sreenivasulu, A. L., Reddy, P. C. (2020). NLDA non-linear regression model for preserving data privacy in wireless sensor networks. *Digital Communications and Networks*, 6(1), 101–107. https://doi.org/10.1016/j.dcan.2019.01.004

[37] Tiloca, M., De Guglielmo, D., DIni, G., Anastasi, G., Das, S. K. (2017). JAMMY: a distributed and dynamic solution to selective jamming attack in TDMA WSNs. *IEEE Transactions on Dependable and Secure Computing*, 14(4), 392–405. https://doi.org/10.1109/TDSC.2015.2467391

[38] Xie, M., Hu, J., Guo, S. (2015). Segment-based anomaly detection with approximated sample covariance matrix in wireless sensor networks. *IEEE Transactions on Parallel and Distributed Systems*, 26(2), 574–583. https://doi.org/10.1109/TPDS.2014.2308198

[39] Yan, K. Q., Wang, S. S., Liu, C. W. (2010). Hybrid intrusion detection system for enhancing the security of a cluster-based wireless sensor network. In Proceedings, 2010 3rd IEEE International Conference on Computer Science and Information Technology, 114–118.

28 Comparison of machine learning, ensemble learning, and deep learning model for predicting ovarian cancer: a data-driven study

Naba Jyoti Sarmah[1,a], Dwipen Laskar[2,b] and Kaushik Kumar Bharadwaj[3,c]

[1]Department of Computer Application, Mahapurusha Srimanta Sankaradeva Viswavidyalaya, Nagaon-782001, Assam, India

[2]Department of Computer Science, Gauhati University, Guwahati-781014, Assam, India

[3]Department of Computational Biology and Biotechnology, Mahapurusha Srimanta Sankaradeva Viswavidyalaya, Nagaon-782001, Assam, India

Abstract

Ovarian Cancer (OC) is still one of the deadliest gynaecological cancers, and patient outcomes are greatly enhanced by early identification. This study aims to assess the predictive abilities of Deep Learning (DL), Ensemble Learning (EL), and Machine Learning (ML) algorithms for OC identification using clinical data. ML algorithms—Support Vector Machine (SVM), Naive Bayes (NB), Logistic Regression (LR), k-Nearest Neighbor (KNN), Decision Tree (DT); ensemble learning techniques—Random Forest (RF), Extreme Gradient Boosting Machine (XGBoost), Light Gradient Boosting Machine (LGBM), and DL architectures—TabNet, Artificial Neural Networks (ANNs), and Convolutional Neural Networks (CNNs) are considered in the experiment. In this work, all the models were trained and tested on publicly available OC clinical datasets containing data from 349 individual patients. This paper investigates the predictive accuracy of these models in identifying early-stage OC using patient-specific clinical data, intending to enable predictive analytics for early diagnosis. Integrated DL, EL, and ML methods are applied and tested for better predictive accuracy. The performance of various models is evaluated based on the scores obtained using different metrics, precision, recall, accuracy, F1-score, and ROC-AUC score. In the results, EL models outperformed the DL and ML models. Among the EL models, RF achieved the maximum accuracy of 91%. Other EL models, LGBM, and XGBoost, performed very well compared to DL and ML models. Among DL models, TabNet notably reached an accuracy of 86%. It can thus be concluded that both EL and DL approaches are quite promising. This work thus contributes to the advancement of data-driven models relevant to the improvement of prophecy in OC and different similar datasets, with implications for personalised medicine and early intervention.

Keywords: Ovarian cancer, prediction, machine learning, ensemble learning, deep learning

Introduction

Ovarian carcinoma is one of the most fatal cancers of the female reproductive system. Because of the elevated mortality rate, early detection has emerged as the challenge of the current time. Inter- and intra-histological heterogeneity, a high incidence of recurrence, and the lack of effective treatments make it even more challenging [15]. GLOBOCAN projections forecast a 56% global rise in OC cases by 2050, with an alarming 75% of diagnoses expected to occur in the advanced stages of the disease [1]. Early detection of this condition significantly improves the five-year survival rate, raising it from 3% at Stage IV to 90% at Stage I [17]. Therefore, it is crucial to identify OC recurrence, because it helps in developing personalised treatment and monitoring strategies, including selecting the most suitable chemotherapeutic alternatives [18]. Although screening tools such as pelvic exams, ultrasounds, CA125 tests, and MRIs are frequently used to identify OC, but these examinations are very costly, and the accuracy of these methods for early diagnosis of

OC is not established. Furthermore, MRI requires specialised knowledge. Therefore, early prophecy and successful intervention depend on designing of precise and timely predictive models. AI offers a promising alternative for efficiently handling large datasets, reduces the data gaps, and enables personalised, cost-effective diagnostics [4]. One type of AI that helps the clinicians and researchers to overcome the challenge of large data processing is Machine Learning (ML). It provides reliable methods for understanding complex data and effectively addresses multiple variables to provide amazingly accurate prediction results. Nowadays, ML algorithms are used for accurate diagnosis of disease and to develop personalised therapeutic strategies [18]. OC affects mainly postmenopausal women but recently also occurred among young women. The unhealthy lifestyle of most individuals has become the cause in recent days, and thus poor survival rate due to late diagnosis so early detection should be ensured, and a recent non-invasive, cost-efficient methods by implementing ML algorithms on blood tests and protein marker

[a]nabajyoti2006@gmail.com, [b]laskardwipen@gauhati.ac.in, [c]kkbhrdwj01@gmail.com

DOI: 10.1201/9781003675235-28

CA-125 and CA-19 [14]. Often denoted to as the 'silent killer,' the symptoms of OC obscure with the symptoms of bloating and abdominal pain, which can cause a delay in diagnosis. There are no standard screening tests for OC, unlike certain other malignancies, and the available indicators, CA-125, are not so sensitive for early identification. The deep pelvic location of the ovaries, as well as the heterogeneity of OC, poses challenges in attaining early detection and treatments. Sufficient patient education, early-stage diagnosis, and standardised diagnostic pathways are a few conditions that need to be satisfied to reduce the problem. Recently, medical researchers with the help of ML techniques promised to improve cancer prediction and diagnosis by analysing large datasets to uncover hidden patterns and develop accurate predictive models [13].

This study aims to check the accuracy of different DL, EL, and ML techniques as predictive tools for OC. Patient data was collected from the dataset, and various DL, EL, and ML approaches are applied to develop models for accurate cancer prediction. Comprehensive evaluations and comparative analyses of the different model's performance are conducted to determine the most effective approach. Specifically, this research is carried out to analyse 49 different parameters of blood test reports collected from 349 different patients, to distinguish between benign ovarian tumours and OC.

Literature review

Cancer is a complex disease to manage and treat, its accurate and early diagnosis is required for proper treatment. Although histopathological analysis is a primary method in the identification of malignancies, it has several challenges. Therefore, developing a stable, non-invasive ML model is essential to move forward with precise medicine and improve cancer management [6]. The rapid progress of high-throughput and ICT technologies for clinical, imaging, and multi-omics data, with predictive and prognostic biomarkers, produces enormous opportunities to design fast and accurate ML models. ML models are used to extract worthwhile information from complex datasets, and they become invaluable in cancer research, especially for early detection and accurate diagnosis. The rapid growth of technologies such as high-throughput methods and information communication technology (ICT) has led to the generation of vast data in cancer research, including clinical, imaging, and molecular data. These data contain critical information, such as molecular biomarkers, as it is important for accurate cancer diagnosis and treatment. For the processing of this type of complex data, ML, and DL techniques played an important role. DL has been found to have special applications in dealing with massive, diverse datasets [8,11]. DL was currently exploited for early detection of OC by utilising data of gene expression, tissue images, and ultrasound scans. However, the methods seem promising towards higher accuracy and the detection of the cancer subtypes [7]. Kawakami et al. designed an ML framework to predict clinical parameters, histotypes, and prognosis in epithelial ovarian cancer (EOC) using biomarkers from peripheral blood tests. The EL models such as Gradient Boosting model and RF achieved high accuracy values and were significant for the customisation of therapeutic approaches in patients with EOC [9].

Lavanya and Subbulakshmi investigated the use of ML and explainable AI (XAI) for OC prediction. They analysed different prediction models such as SVM and EL. They found that EL improved accuracy, while XAI provided greater transparency in decision-making, enhancing its clinical application [13]. Mohi Uddin and the team had developed a new model for prediction of cervical cancer prognosis by using ML techniques. By using EL techniques and addressing the class imbalance in their model, they obtained an excellent accuracy of 99.19% [16], which suggests the usefulness of the model for trustworthy cervical cancer diagnosis [16]. Ahamad and co-workers applied ML models on clinical data from 349 patients to create predictive analytics for the early diagnosis of OC. Their research achieved 91% accuracy in a specific case, thus highlighting the promising role of ML models in early cancer detection [2]. Abuzinadah et al. developed a stacked ensemble model using an OC dataset, achieving a high accuracy of 96.87% for OC detection, whereas for the whole dataset, they achieved an accuracy of 87%. Their model, enhanced with Shapley additive explanations (SHAP)-based explainable AI, with hyperparameter tuning outperformed other models, highlighting its potential for early, reliable diagnosis [1].

Methodology

Dataset used

In this study, we consider a dataset covering blood test reports of 349 individuals, collected at the Third Affiliated Hospital of Soochow University. The dataset included two groups: 178 individuals identified with benign ovarian tumours and 171 individuals identified with OC (https://github.com/martuzaiu/Ovarian_Cancer_Project) [2,12]. A total of 49 features were taken. The dataset has three subgroups: blood routine test, general chemistry, and tumour marker. In our study, we consider the whole dataset for our experimental purpose.

Data processing

The raw dataset of 349 patients was pre-processed by standardising the values, with equation (28.1) [10], centering them on the mean with a unit standard deviation.

$$X' = \frac{X-\mu}{\sigma} \tag{28.1}$$

where μ is the mean and σ is the standard deviation.

Machine-learning models

In this study, we observe the accuracy of different supervised DL, EL, and ML algorithms to predict OC. ML algorithms, such as SVM, LR, NB, DT, k-NN, and EL methods, including RF, LGBM, and XGBoost, are compared with various DL architectures such as TabNet, CNNs, and ANNs. The prime goal of the study was to identify the models with best predictive performance. For programming, Python (version 3.7.12) was used. Python libraries like Pandas and NumPy were employed for basic data processing, while Sklearn was utilised to implement the ML algorithms. Additionally, the Matplotlib library was used to generate all plots and figures, facilitating data visualisation and result interpretation.

The overall study design is depicted in the schematic representation of the methodology shown in Figure 28.1.

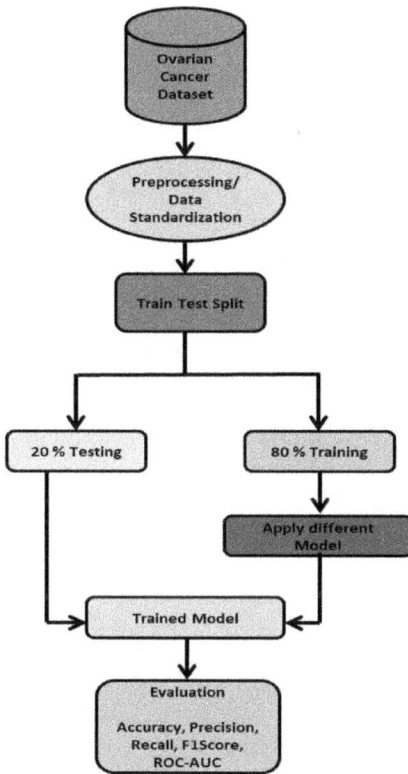

Figure 28.1 A schematic representation of the studied methodology, indicating each step and the flow between them through arrows

Source: Author

Evaluation metrics

In this study, we examined various evaluation metrics, such as accuracy, precision, recall, F1-score, and ROC-AUC score, to check the performance of different classifiers.

Accuracy: Accuracy reflects the overall correctness of a model's predictions [2,3] and it is calculated using the formula:

$$Accuracy = \frac{TP+TN}{TP+TN+FP+FN} \qquad (28.2)$$

where, TP = True positive; FP = False positive; TN = True negative; FN = False negative.

Precision: Precision represents the proportion of correctly identified positive samples amongst all samples predicted as positive [2,3] and is defined as follows:

$$Precision = \frac{TP}{TP+FP} \qquad (28.3)$$

Recall: Recall measures the capability of the classifier to accurately identify all samples within a specific class, as defined below [2,3]:

$$Recall = \frac{TP}{TP+FN} \qquad (28.4)$$

F1-score: The F1-score is used to address the class imbalance in data by balancing Precision and Recall [2,3], and is calculated as follows:

$$F1 = \frac{2*Precision*Recall}{Precision+Recall} = \frac{2*TP}{2*TP+FP+FN} \qquad (28.5)$$

ROC-AUC: ROC-AUC reflects the model's ability to discriminate between classes, illustrating the trade-off between specificity and sensitivity [2,3].

$$True\ Positive\ Rate(TPR) = \frac{TP}{TP+FN} \qquad (28.6)$$

$$False\ Positive\ Rate\ (FPR) = \frac{FP}{FP+TN} \qquad (28.7)$$

The Area Under the Curve (AUC) is called the ROC-AUC Score [2]. It is plotted as TPR (y-axis) against FPR (x-axis) for various threshold values.

Results and discussion

Ovarian tumours, including both benign and malignant cases, from 349 patients were analysed in this study. The models in this study were built by incorporating all 49 features from the dataset. The dataset was split into two parts: 80% training and 20% testing. Evaluation metrics— Accuracy, Precision, Recall, F1-score, and ROC-AUC were used to check the performance of different classifiers. Among all the models, the EL model, RF, achieved the highest accuracy of 91.0%. In the Neural Network and DL models, TabNet exhibited the highest accuracy of 86.0%. For the ML models, NB and DT achieved the highest accuracy of 83.0%. The results are presented in Table 28.1.

Discussion

Early detection of OC is significant for reducing mortality rates by extending patient survival. Here, the effectiveness

Table 28.1 Accuracy, precision, recall, F1-score, and ROC-AUC score of different classification methods.

	RF	LGBM	XGBoost	TabNet	NB	DT	ANN	CNN	LR	SVM	KNN
Accuracy	0.91	0.9	0.89	0.86	0.83	0.83	0.81	0.81	0.8	0.8	0.76
Precision	0.92	0.91	0.89	0.86	0.83	0.83	0.82	0.82	0.81	0.8	0.78
Recall	0.91	0.9	0.89	0.86	0.83	0.83	0.81	0.81	0.8	0.8	0.76
F1-score	0.91	0.9	0.89	0.86	0.83	0.83	0.81	0.81	0.8	0.8	0.75
ROC-AUC	0.95	0.96	0.95	0.9	0.89	0.83	0.85	0.85	0.88	0.88	0.82

(*Source: Author's compilation*)

of eleven different models was examined for the detection of OC on publicly available datasets. These included ML techniques such as SVM, LR, NB, DT, and k-NN; EL techniques like RF, LGBM, and XGBoost; and DL architectures such as TabNet, CNNs, and ANNs. This study analysed the dataset to identify ovarian carcinomas in their early stages. Initially, the raw dataset was pre-processed using standardisation techniques. Various ML methods were then applied individually to the data, with an 80:20 ratio used to divide the dataset into train and test sets. High accuracy scores were obtained from the classification outcomes, demonstrating the prediction model's reliability and applicability in real-world situations. Most models achieved over 80% accuracy, along with strong precision, recall, F1-score, and ROC-AUC values, except for the k-NN algorithm, which showed an accuracy of 76%. However, the extremely limited dataset used in this analysis makes it difficult to draw general conclusions. Despite this limitation, the predictive models developed here have the potential to serve as valuable tools in real-life scenarios. Since early-stage OC is often difficult for practitioners to diagnose, this system could assist in timely and accurate detection, ultimately improving patient outcomes. Our findings indicate that the RF model showed the highest performance with an accuracy of 91.0%. This performance surpasses that of models reported in earlier studies [1,2,5] performed for the whole dataset.

Conclusion

Using clinical datasets, this study assessed the prediction capabilities of ML, EL, and DL models for OC detection. The findings showed that the RF model achieved the highest accuracy of 91%, demonstrating the potentiality of ML algorithms in handling tabular clinical data. The results of TabNet showed an accuracy of 86% and demonstrated its effectiveness among DL models. TabNet highlighted its potential for representing complex data. With an accuracy of 86%, TabNet illustrated its capability among DL models, highlighting its prospective for representing complex data. The findings demonstrate the synergistic benefits of applying ML and DL methods and

show the usefulness of these methods for designing more robust and accurate prediction models. In the direction of data-driven approaches, the present study contributes significantly to OC early diagnosis, which is imperative for taming patient outcomes and making personalised treatment strategies possible. Few studies explored different biomarkers associated with disease occurrence. Future studies can be carried out for selected biomarkers to check the effectiveness of detecting OC. Also, new models with hyperparameter tuning can be developed to achieve higher accuracy rates. In the future, research can be conducted on specific biomarkers for assessing the performance of AI, ML, and DL towards ovarian cancer detection. Research could be carried out by combining larger sets of data and using other biomarkers to strengthen the robustness as well as the generalisability of the models. The welfare of patients and clinical practice may be significantly impacted by the results of this study. By enhancing timely diagnosis and prognosis, it will provide improved care and outcomes for patients from this deadly illness.

References

[1] Abuzinadah, N., Posa, S. K., Alarfaj, A. A., Alabdulqader, E. A., Umer, M., Kim, T., Alsubai, S., Ashraf, I. (2023). Improved prediction of ovarian cancer using ensemble classifier and shaply explainable AI. *Cancers*, 15(24), 5793. https://doi.org/10.3390/cancers15245793

[2] Ahamad, M. M., Aktar, S., Uddin, M. J., Rahman, T., Alyami, S. A., Al-Ashhab, S., Akhdar, H. F., Azad, A., Moni, M. A. (2022). Early-Stage detection of ovarian cancer based on clinical data using machine learning approaches. *Journal of Personalized Medicine*, 12(8), 1211. https://doi.org/10.3390/jpm12081211

[3] Alazab, M., Khan, S., Krishnan, S. S. R., Pham, Q., Reddy, M. P. K., Gadekallu, T. R. (2020). A multidirectional LSTM model for predicting the stability of a smart grid. *IEEE Access*, 8, 85454–85463. https://doi.org/10.1109/access.2020.2991067

[4] Ayyoubzadeh, S. M., Ahmadi, M., Yazdipour, A. B., Ghorbani-Bidkorpeh, F., Ahmadi, M. (2024). Prediction of ovarian cancer using artificial intelligence tools. *Health Science Reports*, 7(7). https://doi.org/10.1002/hsr2.2203

[5] Azar, A. S., Rikan, S. B., Naemi, A., Mohasefi, J. B., Pirnejad, H., Mohasefi, M. B., Wiil, U. K. (2022). Application of machine learning techniques for predicting survival in ovarian cancer. *BMC Medical Informatics and Decision Making*, 22(1). https://doi.org/10.1186/s12911-022-02087-y

[6] Botlagunta, M., Botlagunta, M. D., Myneni, M. B., Lakshmi, D., Nayyar, A., Gullapalli, J. S., Shah, M. A. (2023). Classification and diagnostic prediction of breast cancer metastasis on clinical data using machine learning algorithms. *Scientific Reports*, 13(1). https://doi.org/10.1038/s41598-023-27548-w

[7] El-Latif, E. I. A., El-Dosuky, M., Darwish, A., Hassanien, A. E. (2024). A deep learning approach for ovarian cancer detection and classification based on fuzzy deep learning. *Scientific Reports*, 14(1). https://doi.org/10.1038/s41598-024-75830-2

[8] Hira, M. T., Razzaque, M. A., Sarker, M. (2024). Ovarian cancer data analysis using deep learning: a systematic review. *Engineering Applications of Artificial Intelligence*, 138, 109250. https://doi.org/ 10.1016/j.engappai.2024.109250

[9] Kawakami, E., Tabata, J., Yanaihara, N., Ishikawa, T., Koseki, K., Iida, Y., Saito, M., Komazaki, H., Shapiro, J. S., Goto, C., Akiyama, Y., Saito, R., Saito, M., Takano, H., Yamada, K., Okamoto, A. (2019). Application of artificial intelligence for preoperative diagnostic and prognostic prediction in epithelial ovarian cancer based on blood biomarkers. *Clinical Cancer Research*, 25(10), 3006–3015. https://doi.org/10.1158/1078-0432.ccr-18-3378

[10] Krithikadatta, J. (2014). Normal distribution. *Journal of Conservative Dentistry*, 17(1), 96. https://doi.org/ 10.4103/0972-0707.124171

[11] LeCun, Y., Bengio, Y., Hinton, G. (2015). Deep learning. *Nature*, 521(7553), 436–444. https://doi.org/ 10.1038/nature14539

[12] Lu, M., Fan, Z., Xu, B., Chen, L., Zheng, X., Li, J., Znati, T., Mi, Q., Jiang, J. (2020). Using machine learning to predict ovarian cancer. International *Journal of Medical Informatics*, 141, 104195. https://doi.org/10.1016/j.ijmedinf.2020.104195

[13] M, S. L. J., P, S. (2024). Innovative approach towards early prediction of ovarian cancer: machine learning-enabled XAI techniques. *Heliyon*, 10(9), e29197. https://doi.org/10.1016/j.heliyon.2024.e29197

[14] Nayak, C., Tripathy, A., Parhi, M., Barisal, S. K. (2023). Early stage ovarian cancer prediction using machine learning. In 2021 International Conference in Advances in Power, Signal, and Information Technology (APSIT), 603–608. https://doi.org/10.1109 /apsit58554.2023.10201764

[15] Qian, L., Zhu, J., Xue, Z., Zhou, Y., Xiang, N., Xu, H., Sun, R., Gong, W., Cai, X., Sun, L., Ge, W., Liu, Y., Su, Y., Lin, W., Zhan, Y., Wang, J., Song, S., Yi, X., Ni, M., . . . Guo, T. (2024). Proteomic landscape of epithelial ovarian cancer. *Nature Communications*, 15(1). https://doi.org/10.1038/s41467-024-50786-z

[16] Uddin, K. M. M., Mamun, A. A., Chakrabarti, A., Mostafiz, R., Dey, S. K. (2024). An ensemble machine learning-based approach to predict cervical cancer using hybrid feature selection. *Neuroscience Informatics*, 4(3), 100169. https://doi.org/10.1016 /j.neuri.2024.100169

[17] Vázquez, M. A., Mariño, I. P., Blyuss, O., Ryan, A., Gentry-Maharaj, A., Kalsi, J., . . . Zaikin, A. (2018). A quantitative performance study of two automatic methods for the diagnosis of ovarian cancer. *Biomedical Signal Processing and Control*, 46, 86–93. https://doi.org/10.1016/j.bspc.2018.07.001[24]

[18] Zhou, L., Hong, H., Chu, F., Chen, X., Wang, C. (2024). Predicting the recurrence of ovarian cancer based on machine learning. *Cancer Management and Research*, 16, 1375–1387. https://doi.org/ 10.2147/cmar.s482837

29 Efficient crack classification through transfer learning with pre-trained convolutional neural networks

Thokchom Chittaranjan Singh[a], Romesh Laishram[b] and N. Basanta Singh[c]

Department of Electronics and Communication Engineering, Manipur Institute of Technology, Manipur, India

Abstract

Cracks in concrete structures reduce their safety and durability; therefore, timely detection and accurate assessment of the same are highly essential. Traditional manual inspection methods are time-consuming, resource-intensive, and highly subjective. Image processing techniques can improve this process. In this paper, transfer learning with convolutional neural networks is employed for automated crack detection. This work fine-tuned four pre-trained architectures—MobileNetV2, ResNet50, InceptionV3, and Xception—on the balanced dataset by using the same data pre-processing and augmentation techniques to compare their performances. Among those tested, the MobileNetV2 model achieved 99.9% accuracy, best for resource-constrained environment tasks, while Xception achieved 99.8%, followed by InceptionV3 at 99.7%. ResNet50 was found to be very capable with respect to feature extraction but imposed greater computational overhead. The results underline that transfer learning scales for infrastructure monitoring applications, where MobileNetV2 finds a niche for lightweight and real-time deployment However, for more complex applications needing higher accuracy, Xception or InceptionV3 should be applied. The future direction may focus on implementing lightweight model integration into UAV platforms that can offer promising real-world applications.

Keywords: Crack detection, transfer learning, CNN, MobileNetV2, ResNet50, InceptionV3, Xception, UAV

Introduction

Cracks are visible fractures or separations in materials caused by mechanical stress, environmental factors, or material fatigue. In concrete structures, cracks are particularly worrisome because they can weaken structural integrity, reduce durability, and allow water or chemicals to penetrate. This infiltration accelerates the degradation process, making the structure more vulnerable.

So, maintenance of concrete structures has become essential for the durability, safety, and reliability of infrastructure globally. In spite of its popularity in construction, concrete surfaces often develop cracks that can compromise the integrity of the structure if not identified early. These cracks can escalate into expensive repairs or, in extreme cases, catastrophic structural failures. Early detection of cracks is important for lowering these risks, but standard methods such as manual inspections are hard to do, take a long time, and are prone to mistakes [1]. Furthermore, these approaches are less effective for large or inaccessible structures.

Recent technological breakthroughs have established deep learning as a transformational instrument for automating intricate activities, such as surface fracture identification. To overcome the limitations of manual inspections, UAVs equipped with advanced imaging tools are being adopted to inspect concrete structures. The feasibility of UAV-based fatigue crack detection for bridges has been demonstrated using three different UAV models,

highlighting both their cost-effectiveness and ability to access hard-to-reach areas [2]. Nonetheless, the analysis of the gathered data continues to be a hurdle. Conventional techniques for image processing involve edge detection, thresholding, and morphology, which are rule-based; therefore, they are not very flexible and lack accuracy as per [3] and [4]. Additionally, classical CNN-based models like AlexNet [17], combined with exhaustive search techniques, have proven effective in crack localization. However, their approach requires significant computation, motivating the need for more lightweight and scalable models suitable for deployment in real-time or embedded systems.

In this work, we explored the advanced approach of crack classification in captured images using transfer-learning models of CNNs. Transfer-learning methods based on pre-trained CNN models have gained much attention in the case of limited data and show robust performance, avoiding problems such as overfitting and class imbalance. This study will focus on the application of lightweight pre-trained CNN models that provide improved accuracy of classification, while keeping the computational efficiency high, which is of primary importance for the deployment of such systems in resource-constrained devices like UAVs and portable platforms. Pre-trained models explored in this paper involve MobileNetV2, ResNet50, Inception-V3, and Xception; all of these underwent fine-tuning in correspondence to achieve the best of their capabilities for the

[a]thchittaranjan05@gmail.com, [b]romesh.mit.ece@manipuruniv.ac.in, [c]basanta_n@rediffmail.com

DOI: 10.1201/9781003675235-29

related crack detection task. With the integration of such architectures, we intend to bring a balance between accuracy and efficiency to satisfy practical real-world demands for infrastructure monitoring.

The paper is organised as follows: Section 2 reviews relevant literature, Section 3 describes the research methodology and dataset, Section 4 presents experimental results, and Section 5 concludes the work with potential future directions.

Related work

Machine learning and computer vision have significantly transformed automated crack detection, delivering superior accuracy, efficiency, and scalability when compared to conventional methods. Modern lightweight deep learning models have made it possible to deploy crack detection technologies on devices with limited computational resources, such as drones and portable tools, thereby increasing their functionality in practical scenarios. Early approaches to crack detection relied heavily on manual efforts and simple image processing, which were both time-intensive and error-prone. Hamishebahar et al. [5] emphasized in their research the extent to which deep learning techniques overcome the constraints of traditional methods, particularly in the context of complicated patterns of cracks and a variety of environmental challenges.

The integration of deep learning with UAV technology has transformed crack detection through expedited data collection and processing. Xiang et al. [6] introduced GC-YOLOv5s, a lightweight detector specifically designed for UAV-based road crack detection, which improves precision and reduces model parameters to enable real-time processing on resource-constrained platforms. Omoebamije et al. [7] devised an enhanced deep-learning convolutional neural network for crack identification utilising UAV imagery, tackling issues associated with recognition rates and processing costs.

These lightweight architectures, such as MobileNetV2 and SqueezeNet, are more suitable for resource-constrained environments. Ali et al. [8] have pointed out that in the process of crack detection, lightweight architecture is becoming increasingly important in deep learning. The models also benefit from transfer learning through the pre-trained networks that adapt them to perform particular tasks quicker with fewer data.

Kavitha et al. [9] and Khan et al. [18] examined advanced deep learning techniques, combining transfer learning with pre-trained models such as ResNet and InceptionNet, while including explainable AI methods to enhance the interpretability and practical utility of the results. Aliu et al. [3] employed semantic segmentation techniques incorporating edge identification and heatmaps to find cracks, accomplishing stunning precision and efficiency. The improvement of training methods has also

been a priority. Yazdani and Hashi [10] demonstrated that data augmentation improves model generalization by contrasting techniques such as ResNet152 with customized CNNs, while Xu et al. [14] highlighted the use of GAN-based augmentation to enhance performance with limited data. Gopalakrishnan et al. [4] applied transfer learning using ImageNet-pretrained models to detect pavement cracks, solving challenges like surface texture variation and misleading anomalies.

The proposed work will implement transfer learning by integrating lightweight architectures for UAVs and portable devices. It focuses on robust model improvement with advanced augmentation techniques and attention mechanisms, and the evaluation of real-time performance and energy efficiency for its optimal operation on resource-constrained platforms.

Methodology

Dataset preparation
This study used a labelled dataset of high-resolution concrete surface images from the Kaggle surface dataset [11], divided into two classes: Positive (images with visible cracks) and Negative (images without visible cracks). Each class contains 20,000 images, totalling 40,000 samples, all with dimensions of 227 × 227 pixels and RGB channels. Figure 29.1 illustrates sample images of two different classes.

Pre-processing
All images were resized to 128 × 128 pixels to ensure uniformity for CNN input. Data augmentation techniques, including rotation, flipping, and intensity adjustments, were applied to enhance variability and improve model robustness, reducing overfitting and preparing the model for real-world conditions [12–14]. The dataset was split into 80% for training and 20% for validation, a standard practice in crack detection studies to ensure reliable performance evaluation on unseen data.

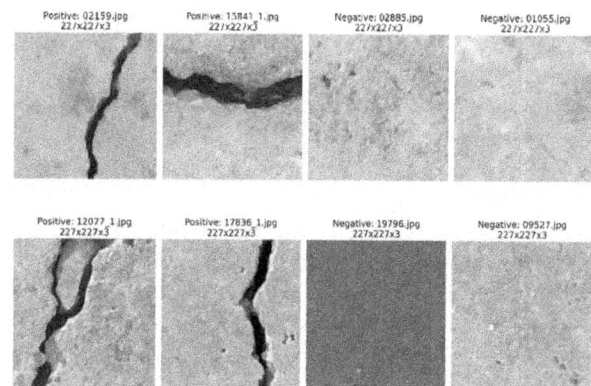

Figure 29.1 Sample of the datasets
Source: Author

Model implementation

To effectively address the challenge of feature extraction in image data, convolutional neural networks (CNNs) were utilised to create a robust image classification model. CNNs were chosen for their proven ability to capture high-level features and enhance classification performance. This study evaluated the performance of four pre-trained CNN architectures: MobileNetV2, ResNet50, InceptionV3, and Xception. These models were fine-tuned using transfer learning to align with the specific needs of crack detection tasks.

Transfer learning approach

Each of the pre-trained CNN models was adapted for binary classification of crack images through the following steps:

1. ***Feature extraction:*** The initial layers of the pre-trained models, designed to identify general patterns like edges and textures, were kept fixed. This step ensured that the valuable knowledge embedded in these layers was retained during training.
2. ***Custom layers:*** Additional layers were introduced to the base model, including convolutional layers with ReLU activation for non-linear transformations, batch normalisation to stabilise the training process, global average pooling to reduce spatial dimensions, and dense layers tailored for binary classification.
3. ***Attention mechanism:*** Squeeze-and-Excitation (SE) blocks were integrated into the architecture to recalibrate feature maps, focusing on the most relevant features while suppressing irrelevant noise.

The training occurred in two phases. First, the additional layers were trained independently with the base model frozen to maintain the pre-trained characteristics during the initial stage. Fine-tuning was the second stage of the training, where the base model's top blocks were unfrozen and trained using a lower learning rate to enhance feature extraction for the detection of cracks. We employed the Adam optimizer for efficient learning with the binary cross-entropy loss function. Figure 29.2 shows a visual summary of the experimental procedure including transfer learning configuration and CNN architectures.

CNN Architectures
MobileNetV2

MobileNetV2 is known for its lightweight architecture, making it ideal for deployment in resource-constrained environments such as drones or mobile devices. The model was fine-tuned using a binary cross-entropy loss function after replacing the classification head with a dense layer and sigmoid activation. Depthwise separable convolutions in MobileNetV2 significantly reduce computational overhead while maintaining high accuracy. This model

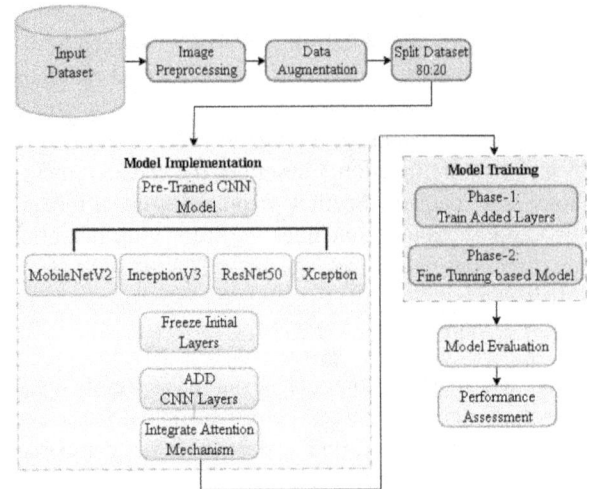

Figure 29.2 Flowchart of the proposed model
Source: Author

achieved state-of-the-art performance in crack detection due to its efficient feature extraction capabilities.

InceptionV3

InceptionV3 utilises inception modules to capture multi-scale spatial features. For crack detection, the model was fine-tuned by freezing the initial layers and customising the final layers for binary output. Auxiliary classifiers in InceptionV3 enhance gradient flow during training, leading to faster convergence. This model is particularly effective in capturing intricate crack details, achieving high validation accuracy.

ResNet50

ResNet50 is a deep convolutional neural network built on the concept of residual learning, featuring 50 layers with skip connections to mitigate vanishing gradient issues. For crack detection tasks, the model was fine-tuned by replacing the final fully connected layer with a binary classifier and using a cross-entropy loss function for optimisation. ResNet50's deeper architecture enables it to capture complex hierarchical features, making it particularly effective on diverse and challenging datasets. Despite its depth, the skip connections ensure efficient training and faster convergence. This model is well-suited for scenarios requiring high accuracy and robustness in feature extraction while balancing computational requirements.

Xception

The Xception neural network architecture surpasses the performance of the InceptionV3 model. While utilising the same number of parameters as InceptionV3, Xception leverages these parameters more effectively, resulting in improved model accuracy. Like InceptionV3, the Xception model is trained on the ImageNet dataset and is capable of classifying images across a wide variety of classes.

Quality metrics

Key metrics such as Accuracy (Acc), Precision (P_r), Recall (R_e), and F1-Score were used to evaluate the model's performance for crack detection. Precision is the measure of how accurately the system classifies images as cracks without including false positives. Recall is the ratio of accurate predictions of cracked images to the total number of sample images with cracks. Accuracy is the fundamental performance parameter, defined as the ratio of accurately predicted cracks to the total number of sample images available. The F1 score is a reliable metric that accurately evaluates the algorithm's performance, particularly by including false positives and false negatives. These metrics are defined as follows:

$$P_r = \frac{T_p}{T_p+F_p} \tag{29.1}$$

$$R_e = \frac{T_p}{T_p+F_n} \tag{29.2}$$

$$F1 = 2 \times \frac{P_r \times R_e}{P_r + R_e} \tag{29.3}$$

where T_p is the number of true positives, F_p is the number of false positives, and F_n is the number of false negatives.

Experimental setup

The crack detection experiments were conducted on a high-performance system featuring a 13th Gen Intel Core i7 processor, 16 GB RAM, and an NVIDIA GeForce RTX 4050 GPU. Transfer learning models, including MobileNetV2, ResNet50, InceptionV3 and Xception were customised for binary classification and trained using TensorFlow and Keras frameworks in Python environment.

Result and discussion

Table 29.1 summarises the performance of the pre-trained CNN models. MobileNetV2 achieved an impressive validation accuracy of 99.9%, with perfect scores across precision, recall, and F1 metrics. Its lightweight architecture and low computational complexity make it ideal for real-time applications in resource-constrained settings, such as mobile devices or drones.

Table 29.1 Evaluation metrics of pre-trained CNN models used.

Model	Acc	Pr	Re	F1
MobileNetV2	99.9	1	1	1
InceptionV3	99.7	0.995	0.995	0.995
Xception	99.8	~1	~1	~1
Resnet50	99.6	~1	~1	~1

Source: Author

InceptionV3 achieved a good accuracy rate of 99.7%, which was nearly perfect in terms of precision, recall, and F1 score because it was able to capture multi-scale spatial features effectively through the inception module. It can detect complicated patterns and is also very powerful for different kinds of crack datasets but does need relatively more computation compared to MobileNetV2.

Xception did slightly better, with an accuracy rate of 99.8%, also performing strongly in terms of precision, recall, and F1. More importantly, though, its efficient use of parameters means it can handle more advanced feature extraction with a balance between high accuracy and moderate resource needs. This makes it perfect for when performance is important, but both need to be considered.

ResNet50 scored 99.6% accuracy, with near-perfect precision, recall, and F1 score. Being deeper in architecture would be best suited for the purpose of retaining complex features. For that, the application could be in problematic datasets. The higher computational demand it has will make it suitable for applications where resources are not a constraint, and robust feature extraction is important.

The confusion metrics were later plotted. Figures 29.3 to 29.6 show the confusion matrices for MobileNetV2, ResNet50, InceptionV3, and Xception, respectively. A descriptive result regarding the classification performance in every matrix is presented. Based on each matrix, the models present high rates for true positives and true negatives, which indicate their strength for correct classification of crack and non-crack images.

Figure 29.7 depicts that training accuracy of MobileNetV2, which steadily increases, surpassing 99.9%, while validation accuracy closely follows, indicating minimal overfitting and strong generalisation to unseen data.

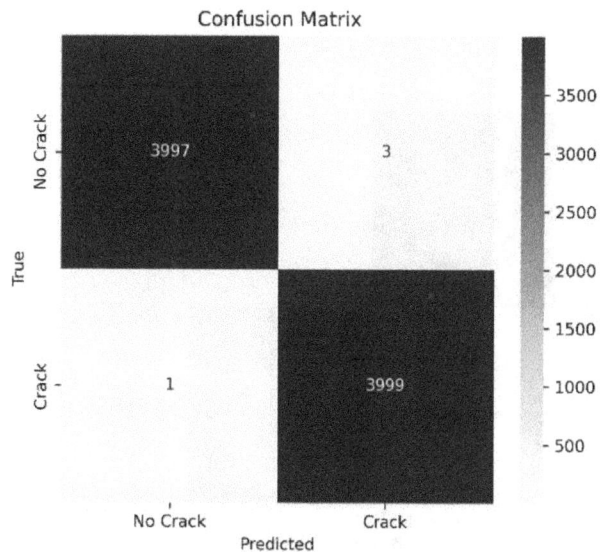

Figure 29.3 Confusion metrics for MobileNetV2

Source: Author

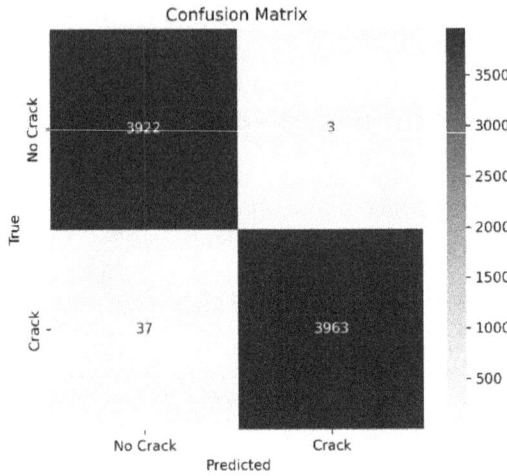

Figure 29.4 Confusion metrics for ResNet50
Source: Author

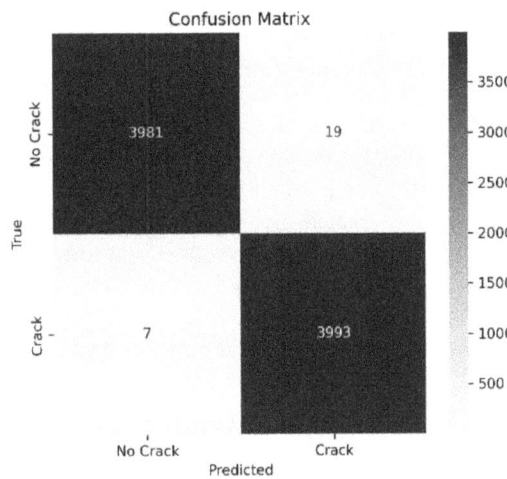

Figure 29.5 Confusion metrics for InceptionV3
Source: Author

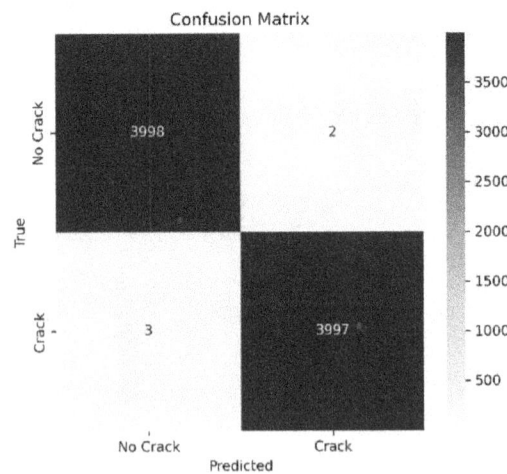

Figure 29.6 Confusion metrics for Xception
Source: Author

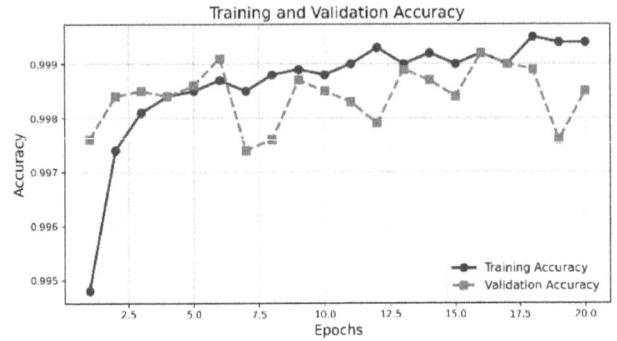

Figure 29.7 Training and validation accuracy for Mobile-NetV2
Source: Author

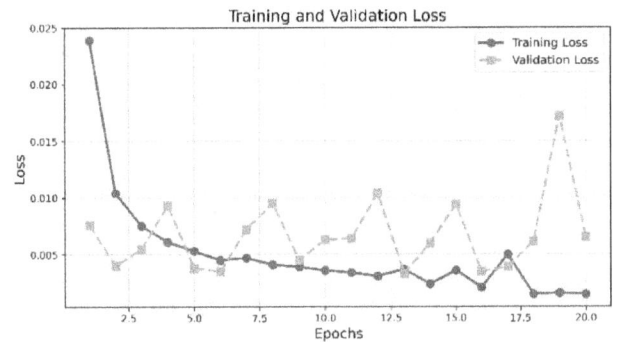

Figure 29.8 Training and validation loss for MobileNetV2
Source: Author

Table 29.2 Comparative analysis of model architectures.

Reference	Architecture	Accuracy (%)
Ren et al [15]	CrackSegNet	98.77%
Chordia et al. [13]	CNN	99.43
Rahai et al. [12]	AlexNet	99.00
Hussain et al. [19]	VGG16	99.52
Our model	MobileNetV2	99.9%

Source: Author

Figure 29.8 highlights a rapid decrease in training loss during the initial epochs, eventually stabilising at a very low level, reflecting effective learning. Validation loss remains consistently low with minor fluctuations, underscoring the model's reliability. These results confirm MobileNetV2's high accuracy and low loss, making it an excellent choice for crack detection tasks.

Table 29.2 demonstrates our proposed model's exceptional performance, MobileNetV2, which attains an accuracy of 99.9%, exceeding all other current methodologies. Hussain et al. [16] showed robust performance using VGG16 at an accuracy of 99.52%, while Chordia et al. attained 99.43% utilising a CNN-based methodology, but our model evidences a distinct enhancement.

Furthermore, AlexNet and CrackSegNet, utilised by Rahai et al. and Ren et al., respectively, achieved slightly lesser accuracies of 99.06% and 98.77%. This comparison highlights the effectiveness and performance of MobileNetV2 in solving the issue, demonstrating its ability to deliver novel outcomes while offering advantages such as less computational complexity relative to more substantial architectures like VGG16.

Conclusion

This study systematically evaluated four transfer learning-based CNN architectures for crack detection in concrete surfaces. All models demonstrated high accuracy and robust performance, with MobileNetV2, Xception, and InceptionV3 standing out for their efficiency and predictive power balance. MobileNetV2, in particular, is recommended for resource-constrained environments, while Xception and InceptionV3 are better suited for applications requiring higher accuracy. These results underscore the potential of transfer learning in automating crack detection, offering a scalable and efficient solution for infrastructure monitoring and maintenance. Future work could focus on real-world deployment and exploring hybrid models to enhance performance.

Future work may focus on implementing lightweight model integration into UAV platforms that can offer promising real-world applications, enabling efficient, automated inspections of infrastructure such as bridges, roads, and buildings.

References

[1] Meng, X. (2021). Concrete crack detection algorithm based on deep residual neural networks. *Scientific Programming*, 2021(1), 3137083

[2] Dorafshan, S., Thomas, R. J., Maguire, M. (2018). Fatigue crack detection using unmanned aerial systems in fracture critical inspection of steel bridges. *Journal of Bridge Engineering*, 23(10), 04018078.

[3] Aliu, A. A., Ariff, N. R. M., Ametefe, D. S., John, D. (2023). Automatic classification and isolation of cracks on masonry surfaces using deep transfer learning and semantic segmentation. *Journal of Building Pathology and Rehabilitation*, 8(1), 28.

[4] Gopalakrishnan, K., Khaitan, S. K., Choudhary, A., Agrawal, A. (2017). Deep convolutional neural networks with transfer learning for computer vision-based data-driven pavement distress detection. *Construction and Building Materials*, 157, 322–330.

[5] Hamishebahar, Y., Guan, H., So, S., Jo, J. (2022). A comprehensive review of deep learning-based crack detection approaches. *Applied Sciences*, 12(3), 1374.

[6] Xiang, X., Hu, H., Ding, Y., Zheng, Y., Wu, S. (2023). GC-YOLOv5s: a lightweight detector for UAV road crack detection. *Applied Sciences*, 13(19), 11030.

[7] Omoebamije, O., Omoniyi, T. M., Musa, A., Duna, S. (2023). An improved deep learning convolutional neural network for crack detection based on UAV images. *Innovative Infrastructure Solutions*, 8(9), 236.

[8] Ali, L., Alnajjar, F., Khan, W., Serhani, M. A., Al Jassmi, H. (2022). Bibliometric analysis and review of deep learning-based crack detection literature published between 2010 and 2022. *Buildings*, 12(4), 432.

[9] Kavitha, S., Baskaran, K. R., Dhanapriya, B. (April 2023). Explainable AI for detecting fissures on concrete surfaces using transfer learning. In 2023 International Conference on Inventive Computation Technologies (ICICT) . IEEE, 376–384.

[10] Yazdani, R., Hashi, E. K. (August 2021). Comparative analysis of deep learning algorithms based on surface image classification. In 2021 International Conference on Science & Contemporary Technologies (ICSCT) . IEEE, 1–6.

[11] Özgenel, Ç. F. (2019). Concrete crack images for classification. *Mendeley Data*, 2, 2019.

[12] Rahai, A., Rahai, M., Iraniparast, M., Ghatee, M. (December 2022). Surface crack detection using deep convolutional neural network in concrete structures. In 2022 IEEE 5th International Conference on Image Processing Applications and Systems (IPAS) . IEEE, 1–5.

[13] Chordia, A., Sarah, S., Gourisaria, M. K., Agrawal, R., Adhikary, P. (September 2021). Surface crack detection using data mining and feature engineering techniques. In 2021 IEEE 4th International Conference on Computing, Power and Communication Technologies (GUCON) . IEEE, 1–7.

[14] Xu, H., Li, C., Rahaman, M. M., Yao, Y., Li, Z., Zhang, J., ... Teng, Y. (2020). An enhanced framework of generative adversarial networks (EF-GANs) for environmental microorganism image augmentation with limited rotation-invariant training data. *IEEE Access*, 8, 187455–187469.

[15] Ren, Y., Huang, J., Hong, Z., Lu, W., Yin, J., Zou, L., Shen, X. (February 2020). Image based concrete crack detection in tunnels using deep fully convolutional networks. *Construction and Building Materials*, 234, 117367.

[16] Hussain, A., Qureshi, K. N., Anwar, R. W., Aslam, A. (February 2024). A novel SCD11 CNN model performance evaluation with Inception V3, VGG16 and ResNet50 using surface crack dataset. In 2024 2nd International Conference on Unmanned Vehicle Systems-Oman (UVS) . IEEE, 1–7.

[17] Li, S., Zhao, X. (2019). Image-based concrete crack detection using convolutional neural network and exhaustive search technique. *Advances in Civil Engineering*, 2019.

[18] Khan, U. S., Ishfaque, M., Khan, S. U. R., Xu, F., Chen, L., Lei, Y. (2024). Comparative analysis of twelve transfer learning models for the prediction and crack detection in concrete dams, based on borehole images. *Frontiers of Structural and Civil Engineering*, 18(10), 1507–1523.

30 Enhancing medical imaging diagnostics with federated learning: overcoming data heterogeneity

Soumyaranjan Panda[1,a], Prayash Tah[2], Vikas Pareek[1] and Sanjay Saxena[2]

[1]Department of Computer Science and IT, Mahatma Gandhi Central University, Bihar-845401, India

[2]Department of Computer Science and Engineering, International Institute of Information Technology, Bhubaneswar, Odisha-751003, India

Abstract

The increased availability of medical imaging data, such as multimodal MRI, X-rays, and CT scans, has expanded the possibilities for machine-driven diagnostic solutions powered by artificial intelligence. However, deploying centralised training models in medical imaging confronts considerable obstacles regarding patient privacy, data security, and regulatory hurdles. Federated learning (FL) has evolved as a potential solution for collaborative training among multiple institutions while protecting private data. Despite its merits, conventional FL often faces inconsistencies in performance due to data distribution variations among clients. To address this challenge and enable effective collaborative learning, we propose the Federated Discrepancy-Based Contrastive Network, which combines model contrastive federated learning with a discrepancy-based weighting technique. By incorporating contrastive learning principles, the model aligns similar representations to refine local training, while utilising discrepancy weights to enhance feature representations and improve global model convergence. This approach ensures a more robust and generalisable model, enabling effective classification of medical conditions from imaging data. We evaluate the proposed method against conventional FL methods to illustrate its potential benefits. The results demonstrate that it can achieve an average of 3% growth across all metrics, including accuracy, Kappa score, recall, precision, and F1 score. This improvement showcases its effectiveness as an essential tool for privacy-preserving, distributed training in medical imaging applications.

Keywords: Deep learning, federated learning, radiographic imaging, supervised contrastive learning

Introduction

The rapid growth in medical imaging and artificial intelligence has significantly improved the capabilities of machine-driven diagnostic solutions [1]. These innovations are critical in diagnosing various diseases, including cancers [2], pneumonia [3], and other abnormalities. However, centralised learning approaches often require collecting data from multiple sources, which poses significant challenges related to patient privacy, data security, and compliance with government regulations such as HIPAA and GDPR [4].

Federated learning (FL) allows multiple parties to collaboratively and distributively train machine learning models without sharing local data, addressing critical concerns of data privacy and security [5,6]. However, a significant challenge in conventional FL is handling the heterogeneity of local data distributions among participating clients, which often affects global model performance. Existing approaches, such as MOON [7] and the Federated Discrepancy method [8], provide valuable insights into mitigating these challenges. MOON introduces model-contrastive federated learning by aligning model representations to correct local training, demonstrating significant performance enhancements on image classification tasks by conducting contrastive learning

at the model level. Meanwhile, the distribution discrepancy-based client weighting method mitigates category distribution heterogeneity using discrepancy-aware aggregation weights. These weights are estimated based on the discrepancies between local and global category distributions, leading to improved model utility, greater modularity, and a tighter theoretical optimisation bound.

Building upon the strengths of these approaches, we present the Federated Discrepancy-Based Contrastive Network (FedDiCoNet), which synergises the contrastive learning principles of MOON and the discrepancy-aware collaboration between clients. FedDiCoNet uses model representation similarity for strong local training correction and discrepancy-aware aggregation weights to manage data heterogeneity completely. By integrating these two complementary approaches, FedDiCoNet enhances feature representations, global model convergence, and overall performance. Extensive studies show that FedDiCoNet outperforms leading federated learning approaches, providing a single solution for privacy-preserving and efficient collaborative model training.

This study evaluates FedDiCoNet for COVID-19 classification using medical images, addressing data heterogeneity. By comparing it with FedAVG, FedProx, and MOON, we highlight its potential for privacy-preserving, robust healthcare diagnostics.

[a]soumya.r.panda01@gmail.com

DOI: 10.1201/9781003675235-30

The organisation of this paper is as follows: **Section 2** provides the background and underlying knowledge needed for this study. **Section 3** describes the recommended methodology. **Section 4** summarises the findings. **Section 5** concludes the paper by summarising significant findings and outlining future research directions.

Literature review

Federated learning

Federated learning (FL) enables multiple clients to collaboratively train a shared model while keeping data private by exchanging only model updates, ensuring security, especially in healthcare with strict privacy regulations [5]. However, FL faces communication inefficiency, non-IID data, and client heterogeneity [9]. FedAvg, the foundational FL algorithm, aggregates locally trained models while preserving privacy [5]. FedProx enhances FedAvg with a proximal term for better handling non-IID data [10].

Privacy-preserving medical imaging: unlocking potential with federated learning

Federated learning enhances diagnostic models while ensuring data security and compliance with privacy regulations. FedMRI [11] is a collaborative learning framework for MR image reconstruction, which combines a shared encoder and client-specific decoder to preserve domain-specific features, thereby improving performance across multi-institutional data. Furthermore, Zhu et al. [12] developed MLA–BIN, a lightweight federated learning framework for medical image segmentation that enhances domain generalisation and achieves superior Dice scores across the prostate, optic disc, and COVID-19 datasets. Moreover, Tölle et al. [16] conducted the largest federated cardiac CT analysis, using a semi-supervised CNN-transformer approach, to enhance predictive accuracy and generalisability across eight hospitals. Huangsuwan et al. [17] enhanced federated learning on the NIH ChestX-ray14 dataset using FedDrip, a diffusion-based synthetic data approach, improving FedAvg, FedDyn, and FedProx performance while addressing non-IID challenges and preserving data privacy. Despite many advancements, research on data heterogeneity in federated learning for medical imaging remains limited, underscoring the need for effective mitigation techniques to ensure robust, generalisable models [13].

Supervised contrastive learning

Supervised contrastive learning extends contrastive learning by using labelled data to define positive and negative pairs, enhancing feature representations through similarity and dissimilarity comparisons [15]. This approach encourages the model to bring representations of the same class closer together in the feature space while pushing those of different classes further apart. Supervised contrastive learning strengthens the discriminative capability of the extracted features by emphasising class-specific relationships, making it particularly effective for classification tasks. We incorporated this technique into federated learning to reduce the effects of data heterogeneity by aligning representations across distributed clients through model-level comparisons [7].

Material and methods

Methodology

In FL, the goal is to train a global DL model that combines the client's parameters ω without sharing data between clients and the server [5]. The optimisation problem that defines FL can be mathematically expressed as follows:

$$\min_{\omega} \left[\sum_{k=1}^{K} \frac{n_k}{n} \mathcal{L}_k(\omega) \right]$$

where ω is the server DL model weights, K denotes the total number of clients (e.g., Hospitals or Devices), n_k is the number of data instances at client k. $n = \Sigma_{k=1}^{k} n_k$ denotes the total number of data instances among all clients. $\mathcal{L}_k(\omega)$ is the local loss function of client k, which measures the performance of the DL model ω on its local data.

The optimisation process involves in local training each client k independently minimises its local loss function: $\mathcal{L}_k(\omega) = \frac{1}{n} \sum_{i=1}^{n_k} l(f(\omega, x_{k,i}), y_{k,i})$

where l is the individual loss function (cross-entropy, mean squared error). In $x_{k,i}$, $y_{k,i}$ the ith data point and its label on client k. $f(\omega, x_{k,i})$ is the model's prediction.

Global aggregation, the server aggregates updates from all clients to update the global model.

$$\omega = \sum_{k=1}^{k} \frac{n_k}{n} \omega_k$$

where ω_k is the locally trained model at client k. The iterative process continues until a predefined stopping criterion is met.

Contrastive loss function for local client

In our case, we use a custom loss function introduced by Li et al. [7], called the model contrastive loss function. This is integrated with the supervised loss function to create a joint loss function for each client. During model training, the client minimises this combined loss function through backpropagation.

$$client\ contrastive\ loss =$$

$$-log\frac{exp\big(sim(h,h_g)\,/\,\gamma\big)}{exp\big(sim(h,h_g)/\gamma\big) + exp\big(sim(h,h_p)\,/\,\gamma\big)}$$

Discrepancy weights for global server
In our study, we used discrepancy weight for the server introduced by Ye et al. [8]. In place of $\frac{n_k}{n}$, we used P_i as client weights that are considered by the server.

$$P_i = \frac{ReLU(n_i - c \cdot d_i + e)}{\Sigma_{j=1}^l ReLU\big(n_j - c \cdot d_j + e\big)}$$

where n_i is the client's relative dataset size and d_i is the client's discrepancy level to server uniform weight. ReLU is used to manage negative value, and c is a hyperparameter to balance n_i and d_i, e is a hyperparameter to adjust weight. This will assign a large weight to clients with large data sizes and lower discrepancy levels.

Federated training framework
Figure 30.1 illustrates the federated training workflow, where each client maintains a local model and computes cross-entropy loss. Each client retains three models: the previously trained model, the global model, and the currently trained model, each with a projection head. Extracted representations from Global_head, Client_current_head, and Client_previous_head are then fed into

the supervised contrastive loss module to calculate the contrastive loss. The total loss, combining cross-entropy and contrastive losses, is minimised through backpropagation. After local epochs, clients send updates to the server for weighted aggregation of model parameters. The process repeats until predefined global rounds is reached. Moreover, before federated training, the server estimates discrepancy weights for each client based on their data size and discrepancy levels.

Data and experimental setup
We used the COVID-19 dataset [14] to evaluate data across 10 clients using a Dirichlet distribution, as shown in Figure 30.2, which depicts client heterogeneity data distribution. The used dataset is available for download at this Kaggle link: https://www.kaggle.com/datasets/anasmohammed-tahir/covidqu. Model performance was assessed using accuracy, recall, precision, F1-score, and kappa score. The federated model was trained for 100 global rounds, with each client performing 10 local epochs using the SGD optimizer. A local learning rate of 0.01 and a batch size of 64 were used. And for the local model, we have used ResNet18.

Results and analysis

Best performances
Table 30.1 shows FedDiCoNet achieving the highest Top-1 accuracy of 91.20% among other FL methods. FedAvg scores 86.15%, while FedProx lags at 78.37%.

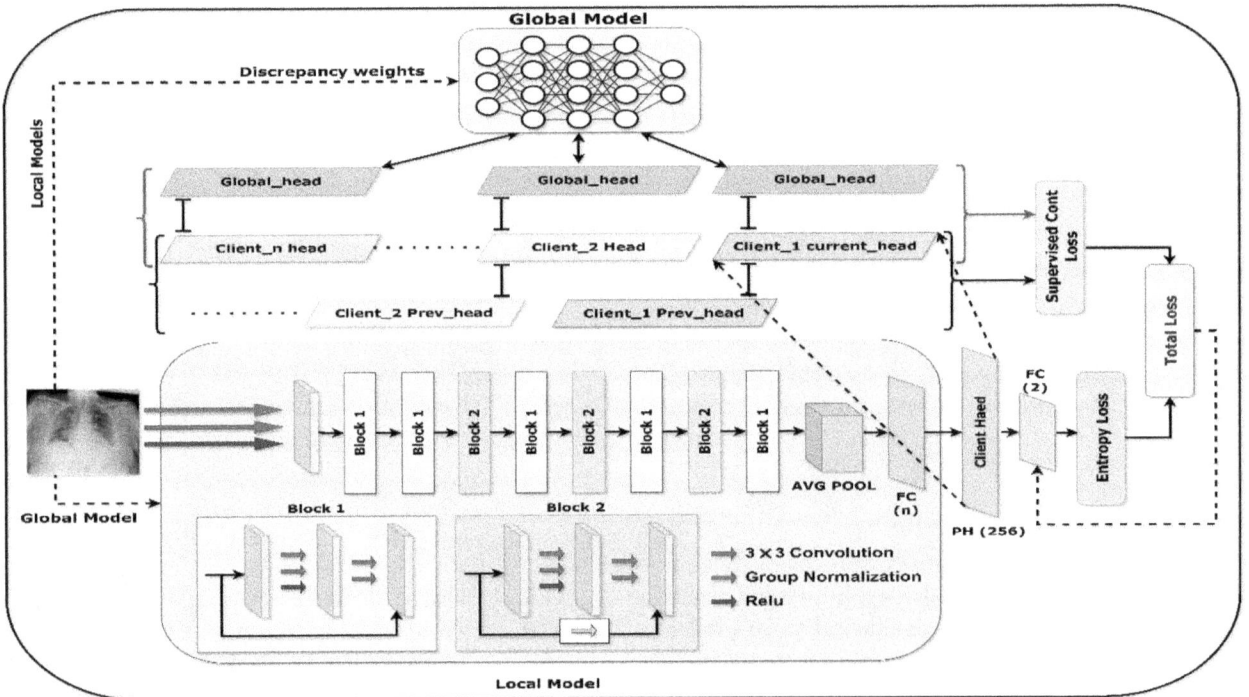

Figure 30.1 The flow of the federated learning training
Source: Author

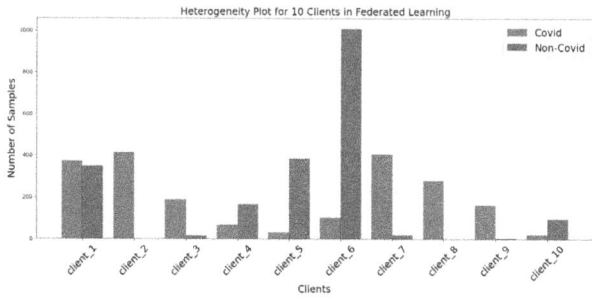

Figure 30.2 Heterogeneity data distribution over 10 clients
Source: Author

MOON slightly surpasses FedAvg with 86.44%. FedDiCoNet's superior performance highlights its robustness in handling data heterogeneity in federated learning. Figure 30.3 presents the t-SNE plot for all four models, where FedDiCoNet demonstrates better class separability between the two classes.

Performance over round and convergence

Table 30.2 shows that FedDiCoNet achieves 79.80% accuracy at 25 rounds, outperforming other FL methods. This early performance demonstrates its superior convergence, making it more efficient with fewer communication rounds in FL. Additionally, FedDiCoNet consistently surpasses other FL methods in every round, highlighting its consistent performances over communication rounds.

Comparative analysis of federated learning models on COVID-19 prediction

Table 30.3 highlights FedDiCoNet's superior performance on COVID-19 data across all metrics. It achieves the highest accuracy of 89.53%, surpassing FedAvg, MOON, and FedProx. Its precision of 90.20% and recall of 89.54% ensure reliable predictions, outperforming FedAvg, MOON, and FedProx. FedDiCoNet's F1-score of 89.49% and Kappa score of 0.79 further confirm its robustness. These results demonstrate FedDiCoNet's effectiveness in federated learning for COVID-19 data, ensuring higher

Table 30.1 Top1 accuracy achieved by various federated learning algorithm.

Methods	Top-1 accuracy
FedAvg	86.15%
FedProx	78.37%
MOON	86.44%
FedDiCoNet	**91.20%**

Source: Author

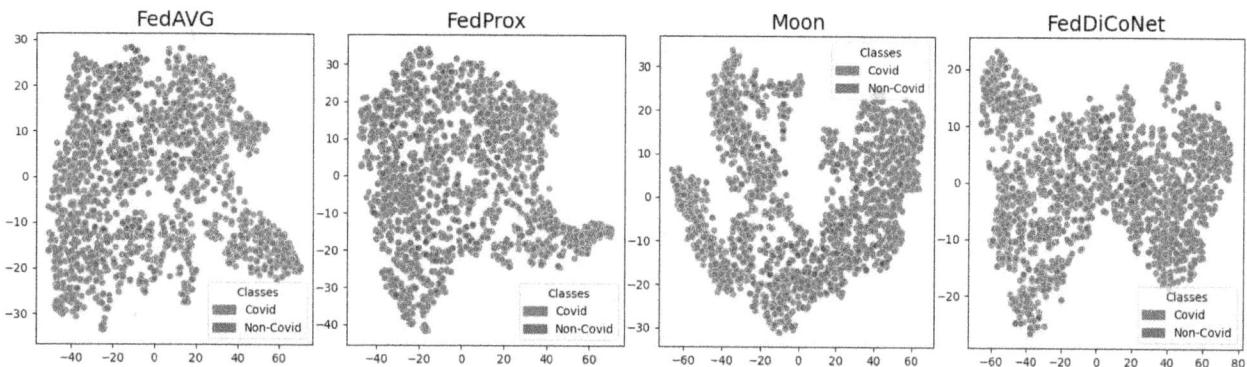

Figure 30.3 t-SNE visualisation for different federated learning models
Source: Author

Table 30.2 Performance across various rounds during training.

Methods	Communication rounds			
	25	50	75	100
FedAvg	51.14%	70.59%	51.09%	86.15%
FedProx	50.00%	67.67%	55.03%	78.14%
MOON	50.00%	50.00%	67.84%	86.44%
FedDiCoNet	**79.80%**	**80.83%**	**88.61%**	**89.53%**

Source: Author

Table 30.3 Various quantitative measures for COVID-19 tasks.

Methods	Accuracy	Precision	Recall	F1-score	Kappa
FedAvg	86.15%	86.28%	86.16%	86.14%	0.72
FedProx	78.14%	82.14%	78.15%	77.45%	0.56
MOON	86.44%	87.02%	86.43%	86.39%	0.72
FedDiCoNet	**89.53%**	**90.20%**	**89.54%**	**89.49%**	**0.79**

Source: Author

accuracy, better sensitivity, and improved agreement between predictions and actual labels.

Conclusion

This study highlights FedDiCoNet's effectiveness in addressing data heterogeneity in federated learning. Integrating model-based contrastive learning and discrepancy-aware aggregation enhances model convergence and performance. On the COVID-19 chest X-ray dataset, FedDiCoNet achieves the best accuracy of 91.20%, outperforming FedAvg, FedProx, and MOON across all metrics. This demonstrates its potential as a robust, privacy-preserving solution for medical imaging applications. Its strong early performance indicates superior convergence, making it more efficient with fewer communication rounds in federated learning. Future work will extend the evaluation to diverse datasets across different medical imaging modalities, such as MRI, CT, and X-ray, to further validate its robustness and applicability.

References

[1] Dayarathna, S., Islam, K. T., Uribe, S., Yang, G., Hayat, M., Chen, Z. (2024). Deep learning based synthesis of MRI, CT and PET: review and analysis. *Medical Image Analysis*, 92:103046.

[2] Saxena, S., Jena, B., Mohapatra, B., Gupta, N., Kalra, M., Scartozzi, M., Saba, L., Suri, J. S. (2023). Fused deep learning paradigm for the prediction of O6-methylguanine-DNA methyltransferase genotype in glioblastoma patients: a neuro-oncological investigation. *Computers in Biology and Medicine*, 153, 106492.

[3] Ali, M., Shahroz, M., Akram, U., Mushtaq, M. F., Carvajal Altamiranda, S., Aparicio Obregon, S., De La Torre Díez, I., Ashraf, I. (2024). Pneumonia detection using chest radiographs with novel EfficientNetV2L model. *IEEE Access*, 12, 34691–34707

[4] AlSalman, H., Al-Rakhami, M. S., Alfakih, T., Hassan, M. M. (2024). Federated learning approach for breast cancer detection based on DCNN. *IEEE Access*, 12, 40114–40138. doi: 10.1109/ACCESS.2024.3374650.

[5] McMahan, B., Moore, E., Ramage, D., Hampson, S., Aguera y Arcas, B. (2017). Communication-efficient learning of deep networks from decentralized data. In Artificial Intelligence Statistics. PMLR, 1273–1282.

[6] Guan, H., Yap, P. T., Bozoki, A., Liu, M. (2024). Federated learning for medical image analysis: a survey. *Pattern Recognition*, 110424.

[7] Li, Q., He, B., Song, D. (2021). Model-contrastive federated learning. In Proc. IEEE/CVF Conf. Comput. Vis. Pattern Recognit., 10713–10722.

[8] Ye, R., Xu, M., Wang, J., Xu, C., Chen, S., Wang, Y. (2023). FedDisco: federated learning with discrepancy-aware collaboration. In Int. Conf. Mach. Learn. PMLR, 39879–39902.

[9] Lu, Z., Pan, H., Dai, Y., Si, X., Zhang, Y. (2024). Federated learning with non-IID data: a survey. *IEEE Internet Things Journal*, 11(11),19188–19209. doi: 10.1109/JIOT.2024.3376548.

[10] Yuan, X., Li, P. (2022). On convergence of FedProx: local dissimilarity invariant bounds, non-smoothness and beyond. *Advances in Neural Information Processing Systems*, 35, 10752–10765.

[11] Feng, C.-M., Yan, Y., Wang, S., Xu, Y., Shao, L. Fu, H. (2023). Specificity-preserving federated learning for MR image reconstruction. *IEEE Transactions on Medical Imaging*, 42(7), 2010–2021.

[12] Zhu, F., Tian, Y., Han, C., Li, Y., Nan, J., Yao, N. Zhou, W. (2024). Model-level attention and batch-instance style normalization for federated learning on medical image segmentation. *Information Fusion*, 107, 102348.

[13] Rauniyar, A., Hagos, D. H., Jha, D., Håkegård, J. E., Bagci, U., Rawat, D. B. Vlassov, V. (2023). Federated learning for medical applications: a taxonomy, current trends, challenges, and future research directions. *IEEE Internet of Things Journal*.

[14] Tahir, A. M., Chowdhury, M. E. H., Khandakar, A., Rahman, T., Qiblawey, Y., Khurshid, U., Kiranyaz, S., et al. (2021). COVID-19 infection localization and severity grading from chest X-ray images. *Computers in Biology and Medicine*, 139, 105002.

[15] Khosla, P., Teterwak, P., Wang, C., Sarna, A., Tian, Y., Isola, P., Maschinot, A., Liu, C., Krishnan, D. (2020). Supervised contrastive learning. *Advances in Neural Information Processing Systems*, 33, 18661–18673.

[16] Tölle, M., Garthe, P., Scherer, C., et al. (2025). Real world federated learning with a knowledge-distilled transformer for cardiac CT imaging. *npj Digital Medicine*, 8, 88. https://doi.org/10.1038/s41746-025-01434-3.

[17] Huangsuwan, K., Liu, T., See, S., Ng, A. B., Vateekul, P. (2025). FedDrip: federated learning with diffusion-generated synthetic image. *IEEE Access*, 13, 10111–10125. https://doi.org/10.1109/ACCESS.2025.3525806.

31 A hybrid deep learning model for accurate brain tumour segmentation and classification using ResNet101 and U-Net

Wu Min Fen[a], Yang Jin[b], Liao Shi Chin[c] and Haobijam Basanta[d]

Department of Electrical Engineering, National Taipei University, Sanxia District, New Taipei City, Taiwan R.O.C

Abstract

Accurate and timely brain tumours diagnosis is crucial for effective treatment; however, it remains a significant challenge worldwide. To address this challenge, this study introduces a novel hybrid deep learning system for automated MRI-based tumour segmentation and classification, which significantly enhances diagnostic speed and precision in clinical settings. The system combines ResNet101, U-Net, and an attention mechanism to extract key features, precisely delineate tumour boundaries, and classify tumour types. Validation on a benchmark dataset yielded an impressive 96% F1-score and 97% recall and sensitivity, outperforming existing methods. These results highlight the system's potential as a highly reliable and accurate tool for improving clinical diagnosis and facilitating early tumour detection.

Keywords: Tumour, deep learning, hybrid model, ResNet101, U-Net, automatic segmentation, MRI scans, attention mechanism, tumour classification

Introduction

Devastating brain tumours pose a major health threat, contributing significantly to mortality and disability. Despite advances in medical imaging, early detection, precise diagnosis, and successful treatment remain significant challenges for clinicians [1]. The inherent variability of tumour presentation, location, and size, coupled with the critical need to distinguish between benign and malignant types, hinders consistent and timely diagnoses [2]. Treatment decisions are further complicated by tumour type, grade, and location, emphasising the necessity for accurate early detection to enable effective therapeutic interventions [3]. The global burden of brain tumours is substantial, with over 300,000 new cases diagnosed annually, according to the World Health Organization [4]. Among these, gliomas, particularly glioblastoma multiforme, are highly prevalent and aggressive. Glioblastomas, characterised by their extreme malignancy, often result in poor patient prognoses, with survival rates typically less than 15 months post-diagnosis, even with advanced treatments [5]. This aggressive nature underscores the urgent imperative for early detection and rapid intervention, which are critical for enhancing survival rates and improving patients' quality of life. Early detection is paramount, as malignant brain tumours proliferate rapidly, leading to irreversible neurological damage. Identifying tumours at an early stage facilitates targeted therapies, significantly improving patient outcomes. However, achieving accurate diagnoses remains a significant challenge. Brain tumours exhibit complex morphologies, indistinct boundaries, and unpredictable growth patterns, complicating precise identification. Furthermore, inter-observer variability among radiologists can lead to inconsistent diagnoses [6]. To address these challenges, automated diagnostic solutions are essential. The overwhelming volume of medical images demands AI-driven tools to enhance diagnostic precision, minimise human error, and expedite reliable clinical decisions [7]. Artificial intelligence, encompassing a spectrum of machine learning techniques, including traditional neural networks and, notably, CNNs, has revolutionised medical image analysis [8,9]. These models excel at processing complex MRI and CT scans, uncovering subtle abnormalities undetectable by human observers, and significantly improving tumour detection accuracy. Integrating these technologies into clinical workflows promises to accelerate diagnosis, enhance consistency, reduce diagnostic time, and ultimately, deliver superior patient outcomes [10]. While extensive research has addressed medical image segmentation [11–14], brain tumour detection remains challenging. These include difficulties in integrating multi-modal data due to variations in resolution, modality-specific characteristics, and alignment. Additionally, deep learning models often struggle to generalise across diverse datasets and imaging protocols, limiting their effectiveness in real-world clinical applications. Specifically, even with progress in medical imaging, accurate and timely brain tumour diagnosis continues to be a significant problem in clinical settings. To address these limitations, we propose the development of an advanced deep learning framework for automated brain tumour segmentation and classification from MRI scans.

[a]wuminfen030@gmail.com, [b]yangjoe20000905@gmail.com, [c]ysesst96036@gmail.com, [d]haobijambasanta@mail.ntpu.edu.tw

DOI: 10.1201/9781003675235-31

Our research centers on the design, implementation, and rigorous evaluation of AI models aimed at significantly improve diagnostic accuracy while reducing processing time. The primary objectives of this study are as follows:

1. Construct a novel hybrid deep learning architecture, combining ResNet101 for robust feature extraction with U-Net for precise tumour segmentation.
2. Extract and integrate discriminative features leveraging ResNet101 to capture high-level features from MRI scans, and then seamlessly integrating them into the U-Net architecture for accurate tumour delineation.
3. Enhance diagnostic precision by incorporating a specialised classification module to differentiate between benign and malignant tumours, and integrating an attention mechanism to prioritise salient tumour regions within MRI scans.
4. Optimise the model for clinical deployment balancing model complexity and performance to guarantee high accuracy while achieving computational efficiency for practical clinical use.

Achieving these objectives will enable the development of AI-powered tools that empower radiologists to deliver more accurate and timely diagnoses, thereby enhancing patient care and clinical outcomes.

Related works

Accurate brain tumour segmentation and identification remain critical challenges in clinical practice, directly impacting treatment planning and patient outcomes. Researchers have explored various algorithmic approaches, particularly in multi-modal MRI-based analyses, to enhance tumour detection and classification. Deep learning has proven to be a valuable technique, offering automated and highly precise segmentation. For instance, Sapra et al. [15] achieved 100% classification accuracy using an enhanced probabilistic neural network, highlighting the potential of specialised architectures, though such high accuracy is rarely observed in more generalised datasets. Amin et al. [16] proposed a three-stage pipeline, emphasising the significance of robust pre-processing and feature extraction for reliable classification. Mathivanan et al. [17] investigated transfer learning strategies for brain tumour classification, testing models such as ResNet152, VGG19, DenseNet169, and MobileNetv3, and implemented a robust validation strategy, dividing the dataset into five distinct subsets, to improve the model's ability to generalise and minimise overfitting. Ranjbarzadeh et al. [18] tackled data bias in multi-modal MRI analysis using Z-score normalisation, ensuring consistent data representation, and incorporated an attention-based cascade CNN to enhance tumour-specific focus. Ali et al.

[19] demonstrated enhanced segmentation generalisation through training data, specifically rotations, reflections, and gamma adjustments, within a hybrid 3D CNN and U-Net architecture. Yogananda et al. [20] utilised the BraTS 2018 dataset with three-fold cross-validation, yet dataset diversity limitations persisted, underscoring the need for broader representation in training data. Recent advancements in deep learning have further refined segmentation accuracy. Zhu et al. [21] introduced an enhanced U-Net with ResNet50 and CBAM, achieving an 87.47% Dice score, demonstrating the efficacy of hybrid architectures in capturing intricate tumour details. Wang et al. [22] highlighted Vision Transformers' ability to model global contextual information in multi-modal segmentation, surpassing traditional CNN-based approaches. Sahoo et al. [23] developed a composite deep learning network, integrating Inception V2, to achieve multi-scale feature analysis and enhance tumour core identification accuracy. Hou et al. [24] introduced MFD-Net, which incorporates diffractive blocks for refined feature extraction and enhanced segmentation accuracy. Çetiner et al. [25] proposed DenseUNet+, which integrates dense connections to improve feature propagation in multi-modal MRI segmentation. The 2024 BraTS challenge has further advanced the field by focusing on post-treatment gliomas, providing a benchmarking platform for evaluating the clinical applicability of emerging deep learning models. However, despite the progress made, significant obstacles remain, hindering the broad integration of deep learning models in clinical settings. A major hurdle is the difficulty in achieving generalisation across diverse patient populations due to inherent dataset constraints. Additionally, capturing the complex structural and textural variations in tumours remains a challenge, particularly in multi-modal imaging scenarios. In response, we introduce a novel deep learning model integrating ResNet101 for strong feature learning, U-Net for accurate tumour delineation, and an attention module to improve tumour-specific feature selection. ResNet101's deep residual structure effectively captures hierarchical tumour features, which are crucial for handling complex tumour morphologies. U-Net's encoder-decoder architecture ensures precise segmentation by integrating local and global features, while an attention mechanism dynamically prioritises key image regions, refining segmentation accuracy. This architecture achieves a synergy between high accuracy and computational efficiency, facilitating seamless clinical integration. Our experimental findings, with F1-score of 96% and 97% recall and precision, significantly surpass current benchmarks. This superior performance, achieved through the integration of advanced architectures and attention mechanisms, presents a promising avenue for improving brain tumour diagnosis, refining clinical decision-making, and enabling personalised treatment strategies. In conclusion,

while artificial intelligence algorithms have yielded substantial advancements in brain tumour segmentation and classification, challenges related to dataset diversity, multi-modal data integration, and model generalisation continue to impede clinical translation. This research overcomes these limitations by presenting a novel hybrid model, combining robust feature extraction, precise segmentation, and attention mechanisms to significantly improve accuracy and clinical applicability.

Methodology

Dataset

This study utilised two distinct datasets: one for training the segmentation task, derived from the TCIA's [26] TCGA low-grade glioma collection, consisting of 7,585 MRI scans with manual segmentation masks from 110 patients; and another for classification, sourced from Kaggle [27], comprising 7,023 MRI scans classified into four categories: glioma, meningioma, pituitary tumour, and no-tumour. Figure 31.1(a) presents sample MRI scans along with their corresponding segmentation masks, while Figure 31.1(b) displays representative images from the classification dataset. However, these datasets exhibit inherent potential biases due to variations in MRI scanner types, acquisition settings, and patient demographics. The TCIA dataset primarily contains data from specific medical institutions, which may not generalise well to diverse clinical settings. Similarly, the Kaggle dataset lacks comprehensive patient metadata, limiting our ability to fully assess its clinical applicability. To mitigate these biases, future work will focus on expanding dataset diversity and

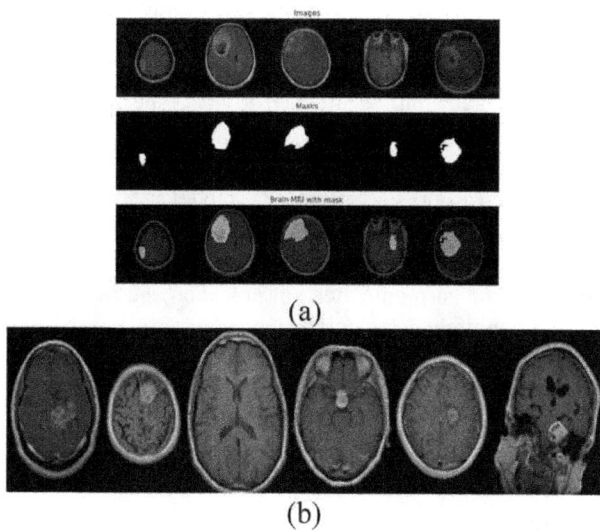

validating the model with external, more varied datasets. Additionally, subsequent studies will incorporate comprehensive patient metadata to improve the evaluation of clinical applicability. Overcoming these challenges is essential for ensuring the model's reliability, adaptability, and fairness in clinical practice.

Hybrid model architecture and design

This work proposes a unique hybrid deep learning framework for the combined task of brain tumour classification and segmentation in MRI scans. Through the integration of ResNet101 for effective feature extraction, U-Net for exact segmentation, and CAM and SAM for optimised feature representation, the resulting architecture offers a significant advancement in clinical diagnostic tools (Figure 31.2).

To achieve superior tumour localisation, CAM and SAM are strategically implemented along skip connections before the concatenation with the upsampled features, amplifying relevant features and suppressing noise. Specifically, CAM dynamically refines feature importance across channels by weighting feature maps based on their segmentation process, while SAM enhances spatial attention by highlighting tumour-related pixels through the creation of a spatial attention map that emphasises tumour regions. This dual-attention strategy significantly improves segmentation accuracy, particularly for small or complex tumours. The process begins with a $65 \times 65 \times 65$ MRI input volume, representing spatial dimensions and channels. Utilising its residual connections, ResNet101 robustly derives deep, informative features, overcoming the vanishing gradient problem and enabling the capture of intricate tumour structures. These features are then routed to the U-Net decoder through skip connections, ensuring the preservation of both fine-grained spatial and high-level semantic information. This integration aids in restoring spatial resolution, resulting in precise pixel-level segmentation masks. Subsequently, the segmented output is flattened (converted to a 1-dimensional vector) and passed to a fully connected layer for tumour classification, categorising tumours as glioma, meningioma,

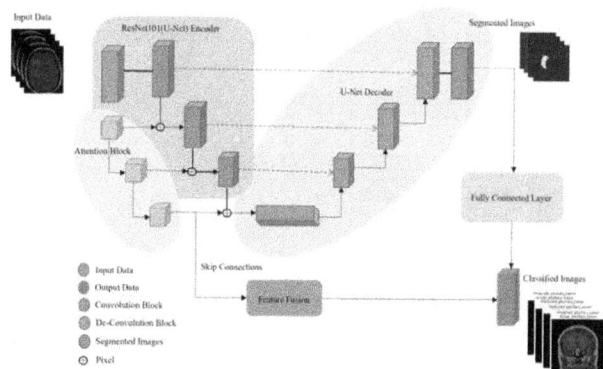

(a)

(b)

Figure 31.1 Datasets used in this study. (a) Sample MRI scans from the segmentation dataset with ground truth tumour regions. (b) Representative images from the classification dataset

Source: Author

Figure 31.2 Overview of the hybrid model architecture

Source: Author

pituitary tumours, and no-tumour. This classification relies on the integrated and refined features from ResNet101 and U-Net, potentially leading to reliable diagnostic outcomes. Figure 31.3 depicts the hybrid deep learning model's processing pipeline.

Model optimised framework for training and evaluation
A rigorous assessment of training efficacy, inference speed, and computational resource utilisation was performed to validate the clinical viability of our brain tumour segmentation and classification architecture. Our hybrid U-Net architecture, incorporating a ResNet101 encoder, was trained on an NVIDIA RTX 3060 GPU with 12GB of VRAM, and converged within 200 epochs, averaging 20 minutes per epoch. Notably, a single MRI image was processed in just 0.18 seconds, demonstrating its potential for real-time clinical diagnostic applications. For robust model evaluation, the dataset was partitioned into 70% training, 20% validation, and 10% testing subsets. This structured partitioning enabled precise parameter optimisation during training, fine-tuned hyperparameter adjustments through validation, and an unbiased final accuracy evaluation using the test set. To further enhance generalisation and robustness, the training data was diversified through geometric transformations, including random flipping and rotation up to 15 degrees. These specific augmentations were chosen to simulate common variations in MRI scans. The training was optimised using the Adam optimizer and a ReduceLROnPlateau scheduler, which decreased the learning rate by 10% when the validation loss plateaued, ensuring smooth convergence. To prevent overfitting, an early stopping mechanism was implemented, halting training after 10 epochs without validation improvement. ResNet101 was selected as the encoder for its robust feature extraction capabilities, and U-Net as the decoder for its proven effectiveness in semantic segmentation. The model's computational efficiency was evaluated through key metrics, including memory usage, FLOPs, and runtime. The U-Net + ResNet101 framework exhibited a computational load of approximately 1.06 billion FLOPs per forward pass, with a peak GPU memory utilisation of 10.5 GB during training. Compared to traditional models, our hybrid approach offers a superior balance of computational efficiency and accuracy, making it suitable for practical use in clinical environments with high-performance hardware. For performance evaluation, a composite loss function was adopted to optimise both segmentation and classification, with Dice and BCE losses addressing class imbalance. The Dice loss was calculated as:

$$L_{Dice} = 1 - \frac{2 \sum_{i=1}^{N} pi * gi}{\sum_{i=1}^{N} pi^2 + \sum_{i=1}^{N} gi^2} \tag{1}$$

where pi and gi denote the predicted pixel and ground truth mask intensities. The loss function is designed to enhance tumour region coverage during segmentation.

For tumour type classification, cross-entropy loss is employed, which is mathematically expressed as:

$$L_{BCE} = -\sum_{i=1}^{C} yi \log (\hat{y}i) \tag{2}$$

where yi represents the actual class distribution and denotes the predicted probability for class i.

The final unified loss function is obtained by merging both components and is calculated as:

$$L_{total} = \alpha * L_{Dice} + \beta * L_{BCE} \tag{3}$$

where α and β are adjustable weights that balance the contributions of the segmentation and classification tasks. During validation, these hyperparameters are optimised to maximise the overall model performance. Fine-tuning these weights ensures balanced learning and improved accuracy. This refined composite loss function enables the hybrid model to excel in multiple tasks, providing a robust and efficient approach for brain tumour diagnosis. Furthermore, to comprehensively evaluate classification performance, standard evaluation metrics were utilised, including precision, recall (sensitivity), specificity, and F1-score. Precision quantifies the percentage of correctly identified tumour cases among all predicted positives. Recall (sensitivity) assesses the model's effectiveness in detecting actual tumour cases. Specificity indicates how well the model distinguishes non-tumour cases. The F1-score, the harmonic mean of precision and recall, provides a comprehensive performance measure, particularly valuable for handling imbalanced datasets. These metrics are mathematically defined as:

$$\text{Precision} = \frac{TP}{TP+FP} \tag{4}$$

$$\text{Recall} = \frac{TP}{TP+FN} \tag{5}$$

$$\text{Specificity} = \frac{TN}{TN+FP} \tag{6}$$

$$\text{F1} - \text{Score} = \frac{2 * Precision * Recall}{Precision + Recal} \tag{7}$$

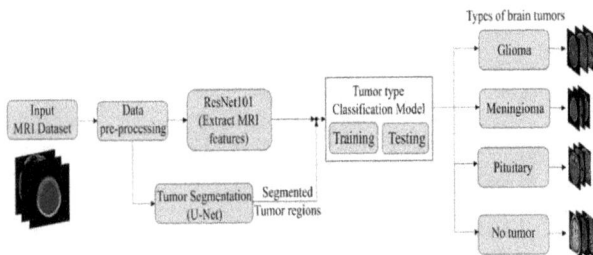

Figure 31.3 Architectural overview of the hybrid deep learning model
Source: Author

where TP (True Positive) refers to correctly identified positive tumour cases, FP (False Positive) denotes incorrectly predicted positive tumour cases, FN (False Negative) represents incorrectly predicted negative tumour cases, and TN (True Negative) refers to correctly identified negative tumour cases.

Empirical results

Hybrid model performance evaluation

The hybrid model's classification performance across glioma, meningioma, pituitary tumours, and no-tumour cases was assessed using a confusion matrix (Figure 31.4). This matrix effectively visualised correct (diagonal) and incorrect (off-diagonal) predictions. The model achieved high accuracy for meningioma (159 correct) and pituitary tumours (152 correct), with few misclassifications. However, distinguishing between glioma (170 correct) and meningioma presented a challenge, evidenced by notable misclassifications (3 glioma as meningioma, 3 meningioma as glioma). Additional misclassifications included 73 no-tumour correct, 1 no-tumour as meningioma, 3 no-tumour as pituitary, 5 meningioma as pituitary, and 4 pituitary as meningioma. These observed errors, particularly misclassifications between glioma and meningioma, highlight potential weaknesses in feature discrimination. Additionally, class imbalance within the dataset likely affected overall performance and the confusion matrix, necessitating future balancing strategies to address this issue. Clinically, these misclassifications could lead to inappropriate treatment decisions, emphasising the necessity of improved feature discrimination and balanced datasets to enhance diagnostic accuracy. Beyond classification, the model demonstrated strong segmentation capabilities. Figure 31.5 illustrates accurate tumour boundary delineation across various MRI scans, highlighting the model's precision. Figure 31.6 further validates its clinical potential by showcasing precise tumour type and location predictions.

Comparative performance evaluation with existing models
The performance of our hybrid model, U-Net + ResNet101, was thoroughly evaluated and compared with several other models across various key metrics. In this comparative performance analysis, our hybrid U-Net + ResNet101 model achieved the highest F1-Score of 0.96, outperforming VGG16 (0.92), ResNet101 (0.95), DenseNet121 (0.95), and Xception (0.94). This highlights its superior tumour detection accuracy, effectively minimising both false positives and false negatives. Regarding recall/sensitivity, U-Net + ResNet101 scored 0.97, highlighting its strong ability to capture true positives and minimise missed tumour detections, which is crucial in medical imaging. VGG16 had a lower recall/sensitivity of 0.93, while DenseNet121 and Xception both scored 0.96, trailing slightly behind U-Net + ResNet101. For specificity, U-Net + ResNet101 scored 0.97 in specificity, indicating that it effectively excludes no-tumour cases. The other models had specificity values of 0.95 or higher, but none matched U-Net + ResNet101's ability to avoid false positives as effectively. Finally, in terms of precision, U-Net + ResNet101 achieved a precision of 0.96, ensuring a high

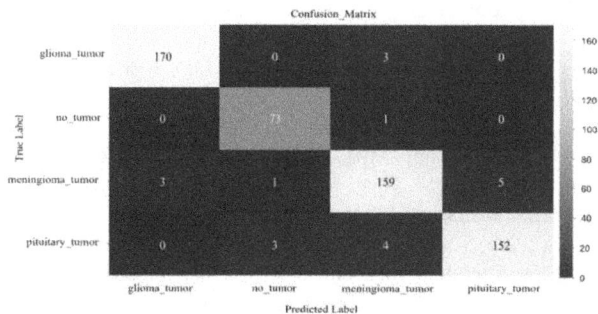

Figure 31.5 Brain tumour segmentation results
Source: Author

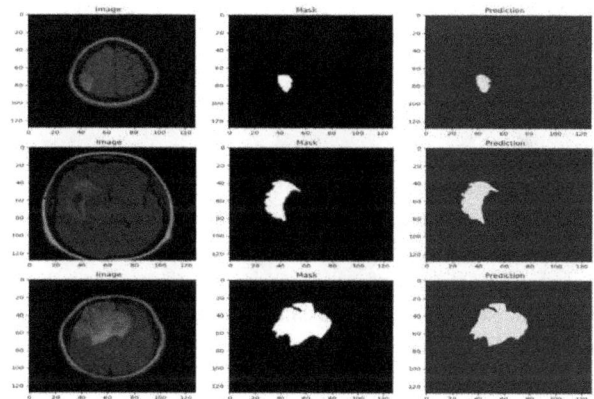

Figure 31.4 Confusion matrix for brain tumour classification
Source: Author

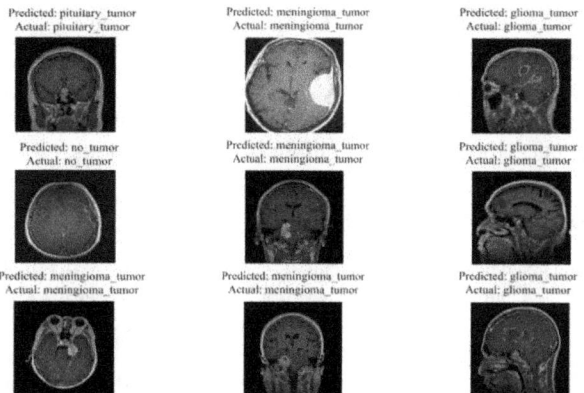

Figure 31.6 Brain tumour prediction results
Source: Author

Table 31.1 U-Net + ResNet101 performance vs. leading models.

Metric	U-Net + ResNet101	VGG 16	Res Net101	Dense Net121	Xcep-tion
F1-score	0.96	0.92	0.95	0.95	0.94
Sensitivity	0.97	0.93	0.95	0.96	0.96
Specificity	0.97	0.96	0.97	0.96	0.96
Precision	0.96	0.90	0.94	0.95	0.95

Source: Author

rate of correctly predicted tumour cases, minimising false positives. ResNet101 and DenseNet121 scored slightly lower at 0.94 and 0.95, respectively, while Xception also had a precision of 0.94, showing that U-Net + ResNet101 outperformed other models in this aspect as well. Notably, the integration of attention mechanisms yielded a 2.5% increase in segmentation Dice score and a 1.8% improvement in classification F1-score over the baseline. As a result, the hybrid model consistently exhibited superior performance across all evaluation metrics, confirming its efficacy in both tumour segmentation and classification, as detailed in Table 31.1. This superior performance stems from the hybrid model's ability to integrate ResNet101's feature extraction, U-Net's segmentation precision, and the attention mechanism's focus on critical image regions.

Conclusion

This research introduces a novel hybrid deep learning architecture, combining ResNet101's robust feature extraction with U-Net's precise segmentation, enhanced by an attention mechanism that focuses on critical image regions. Comprehensive evaluation confirms its superior performance in both tumour segmentation and classification compared to leading methods. This model holds significant potential to revolutionise clinical practice by enabling rapid, reliable diagnoses and facilitating timely, personalised patient care. Ultimately, we aim to improve patient outcomes through earlier and more accurate diagnoses, leading to enhanced treatment strategies. Future work will concentrate on refining the attention mechanism, exploring 3D Convolutional Neural Networks for volumetric analysis, integrating multi-modal data, validating the model on external datasets to ensure generalisation, and developing explainable AI models to foster clinical trust and support informed decision-making.

References

[1] Muhammad, S., Syeda, F. S., Izza, T., Aly, H. (2024). Challenges to early detection of brain tumors in low- and middle-income countries: a systematic review. *World Neurosurgery*, 191, 68–80.

[2] Bi, W. L., Hosny, A., Schabath, M. B. (2019). Artificial intelligence in cancer imaging: clinical challenges and applications. *CA Cancer Journal for Clinicians*, 69(2), 127–157.

[3] Zhou, Y., Tao, L., Qiu, J., et al. (2024). Tumor biomarkers for diagnosis, prognosis and targeted therapy. *Signal Transduction and Targeted Therapy*, 9, 132.

[4] Louis, D. N., Perry, A., Wesseling, P., et al. (2021). The 2021 WHO classification of tumors of the central nervous system: a summary. *Neuro-Oncology*, 23(8), 1231–1251.

[5] Rong, L., Li, N., Zhang, Z. (2022). Emerging therapies for glioblastoma: current state and future directions. *Journal of Experimental & Clinical Cancer Research*, 41, 142.

[6] Craig, A., Beam, E. F., Conant, E. A. (2002). Sickles, factors affecting radiologist inconsistency in screening mammography. *Academic Radiology*, 9(5), 531–540.

[7] Pinto-Coelho, L. (2023). How artificial intelligence is shaping medical imaging technology: a survey of innovations and applications. *Bioengineering (Basel)*, 10(12), 1435.

[8] Chen, Z. H., Lin, L., Wu, C. F., Li, C. F., Xu, R. H., Sun Y. (2021). Artificial intelligence for assisting cancer diagnosis and treatment in the era of precision medicine. *Cancer Communications (London)*, 41(11), 1100–1115.

[9] Archana, R., Jeevaraj, P. S. E. (2024). Deep learning models for digital image processing: a review. *Artificial Intelligence Review*, 57, 11.

[10] Jiang, X., Hu, Z., Wang, S., Zhang, Y. (2023). Deep Learning for medical image-based cancer diagnosis. *Cancers (Basel)*, 15(14), 3608.

[11] Taha, A. A., Hanbury, A. (2015). Metrics for evaluating 3D medical im age segmentation: analysis, selection, and tool. *BMC Medical Imaging*, 15(1), 1–28.

[12] Norouzi, A., Rahim, M. S. M., Altameem, A., Saba, T., Rad, A. E., Rehman, A., Uddin, M. (2014). Medical image segmentation methods, algorithms, and applications. *IETE Technical Review*, 31(3), 199–213.

[13] Ronneberger, O., Fischer, P., Brox, T. (2015). U-net: convolutional networks for biomedical image segmentation. In Medical Image Computing and Computer-Assisted Intervention–MICCAI 2015: 18th International Conference, Munich, Germany, October 5–9, 2015, Proceedings, Part III 18. Springer International Publishing, 234–241.

[14] Diakogiannis, F. I., Waldner, F., Caccetta, P., Wu, C. (2020). ResUNet-a: a deep learning framework for semantic segmentation of remotely sensed data. *ISPRS Journal of Photogrammetry and Remote Sensing*, 162, 94–114.

[15] Sapra, P., Singh, R., Khurana, S. (2013). Brain tumor detection using neural network. *International Journal of Science and Modern Engineering (IJISME)*, 1(9), 2319–6386.

[16] Amin, J., Sharif, M., Yasmin, M., Fernandes, S. L. (2020). A distinctive approach in brain tumor detection and classification using MRI. *Pattern Recognition Letters*, 139, 118–127.

[17] Mathivanan, S. K., Sonaimuthu, S., Murugesan, S., et al. (2024). Employing deep learning and transfer learning for accurate brain tumor detection. *Scientific Reports*, 14, 7232.

[18] Ranjbarzadeh, R., Bagherian Kasgari, A., Jafarzadeh Ghoushchi, S., et al. (2021). Brain tumor segmentation

based on deep learning and an attention mechanism using MRI multi-modalities brain images. *Scientific Reports*, 11, 10930.

[19] Ali, M., Gilani, S.O., Waris, A., Zafar, K., Jamil, M. (2020). Brain tumour image segmentation using deep networks. *IEEE Access*, 8, 153589–153598.

[20] Yogananda, C. G. B, Shah, B. R., Vejdani-Jahromi, M., et al. (2020). A fully automated deep learning network for brain tumor segmentation. *Tomography*, 6(2), 186–193.

[21] Zhu, J., Zhang, R., Zhang H. (2023). An MRI brain tumor segmentation method based on improved U-Net. *Mathematical Biosciences and Engineering*, 21(1), 778–791.

[22] Wang, P., Yang, Q., He, Z., Yuan, Y. (2023). Vision transformers in multi-modal brain tumor MRI segmentation: a review. *Meta-Radiology*, 1(1), 1–14.

[23] Sahoo, A. K., Parida, P., Muralibabu, K., Dash, S. (2023). An improved DNN with FFCM method for multimodal brain tumor segmentation. *Intelligent Systems with Applications*, 18, 1–12.

[24] Hou, P., Peng, Y., Wang, Z., Wang, J., Jiang, J. (2023). MFD-Net: MFD-Net: modality fusion diffractive network for segmentation of multimodal brain tumor image. *IEEE Journal of Biomedical and Health Informatics*, 27(12), 5958–5969.

[25] Çetiner, H., Metlek., S. (2023). DenseUNet+: a novel hybrid segmentation approach based on multi-modality images for brain tumor segmentation. *Journal of King Saud University - Computer and Information Sciences*, 25(8), 1–15.

[26] Pedano, N., Flanders, A. E., Scarpace, L., Mikkelsen, T., et al. (2016). The cancer genome atlas low grade glioma collection (TCGA-LGG) (Version 3) [Data set]. *The Cancer Imaging Archive*.

[27] Nickparvar, M. (2020). Brain tumor MRI dataset. Kaggle. Available at https://www.kaggle.com/datasets/masoud-nickparvar/brain-tumor-mri-dataset

32 Evaluating machine learning models for precision disease prediction

Pallabi Patowary[1,a] and Dhruba K Bhattacharyya[2,b]

[1]Department of Computer Science and Engineering, Golaghat Engineering College, Assam, 785621, India

[2]Department of Computer Science and Engineering, Tezpur University, Assam, 784028, India

Abstract

Accurate disease prediction is critical in healthcare for timely intervention and effective decision-making. The evaluation in this study focuses on seven Machine Learning (ML) models, including Support Vector Machine, Decision Tree, k-Nearest Neighbors, Naïve Bayes, Logistic Regression, Random Forest, and Multi-Layer Perceptron. The evaluation is conducted using two datasets: a symptom-based dataset and a biomarker-based dataset. Each model is optimised using GridSearchCV for hyperparameter tuning and evaluated with the 5-fold stratified cross-validation technique. Performance metrics, such as accuracy, recall, precision, F1 score, AUC-ROC, overfit gap, and runtime are used to assess model reliability and generalisability. This systematic evaluation highlights Random Forest as the most effective algorithm for disease diagnosis in healthcare applications, with SVM offering a reliable secondary option. The Multi-Layer Perceptron, in particular, provides a balance between accuracy and computational efficiency. The findings emphasise the importance of using diverse metrics and datasets for comprehensive model assessment in medical ML applications.

Keywords: Multiple disease dataset, machine learning, early detection, accuracy score, disease prediction

Introduction

The healthcare sector is increasingly adopting data-driven approaches to improve diagnostic accuracy and patient outcomes. Machine Learning (ML) has shown promising results in predictive modelling, particularly for disease diagnosis. Predictive models help clinicians make informed decisions and allow for timely interventions, especially in complex, multi-disease scenarios. This study compares the performance of several popular ML models on two distinct datasets. The first dataset (symptom-based) includes symptoms commonly reported in medical screenings, while the second (biomarker-based) comprises clinical biomarkers linked to conditions such as heart disease, diabetes, and anaemia. The main objective of our work is to evaluate the performance of each ML model and determine which are the best suited for disease prediction tasks across different types of healthcare data which can contribute to the development of robust, scalable, and accessible diagnostic tools or applications.

Background

Grampurohit et al. [1] studied ML classifiers i.e., Decision Tree (DT), Random Forest (RF), and Naive Bayes (NB) to create an effective disease prediction system based on patient symptom data. Here, the authors developed the model by considering 95 symptoms out of 132, but it is unclear what data reduction method they used. They only reported the accuracy score. Arumugam et al. [2] explored the development of a ML-based system using three ML models i.e., DT, NB, and Support Vector Machine (SVM) for predicting multiple diseases. In this work, the DT model performed better than NB and SVM. Raj et al. [3] proposed a disease prediction model based on only RF. Rout et al. [4] performed a comparative analysis for disease prediction using ML and found the same prediction accuracy for DT, RF, NB, SVM, and Logistic Regression (LR). Here, the authors have only presented the accuracy score but have not shown other details of the confusion matrix and have not applied cross-fold validation. Thus, existing work often focused on single-dataset or single-algorithm approaches, leaving gaps in understanding the comparative performances of multiple algorithms across diverse datasets using various predictive models.

Proposed method

Dataset used

Our analysis is conducted on two datasets, and these datasets are downloaded from the Kaggle website (https://www.kaggle.com/datasets). They are as follows:

(a) *Dataset 1 (symptom-based):* This dataset contains 4,961 records with 132 features (symptoms) and 41 diseases. Commonly reported symptoms by patients in initial screenings are itching, joint pain, fatigue, headache, and so on, and associated diseases which are due to the combinations of various symptoms such as Vertigo, AIDS, Arthritis, Bronchial Asth-

[a]pallabi.cse09@gmail.com, [b]dkb@tezu.ernet.in

DOI: 10.1201/9781003675235-32

ma, Cervical spondylosis, Chicken pox, Chronic cholestasis, Common Cold, Dengue, Diabetes, and so on.

(b) *Dataset 2 (biomarker-based):* This dataset comprises 2,837 records with 24 features, focusing on physiological and biochemical parameters such as glucose, cholesterol, haemoglobin, BMI, blood pressure, and so on. The diseases include anaemia, diabetes, heart disease, and thalassemia.

Pre-processing

The framework for multi-disease prediction using ML, as depicted in Figure 32.1, is structured into several steps. The process begins with the collection of multiple disease datasets. In the next step, these datasets undergo comprehensive pre-processing steps. The datasets are pre-processed, including data cleaning, where missing or irrelevant data is removed to ensure data quality; feature selection, where the most relevant attributes are chosen to enhance model performance; and feature scaling, which standardises data values to ensure compatibility across algorithms. For the selection of features, the random forest classifier has been applied.

Model training and testing

Once pre-processed, the data is split into training and testing sets, where the training set is used to build the ML models, and the testing set is reserved for evaluation purposes. In this work, seven well-established ML algorithms such as DT [6], k-NN [5], SVM [8], NB [9], LR [7], RF [10], and MLP [11] models have been used for training

and testing to predict diseases based on some symptoms or patient's health parameters. For each model, hyperparameter tuning is performed using GridSearchCV to identify the best possible set of hyperparameters. Cross-validation with 5 splits (StratifiedKFold) is used to assess model performance more reliably. This technique allows us to better generalise the model's predictive power and provides a comprehensive assessment of its accuracy, especially important in healthcare applications where reliable predictions are crucial.

Model evaluation

We have used evaluation metrics such as Accuracy (Acc), Recall (R), Precision (P), F1 Score (F1), AUC-ROC (AUC), cross-validation score (CV), Overfitting Gap (OG), and run-time instead of accuracy alone, as they better reflect performance across imbalanced datasets. Accuracy defines the proportion of correct predictions. AUC (Area Under the Curve) measures the model's ability to distinguish between classes. The overfitting gap, which measures the difference between training and testing accuracy, is also calculated.

This framework concludes with disease prediction and comparisons, where the model's predictions are analysed, and their performances are compared among the seven popular ML models using above mentioned metrics to determine the most effective models for disease diagnosis. This systematic approach ensures reliable and interpretable disease predictions, with potential applications in healthcare decision-making.

Results and discussion

To ensure data quality and improve model performance, several pre-processing steps are carried out. Missing values are imputed using the mean to preserve all records in the dataset. Feature selection is done using the Random Forest algorithm, which helps identify the most important features while removing less relevant ones to reduce noise. The class distribution is examined and features are scaled using MinMaxScaler to ensure compatibility with algorithms sensitive to feature magnitudes. Finally, the data is divided into training and testing sets with an 80:20 split, and 5-fold stratified cross-validation is used to ensure balanced class representation and reliable evaluation. The pre-processed Dataset 1 consists of 38 symptoms and Dataset 2 contains 20 features.

Table 32.1 compares the performance of various ML models on Dataset 1 i.e., for symptom-based dataset. The results indicate that RF, SVM, LR, and NB consistently demonstrate exceptional performance across all metrics, with a high-test accuracy (0.9698), precision (0.9742), recall (0.9698), F1-score (0.9704), and near-perfect AUC-ROC values (0.9998 for RF, LR, and NB; and 0.9993 for SVM). Additionally, these models exhibit negligible OG

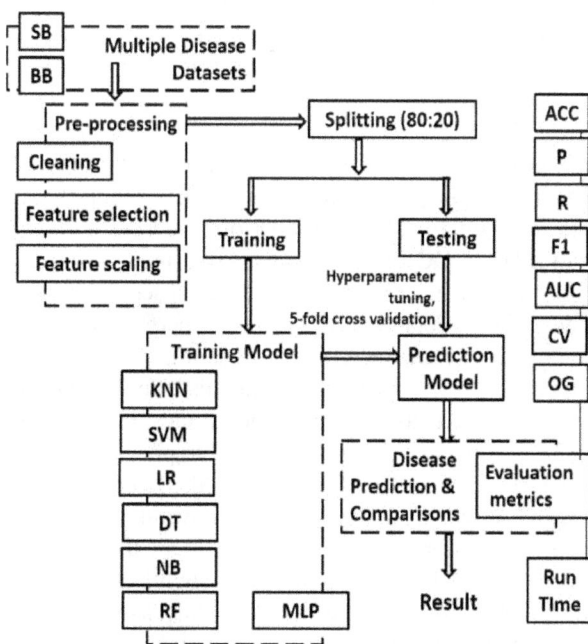

Figure 32.1 Proposed framework for disease prediction
Source: Author

Table 32.1 Performance metrics for various ML models on symptom-based dataset (Dataset 1).

Model	Accuracy	Precision	Recall	F1 score	AUC-ROC	CV score	Overfit gap	Run time (Sec)
Random forest	0.9698	0.9742	0.9698	0.9704	0.9998	0.9675	0.0000	53.5979
SVM	0.9698	0.9742	0.9698	0.9704	0.9993	0.9675	0.0000	10.3532
Logistic regression	0.9698	0.9742	0.9698	0.9704	0.9998	0.9675	0.0000	3.3247
Naïve Bayes	0.9698	0.9742	0.9698	0.9704	0.9998	0.9665	0.0000	0.0641
k-NN	0.9668	0.9712	0.9668	0.9670	0.9968	0.9670	0.0153	4.4762
Decision tree	0.9557	0.9715	0.9557	0.9596	0.9993	0.9451	0.0051	1.6804
MLP	0.9607	0.9684	0.9607	0.9616	0.9996	0.9682	0.0087	41.221

Source: Author

Table 32.2 Performance metrics for various ML models on biomarker-based dataset (Dataset 2).

Model	Accuracy	Precision	Recall	F1 score	AUC-ROC	CV score	Overfit gap	Run time (Sec)
Random forest	0.9613	0.9624	0.9613	0.9604	0.9980	0.9765	0.0387	14.9446
SVM	0.9489	0.9406	0.9489	0.9441	0.9973	0.9773	0.0311	22.4919
Decision tree	0.9384	0.9424	0.9384	0.9400	0.9556	0.9545	0.0612	8.1932
k-NN	0.9014	0.9263	0.9014	0.9102	0.9546	0.9508	0.0986	6.8372
Logistic regression	0.8504	0.8666	0.8504	0.8516	0.9419	0.8906	0.0236	0.9252
Naïve Bayes	0.8345	0.8419	0.8345	0.8313	0.9654	0.8703	0.0179	0.0044
MLP	0.9384	0.9370	0.9383	0.9373	0.9874	0.9278	0.0387	15.886

Source: Author

and stable CV scores, with NB achieving a CV score of 0.9665 and RF, SVM, and LR slightly higher at 0.9675. k-NN, while achieving competitive performance, falls slightly behind with a test accuracy of 0.9668, lower AUC-ROC, and a noticeable OG, suggesting room for improvement in generalisation. DT is the least effective model, with the lowest test accuracy (0.9557), a slightly higher OG (0.0051), and a lower CV score (0.9451), indicating reduced robustness and reliability. In terms of run time, RF took the longest (53.59 seconds), while NB was the fastest (0.0641 seconds), making it an efficient yet effective model. MLP demonstrates a moderate performance but its AUC-ROC value is close to the top-performing models. Additionally, MLP has a high computational cost, with a run time of 41.22 seconds, making it one of the slower models after RF. Overall, RF, SVM, LR, and NB emerge as reliable models, combining strong predictive performance with minimal overfitting and excellent generalisability. For real-world applications, where accuracy is critical (like disease prediction), RF and SVM are justified choices despite their computational cost. MLP and KNN demonstrate competitive but slightly lower performance, while DT shows the least effectiveness in this dataset.

Table 32.2 evaluates the performance of seven ML models on a disease prediction Dataset 2, i.e., biomarker-based dataset. The performance of the models on Dataset 2 reveals clear distinctions in terms of accuracy, precision, recall, and overall generalisation capabilities.

RF emerges as the top performer with the highest test accuracy (0.9613), precision (0.9624), and recall (0.9613), alongside a high AUC-ROC score of 0.9980, indicating its robust capability to distinguish between classes. It also exhibits a relatively low OG, suggesting good generalisation on unseen data. Following closely, SVM also performs strongly, with a test accuracy of 0.9489 and slightly lower precision and recall compared to RF, but its CV score of 0.9773 indicates strong model stability across folds. MLP performs competitively, with an accuracy of 0.9384, precision of 0.9370, and recall of 0.9383, placing it slightly below SVM and RF. DT, while slightly less accurate (0.9384), still performs decently, though with a larger OG of 0.0612, indicating more fluctuation in performance across different data subsets. The k-NN model shows notable deterioration in performance, with the lowest test accuracy (0.9014) and a larger overfit gap (0.0986), which implies poor generalisation. LR and NB are the least effective, with test accuracies of 0.8504 and 0.8345, respectively, as well as lower precision, recall, and AUC-ROC scores. These models also demonstrate the lowest CV scores, pointing to weaker overall stability and predictive power. Overall, RF and SVM are the most consistent and reliable models, with RF slightly outperforming SVM across most metrics. Meanwhile, k-NN, LR, and NB show progressively weaker performance, particularly in terms of accuracy and generalisation, with k-NN struggling the most with an elevated OG. These results suggest

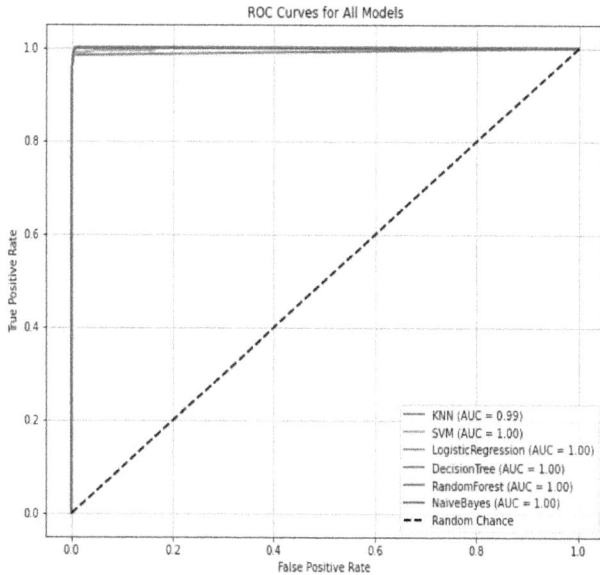

Figure 32.2 AUC-ROC curves for various machine learning. Models on symptom-based Dataset (Dataset 1) [The legend specifies the model name followed by its corresponding AUC score]

Source: Author

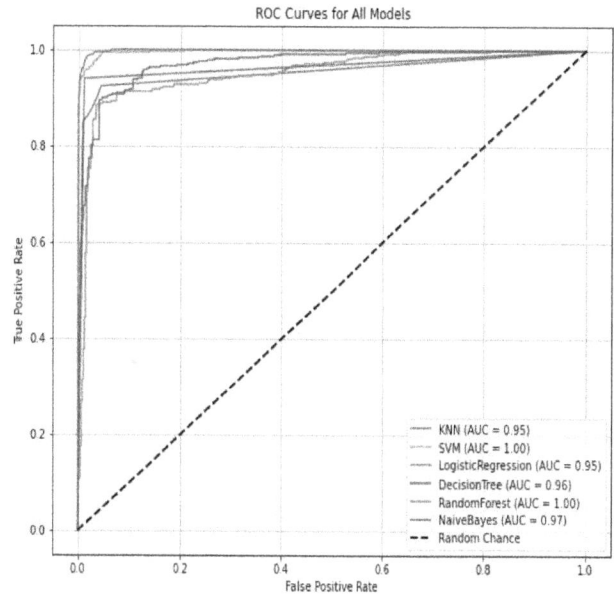

Figure 32.3 AUC-ROC curves for various machine learning models on biomarker-based dataset (Dataset 2)

Source: Author

that for this particular dataset, RF is the preferred model for optimal performance, while SVM is a close alternative for slightly lower computational demands. In terms of computational efficiency, RF and SVM are slower than some models, but they offer superior performance. MLP has a comparable runtime to RF but is slightly less accurate. DT and k-NN execute faster, but their lower accuracy and generalisation issues make them less favourable. LR and NB are the fastest models, but at the cost of significantly lower accuracy and AUC values.

AUC-ROC curves for Dataset-1 and Dataset-2 are presented in Figures 32.2 and 32.3. These figures illustrate the performances of multiple ML models used in our study for disease prediction. The ROC curves are generated using a micro-average approach, which aggregates the contributions from all disease classes. Each curve represents the trade-off between the True Positive Rate (TPR) (or Sensitivity) and the False Positive Rate (FPR) across various threshold values. The AUC (Area Under the Curve), shown in the legend, quantifies the model's ability to differentiate between classes, with higher AUC values indicating better performance. Here, the X-axis (False Positive Rate – FPR) represents the proportion of negative cases incorrectly classified as positive, the Y-axis (True Positive Rate – TPR) represents the proportion of positive cases correctly classified. Each curve corresponds to the ROC curve of a specific model, allowing for a direct comparison of how each classifier performs across various thresholds. The dashed diagonal line in both figures represents the random chance baseline (AUC = 0.50). The ROC curves positioned significantly above this line

confirm the models' predictive capabilities. In Figure 32.2, all models show nearly perfect performance, with most achieving an AUC of 1.00 and k-NN slightly lower at 0.99. This reflects excellent model performance. In Figure 32.3, it is seen that the models exhibit more variance in their performance. While SVM and RF maintain perfect AUC values of 1.00, other models such as KNN, LR, DT, and NB achieve AUCs ranging from 0.95 to 0.97. This indicates slightly reduced but still strong performance across all models.

In conclusion, RF has been found to be the most consistent high performer across both datasets, achieving the best balance of accuracy, recall, precision, and AUC-ROC scores. It demonstrates strong predictive power with minimal overfitting, making it the best choice for both Dataset 1 (symptom-based) and Dataset 2 (biomarker-based). SVM is another solid performer, closely following RF, particularly in Dataset 2, but it slightly lags in terms of generalisation. Other models, such as DT, KNN, LR, and NB, show weaker results, with k-NN and NB especially underperforming. Overall, for robust performance across both datasets, RF is recommended, with SVM as a reliable alternative. MLP also offers a balance between performance and runtime efficiency.

Conclusion

This study shows that machine learning models, particularly Random Forest and SVM, are effective at predicting multiple diseases using both symptom-based and biomarker-based data. Both datasets showed high accuracy, making these models useful in real-world healthcare applications. In the future, researchers could explore using more advanced deep-learning techniques and more disease data to improve the models. Additionally, these models could be turned into web applications, allowing healthcare providers to quickly enter patient information and get real-time disease risk predictions, thereby helping them make faster and more accurate decisions.

References

[1] Grampurohit, S., Sagarnal, C. (2020). Disease prediction using machine learning algorithms. In 2020 International Conference for Emerging Technology (INCET). IEEE, 1–7.

[2] Arumugam, K., Naved, M., Shinde, P. P., Leiva-Chauca, O., Huaman-Osorio, A., Gonzales-Yanac, T. (2023). Multiple disease prediction using machine learning algorithms. *Materials Today: Proceedings*, 80, 3682–3685.

[3] Raj, M. K., Malardhas, J. P., Devapriya, I. (2024). Machine learning approach to predict multiple diseases based on symptoms. In 2024 10th International Conference on Communication and Signal Processing (ICCSP). IEEE, 1195–1199.

[4] Rout, A. K., Sethy, A., Rani, V. V. (2024). Predicting disease risk with machine learning: a comparative study of classification algorithms. In 2024 International Conference on Advances in Modern Age Technologies for Health and Engineering Science (AM ATHE). IEEE, 1–7.

[5] Peterson, Leif E. (2009). K-nearest neighbor. *Scholarpedia*, 4(2), 1883.

[6] Song, Yan-Yan, L. U. Ying. (2015). Decision tree methods: applications for classification and prediction. *Shanghai Archives of Psychiatry*, 27(2), 130.

[7] Kleinbaum, David G., et al. (2002). *Logistic Regression*. New York: Springer-Verlag.

[8] Hearst, M. A., Dumais, S. T., Osuna, E., Platt, J., Scholkopf, B. (1998). Support vector machines. *IEEE Intelligent Systems and their Applications*, 13(4), 18–28.

[9] Rish, Irina. (2001). An empirical study of the naive Bayes classifier. In IJCAI 2001 Workshop on Empirical Methods in Artificial Intelligence, 3(22).

[10] Breiman, Leo. (2001). Random forests. *Machine Learning*, 45, 5–32.

[11] Gardner, Matt W., Dorling, Stephen R. (1998). Artificial neural networks (the multilayer perceptron)—a review of applications in the atmospheric sciences. *Atmospheric Environment*, 32, 14–15, 2627–2636.

33 Modelling stock prices using XGBoost, Prophet, and ARIMA: a comprehensive evaluation

K. B. Vaisshnavi[a] and A Padmavathi[b]

Department of Computer Science and Engineering, Amrita School of Computing, Amrita Vishwa Vidyapeetham, Chennai, India

Abstract

Accurate stock price prediction is necessary for analysts and investors, supporting well-informed decisions in the evolving markets. This research compares three models such as XGBoost, Prophet, and ARIMA for predicting stock values based on previous years data, which includes volume, date, and opening and closing prices. Addressing the missing values, misfits, and ensuring stationarity is required in data pre-processing. The algorithms used are very useful in capturing non-linear relationships, model seasonality, and handling linear trends. The strengths and weaknesses of each model are highlighted in this study including the vital importance of model selection for stock price prediction. The research insight intends to predict accuracy and improve financial forecasting using recent machine learning and statistical techniques.

Keywords:Stock price prediction, model selection, XGBoost, Prophet, ARIMA

Introduction

Forecasting stock prices is an important activity in the financial market that allows the investors, analysts, and institutions to make the right decisions regarding trading, portfolio management, and risk evaluation. In today's data-driven world, accurate forecasting can help with identifying opportunities for profitable investment, reducing uncertainty, and allocating resources efficiently. The high volatility of stock markets, caused mainly by economic, political, and social factors, makes precise forecasting a challenging task. Market fluctuations come owing to global economic changes, corporate earnings announcements, geopolitical events, and even public sentiment driven from news and social media. Because these factors are unexpected, sophisticated modelling methods that can capture both linear and non-linear stock price movements are required.

Volatility of the stock market arises from unpredictable economic, social, and political conditions such as inflation, interest rates, company profits, and opinion. Sudden changes due to events such as financial crises and geopolitical realignments complicate short- and long-term predictions. The integration of structured data (prices, quantities) and unstructured data (news, social media) also complicates the accuracy in predictions. Sophisticated machine learning and time-series algorithms aid in detecting underlying patterns, evolving to capture the dynamic nature of financial markets.

Conventional methods like linear regression and moving averages have been applied extensively in stock market predictions. Nevertheless, the models tend to miss advanced interdependencies, abrupt market changes, and non-linear financial patterns. Moreover, although research has utilised machine learning algorithms like neural networks, they are computationally intensive and overfit. To fill these gaps, our research compares three strong prediction models—XGBoost, Prophet, and ARIMA—each of which is built to handle a different side of stock price prediction.

The motivation for this work comes from the growing need for more interpretable and accurate financial forecasting models. XGBoost, a gradient boosting method, is able to capture non-linear dependencies and feature relationships well, so it is ideally suited for unpredictable market environments. Prophet, the time series model developed by Facebook, attempts to capture long-term trends along with the seasonal patterns. ARIMA is the traditional statistical model for time-series forecasting with linear trends. By comparing these three models, this study will provide real recommendations to traders, investors, and even financial analysts about choosing the best model, depending on the market conditions and the forecasting needs.

In contrast to previous research involving a single model or deep learning with excessive computational requirements, this study compares various algorithms based on data pre-processing, performance, and interpretability. We examine how RMSE, MAE, and R^2 indicate model accuracy in both stable and volatile markets. We also propose a hybrid model incorporating XGBoost, Prophet, and ARIMA to enhance accuracy. This combination utilises a powerful feature of machine learning and interpretability of statistical models, creating a scalable, flexible stock price forecasting system for real-world applications.

In this study, we aim at a more exhaustive review of these models so as better-informed decision-making

[a]kbvaisshnavi@gmail.com, [b]padmavathi@ch.amrita.edu

DOI: 10.1201/9781003675235-33

by market players. On the same ground, it shall form a basis for future developments in stock market prediction methods.

Related works

The ARIMA (AutoRegressive Integrated Moving Average) model will be the most suitable model that can predict the stock prices based on previous time series data. Such models are indeed useful as they identify linear patterns and trends in univariate datasets and will be useful in forecasting. Its advantages include simplicity and accuracy in the modelling of linear data, while the disadvantages may involve its inability to treat non-linear relationships and to require much care in the tuning of parameters [1]. The stock price forecasting uses a methodology, which integrates CNN, BiLSTM, and attention mechanisms. Spatial heterogeneity is captured through CNN, temporal dependencies are modelled through BiLSTM, while attention highlights the important patterns. It is a method for handling datasets that are complex but resource intensive [2]. Neural networks are applied to stock price prediction by learning patterns from historical data. They excel at modelling non-linear relationships and adapting to complex market dynamics. However, they require extensive data and computational power for accurate predictions [3]. A review of machine learning methodologies and algorithms for stock price prediction would lay bare the analysis of historical data and show predictive patterns on the market. They are flexible and accurate in adjusting to varying conditions. Their demerit, however, will emanate from extremely high data variability and size requirements for the dataset [4].The study on the history of the market has been pursued with an aim to identify trends and patterns that have consequently succeeded in increasing stock price prediction accuracy. It processes large amounts of complex data, determines hidden dependencies, and generally enhances prediction accuracy. Though this has its advantages, these can be outweighed by challenges such as handling changing behaviours, complex computational processes, and even the pitfalls of overfitting [5]. Prediction of stock prices or sales in the foreign exchange market is done using deep learning and hybrid neural network models by analysing market indicators along with historical data. The use of neural network-based methods, along with deep learning techniques, would help improve accuracy [6,7]. Benefits include improved accuracy, while quickly adapting to changes in a market, and efficiently handling of complex datasets. However, challenges involve the "heavy" processing requirement, as well as a high level of sensitivity towards data quality, not to mention the risk of overfitting with complex models [8,23].

Neural networks in hybrid modes, such as moving averages and relative strength index (RSI), are used to predict stock prices. This historical stock data is used to train the model neural network with many technical indicators, which is aimed at improving accuracy in the prediction. It helps consider the complex patterns that appear in stock markets. The model's hybrid approach aims at improving forecasting performance through the use of historical and market data [9]. Forecasting stock prices is done through the integration of neural networks and hybridised market indicators such as moving averages and relative strength index (RSI). The networks then use these technical indicators in conjunction with neural networks to provide better forecasting accuracy. Improved prediction and the ability to identify complex trends in the market are some of the defining advantages. However, the data requirement turns out to be huge and thus computational costs are high; overfitting is also one other aspect of this approach [10,26]. The advanced deep learning models on stock price prediction are LSTM, RNN, and CNN, which use its sliding window approach for capturing both time and space dependency. Deep learning technologies have also applied complex patterns into models from past stock data [11,27]. These models identify complicated market trends, which improve prediction accuracy, but they require large datasets and huge computational power along with introducing overfitting in training if not properly tuned or regularised [12]. Reinforcement learning is applied to stock price prediction by training an agent to make decisions based on market conditions and historical data. The algorithm acquires knowledge from performance evaluation and subsequently makes itself improved at predicting future stock prices. For this reason, this can be adaptive in real-time decision-making in moving market environments. However, plentiful data are needed, and high computation is one of the disadvantages with possible performance failure if not properly trained [13,24]. In doing so, an initial portion of this model will include familiarity with the markets to foster an increased model accuracy by employing historical knowledge. Besides, text-based-models use textual data, such as news articles or accounts of financial reports, to help them predict stock price movements through analysing the sentiment extracted from such texts [14,29]. These approaches improve prediction accuracy but require high-quality data and are computationally expensive, making it inherently challenging to handle noisy and missing data information [15].

Methodology

The research methodology will be set out to introduce and statistically apply three models—XGBoost, Prophet, and ARIMA—for price prediction in stock. It starts from the data preparation, wherein historical stock price observations contain factors like opening price, closing price, volume, and date. Then, this data was divided into training and testing portions in a way that the models could be trained on past data and tested on unseen data for reliable evaluation. After preparing the dataset, data pre-processing is performed to remove missing values and detect outliers, as well as to provide the data with properties of stationarity for time series analysis. Each model is implemented according to its peculiar capabilities: XGBoost takes care of all those complex non-linear relationships; Prophet handles seasonality and trends; and ARIMA gives linear time series forecasts. Finally, a comparative analysis is carried out among the models with these metrics: Root Mean Squared Error (RMSE), Mean Absolute Error (MAE), and R-squared (R²), and this will highlight the merits and demerits of different models under different market conditions. This structured methodology guarantees a holistic view in the evaluation of the performance of the models in turbulent financial markets.

Dataset

In this research, the publicly available financial data repository has multiple datasets that provide complete historical stock price data. Each data sample has an entry capturing the date, opening price, closing price, highest price, lowest price, and trading volume; these capture critical and elaborate activities in the stock market for analysis and forecast purposes. The data is structured in tabular form where rows correspond to daily observations of stocks and columns represent features critical to prediction efforts. Number of rows (data points): Stock price datasets usually consist of 1,000 to 50,000 rows depending on the interval (daily, weekly, monthly). Number of columns (features): Critical features: Date, Open Price, Close Price, High, Low, and Volume. The dataset has been divided into two parts: the training and testing parts, most of which has representation in historical data for model training and validation. Also, the training set is used to fit models while the testing set is used to evaluate their predictive performance on unseen data.

Data pre-processing and analysis

Data pre-processing is one of the primary stages that prepares the entire dataset to model the task of prediction rightly and also ensures accuracy and reliability in the predictions. The historical stock price dataset of the project contains different features such as date, open price, close price, high price, low price, and trading volume. The pre-processing steps were structured as follows:

- Data conversion and indexing: The Date column was transformed into date–time format so that it could be analysed for time series data. This column was also set as index so that trends in time can be easily determined using the likes of Prophet and ARIMA.
- Feature selection: Relevant features in this dataset, such as the open, close, high, low prices, and volume, were selected as they played an integral part in pre-

dicting stock prices. All irrelevant or duplicate columns were removed to simplify the analysis.

- Handling missing values: Lost data points were detected and addressed using interpolation procedure within series data to maintain continuity. The method ensured that the model was not disrupted.
- Outlier detection and removal: Box plots were used to visualise potential outliers in stock price and volume data. Excluding these anomalies made certain that the models were not biased/misled due to extreme values.
- Stationarity check: As far the stationarity of a time series is concerned, tests such as the Augmented Dickey-eyFuller (ADF) test are used to check it before applying both differencing and moving averages to stabilise the mean and variance of incoming data. This will help the data meet the assumptions required for an ARIMA model. Feature scaling: So, for stock prices and volumes, scaling was done to ensure numerical stability for models like XGBoost and so forth, allowing improvements in convergence or performance of the model.

Exploratory data analysis (EDA): The analysis will be used to show all pattern, trends, and seasonality in stock price dataset. Line graphs would show the time-based trends, while correlation heatmaps can show relationships between different attributes, such as volume and price changes. The patterns from the data are extracted using seasonal decomposition, which will enable models like Prophet to capture the trends better. These pre-processing and analytical procedures were actually optimised for such dataset for implementing XGBoost, Prophet, and ARIMA as making a good foundation for accurate forecasting on stock prices.

Comparison of other algorithms

XGBoost, Prophet, and ARIMA models were analysed using metrics such as RMSE, MAE, and R-squared for performance in predicting; hence, the results offer an

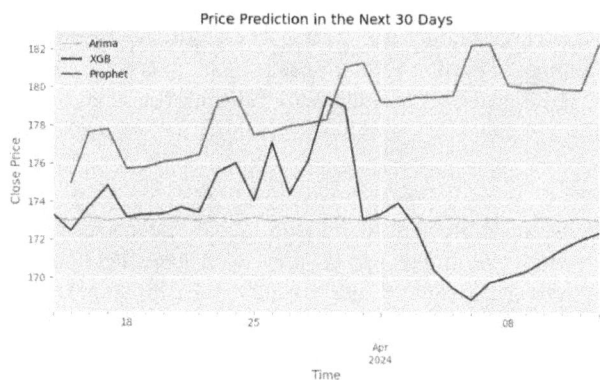

Figure 33.1 Stock price prediction
Source: Author

insight into each model's accuracy, precision, and the ability to generalise. These are put through their strengths and weaknesses regarding performance during periods of increased volatility, seasonal trends, and flat phases. By using systematic comparative analysis, the research determined the best model for predicting stock prices under various conditions and hence provides a useful insight in financial forecasting and helps to develop reliable and robust prediction techniques in dynamic financial markets.

Results and discussion

The results of this study provide an in-depth evaluation of the predictive capabilities of the XGBoost, Prophet, and ARIMA models in forecasting stock prices. Each algorithm's performance was assessed based on metrics such as Root Mean Squared Error (RMSE), Mean Absolute Error (MAE), and R-squared (R^2) through comprehensive data pre-processing and model training.

Model performance
The highest accuracy was obtained by the XGBoost model, with an accuracy of 98.24%, showing its high capability to capture complex non-linear relationships in the dataset and making it a good indicator of robustness of models in utilising features such as trading volume, opening and closing prices for accurate prediction, thereby suitable for dynamic volatile market conditions. ARIMA and Prophet models performed similarly, as both achieved accuracies of 93.85% and 93.17%, respectively. ARIMA was good in stable conditions because it can deal with linear relationships and stationary time series data. However, its inability to capture the non-linear trends led to less accuracy than XGBoost. Prophet was excellent for modelling seasonality and long-term trends, especially where the periodic pattern is quite clear, but its performance decreased in highly volatile situations.

Comparative analysis
The comparative analysis showed that XGBoost has the maximum predictive ability and adaptability by reaching an accuracy rating of 98.24%. The ability to understand complex non-linear relationships and integrate all the features, such as opening price, closing price, and trading volume, adds to the effectiveness of that algorithm while at play in dynamic and volatile market conditions. Along with its other advantages, XGBoost's gradient boosting framework helps with viable feature selection, reduces overfitting, and assures good generalisation on unseen data. ARIMA and Prophet, for example, achieved 93.85% and 93.17%, respectively, but were useful in a very narrow space, unable to generalise into the unpredictable behaviour domain of the markets. ARIMA did well in stable, linear trends and poorly in non-stationary,

non-linear patterns, while Prophet did well on seasonality and long-term trends, whereas there were limitations regarding modelling sudden shifts in the market. Thus, XGBoost turns out to be the champion of flexibility and cutting-edge features, proving to be an excellent choice for stock price prediction.

Discussion

The results highlight the importance of narrowing the model selection to specific characteristics of the market type and the forecasting objectives. In this regard, XGBoost, with an unmatched accuracy of 98.24% regarding the handling of non-linear relationships as well as the insightfulness from it, stands out as a very good choice in highly dynamic and volatile markets, where changes in stock prices occur rather fast, thus requiring a much deeper adaptability. Multiplicity of features integration, for instance, trading volume and historical prices, adds potency to its prediction. Prophet, on the contrary, shows good performance in datasets with clear, seasonal cycles and long-term trends, and it may be a very simple implementation for periodic behaviour time series. Its performance is reduced substantially, however, when confronted with an abrupt market transition or very high volatility. ARIMA has little flexibility considering dynamic conditions, but it is best suited to a stable market with a linear approach, given the performance guarantees from strict reliance upon stationarity. These insights demand a clear-cut approach for effective model selection because each method has its unique strengths in respective aspects of market consideration.

XGBoost performed better than Prophet and ARIMA because it could retain non-linear relationships, deal with volatility in the market, and capture multiple features, such as trading volume and past prices, and reached 98.24% accuracy. Prophet did not handle abrupt price movements well since it assumes seasonality, whereas ARIMA was

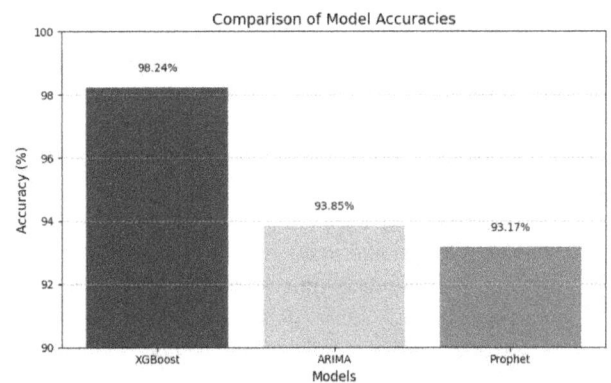

Figure 33.2 Graph representation of the comparison of algorithms
Source: Author

restricted by its stationarity condition and weak non-linear trend handling. XGBoost's gradient boosting framework and regularisation methods rendered it more versatile to respond to dynamic stock price fluctuations, imparting to it a considerable advantage over the remaining models.

Future research may involve hybrid models that mix the benefits from all these three systems: XGBoost, Prophet, and ARIMA, to become an all-inclusive predictive framework. The great potential lies in the integration of the non-linear adaptability of XGBoost with seasonal trend modelling from Prophet and that ARIMA could model the presence of linear relationships, tethering to an even higher predictably operating from different spaces. This will add some outside perspective macro-economic indicators, geopolitical, and news sentiments to provide a more comprehensive view of the market. Real-time and adaptive learning involve models much better in catching sudden market idiocies making these models very much more applicable in practice. Future works opened up to ensemble strategies and machine learning advancements may someday pave the road to more robust, scalable, and trustworthy forecasting tools in financial markets.

Conclusion

The present research provides comparative studies of the performances of the three major prediction methodologies that include XGBoost, Prophet, and ARIMA, with respect to stock price prediction as evaluated through well-known metrics of prediction error, namely RMSE, MAE, and R-squared. The results finally indicated that XGBoost had better predictions compared to former models, achieving accuracy levels of up to 98.24%, thus making it the most relevant choice for dynamic and volatile markets. Prophet and ARIMA were found to achieve accuracies of 93.17% and 93.85%, respectively, while they individually will also work well under specific situations whereby Prophet is the best towards a pattern with seasonality and tendencies, whereas ARIMA gives good results when conditions are quite constant with linear trends. These results indicate the utmost primacy in model selection following market-determined properties and forecasting requirements. XGBoost is versatile and very efficient and can, therefore, be accepted as a good tool for extending applications in areas of financial forecasting-broadly, especially in ever-changing markets. At the same time, Prophet and ARIMA have their teams and benefits complementary to their capability when a seasonal trend or linear stability is at work. Future studies may consider combined model building that puts together all unique strengths of these models, deserves the guide for future work, and accompanies external data from news sentiment and macroeconomic variables towards improved forecasting reliability. They may be manipulated with real-time data and dynamic learning technologies for an improved responsiveness of the models to sudden market shifts, thus making them more relevant in dynamic financial environments. This study contributes to the progress of financial forecasting by providing guidance in model selection and optimisation, setting the stage for the development of more robust and accurate methodologies for stock price forecasting.

References

[1]   Ariyo, Adebiyi A., Adewumi, Adewumi O., Ayo, Charles K. (2014). Stock price prediction using the ARIMA model. In 2014 UKSim-AMSS 16th International Conference on Computer Modelling and Simulation. IEEE.

[2]   Lu, Wenjie, Jiazheng Li, Jingyang Wang, Lele Qin. (2021). A CNNBiLSTM-AM method for stock price prediction. *Neural Computing and Applications*, 33(10).

[3]   Schoneburg, Eberhard. (1990). Stock price prediction using neural networks: a project report. *Neurocomputing*, 2(1).

[4]   Obthong, Mehtabhorn, No ngnuch Tantisantiwong, Watthanasak Jeamwatthanachai, Gary Wills. (2020). A survey on machine learning for stock price prediction: algorithms and techniques.

[5]   Leung, Carson Kai-Sang, Richard Kyle MacKinnon, Yang Wang. 2014). A machine learning approach for stock price prediction. In Proceedings of the 18th International Database Engineering & Applications Symposium.

[6]   Yu, Pengfei, Xuesong Yan. Stock price prediction based on deep neural networks. *Neural Computing and Applications*, 32(6) (2020)

[7]   Soni, Payal, Yogya Tewari, Deepa Krishnan. (2022). Machine learning approaches in stock price prediction: a systematic review. *Journal of Physics: Conference Series*, 2161(1). IOP Publishing.

[8]   Hu, Zexin, Yiqi Zhao, Matloob Khushi. (2021). A survey of forex and stock price prediction using deep learning. *Applied System Innovation*, 4(1).

[9]   Adebiyi, Ayodele A., et al. (2012). Stock price prediction using neural network with hybridized market indicators. *Journal of Emerging Trends in Computing and Information Sciences*, 3(1).

[10] Adebiyi, Ayodele A., et al. (2012). Stock price prediction using neural network with hybridized market indicators. *Journal of Emerging Trends in Computing and Information Sciences*, 3(1).

[11] Selvin, Sreelekshmy, et al. (2017). Stock price prediction using LSTM, RNN and CNN-sliding window model. In 2017 International Conference on Advances in Computing, Communications and Informatics (ICACCI). IEEE.

[12] Ji, Xuan, Jiachen Wang, Zhijun Yan. (2021). A stock price prediction method based on deep learning technology. *International Journal of Crowd Science*, 5(1).

[13] Lee, Jae Won. (2001). Stock price prediction using reinforcement learning. In 2001 IEEE International Symposium on Industrial Electronics (ISIE) Proceedings (Cat. No. 01TH8570). vol. 1. IEEE.

[14] Kohara, Kazuhiro, et al. (1997). Stock price prediction using prior knowledge and neural networks. *Intelligent Systems in Accounting, Finance & Management*, 6(1).

[15] De Fortuny, Enric Junque, et al. (2014). Evaluating and understanding text-based stock price prediction models. *Information Processing & Management*, 50(2).

[16] Cakra, Yahya Eru, Bayu Distiawan Trisedya. (2015). Stock price prediction using linear regression based on sentiment analysis. In 2015 International Conference on Advanced Computer Science and Information Systems (ICACSIS). IEEE.

[17] Hossain, Mohammad Asiful, et al. (2018). Hybrid deep learning model for stock price prediction. In 2018 IEEE Symposium Series on Computational Intelligence (SSCI). IEEE.

[18] Mohan, Saloni, et al. (2019). Stock price prediction using news sentiment analysis. In 2019 IEEE Fifth International Conference on Big Data Computing Service and Applications (BigDataService). IEEE.

[19] Ghosh, Achyut, et al. (2019). Stock price prediction using LSTM on Indian share market. In Proceedings of 32nd International Conference, vol. 63.

[20] Zhang, Jing, et al. (2018). A novel data-driven stock price trend prediction system. *Expert Systems with Applications*, 97.

[21] Zhang, Liheng, Charu Aggarwal, Guo-Jun Qi. (2017). Stock price prediction via discovering multi-frequency trading patterns. Proceedings of the 23rd ACM SIGKDD International Conference on Knowledge Discovery and Data Mining.

[22] Khan, Zabir Haider, Tasnim Sharmin Alin, Md Akter Hussain. (2011). Price prediction of share market using artificial neural network (ANN). *International Journal of Computer Applications*, 22(2).

[23] Padmavathi, A. , Sarker, D., (2023). RecipeMate: a food media recommendation system based on regional raw ingredients. In 2023 14th International Conference on Computing Communication and Networking Technologies (ICCCNT), Delhi, India, 1–6. doi: 10.1109/ICCCNT56998.2023.10307728.

[24] Padmavathi, A., Kumar, D. P., Reddy, K. N. K. (March 2024). Gaze based online examination cheat detection with future integration of deep learning models. In 2024 11th International Conference on Reliability, Infocom Technologies and Optimization (Trends and Future Directions) (ICRITO). IEEE, 1–5.

[25] Padmavathi, A., Amrutha, G., Sah, R. K., Chapagain, B., Manasa, A. (2024). Performance evaluation of movie-based recommendation systems using hybrid machine learning models. In 2024 5th International Conference on Mobile Computing and Sustainable Informatics (ICMCSI), Lalitpur, Nepal, 195–201. doi: 10.1109/ICMCSI61536.2024.00035.

[26] Mailni, M., Srivastava, P., Soni, H., Hemprasanna, Pillai, A. S. (2024). Towards suicide prevention: a natural language processing and machine learning approach integrated with chatbot. In 2024 International Conference on Automation and Computation (AUTOCOM), Dehradun, India, 181–186. doi: 10.1109/AUTOCOM60220.2024.10486154.

[27] Padmavathi, A., Gupta, A., Prakash, K. B. S. (2024). Crop recommendation and yield prediction using machine learning based approaches. In 2024 5th International Conference on Recent Trends in Computer Science and Technology (ICRTCST), Jamshedpur, India, 302–309. doi: 10.1109/ICRTCST61793.2024.10578531.

[28] Padmavathi, A., Kavyasree, N., Sanjana, N., Devi, M. G.. (2024). Comparative analysis of convolutional neural network architectures for pneumonia diagnosis from chest x-ray images. In 2024 15th International Conference on Computing Communication and Networking Technologies (ICCCNT), Kamand, India, 1–7. doi: 10.1109/ICCCNT61001.2024.10725068.

[29] Kumar, S., Muthaiah, U., Roy, R. V. Harnessing machine learning-based classification techniques to optimize crop and fertilizer recommendations. *IFIP Advances in Information and Communication Technology*, 718, 265–275. doi: 10.1007/978-3-031-69986-320.

34 Advancements in Sugarcane Disease Detection and Prediction

Ankita Goyal Agarwala[1,a] and Mirzanur Rahman[2,b]

[1]Research Scholar, Department of IT, Gauhati University & Assistant Professor, Department of CSE, The Assam Royal Global University, India

[2]Assistant Professor, Department of IT, Gauhati University, India

Abstract

The economic growth of a country like India predominantly relies on its agricultural sector. India ranks second globally in sugarcane production, a vital cash crop with significant demand due to its by-products. However, various diseases reduce the yield and result in substantial economic losses. Manual identification of these diseases is time-consuming and often inefficient. This paper reviews the work that has been carried out on sugarcane disease identification, identifies the lapses of the existing system, and proposes some novel hybrid architectures that combine the advantages of typical convolutional neural networks along with SVMs and Vision Transformers.

Keywords: Sugarcane diseases, image analysis methods, artificial algorithms, deep learning

Introduction

India's farming industry has experienced substantial expansion, contributing substantially to its economy. Two-thirds of the population relies on agriculture for their livelihood [1]. Sugarcane, being a major cash crop, supports the production of sugar, jaggery, molasses, biofuels, and livestock feed [2]. Despite its importance, various diseases drastically affect sugarcane yield, causing economic setbacks. Traditional manual approaches for identifying diseases are labour-intensive and lack accuracy. Hence, an urgent need for automated, efficient systems for early disease detection is required.

Motivation

Apart from producing milk, pulses, and jute, India stands second at producing crops like sugarcane, wheat, rice, amongst others [3]. Sugarcane contributes significantly to India's economy due to its versatile applications. Despite advancements in agriculture, diseases such as red rot, smut, and yellow leaf disease continue to hinder its production, thereby affecting the overall economic growth. Early and accurate detection systems are imperative to mitigate these challenges and ensure sustainable yield.

Common Types of Sugarcane Diseases

Table 34.1 summarises the prevalent sugarcane diseases [4], their symptoms, and favourable conditions for proliferation.

The diseases discussed in the above table are depicted in Figure 34.1. Figure 34.2 depicts the healthy plants so as to differentiate with the diseased ones.

Objectives

The primary objectives of this work include:

- To perform a detailed analysis of the work done so far in the field.
- To identify the gaps in the existing systems.
- To analyse the behaviour of a model that will be found best from the survey for sugarcane disease prediction
- To propose some hybrid architectures for the same.

Background Study

To detect the diseases of sugarcane at an early stage, various computational techniques are being used. Starting from thresholding techniques, template matching techniques to typical machine learning algorithms, deep learning, and expert system techniques, researchers have till date tried every possible method of prior disease recognition and prediction. The datasets utilised for this purpose differs from numerical or ordinal data to typical RGB images to hyperspectral images. A lot of advancements have been achieved in this regard. An overview of all such works is reported below:

In 2014, Ratnasari et al. [6] proposed a leaf-based disease identification technique that uses 30 segmented spot images and SVM was employed for categorisation with ROI identification using L*a*b colour space and GLCM for texture feature extraction. Diseases like rust spot, yellow spot, and ring spot were identified with an accuracy of 80%. However, the methodology seems outdated with a low accuracy rate, and the dimension of the dataset was insufficiently large.

[a]ankita89.goyal@gmail.com, [b]mirzanurrahman@gmail.com

DOI: 10.1201/9781003675235-34

Table 34.1 Common sugarcane diseases and their symptoms.

Disease	Symptoms	Favourable conditions
Red rot	Dried spindle leaves, smudged stalks	Air, rain splash, and soil transmission
Wilt	Yellowing leaves, thin canes	Soil, wind, rain, and irrigation water
Grassy shoot	Slender, yellowing leaves	Diseased seeds, insects, root parasites
Smut	Belt-like structures, thin internodes	Diseased seeds, spores in soil
Red stripe	Dark red lesions, brown stripes	Wind and rain transmission
Mosaic	Chlorotic, mosaic-patterned leaves	Diseased seed material
Rust	Small yellowish spots	Wind and water splash
Yellow disease	Yellowing of leaf midrib	Diseased seeds
Pokkah boeng	Wrinkling, twisting leaves, dead spindles	Air currents, infected setts

Source: Author

Figure 34.1 Visual representation of sugarcane diseases
Source: Author

Figure 34.2 Healthy sugarcane plants
Source: Author

Eaganathan et al. [7] used a hybrid model for the identification of sugarcane leaf scorch disease with typical image processing techniques. After pre-processing the images for noise removal and smoothening, it uses K-means clustering for segmentation purpose and k-NN was used for feature extraction. This method reported an accuracy of 95%. More hybrid models with different algorithms can be tested for better results. However, only a single disease was considered for identification.

In 2016, Mohanty et al. [8] proposed a deep learning technique to identify 14 crops and 26 diseases. Different sample sizes were trained and tested. AlexNet and GoogleNet architectures of deep learning were used, which reported an accuracy of 85.53% and 99.34%, respectively. However, the system was unable to consider non-homogeneous backgrounds.

Sathiamoorthy et al. [9] used data mining techniques to detect sugarcane disease. With the help of WEKA tool, algorithms like Multi-layer Perceptron, J48, and K-means were used to detect diseases such as rust spot, yellow spot, and ring spot. Numerical dataset having 130 samples were used and K-means performed better out of the three. The dataset's size was too small, and lacked information on accuracy and precision scores.

In 2019, Militante et al. [10] introduced a different deep learning approach for identifying sugarcane diseases. Diseases like grassy shoot, smut, rust, and yellow leaf disease were recognised through 13,842 images using the CNN algorithm, in which the RELU activation function was used. This method had shown an accuracy of 95%.

Using Hidden Markov Model and anisotropic diffusion algorithm, diseases like rust spot, yellow spot, and ring spot were detected with the CNN classification algorithm by Pawar and Talukdar [11]. This work also developed an application where disease prediction was carried out at the backend. However, details regarding the dataset type, as well as accuracy and precision scores, are not provided.

On a limited sized image dataset, Kumar et al. [12] implemented a hybrid model for disease identification with the conventional image analysis methods. After pre-processing, the K-means algorithm was employed for segmentation, and multi-SVM for classification. Diseases like mosaic disease, red rot, and eye spot were detected with an accuracy of 97%.

In 2019, Bao et al. [13] detected smut disease in sugarcane by using hyperspectral images and deep neural networks (DNN). In addition to this, a Dual Self Attention Block (DSAB) module with ResNet architecture was embedded with DNN to detect even invisible smut symptoms. This combination of spectral and spatial components for identifying diseases gave considerably better results. However, the dataset dimension was relatively smaller, and the work could have been tested for other diseases as well.

Daphal and Koli [14] detected leaf diseases such as rust, eye spot, mosaic, and banded chlorosis using various deep learning techniques. Multispectral images were used, and modifications to certain parameters in VGGNET architecture of CNN model were done. The proposed model achieved an accuracy near that of other models. However, there was no clear mention of the accuracy scores.

Thilagavathi et al. [15] detected diseases like ring spot, mosaic, red rot, eye spot, yellow leaf, and leaf scald using image processing techniques, and a web application was also created for detecting these diseases. They applied a standard image processing method combined with K-means for segmentation and SVM for classification. An overall accuracy of 95% was reported. Different techniques, instead of conventional ones, could have been utilised for segmentation and classification.

In 2020, Srivastava et al. [16] implemented multiple deep learning models, such as neural networks and hybrid AdaBoost, and were compared with algorithms by analysing its leaves, stem, colour, etc. A precision of 90.2% was obtained using VGG-16 as the feature extraction method and SVM as the classifier. A comparison with its counterparts could have well established the method.

D. Dutta [17] presented the recent trends of sugarcane diseases in Assam. The study claimed that fungal infections are more common in Assam compared to viral and bacterial diseases. Yellow leaf disease is the most common disease, and there is an increased trend in diseases like red rot and pokkahboeng. Other diseases like wilt, ring rot, banded sclerotial disease occur in trace amount. The given assertion has already been verified by Begum et al. [18].

Kumar et al. [19] tried collecting images in three different ways, i.e., manually, from Kaggle and UCI machine learning repository, and applied deep learning methods like AlexNet and GoogleNet on those datasets with respect to the CNN model. Algorithms like Quicker RCNN, RCFN, and SSD are also used for object identification. The accuracy for this methodology came around 93%. Though the methodology was robust, the results were not well presented.

In 2021 Rajesh and Gowri [20] developed a system that used the rainfall dataset of Maharashtra and predicted sugarcane diseases with the help of SVM and k-NN algorithms, and also highlighted the disease concerned with that locality which will help the farmers take preventive measures to improve the yield quality. However, limited information was provided about the dataset used, and the reported accuracy scores were subpar.

A comprehensive analysis of the different approaches utilised thus far for identification and prediction of sugarcane diseases was presented by Manavalan [5].

Prithvi R et al. [21] captured 13,842 images for seven different classes manually and applied the CNN model with ReLU activation function. Techniques such as Canny edge detection and thresholding-based segmentation were employed. This work also presents a pesticide recommendation system. The results, however, were presented in a very haphazard manner.

In 2022, Kumpala et al. [22] detected the red stripe disease using the YOLO algorithm of the CNN model. They developed a disease diagnosis and detection system, apart from a system for displaying responses. The work was carried out on 4,000 image samples, out of which 2,000 were diseased and 2,000 were healthy. The system reported a precision of approximately 96%. A unique characteristic of the work was that manual samples were submitted for laboratory testing so as to conform to the results obtained computationally. The system could have been more robust if additional diseases were included.

Narmilan [23]used unmanned aerial vehicle (UAV) multispectral images and attempted to detect white leaf disease using algorithms such as XGBoost, decision tree, random forest, and k-NN. An accuracy of 94% was reported. The concept of vegetation index was used to detect white leaf disease-infected plants from multispectral images.

Gap Analysis

After performing a detailed study of the various approaches that have been utilised so far in the identification of sugarcane diseases, the following analysis was made:

- Firstly, the efficiency of the overall disease identification process depends on building a strong dataset. The major factors to be considered for data collection include the source from which the data is to be collected, the weather/climatic conditions, the device used, the different angles, and the source of light, among others. For better precision, the dataset size has to be very large. Also nowadays, hyperspectral or multispectral images are used for the purpose, which gives nearly accurate results.

- The pre-processing and feature extraction from the raw data has to be done very carefully. Different combinations of features like stem, colour, rainfall, soil quality, texture, moisture, etc. can be extracted and compared for more accurate disease prediction.

- The process of disease identification or recognition depends on the techniques being used, such as data mining methods, machine learning techniques, deep learning etc. It has been noted that out of the mentioned techniques, deep learning techniques implemented with different architectures like AlexNet, GoogleNet, ResNet, VGG, etc. have proved to be the most accurate models for prediction.

- None of the works have tried to highlight the prevention of these diseases through automated systems, and the effect of pesticides can also be studied in this regard.

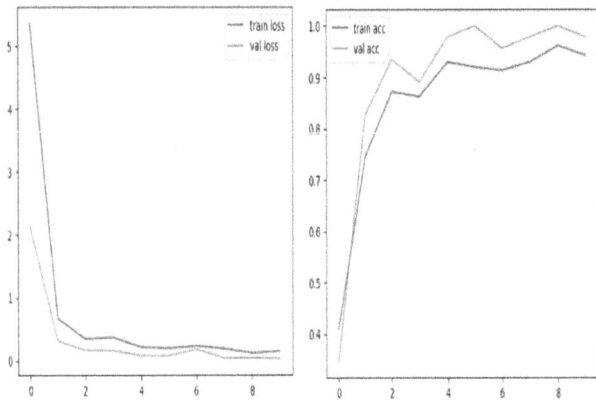

Figure 34.3 Comparative evaluation of accuracy and loss
Source: Author

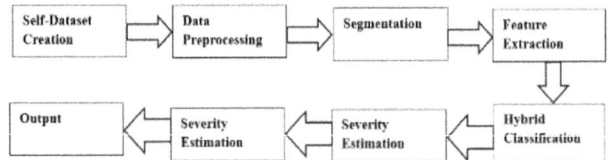

Figure 34.4 Proposed CNN-SVM architecture
Source: Author

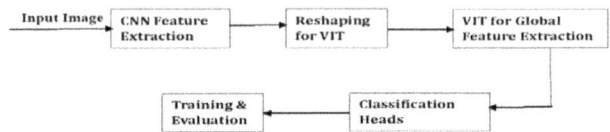

Figure 34.5 Proposed CNN-Vit architecture
Source: Author

- Though Assam is among the primary producers of sugarcane, an image dataset on the varied diseases obtained from the fields is not yet available, and no work on automated disease detection of sugarcane has been reported so far.

Sample Execution

A sample execution was conducted on a downloaded dataset from Kaggle [24] having three classes (Red Rot, Red Rust, and Healthy) using the ResNet50, Inception V3, and VGG16 architectures for 10 epochs. The dataset was already pre-processed. However, the results for Resnet 50 are presented, as it seemed to be the more effective architecture out of the three, with higher efficiency, lower computational costs and a higher learning rate.

A comparative evaluation of accuracy vs loss is shown below in Figure 34.3. It has been noted from the curve that the accuracy is gradually increasing, while the loss is decreasing. Also, the model gradually learns to generalise the data well, and since both the training and validation loss minimises gradually, this indicates that the model has learned meaningful patterns from the training data without overfitting, and can effectively adapt to new, unseen data.

Some Proposed Hybrid Architectures

Creating a novel combined architecture for sugarcane plant disease detection using RGB image datasets can involve combining classical image processing techniques with modern deep learning approaches. Some of the proposed techniques are discussed below:

Proposed cascade-CNN-SVM (CCSNet)
As depicted in Figure 34.4, in the proposed CCSNet architecture, after creating one's own local dataset and pre-processing it, techniques like superpixel segmentation,

thresholding, and masking can be utilised to obtain the ROIs. Further, classical features like texture, colour, and shape can be extracted, along with deep features using lightweight CNN models like MobileNetV3 to extract feature vectors from the segmented ROIs. To create a robust feature set, hybrid classification can be performed using classical SVM with a radial basis function (RBF). Finally, after severity estimation using the conventional methods, the disease class can be predicted as the output with the severity percentage.

Proposed cascade CNN-ViT architecture
Figure 34.5 depicts the architecture involving Vision Transformers (ViTs). ViTs are nowadays extensively used in image processing and have produced outstanding results. In the proposed architecture, a CNN model can be used for extracting features, and the extracted features, after reshaping, can be fed into a ViT model for deep feature extraction, followed by classification and evaluation. It is hoped that this model will prove to be a fruitful one.

Conclusion

Since the Indian economy depends primarily on agriculture and a lot of crops are being wasted because of the presence of various kinds of diseases within them, so there is an urgent need to limit the occurrence of diseases or infections in plants. Among all crops, sugarcane is the most commercially significant. So, early detection and prevention of sugarcane diseases through various automated systems is an essential need for the farmers.

A lot of systems were developed but with various limitations. Therefore, an efficient hybrid system which can help the farmers in real-time to detect infected plants/crops is in high demand. Addressing these lapses, this paper proposes some hybrid methods which can definitely produces superior results compared to similar methods. The future scope will focus on implementing the proposed models.

References

[1] Annapoorna, Shankar. (2023). Invest India. www.investindia.gov.in.

[2] Ministry of Agriculture & Farmers Welfare. (2023). Press Information Bureau. www.pib.gov.in.

[3] Food and Agriculture Organisation. (2023). www.fao.org.

[4] Vikaspedia: Sugarcane Diseases and Symptoms. www.vikaspedia.in.

[5] Manavalan, R. (2021). Efficient detection of sugarcane diseases through intelligent approaches: a review. *Asian Journal of Research and Review in Agriculture*, 27–37.

[6] Ratnasari, E. K., Mentari, M., Dewi, R. K., Ginardi, R. V. H. (2014). Sugarcane leaf disease detection and severity estimation based on segmented spots image. In Proc 2014 Int Conf Information, Commun Technol Syst ICTS 2014, 93–98.

[7] Umapathy Eaganathan, J. S., Lackose, V., Benjamin, F. J. (2014). Identification of sugarcane leaf scorch disease using K-means clustering segmentation and KNN based classification. *International Journal of Advances in Computer Science and Technology (IJACST)*, 3(12), 11–16.

[8] Mohanty, S. P., Hughes, D. P., Salathé, M. (2016). Using deep learning for image-based plant disease detection. *Frontiers in Plant Science*, 7, 1419.

[9] Sathiamoorthy, S., Ponnusamy, R., Natarajan, M. (2018). Sugarcane disease detection using data mining techniques. *International Journal of Research in Advent Technology, Special Issue*, 296–301.

[10] Militante, S. V., Gerardo, B. D., Medina, R. P. (October 2019). Sugarcane disease recognition using deep learning. In 2019 IEEE Eurasia Conference on IOT, Communication and Engineering (ECICE). IEEE, 575–578.

[11] Dr Jyotismita Talukdar, Ms Snehal Pawar. (Apr 2016). Sugarcane leaf disease detection. *International Research Journal of Engineering and Technology (IRJET)*, 06(04).

[12] Kumar, A., Tiwari, A. (2019). Detection of sugarcane disease and classification using image processing. International Journal of Research in Applied Science and Engineering Technology, 7(5), 2023–30.

[13] Bao, D., Zhou, J., Bhuiyan, S. A., Zia, A., Ford, R., Gao, Y. (December 2021). Early detection of sugarcane smut disease in hyperspectral images. In 2021 36th International Conference on Image and Vision Computing New Zealand (IVCNZ). IEEE, 1–6.

[14] Dadabhau Daphal Swapnil, Koli S. M, (February 2020). Identification of sugarcane foliar diseases: methods and datasets. *International Journal of Engineering and Advanced Technology (IJEAT)*. ISSN-2249-8958.

[15] Thilagavathi, K., Kavitha, K., Praba, R. D., Arina, S. V., Sahana, R. C. (2020). Detection of diseases in sugarcane using image processing techniques. *Bioscience Biotechnology Research Communications, Special Issue*, (11), 109–115.

[16] Srivastava, S., Kumar, P., Mohd, N., Singh, A., Gill, F. S. (2020). A novel deep learning framework approach for sugarcane disease detection. *SN Computer Science*, 1(2), 1–7.

[17] Dutta, D. (2020). Recent trends of sugarcane diseases in Assam. *Tropical Plant Research*. ISSN (Online/Offline): 2349-1183/ 2349-9265

[18] Begum, M., Singha, D. D., Bordoloi, B. C. (2016). Trend of sugarcane and jaggery production in Assam and associated problems and prospects. *International Journal of Agriculture & Environmental Science*, 3(6), 15–23.

[19] Kumar, P. (2021). Research paper on sugarcane disease detection model. *Turkish Journal of Computer and Mathematics Education (TURCOMAT)*, 12(6), 5167–5174.

[20] Rajesh, T., Gowri, V. (2021). Enhanced approach for disease prediction in sugarcane crop with the support of advanced machine learning strategies. *Annals of the Romanian Society for Cell Biology*, 16805–16814.

[21] Prithvi, R., Priyanka, S., Keerthana, S., Yashaswini, A. R. (2022). Sugarcane leaf disease prediction and pesticides recommendation system. *International Research Journal of Modernization in Engineering Technology and Science*, 4. e-ISSN: 2582-5208.

[22] Kumpala, I., Wichapha, N., Prasomsab, P. (2022). Sugar cane red stripe disease detection using YOLO CNN of deep learning technique. *Engineering Access*, 8(2), 192–197.

[23] Narmilan, A., Gonzalez, F., Salgadoe, A. S. A., Powell, K. (2022). Detection of white leaf disease in sugarcane using machine learning techniques over UAV multispectral images. *Drones*, 6(9), 230.

[24] Roshita, B. (2023). Sugarcane Leaf Disease Dataset. https://www.kaggle.com/datasets/roshitab/sugarcane-leaf-disease-dataset

35 Impact of machine translation evaluation metrics on low resource languages

Sonithoi Ningombam[a], N Donald Jefferson Thabah[b], Arindam Roy[c] and Bipul Syam Purkayastha[d]

Department to Computer Science, Assam University Silchar, India

Abstract

Machine translation (MT) is one of the most prominent fields in computational linguistics and artificial intelligence (AI). Evaluating MT systems is essential, as it provides key insights into how translation systems and fine-tuned models have improved. While traditional human evaluations provide thorough insights, they are often resource-intensive and costly. To address these limitations, researchers have developed various automatic evaluation metrics that offer fast, cost-effective, and reliable assessments of translation quality. These metrics can be broadly classified into two categories: language-independent and language-dependent, with the latter emphasising the semantic context of sentences. This study examines prominent automatic evaluation metrics, including the language-independent BLEU and chrF, as well as the language-dependent BERT, RoBERTa, and S-BERT, on low-resource Indian Languages (namely: Meiteilon and Khasi). The discussion highlights their methodologies, strengths, weaknesses, and correlations with human evaluations.

Keywords: Machine translation, automatic evaluation, BERT, S-BERT, RoBERTa, BLEU, chrF

Introduction

With technological advancements, people around the world are increasingly getting interconnected via the Internet. Cultural exchanges and communication between countries are becoming more common as people around the world become more interconnected. The linguistic divide across diverse cultures serves as an impediment when individuals from various backgrounds seek to communicate their ideas. Machine translation (MT) is very essential in this context. MT, a branch of computational linguistics, uses computers to convert texts from one language to another. Yet, the output of this MT system should maintain quality standards. So, reliable evaluation methods are established to assess the output quality, which can be performed manually, i.e., by human evaluation/linguist expert or by automatic evaluation (AE). Human evaluation approaches are costly [9]. Thus, researchers and developers resort to automatic evaluation to pursue a more effective and convenient method for assessing MT systems.

MT evaluation metrics, such as BLEU [10], METEOR [1], and the related National Institute of Standards and Technology (NIST) (Doddington, 2002) metric, are becoming more and more important in MT research and development. ROUGE, which was originally designed to evaluate summarisation systems, has now been repurposed for MT evaluation. It assesses the n-gram overlap between the generated and reference summaries, with a focus on recall [13]. ROUGE variations, such as ROUGE-1 and ROUGE-2, evaluate various levels of n-gram matching, whereas ROUGE-L looks at the longest common subsequence. Popović [11]

created chrF, a character n-gram F-score metric that assesses translation quality through character-level matching. chrF is intended to be language- and tokenisation-independent, making it a useful tool for evaluating translations across several languages. Sentence Embeddings using Siamese BERT-Networks [12] makes an important contribution to NLP. It is a modified BERT architecture designed to generate semantically meaningful sentence embeddings. However, S-BERT is only available in a small number of languages, making it difficult to rely on in some situations. In such circumstances, models such as RoBERTa [7] or BERT [2] can be employed to determine sentence semantic similarity.

Since the evaluation has an immense effect on the improvement and fine-tuning of translation models, this study will discuss AE metrics, primarily for low-resource languages such as Manipuri/Meiteilon and Khasi, which belong to Indic languages.

Literature review

About the languages

The native language of Manipur, a state in Northeast India, is Meiteilon, also known as Manipuri. It is a language of the Tibeto-Burman family and is distinguished by its rich morphological features, which include a complex tonal phonological system, an agglutinative structure, and a Subject–Object–Verb (SOV) word order [14]. There are two scripts used in Meiteilon: Meitei Mayek and Eastern Nagari (Bengali) [5]. This study will use the Eastern Nagari script for the AE of the MT system.

[a]sonithoiningombam@gmail.com, [b]jefson08@gmail.com, [c]arindam_roy74@rediffmail.com, [d]bipulsyam@gmail.com

DOI: 10.1201/9781003675235-35

The Khasi language, primarily spoken in Meghalaya, India, is deeply embedded within the region's cultural and social fabric. This language belongs to the Austro-Asiatic family, specifically the Mon-Khmer branch, setting it apart linguistically from other Indian languages. The morphological systems of this language deal with how words are formed and structured in a language, using roots, prefixes, and suffixes to create meaning or show grammatical relationships [16].

Background of the metrics evaluation

Non-contextual metrics evaluation
BLEU

The BLEU approach emerged and adopted in the investigation by Papineni et al. [10] to automatically analyse the efficiency of machine translation systems. The measure of it is made up of two key components: the shortness penalty and the n-gram overlap, as seen in equation (35.1). The n-gram overlap assesses the alignment of 1-grams (individual words), 2-grams (word pairs), and up to 4-grams between the candidate and reference texts. The 1-gram overlap checks for correct word usage, whereas the 4-gram overlap assures coherence in the candidate translation. The Brevity Penalty reduces the score of translations that are shorter than the reference text. This is because n-gram overlap accuracy may yield unreasonably high ratings for short translations that include a major chunk of the reference text.

$$BLEU = Brevity\ Penalty \times n\text{-gram Overlap chrF} \quad (35.1)$$

Unlike conventional word-based metrics like BLEU, chrF evaluates at the character level, providing greater robustness for morphologically rich and low-resource languages. It is computed as the F-score (harmonic mean) of precision and recall over character *n*-grams, as indicated in equation. (35.2):

$$chrF = \frac{(1+\beta^2).P.R}{\beta^2.P+R} \quad (35.2)$$

where, P is precision, which tells how much of the candidate text matches the reference text, R is recall which shows how much of the reference text is matched by the candidate text, and β is a weighting factor.

Contextual metrics evaluation

Bidirectional encoder representations from transformers (BERT) and RoBERTa
BERT is a transformer-based language model that incorporates consideration of both the left and right contexts of words in a sentence, thus facilitating a bidirectional comprehension of context. Transformers predominantly

depend on attention mechanisms for capturing the relationships between words. BERT exclusively employs the encoder component of the transformer architecture. During training, random words in the input text are substituted by tokens, while the model's goal aims to predict these masked tokens [2]. This method enables BERT to learn the connections between words in a sentence and get a better comprehension of the language as a whole. For finding semantic score, the pre-trained BERT model receives the candidate/reference sentences to generate a sentence embedding. Furthermore, the final layer of the encoder layer yields the contextual embeddings of dimension (n × embedding_size), where n is the sentence length. This contextual embedding is then passed to an aggregate pooling function, aggregating all the individual word embeddings to obtain a single vector of dimension (1 × embedding_size), as shown in Figure 35.1. A Robustly Optimised BERT Pretraining Approach (RoBERTa) [7] architecture and process are similar to BERT, but RoBERTa is optimised for more efficient training, aiming to achieve comparable or superior results within shorter training times.

To find the semantic scores, both candidate and reference sentences are passed to the BERT/RoBERTa language model to produce two sets of vectors corresponding to the candidate and reference, respectively. Using cosine similarity, the score is derived. If the score is close to 1, it indicates high similarity between the two sets of sentences.

Sentence BERT(S-BERT)
S-BERT employs pre-trained embeddings, which are vector representations of text that include semantics. Generating sentence embeddings is a document processing approach that converts sentences into real-valued vectors. These vectors represent meaning and are optimised for processing in NLP tasks[12].

To create a fixed-sized sentence embedding, S-BERT applies a pooling technique to the output of a BERT/RoBERTa model. The pooling approach is used to calculate the mean of all output vectors, and the resulting mean is the sentence embedding.

The process of finding similarity is similar to that of the BERT/RoBERTa, as described earlier. In general, two

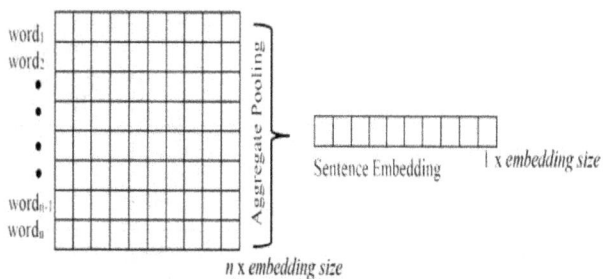

Figure 35.1 Sentence embedding aggregation
Source: Author

highly similar vectors have a cosine similarity of 1, two statistically independent vectors have a similarity of 0, along with two opposing vectors have a similarity of −1. In this evaluation, the cosine similarity result is in a positive space that is neatly bounded by the interval [0,1].

Analysis of metrics evaluation

This empirical study carried out the AE for MT systems. The study analysed the BLEU, chrF, Bert, RoBERTa, and S-BERT for the Indic languages, namely Meiteilon to English, Khasi to English, and vice versa. The data for this analysis was collected from the AI4Bharat[a] for the Meiteilon language, and for Khasi, the data is utilised from Thabah [15]. The collected data further proceeds to the fine tuning from the existing Indic [4] MT system.

Table 35.1 show this study's metric evaluation results. It was observed that the contextual metrics give high scores as compared to the non-contextual metrics. For Indic to English, RoBERTa yields high scores for both languages, i.e., Meiteilon and Khasi. RoBERTa provides the highest

[a]https://ai4bharat.iitm.ac.in/

score for the Meitelon language when translating from English to Indic, but BERT gets the best result for Khasi. S-BERT is not available for Meitelon and Khasi language.

Tables 35.2 and 35.3 demonstrate the scores of various metrics with examples. Figures 35.2 to 35.5 show the 2D t-SNE visualisation of word embeddings for two sentences (reference and predicted) contextual evaluations. The red and blue dots overlap, suggesting that the words in sentence 1 (reference) and sentence 2 (predicted) have comparable embeddings. This shows that the two sentences have high semantic alignment. Words further away in the embedding space (for example, solitary blue or red dots) may reflect semantic or syntactic differences between the two texts. [PAD] and other tokens visible in the visualisation are padding tokens or markers used to process text sequences. These tokens are also integrated and shown here.

In Table 35.2, first row describe the semantic similarity for Meiteilon that predicts "প্রধান মন্ত্রী শ্রী নরেন্দ্র মোদীনা ইদকী কুমহৈদা মহাক্কী য়াইফ পাউজেলে পীখ্রে" translated into English as *"the Prime Minister Shri Narendra Modi has extended his greetings on the occasion of Eid"* while the reference sentence for

Table 35.1 Indic-English and English – Indic metrics.

From (Language)	To (Language)	BLEU	ChrF	BERT	RoBERTa	SBERT
Meiteilon	English	0.28	0.55	0.89	0.98	0.80
Khasi	English	0.35	0.54	0.83	0.98	0.66
English	Meiteilon	0.21	0.53	0.68	0.74	
English	Khasi	0.34	0.51	0.97	0.81	

Source: Author

Table 35.2 Example 1 sentence similarity and metrics score for BERT, S-BERT, RoBERTa, chrf and BLEU (Meiteilon language).

Reference Sentences	Predicted Sentences	BERT	RoBERTa	ChrF	BLEU
ইদকী কুমহৈনা লাকপদা প্রধান মন্ত্রী শ্রী নরেন্দ্র মোদীনা মহাক্কী ওইবা য়াইফা পাউজেলে পীখ্রে	প্রধান মন্ত্রী শ্রী নরেন্দ্র মোদীনা ইদকী কুমহৈদা মহাক্কী য়াইফ পাউজেলে পীখ্রে	0.94	0.96	0.71	0.35
মসগী অপুনবা থবকশংদা য়ু.পি.গী ইমুং খরদি য়ামনা কান্নখ্, শঞ্জা-থুমজা ফজনা তৌনখ্, অদুগা পুর্বাঞ্চলগী অমসুং য়ু.পি.গী মীচম কয়াদি রানা খাঙদুনা অদুমল লনেখ্ৗ	হায়রবি গমে অসদি দাইনাস্তী খরদি চাউখৎখ্ অমসুং শঞ্জা থুমজাগী সাইকল অদু মখা চতথখ্, অদুবু পুর্বাঞ্চল অমসুং য়ু.পি.গী লায়রবা ইমুংশং অসি মাঙখ্	0.96	0.59	0.32	0.08

Source: Author

Table 35.3 Example 2 sentence similarity and metrics score for BERT, S-BERT, RoBERTa, chrf and BLEU (Khasi language).

Reference sentences	Predicted sentences	BERT	RoBERTa	ChrF	BLEU
Phi dei Khynnah Skul ?	Phi dang khynnah?	0.93	0.80	0.34	0.14

Source: Author

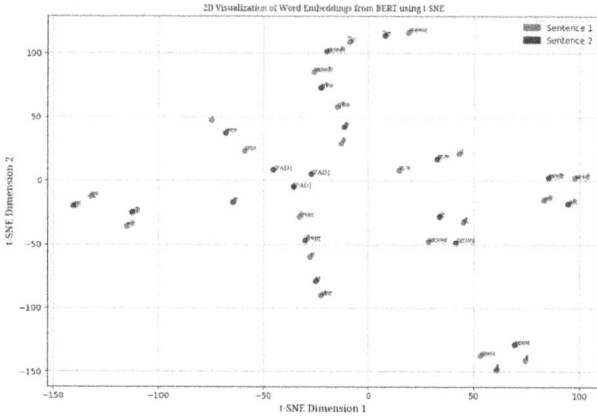

Figure 35.2 2D Visualization of example 1 Word Embedding from using Meiteilon-BERT using t-SNE
Source: Author

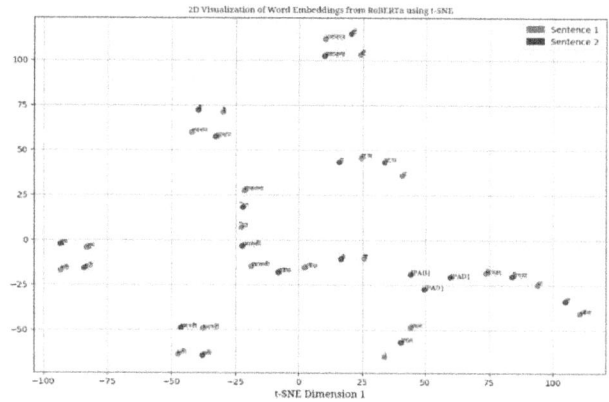

Figure 35.3 2D Visualization of example 2 Word Embedding from using Meiteilon-RoBERTa using t-SNE
Source: Author

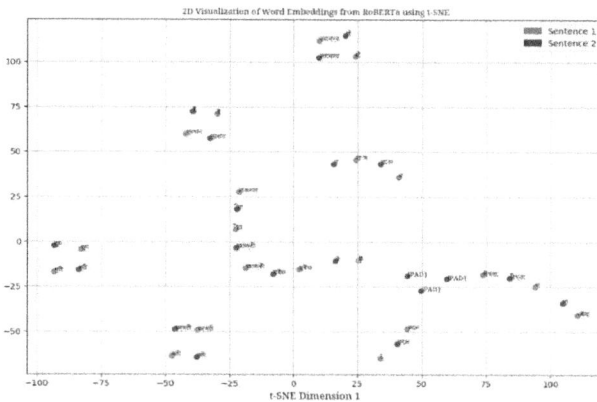

Figure 35.4 2D Visualization of example 3 Word Embedding from using Khasi-BERT using t-SNE
Source: Author

Meiteilon is "ইদকী কুমহৈনা লাকপদা পুরধান মন্ত্রী শ্রী নরন্দ্র মোদীনা মহাক্কী ওইবা য়াইফা পাউজলে পীখ্রে" translated as "*the Prime Minister Shri Narendra Modi has conveyed his greetings on the festival of Eid*".

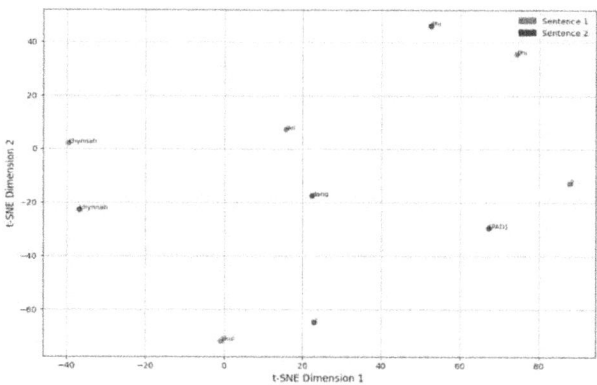

Figure 35.5 2D Visualization of example 3 Word Embedding from using Khasi-RoBERTa using t-SNE
Source: Author

Here, both the predicted and reference sentences convey the semantic similarity. Figures 35.2 and 35.3 show that the equivalent words are projected very close to each other. While Table 35.2, second row shows high evaluation score for the contextual metrics but it sometimes may generate a sentence that does not convey semantically similar meaning. In this table, words like "লনেখি" in the reference are predicted as "মাঙখি" which creates an opposite meaning of the sentence.

Table 35.3 shows the semantic similarity for the Khasi predicted sentence "*Phi dang khynnah?*" translated into English as "*Are you a boy?*" and the reference sentence "*Phi dei khynnah skul?*" translated as "*Are you a school boy?*". Figure 35.4 projects "*Phi*" to be very close, while the rest of the words have a scattered projection. In contrast, Figure 35.5 shows "*Phi*" and "*khynnah*" to be very close in their projection. Semantically, the two sentences are somewhat similar, as they share two common words ("*phi*" (you) and "*khynnah*"(boy)), which the RoBERTa model demonstrates through close projections.

Conclusion

In this work, we investigate sentence similarity using both contextual and non-contextual evaluation metrics, with an emphasis on low-resource languages such as Meiteilon and Khasi. This study observed that contextual BLEU and chrF analyse non-contextual sentences using *n*-gram overlapping, penalising short or incorrect word ordering. Word-level overlaps might differ based on tokenisation and punctuation. BERT and RoBERTa metrics are used to measure similarity in an embedding space, which employ more context and semantics and may provide different scores than BLEU and chrF. And in the context of S-BERT model, it provides the best semantic score because of specifically trained for the task of determining semantic similarity between sentences. However, S-BERT is not available for low-resource languages like Khasi and

Manipuri. Secondly, the RoBERTa model demonstrated better semantic similarity compared to BERT based on vector projections, while BERT achieved a higher semantic score.

The contextual metrics are beneficial when the model generates synonyms during translation but can negatively impact the evaluation if antonyms are predicted. Therefore, a combination of contextual and non-contextual metrics can provide a more balanced and comprehensive analysis and evaluation of translation quality for a model. So, the future research should focus on improving these measures to better address the particular issues given by low-resource languages, as well as developing more rigorous techniques for analysing semantic similarity across different linguistic settings.

References

[1] Banerjee, S., Lavie, A. (2005). METEOR: an automatic metric for MT evaluation with improved correlation with human judgments. In Goldstein, J., Lavie, A., Lin, C.-Y., Voss, C. (eds.), Proceedings of the ACL Workshop on Intrinsic and Extrinsic Evaluation Measures for Machine Translation and/or Summarization. Association for Computational Linguistics, 65–72. https://aclanthology.org/W05-0909.

[2] Devlin, J. (2018). Bert: pre-training of deep bidirectional transformers for language understanding. *arXiv Preprint arXiv:1810.04805.*

[3] Doddington, G. (2002). *Automatic evaluation of machine translation quality using n-gram co-occurrence statistics.* 138–145.

[4] Freitag, M., Rei, R., Mathur, N., Lo, C., Stewart, C., Foster, G., Lavie, A., Bojar, O. (2021). Results of the WMT21 metrics shared task: evaluating metrics with expert-based human evaluations on TED and news domain. In Barrault, L., Bojar, O., Bougares, F., Chatterjee, R., Costa-jussa, M. R., Federmann, C., Fishel, M., Fraser, A., Freitag, M., Graham, Y., Grundkiewicz, R., Guzman, P., Haddow, B., Huck, M., Yepes, A. J., Koehn, P., Kocmi, T., Martins, A., Morishita, M., Monz, C. (eds.), Proceedings of the Sixth Conference on Machine Translation. Association for Computational Linguistics, 733–774. https://aclanthology.org/2021.wmt-1.73

[5] Gala, J., Chitale, P. A., AK, R., Gumma, V., Doddapaneni, S., Kumar, A., Nawale, J., Sujatha, A., Puduppully, R., Raghavan, V., Kumar, P., Khapra, M. M., Dabre, R., Kunchukuttan, A. (2023). IndicTrans2: Towards high-quality and accessible machine translation models for all 22 scheduled Indian languages (No. arXiv:2305.16307). arXiv. https://doi.org/10.48550/arXiv.2305.16307

[6] Huidrom, R., Lepage, Y. (2022). Introducing EM-FT for Manipuri-English neural machine translation. In Jha, G.

N., S. L., Bali, K., Ojha, A. Kr. (eds.). In Proceedings of the WILDRE-6 Workshop within the 13th Language Resources and Evaluation Conference. European Language Resources Association, 1–6. https://aclanthology.org/2022.wildre-1.1

[7] Lee, S., Lee, J., Moon, H., Park, C., Seo, J., Eo, S., Koo, S., Lim, H. (2023). A survey on evaluation metrics for machine translation. *Mathematics*, 11(4), Article 4. https://doi.org/10.3390/math11041006

[8] Liu, Y., Ott, M., Goyal, N., Du, J., Joshi, M., Chen, D., Levy, O., Lewis, M., Zettlemoyer, L., Stoyanov, V. (2019). RoBERTa: a robustly optimized BERT pretraining approach (No. arXiv:1907.11692). arXiv. http://arxiv.org/abs/1907.11692

[9] Ma, Q., Wei, J., Bojar, O., Graham, Y. (2019). Results of the WMT19 metrics shared task: segment-level and strong MT systems pose big challenges. In Proceedings of the Fourth Conference on Machine Translation (Volume 2: Shared Task Papers, Day 1), 62–90. https://doi.org/10.18653/v1/W19-5302

[10] Mondal, S. K., Zhang, H., Kabir, H. D., Ni, K., Dai, H.-N. (2023). Machine translation and its evaluation: a study. *Artificial Intelligence Review*, 56(9), 10137–10226.

[11] Papineni, K., Roukos, S., Ward, T., Zhu, W.-J. (2002). Bleu: a method for automatic evaluation of machine translation. In Isabelle, P., Charniak, E., Lin, D. (eds.). In Proceedings of the 40th Annual Meeting of the Association for Computational Linguistics. Association for Computational Linguistics, 311–318. https://doi.org/10.3115/1073083.1073135

[12] Popović, M. (2015). chrF: character n-gram F-score for automatic MT evaluation. In Bojar, O., Chatterjee, R., Federmann, C., Haddow, B., Hokamp, C., Huck, M., Logacheva, V., Pecina, P. (eds.). In Proceedings of the Tenth Workshop on Statistical Machine Translation. Association for Computational Linguistics, 392–395. https://doi.org/10.18653/v1/W15-3049

[13] Reimers, N., Gurevych, I. (2019). Sentence-BERT: sentence embeddings using Siamese BERT-networks. *arXiv Preprint arXiv:1908.10084.*

[14] Rouge, L. C. (2004). A package for automatic evaluation of summaries. 5.

[15] Singh, C. Y. (2000). Manipuri Grammar. Rajesh Publications.

[16] Thabah, N. D. J. (2024). An enhanced neural machine translation with pre-trained contextual encoding knowledge and data augmentation for low-resource Khasi language. *Journal of Electrical Systems*, 20(10s), 6712–6725.

[17] Thabah, N. D. J., Mitri, A. M., Saha, G., Maji, A. K., Purkayastha, B. S. (2022). A deep connection to Khasi language through pre-trained embedding. *Innovations in Systems and Software Engineering*. https://doi.org/10.1007/s11334-022-00497-

36 Artificial intelligence and entrepreneurship in the context of current socioeconomic growth in India – an examination of the literature

Th. John Lerphangam Monsang[a] and Sanasam Somokanta Singh[b]

Institute of Management Studies (MIMS), Manipur University, India

Abstract

AI has the potential to incomparably influence entrepreneurship through the development of new business version, consumer understanding, increased creativity, and trend forecasting. It can also produce fascinating product varieties and streamline processes. Entrepreneurs may also use AI to customise sales tactics, develop messaging based on consumer input, build AI-generated product roadmaps, and reduce human errors. The moral values, job displacement, and the equitable distribution of benefits are some of the challenges raised by AI. This paper explores how AI empowers entrepreneurs, disrupts traditional business models, and influences economic and social dynamics, while addressing challenges such as inequality and ethical risks.

Keywords: Artificial intelligence, machine learning, entrepreneurship, economic growth

Introduction

As global economies recover from pandemic disruptions and struggle with digital acceleration, AI-driven entrepreneurship is at the forefront of socioeconomic evolution. Artificial Intelligence (AI) has become a transformative force in the 21st century, redefining entrepreneurial landscapes and industries. AI refers to the intelligence that machines, especially computer systems, display by mimicking human learning, comprehension, problem-solving, decision-making, creativity, and autonomy [1]. In computer science, it is the study and development of techniques and software that allow machines to sense their surroundings and employ intelligence and learning to act in ways that increase the likelihood that they will accomplish predetermined objectives [2]. Artificial Intelligence is utilised in several areas [3]. , including machine vision, natural language processing, expert systems, and speech recognition [4]. The capacity to start new businesses necessitates identifying new possibilities [5]. An entrepreneur is a person who plans, oversees, and takes on the risks of a company or organisation. The process of finding novel methods to combine resources is called entrepreneurship. An entrepreneur makes money when the market value created by this novel combination of resources is higher than the market value that these resources could produce individually or in another combination [6]. Entrepreneurs must balance their understanding of business operations with a drive for innovation. They must possess curiosity, flexibility, adaptability, persistence, passion, a willingness to learn, vision, and motivation [7]. The endeavour to start a new company, sometimes incorporating an invention, is known as entrepreneurship.

All industries, from big companies to small enterprises, gain from entrepreneurs. Retail, business services, and food & restaurant operations rank as the top three sectors for small business startups. A recent survey from *Guidant Financial* found that these industries are crucial for entrepreneurs.

AI automates data collection, warehouse stock-picking, and manufacturing, thereby reducing human errors and dangerous tasks. It is used in various industries like customer experience, fraud detection, and recruitment, with chatbots handling customer inquiries and recruitment platforms. However, adopting and maintaining AI workflows comes with challenges and risks, including data, model, operational, ethics, and legal risks. Prioritising safety and ethics ensures that AI-related systems align with societal values [1]. Businesses may use AI as a potent tool to overcome obstacles and accomplish their objectives. It can be used for problem-solving, cultivating creativity, strategic planning, expanding horizons, breaking barriers, and enhancing design thinking. With the ability to analyse enormous volumes of data and spot trends, AI may offer insights that help guide strategic choices. It can also cultivate creativity by analysing existing data, identifying gaps, and proposing innovative solutions. AI-powered tools for idea generation, concept development, and brainstorming sessions can help people become strategically brilliant. AI-powered solutions may assist business owners in making well-informed decisions regarding budget allocation, marketing tactics, and product development. It can also help break barriers by identifying emerging trends, unmet consumer needs, and market gaps. AI can also be used in design thinking to enhance product development and customer experience by analysing user feedback, preferences, and behaviour. By

[a]monsangjohn@gmail.com, [b]sansom68@gmail.com

DOI: 10.1201/9781003675235-36

integrating AI into their DNA, startups can build a foundation for sustained growth. Self-reliance or Atmanirbhar, a self-sustained economy, could be realised.

Literature review

AI revolutionises entrepreneurship by enhancing startup conceptualisation, saving time, and enhancing efficiency. It is transforming the future of business by improving productivity, process innovation, and decision-making with real-time insights, making it a crucial tool for entrepreneurs [9]. AI can assist entrepreneurs in leveraging customer feedback from various sources, such as social media, product reviews, and support tickets, for automated analysis [10]. The combination of entrepreneurship with AI is known as 'AI entrepreneurship,' which involves innovative individuals utilising AI technology to build companies and organisations that seize new possibilities and tackle challenging issues. The possibilities are boundless, especially in a rising nation like India, ranging from AI-powered healthcare solutions to driverless cars and smart cities [11]. AI has been penetrating every sphere of human activities. The discipline of AI has expanded into every field. People from many walks of life enjoy AI's advantages without ever recognising it. We are all essentially being helped by artificial intelligence whenever you use your phone to book a cab, navigate through social media apps on your smartphone, use Google or Apple Maps, or control your smart devices at home [12].

According to the authors in Ref. [13], AI significantly impacts entrepreneurship, benefiting entrepreneurs in areas such as opportunity, performance, decision-making, and research and education. This fusion drives economic growth, innovation, and professional advancements; encourages workforce expansion and societal transformation; and revises industries and the business landscape. AI-powered entrepreneurship uses AI's capabilities to analyse vast data, generate insights, and choose wisely. This technology can revolutionise healthcare, finance, e-commerce, and marketing. AI can enhance decision-making, streamline operations, personalise customer experiences, and create innovative solutions like healthcare diagnostic tools. However, entrepreneurs must navigate moral deliberations and never stop learning and improving [14]. Entrepreneurs create, manage, and bear financial risks in the latest business development or startups, introducing new ideas, products, services, and innovations that enhance competitiveness and societal progress, focusing on growth and scalability. To meet social needs, solve problems, or give peculiar solutions that set them apart from more established companies, they are distinguished for bringing new and inventive goods, services, or technologies to the market. AI integration in startups can disrupt traditional industries and create new ones, driving innovation across healthcare, finance, transportation, and education.

As AI's skills grow, moral and legitimate queries arise. As entrepreneurs and businesses properly utilise AI's assurance, one can look forward to a future characterised by exceptional inventiveness, economic uprising, and development in everyday life. Businesses can find flexibility in the AI ecosystem, which offers opportunities for AI consulting, AI-as-a-service, and AI hardware development. One must stay up to speed on the latest advancements in AI and continuously adapt to the shifting environment to flourish in this fast-paced sector. Confidentiality of the data, talent scarcity, early financing, and ethical and moral issues are some of the difficulties AI-driven companies face in the contemporary context. Yet, businesses, startups, and AI have a mutually beneficial connection in which startups act as incubators for AI research and AI technologies enable startups to overcome markets and effect change with the needs of the varied social status of the populace. Applications and software that use AI in fields such as healthcare, economics, manufacturing, e-learning, natural language processing (NLP), computer vision, agriculture, sustainability, ethics/morality, and compliance, AI hardware, Internet of Things, cybersecurity, and consulting and services are all examples of commercial openings in AI. These domains play a pivotal role for future generations. To improve economic values in this area, it is crucial to stay up-to-date with the latest advancements in AI, understand the unique needs of a target market, and be cognizant of moral and legal issues of the present and future technical domain. AI-powered businesses or firms are changing and transforming varied sectors of market demands with their creative and innovative solutions. These include chatbots, virtual assistants, and self-governing gadgets. Applications of AI include varied and wide areas of human needs as well as the market demands such as in the field of healthcare, finance, education, entertainment, navigation, e-commerce, transportation, cybersecurity, personalised marketing, manufacturing, agriculture, legal technology, real estate, human resources, mental health, gaming, financial planning, smart cities, robotics, and natural language processing (NLP). The agricultural domain, in particular, has been the main areas of the businesses world. AI can enhance social economy development by aiding vulnerable groups, promoting environmental protection, and utilising advanced technological solutions [15]. Collaborating the technical know-how, market knowledge, and flexibility in response to changing trends and technology is necessary for success in AI-driven business. A thorough understanding of the AI technology, a well-defined business model, and an emphasis on addressing issues or adding customer value are all necessary for the success of AI-based business. AI-powered enterprises have to embrace with the problems such as moral values, data protection, and the requirement for constant innovation as it has been the main domain of the study. In the process of forming AI-based goods or services, entrepreneurs must keep in mind the core areas of going hand-in-hand to the recent developments, and most importantly understanding the domain legal landscape and the ethical values [16].

Methodology

This research study is based on a survey of the literature and uses secondary data. The pertinent data is derived from the issue, which is the growth of entrepreneurs in the setting of artificial intelligence. Journals, official government website, and websites like Google and Google Scholar are the sources used to gather pertinent data.

Discussion and results

AI and socioeconomic development

AI and the social economy are increasingly used in sustainable socioeconomic development studies. Although the implications of AI and the economy differ between nations and governments, social innovations might help the social economy and lessen negative impacts on well-being. AI can support vulnerable groups, promote environmental protection, and apply technological solutions, contributing to efficiency, innovation, sustainability, social responsibility, and inclusion [17]. These aspects have been the focus areas for the Government of India in emphasising the cashless economy. The Landscape Report on Artificial Intelligence in Social Innovation, based on data from 300 social innovators across 50 countries, highlights the widespread use of AI across both high- and low-income regions. It highlights AI's potential to address diverse societal issues [18]. AI could be a channel of developing in bringing forth by creating and inculcating the entrepreneurial-based AI business, even to the far flung areas. Among the numerous reports highlighted, AI has made significant contributions in sectors such as airbuses business, hospital services, environmental sustainability, and economic empowerment to many demographics of the world. In agriculture, AI plays a vital role in addressing climate and seasonal changes and productivity to many domains of the entrepreneurial domain. Hence, Machine Learning (ML) is the most deployed AI capability, with Natural Language Processing (NLP) being used by 15% of social innovators. The most common combination is ML and NLP [18]. Thus, these findings have highlighted the essentiality of AI in the domain of businesses world.

The application of AI in the contemporary context can be witnessed from the humble application of smartphones to the varied domains of today's world. This goes along with AI, from simple usage of Google Maps to rocket launches. And it has become indispensable to the world of humans as without the technical understanding and its usage, it shall be incapable for a human to associate with the business world. On the other hand, the application of AI never distinguishes between the have and have not, from the banking application to the simple placing of an order of a product from Amazon, Myntra, and Flipkart, to the various marketing apps. It is inevitable for a man to know the need of the technical world of AI. Not only within the domain of the business world, but also there is transparency in the economic activities for both entrepreneurial and government sectors. AI has become indispensable to the world of business's domain, from family businesses to high profile businesses. However, concerns arise about its negative impact on society, such as social changes, decreased human closeness, unemployment, and wealth inequality. AI can also lead to racial bias and egocentrism and to sedentary ways of life. On the other hand, AI can positively impact healthcare by providing reliable medical diagnosis and treatment systems—for example, medical reports could be accessed via associate website of the diagnostic centre and one can have medical treatment through internet domain of services. Such are the positive ways of development enhance by the AI, for which social therapeutic robots and cyborgs can enhance the quality of life for seniors and physically challenged individuals [19]. As we proceed to the business and markets, AI has become a focal point for today's world and future generations' development. As it plays an important part for the numerous social prospects, mainly in the entrepreneurial business proceed and its prospect. AI has become a critical line of development for all enterprises say for example in cashless transactions, advertisement, accounting, and bookkeeping, the need for AI is increasing to maximum high for the economic growth, and improving productivity. Social governance, Industry 4.0, labour market changes, and intelligent decision-making are important spheres of influence [20].

AI and the entrepreneurial prospect of India

The Indian businesses or the startup ecosystem is experiencing rapid growth, driven by technological advancements [21]. A growing young population, which is considered as one of the immense potentials for the growth of the Indian economy along with the support of the government policies, have fuelled the startup ecosystem in India. India has the potential of enormous growth through various technical expertise and associating with renowned companies such as in Google, Facebook, Amazon, SpaceX, and too many high-profile companies. With this high knowledge at hand, the youths in association with latest AI play an inevitable and significant role in the growth and development of market and human needs which is easily assessable to both parties. Researchers found that 77% of businesses/startups are investing in cutting-edge technologies like blockchain, AI, ML, and IoT. The AI market is expected to grow at a Compound Annual Growth Rate (CAGR) of 25%–35% between 2024 and 2027, with India being one of the leading competitors. AI has the potential to improve productivity and strength by offering data-driven insights and facilitating customised consumer experiences for Indian entrepreneurs. Additionally, it may automate repetitive processes, which lowers operating costs and may increase the profit that businesses and markets require.

AI tools for startups in India

ChatGPT, Gemini, Midjourney, Notion AI, Rewind, Albert, GitHub Copilot, Perplexity, Salesforce Einstein, and Murf are among the top AI tools for startups in India. These are a few AI tools playing a pivotal role in the development of entrepreneurial prospects in the domain of India's contemporary prospects.

AI adoption in large Indian corporations: The following renowned companies are associated with AI at the front desk level for marketing and prospects. Companies like Cognizant, Genpact, and Infosys leverage AI to boost productivity and upskill developers. Zomato, Myntra, and Meesho leverage Azure OpenAI services to streamline operations and enhance customer experiences.

Government initiatives boosting AI adoption

The Indian government has launched the National AI Strategy, AI for All, and AI-Powered Public Services to promote AI adoption.

Challenges and opportunities in AI in startups in India

High Development Costs: Building AI models requires significant technological and infrastructure investments. **Resource Barriers**: High compute costs and data scarcity hinder startups in low-income regions. **Data Quality and Privacy**: Startups must navigate complex regulations and ensure ethical use of customer data. **Shortage of Talent**: The demand for AI professionals exceeds the supply, leading to a talent gap.

Moral values

AI pictures moral domains like bias and discrimination. **Ethical risks**: Bias in facial recognition systems and ChatGPT's misinformation potential underscore the need for ethical frameworks.

AI's future in Indian startups

Deep tech integration: Startups will integrate AI with emerging technologies like IoT, blockchain, and quantum computing. **Global expansion**: AI-driven innovations will expand Indian startups' footprint in international markets. **Social impact**: AI will be pivotal in addressing societal challenges.

AI's impact on entrepreneurship [22]

AI can potentially replace some jobs and create new ones such as AI validation and verification. AI is used in consulting, big data analytics, coding, and assisting with grant applications/funding requests. AI-related resources can effectively save time, energy, and money, thereby boosting productivity. Efficiency level of production could be enhanced, while the error and losses in business can be reduced.

Additional benefits of leveraging AI

AI enhances performance in complex work tasks and acts as a leveller by supporting lower performers. It helps identify broad market possibilities and capitalise on diverse products and services. It bridges literacy gaps and enhances skills development in marketing, finance, human resources, and logistics. It simplifies complex tasks like market research, business planning, pricing, inventory management, and social media management. Accounting and booking could be well established.

The geographical and demographic influence of AI

India's early moves in AI

AI is a key driver of the Fourth Industrial Revolution, with applications in hospital services, healthcare, education, finance, manufacturing, etc. [23]. India has an early-mover advantage in AI, with the Niti Aayog's National Strategy for AI, published in June 2018. The strategy focuses on 'AI for All' with five priority sectors: healthcare, education, agriculture, smart cities, and smart mobility. NetApp's report shows a stark division in AI adoption, with India, Singapore, the UK, and the US emerging as frontrunners. Open AI's VP and Nvidia's Head of AI Research and Developer Relations to support India's AI mission. These has been a channel through which it has penetrated to the diverse culture of the state, giving hope to many in realising ones potential in building and creating hopes to reality.

India AI mission launch and aims

The India AI mission was launched on March 7, 2024, with an expense of Rs. 10,372 crore for five years [23]. The mission aims to foster AI innovation by democratising computing access, enhancing data quality, developing indigenous AI capabilities, attracting top AI talent, enabling industry collaboration, providing startup risk capital, ensuring socially impactful AI projects, and promoting ethical AI. This has led many to take up entrepreneurial traits in developing and associating along with the AI domain world.

Seven pillars drive the responsible and inclusive growth of India's AI ecosystem

Here are some of the focus points in the development of business and AI in India. **India AI compute capacity**:

with more than 10,000 GPUs, it builds a scalable AI computing infrastructure. **Indian AI innovation center**: creates and implements large multimodal, domain-specific, and indigenous fundamental models. **India AI datasets platform**: it makes it easier to obtain high-quality, non-personal datasets for AI research. **Enhancing the use of AI applications**: enhancing the analytical area of various sectors is the goal of the India AI Application Development Initiative. **India AI future skills**: Expands AI courses at the undergraduate, graduate, and doctoral levels. **India AI startup financing**: Ssupporting and accelerating deep-tech AI businesses are. **Secure and reliable AI**: Guarantees conscientious AI development, implementation, and uptake.

AI and job opportunities

AI's impact on the Indian economy and job market
AI's Impact on the Indian Economy in the build-up for business world as without the entrepreneurial acumen, there will be heavy load on the part of government to accommodate large number of educated youths of India. To address this, following are the steps proceeded by the Government of India. Firstly, by 2035, artificial intelligence will contribute $967 billion to the Indian economy. 'The "Future of Jobs" report from the World Economic Forum forecasts major shifts in the global labour sector.' According to the poll, 75% of businesses will use AI, which will significantly alter the labour market [24]. A survey in India reveals that 85% of organisations anticipate AI will create new jobs within the next 1–5 years and enhance the quality of existing roles. 77% of employers are optimistic that AI will improve job security and foster career development.

Challenges in the job market

One of the biggest challenges facing the labour market is the disconnect between the changing demands of the business and the existing level of STEM (science, technology, engineering, and mathematics) education and skill development. This might lead to healthy growth and increase demand if properly implemênted. There is a noticeable lack of skill in STEM fields, and the rise of STEM graduates is not in line with their employability. Though there are numerous youth who have learned, yet there are many who lack in skill development, such as critical thinking, deep understanding of scientific principles, and technological acumen in the present market domain. Hence, it is very much vital to embrace the required skill development of various scientific principles and their application to align the workforce demand of the worldwide labour market and the skills needed for upcoming technology developments.

The need for specialised talent in the Indian job market

The Indian job market will need diverse specialised talent, including technical proficiency in Product Management, Backend and Frontend Development, Full-Stack Capabilities, Data Science, AI/ML, Cybersecurity, Software Reliability, and UI/UX Design. Expertise in cloud service platforms, cloud application development, DevOps, and Big Data will increasingly become vital to meet the current and future demands of the business market.

Addressing the skills gap

The National Education Policy (NEP) 2020 advocates a holistic, multifaceted strategy to address the skills gap. Significant changes are required in vocational education and industrial training facilities, primarily through the broad implementation of apprenticeship programs. Entrepreneurship has significant potential in bridging the skills gap. The Central University has the provision to groom youths by giving them the privilege of pursuing vocational studies with hands-on training, which is a positive note for the development of entrepreneurial prospect of many youths in India and across the states. There is an urgent need to nurture an array of essential professional and life skills beyond technical proficiency to build adaptability in the workforce.

In the progress and development of AI and entrepreneurship prospects, AI development companies are bringing a massive change by pushing to the periphery of AI technology such as transforming hospital services, healthcare field, finance, transportation, education, and, most importantly, the business world. Not only within the purview of the varied contour of businesses domain, but also it has wonderfully brought a big leap to the transportation with autonomous vehicles, reducing traffic congestion and human error. It is also enhancing medical diagnosis and treatment using machine learning and natural language processing, potentially saving lives and improving patient care. AI has made a large imprint on educational institutions by creating adaptive learning systems that provide personalised instruction, feedback, and assessment. Through data analysis and pattern recognition, AI enlightened both students and teachers. AI development companies create educational institutions by creating adaptive learning systems that provide personalised instruction and feedback. They also enhance fraud detection in the financial sector by developing sophisticated systems that detect fraudulent transactions in real-time. AI generates innovative agricultural and allied agricultural business by bringing massive development using machine learning algorithms and satellite images to give farmers and cultivators with realistic information about natural disasters and timely available of the weather forecast for better harvests and business profits. These solutions help farmers make informed decisions about crop selection, watering schedules, and fertilisation, promoting sustainable farming practices and global food security. By creating autonomous robots that lower costs, simplify procedures, and boost productivity, AI and robotics are redefining sectors like manufacturing, retail, and logistics, altering how companies function and adapt to quickly shifting needs [25]. In the process of development of the business world from the time of the Industrial Revolution in the 18th century [26] to the contemporary domain of the technological era, where the means without AI becomes inevitable. Hence, adopting AI is indispensable—from applications like Dish-TV for leisure time to the mass production of vehicles and varied area usage. AI's presence and related technologies play an essential role in the domain of the business world, and to the varied, complex social strata.

Conclusion

The future of entrepreneurship is intertwined with the evolution of AI, and entrepreneurs who understand and embrace its transformative potential can better navigate challenges and seize opportunities. By leveraging AI capabilities, entrepreneurs can turn data-driven insights into tangible results, streamlining operations, optimising marketing campaigns, and enhancing overall efficiency [8]. AI should not be viewed as a substitute for human creativity but rather as a force that enhances human potential and lessens human mistakes and labour. Entrepreneurs may unleash new dimensions of problem-solving, inventiveness, and strategic planning by cultivating a collaborative connection between AI and human creativity. This will propel innovation and growth in the corporate environment. Startups driven by AI encourage innovation, push business, and market establishment, and moving to the various fields of business development. They tackle complex problems in many fields, automate chores and procedures, provide individualised experiences, and propel advancements in business acumen. Hence, the development of AI-based goods and services requires businesses to keep up with developments in this technological domain, comprehend the legitimate portion, and consider moral thought as a value in the long-term of the AI-business world.

Future studies and suggestions

Indian companies are positioned to grow substantially as they solve societal issues, increase their global footprint, and incorporate AI with cutting-edge technologies like blockchain, IoT, and quantum computing, and many other technological know-hows. The paper is a humble approach to understand AI, its usage in the entrepreneurial domain, and

its influence of change to the positive side of the varied societal stratum. The paper relies more on a literature review and needs a pragmatic approach and deeper understanding of the studies, as AI and entrepreneurial domain are indispensable for the varied areas of development of the vast populace of world and for the future generation down the line.

References

[1]   Stryker, C., Kavlakoglu, E. (August 2024). What is AI?, 16.
[2]   Artificial intelligence. *Wikipedia*.
[3]   Copeland, B. (28 November 2024). Artificial intelligence. *Britannica*.
[4]   Craig, L. What is AI? Artificial intelligence explained.
[5]   Entrepreneurship.
[6]   Sobel, R. S. Entrepreneurship. *ECONLIB*.
[7]   What is entrepreneurship?
[8]   Clark, B. (February 2024). Catalyzing creativity: the role of AI in entrepreneurship. *MicroVentures Blog*, 6.
[9]   (22 January 2024). How AI will change the face of entrepreneurship in the future.
[10] Noyes, E. (9 August 2024). How entrepreneurs can use AI to better understand their target customers.
[11] Surapaneni, M. (2 May 2024). Embracing the era of AI entrepreneurship: a guide for professionals.
[12] Kurtianyk, O. (23 February 2024). Intelligent ways entrepreneurs can leverage artificial intelligence.
[13] Giuggioli, G., Pellegrini, M. M. (4 May 2023). Artificial intelligence as an enabler for entrepreneurs: a systematic literature review and an agenda for future research.
[14] (2 June 2024). The rise of AI powered entrepreneurship: what you need to know.
[15] Meņšikovs, V., Simakhova, A., Šipilova, V. (2024). Harnessing artificial intelligence for socio-economic development, 13(3).
[16] Jha, R. (7 November 2023). Entrepreneurship, startups, and artificial intelligence | entrepreneurial opportunities in AI | rise of AI in entrepreneurship.
[17] Meņšikovs, V., Anastasiia, S., Viktorija, Š. (2024). Harnessing artificial intelligence for socio-economic development. *European Journal of Sustainable Development*, 13(3).
[18] Artificial intelligence for social innovation.
[19] Tai, M. C.-T. (14 Aug 2020). The impact of artificial intelligence on human society and bioethics, 32(4), 339–343.
[20] Yuan, C., Tang, J., Cao, Y., Wei, T., Shen, W. (October 2024). The impact of artificial intelligence on economic development: a systematic review. *International Theory and Practice in Humanitiesand Social Sciences*, 1(1), 18.
[21] (9 December 2024). The role of AI in Indian startups: trends and future prospects. *The United Indian*.
[22] King, K., Ganguli, A. (20 March 2024). Impact of AI on entrepreneurship.
[23] Rathi, S. (12 August 2024). India and AI: achievements, ongoing efforts and future prospects.
[24] Kant, A. (18 January 2024). Creating jobs in the new dawn of AI.
[25] Pioneering progress: remarkable AI achievements shaping our future.

37 An extended SMOTE-based class balancing technique to improve classification accuracy on imbalanced datasets

P. Sendash Singh[1,a], Ksh. Nilakanta Singh[2,b], Bishambhor Th[2,c], M. Surchand Singh[2,d], Joychandra N.[2,e] and Khumukcham Robindro Singh[2,f]

[1]Department of Computer Science and Application, Dr. Harisingh Gour Vishwavidyalaya, Sagar, M.P. 470003, India

[2]Department of Computer Science, Manipur University, India

Abstract

Class imbalance is a common issue in real-time data collection, where certain classes are significantly overrepresented compared to others, often negatively impacting predictive model performance. The Synthetic Minority Oversampling Technique (SMOTE) is a widely used method for addressing this imbalance by generating synthetic samples to enhance minority class representation. However, SMOTE may inadvertently introduce noisy samples, which can degrade the performance of predictive models. Additionally, the generated large datasets present computational challenges, highlighting the need for effective data reduction strategies. This paper proposes a novel data balancing approach that integrates an extended version of SMOTE with a distance filtering technique to upsample the minority class instances judiciously. This method minimises dataset expansion by selectively incorporating synthetic instances into the original data. An empirical study is conducted to evaluate the effectiveness of the proposed method compared to existing techniques across several datasets. The results demonstrate its superiority in managing imbalanced datasets, leading to improved classifier performance.

Keywords: Data imbalance, synthetic minority oversampling technique (SMOTE), filtering techniques, sampling techniques

Introduction

Class imbalance is a significant challenge for machine learning due to the unequal distribution of samples across classes in the training dataset. This imbalance is a common issue because predictive models tend to favour the majority class, leading to misclassification of the minority class. It is frequently encountered in domains like fraud detection, medical diagnosis, sentiment analysis, etc., where the occurrence of specific instances is infrequent in real-world scenarios. In such cases, the majority class might account for 99% of the dataset, while the minority class represents only 1%. For example, the number of fraudulent transactions is typically much lower than legitimate ones in fraud detection. Similarly, rare diseases have fewer cases than common ailments in medical diagnosis. This inherent skew in datasets often leads to suboptimal model performance. The issue of class imbalance is an ongoing challenge with wide-ranging implications for model performance and reliability. To address this challenge, researchers and practitioners have explored various strategies. SMOTE is a strategy that balances class distribution by generating synthetic instances of the minority class. While this technique shows promise, it can introduce noise into the data as well as increase the dataset size, which potentially affects model performance negatively.

This paper proposes a novel class-balancing technique that combines extended SMOTE with distance filtering methods. The SMOTE approach increases the number of minority instances in the dataset, while the distance filtering techniques selectively reduce the sample size by retaining only the most important instances from the augmented dataset. A comprehensive evaluation of the proposed method is conducted to show its ability to generate synthetic data effectively. The methodology incorporates various processes, including noise filtering, model training, and performance evaluation. The primary objective is to assess the proposed class-balancing technique on diverse imbalanced datasets, using random forest models as the baseline machine learning algorithm. The evaluation compares the proposed technique with other variants, employing five distinct filtering methods: Z-Score/Standard Deviation, Median Absolute Deviation, Isolation Filtering, Distance Filtering, and Gaussian Filtering. The results are analysed using multiple performance metrics, including accuracy, precision, recall, F1-score, and the area under the receiver operating characteristic curve

[a]sendashpangambam@gmail.com, [b]nilakanta.kakching@gmail.com, [c]bishambor@gmail.com, [d]mayanglambamhundung@gmail.com, [e]joy140nong@gmail.com, [f]rbkh@manipuruniv.ac.in

DOI: 10.1201/9781003675235-37

(AUC-ROC). The proposed technique demonstrates superior performance across these metrics, highlighting its effectiveness in addressing class imbalance problems in real-world scenarios.

Literature review

Several techniques have been developed to address the issue of class imbalance in machine learning. Chawla et al. [1] introduced the Synthetic Minority Over-sampling Technique (SMOTE), which generates synthetic minority class samples by interpolating between existing minority class samples, improving classification performance on imbalanced datasets. Han et al. [2] proposed Borderline-SMOTE, which focuses on generating synthetic instances for minority class instances near the decision boundary, while He et al. [3] introduced ADASYN, which generates synthetic samples based on the difficulty of learning certain minority class instances. Barua et al. [4] developed the Majority Weighted Minority Oversampling Technique (MWMOTE), which assigns weights to minority class instances based on their proximity to the decision boundary, further enhancing performance. Mullick et al. [5] proposed Generative Adversarial Minority Oversampling (GAMO), which uses a Generative Adversarial Network (GAN) to create high-quality synthetic minority class instances. Additionally, Fernández et al. [6] analysed the combination of SMOTE and evolutionary under-sampling methods, showing that SMOTE-based oversampling is effective when paired with undersampling techniques. Liu et al. [9] introduced EasyEnsemble and BalanceCascade, two algorithms that address undersampling by creating balanced subsets of the majority class or sequentially removing correctly classified majority class examples. Galar et al. [8] reviewed ensemble methods, highlighting the effectiveness of bagging, boosting, and hybrid approaches in addressing class imbalance. Wang et al. [10] combined online learning with class imbalance learning in data streams, proposing two new ensemble methods, WEOB1 and WEOB2, which adaptively weigh predictions for improved robustness. These methods have been compared and evaluated for their performance, with various techniques demonstrating differing levels of effectiveness in handling class imbalance.

Methodology

Extended SMOTE
SMOTE is an algorithm that creates synthetic samples through interpolation to enhance the representation of minority class instances in a dataset. Unlike random oversampling, which duplicates instances and risks overfitting, SMOTE generates plausible new instances by leveraging the existing data distribution. It identifies minority class instances and determines the k nearest neighbors for each,

which also belong to the minority class. Synthetic samples are then generated by interpolating between an original instance and one of its randomly selected neighbours. This process is mathematically represented in Equation 37.1:

$$\chi_{new} = \chi_{ori} + \lambda \times (\chi_{neig} - \chi_{ori}) \qquad (37.1)$$

The algorithm continues generating synthetic samples and adds them to the dataset until the desired class balance is achieved. Since SMOTE operates in the feature space, the synthetic samples reflect the underlying data distribution of the minority class. However, SMOTE does not distinguish between informative and noisy minority samples, which can propagate errors. In other words, SMOTE can generate samples near class boundaries, potentially leading to class overlap and misclassification. Additionally, assuming linear interpolation will produce valid samples may not hold for all datasets. Extended SMOTE approaches have been developed to address the aforementioned issues. In these methods, minority class instances are categorised into groups, such as those near the decision boundary, those near majority instances, and those near minority instances. Synthetic samples are generated primarily for instances whose k nearest neighbors are classified as near minority class instances. This approach avoids noisy regions, improves decision boundaries, and reduces noise propagation, resulting in a more accurate dataset representation.

Distance filtering methods
The distance filtering method employs distance metrics such as Euclidean, Manhattan, Cosine similarity, or Mahalanobis distance to assess relationships between data instances. Based on these metrics, instances that fail to meet specific criteria are excluded. The method typically utilises centroids or local neighbourhoods as reference points to determine which instances are retained. Initially, a set of reference points is selected, and the distance of each data instance to these reference points is computed. Instances with distance that satisfy the threshold set by the filtering criteria are retained, while others are removed. It is shown in Equation 37.2.

$$d(p, p_{ref}) \leq r \qquad (37.2)$$

This process is iteratively repeated, updating the reference points in each iteration. This technique is simple and versatile, relying on computationally straightforward metrics like Euclidean distance and cosine similarity. It is applicable across various datasets and problem domains.

Conceptual framework
This section presents a conceptual framework for the experimental evaluation of the proposed ESMOTE method for class balancing technique using extended SMOTE

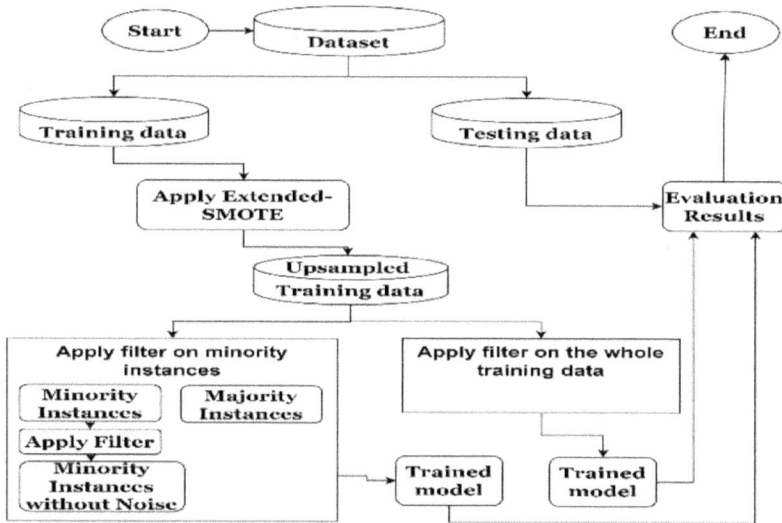

Figure 37.1 Conceptual framework for class balancing technique
Source: Author

to upsample minority classes and distance filtering to remove noise in machine learning datasets. The framework involves several steps, as illustrated in Figure 37.1. Initially, a dataset containing both majority and minority class samples is obtained and split into training and test datasets for model training and evaluation, respectively.

Extended-SMOTE is then applied to the training data to generate synthetic samples for the minority class, ensuring a balanced class distribution. Following this, noise filtering is performed in two scenarios: (i) filtering is applied only to the oversampled minority class, and (ii) filtering is applied to the entire oversampled dataset, encompassing both majority and minority classes. The resulting filtered data forms an optimal, balanced, and noise-free training dataset. This dataset is subsequently used to train a Random Forest (RF) model, which is evaluated on the test dataset using various performance metrics.

Experimental setup
The experimental setup for evaluating the proposed class balancing technique is detailed in this section. The experiments were performed on Intel i5 processor with 8GB of RAM desktop system. All competing models were implemented in Python using the Jupyter Notebook environment. Multiple datasets from the UCI Machine Learning Repository and Kaggle were used to assess the performance of various balancing methods. These datasets, described in Table 37.1, were selected to represent different levels of class imbalance across diverse domains, including credit card fraud detection, breast cancer prediction, and diabetes detection. The evaluation process utilised metrics such as accuracy, F1-score, and the area under the curve (AUC) to measure performance effectively. The proposed ESMOTE method was compared against other techniques that utilise various filtering methods, including Z-score/

standard deviation, median absolute deviation, isolation filtering, and Gaussian-based filtering techniques.

Experimental results
This section discusses the results of the evaluation of various class balancing techniques across various datasets. Table 37.2 summarises the performance of competing class balancing techniques when applied to SMOTED minority samples, while Table 37.3 focuses on the technique applied to all SMOTED samples. The analysis in Table 37.2 reveals that applying filters to SMOTED minority samples consistently improves performance across most datasets. Among the filters evaluated—Z-Score/Standard Deviation Filter, Median Absolute Deviation Filter, Isolation Filter, Distance Filtering, and Gaussian Filtering—the Distance Filtering technique consistently achieves superior results. It delivers the highest accuracy, F1-Score, and AUC for datasets such as Foetal Health (94.98%) and Forest Cover (99.49%). In datasets such as Foetal Health, Forest Cover, and Phoneme, filters applied to the SMOTED minority samples outperform those applied to all SMOTED samples. For instance, most filters, except Gaussian filtering, show significant performance improvements on the Heart and WDBC datasets. When filters are applied to all SMOTED samples, performance tends to be lower compared to filtering the SMOTED minority samples. Gaussian filtering often underperforms, achieving the lowest AUC and accuracy, with results as low as 63.15% in the WDBC dataset. However, an exception is observed in the UCI Credit Card dataset, where applying filters to all SMOTED samples yields slightly better results than filtering only the SMOTED minority samples. Notably in this case, the Median Absolute Deviation Filter outperforms Distance Filtering. Overall, the results suggest that applying filters to SMOTED minority samples generally

Table 37.1 Description of the dataset.

Sl. No.	Dataset	No. of instances	No. of attribute	No. of class	Imbalance ratio	Data type	Source
1	Foetal health	2,126	22	2	3.51	Numerical, Real	Kaggle
2	Forest cover	45,247	56	2	3.77	Numerical	Kaggle
3	Heart	619	13	2	4.16	Numerical, Real	Kaggle
4	Phoneme	5,404	6	2	2.41	Numerical, Real	OpenML
5	WDBC	569	31	2	1.68	Numerical, Real	UCI

Source: Author

Table 37.2 Result obtained on smotted minority sample.

Dataset	Performance metric	Filter applied only on smotted minority samples				
		Z-score /standard deviation filter	Median absolute deviation filter	Isolation filter	Distance filtering	Gaussian filtering
Foetal health	Accuracy	94.35	93.73	94.51	94.98	77.74
	F_1 score	94.35	93.68	94.52	94.98	68
	AUC	91.84	90.18	92.19	92.75	50
Forest cover	Accuracy	99.58	99.32	99.44	99.49	78.96
	F_1 score	99.58	99.32	99.44	99.49	69.68
	AUC	99.34	98.67	99.02	99.11	50
Heart	Accuracy	100	100	100	100	81.72
	F_1 score	100	100	100	100	73.5
	AUC	100	100	100	100	50
Phoneme	Accuracy	89.88	88.03	89.64	89.88	85.57
	F_1 score	89.79	88.03	89.81	90.04	84.86
	AUC	89.69	85.43	89.29	89.46	78.62
WDBC	Accuracy	97.66	97.07	97.66	97.66	63.15
	F_1 score	97.66	97.08	97.66	97.66	48.89
	AUC	97.48	97.02	97.48	97.48	50

Source: Author

enhances performance compared to filters applied to all SMOTE upsampled instances.

Conclusion and future work

This study addresses the persistent challenge of data imbalance in real-time data collection by proposing an Extended SMOTE with a distance filtering-based class balancing technique. Through extensive empirical analysis, the limitations of existing methods for handling class imbalance are highlighted, and the effectiveness of the proposed approach is demonstrated across diverse datasets, including those from the UCI Machine Learning Repository, Kaggle, and gene expression studies. The results reveal that applying filters to SMOTED minority samples generally outperforms filtering all SMOTED samples, emphasising the importance of mitigating noisy samples generated during synthetic data creation. This work provides valuable insights for practitioners and researchers, offering guidance on selecting appropriate methods to address data imbalance while reducing computational overhead in large-scale datasets. A deeper exploration of the exceptions observed in filter performance and dataset interactions could lead to more robust and adaptive methodologies. Investigating the sources of noisy samples during synthetic data generation is critical, with a focus on enhancing the extended SMOTE framework to minimise their impact. Future work could also explore integrating additional data reduction techniques or refining filter selection tailored to specific dataset characteristics.

Table 37.3 Result obtained on smotted majority sample.

Dataset	Performance metric	Filter applied only on smotted majority samples				
		Z-score/standard deviation filter	Median absolute deviation filter	Isolation filter	Distance filtering	Gaussian filtering
Foetal health	Accuracy	94.15	93.6	94.45	94.85	77.89
	F_1 score	94.15	93.6	94.45	94.85	68.25
	AUC	91.6	90	92	92.6	49.75
Forest cover	Accuracy	99.4	99.2	99.4	99.45	78.7
	F_1 score	99.4	99.2	99.4	99.45	69.9
	AUC	99.2	98.4	98.9	99	49.75
Heart	Accuracy	99.9	99.9	99.9	99.9	82.2
	F_1 score	99.9	99.9	99.9	99.9	74.2
	AUC	100	100	100	100	50.25
Phoneme	Accuracy	89.65	87.9	89.4	89.65	85.25
	F_1 score	89.55	87.9	89.6	90	84.6
	AUC	89.4	85.1	89	89.2	78.1
WDBC	Accuracy	97.45	96.85	97.45	97.45	63.65
	F_1 score	97.45	96.9	97.45	97.45	49.15
	AUC	97.3	96.8	97.3	97.3	50.25

Source: Author

References

[1] Chawla, N. V., Bowyer, K. W., Hall, L. O. Kegelmeyer, W.P. (2002). SMOTE: synthetic minority over-sampling technique. *Journal of Artificial Intelligence Research*, 16,321–357.

[2] Han, H., Wang, W. Y., Mao, B.H. (2005). Borderline-SMOTE: a new over-sampling method in imbalanced data sets learning. *International Conference on Intelligent Computing*, 878–887.

[3] He, H., Bai, Y., Garcia, E. A., Li, S. (2008). ADASYN: adaptive synthetic sampling approach for imbalanced learning. In IEEE International Joint Conference on Neural Networks (IEEE World Congress on Computational Intelligence), 1322–1328.

[4] Barua, S., Islam, M. M., Yao, X., Murase, K. (2014). MWMOTE-majority weighted minority oversampling technique for imbalanced data set learning. *IEEE Transactions on Knowledge and Data Engineering*, 26(2), 405–425.

[5] Sankha Subhra Mullick, Shounak Datta, Swagatam Das (2018). Generative adversarial minority oversampling. In Proceedings of the IEEE/CVF International Conference on Computer Vision, 1–7.

[6] Fernández, A., García, S., Herrera, F., Chawla, N. V. (2011). Addressing data complexity for imbalanced data sets: analysis of SMOTE-based oversampling and evolutionary undersampling. *Soft Computing*, 15(10), 1909–1936.

[7] Batista, G. E. A. P. A., Prati, R. C., Monard, M. C. (2004). A study of the behavior of several methods for balancing machine learning training data. *ACM SIGKDD Explorations Newsletter*, 6(1), 20–29.

[8] Galar, M., Fernandez, A., Barrenechea, E., Bustince, H., Herrera, F. (2012). A review on ensembles for the class imbalance problem: bagging-, boosting-, and hybrid-based approaches. *IEEE Transactions on Systems, Man, and Cybernetics, Part C (Applications and Reviews)*. 42(4), 463–484.

[9] Xu-Ying Liu, Jianxin Wu, Zhi-Hua Zhou. (2009). Exploratory undersampling for class-imbalance learning. *IEEE Transactions on Systems, Man, and Cybernetics, Part B (Cybernetics)*, 39(2), 539–550.

[10] Wang, S., Minku, L. L., Yao, X. (2015). Resampling-based ensemble methods for online class imbalance learning. *IEEE Transactions on Knowledge and Data Engineering*, 27(5), 1356–1368.

38 AI-powered personalised learning and intelligent recommendation systems

Sri Sai Harshita Gadavarthi[a], Sanya Gulati[b], Sanika Jagiasi[c], Anushka Jaint[d], Kashish Jindal[e] and Soni Sweta[f]

Department of Computer Engineering, Mukesh Patel School of Technology Management & Engineering, SVKM's Narsee Monjee Institute of Management Studies (NMIMS) Deemed-to-University, Mumbai-400056, India

Abstract

This paper explores the transformative role played by personalised learning in enhancing education through a data-driven approach and intelligent education platforms. It introduces a personalised quiz system that dynamically adjusts questions based on the learner's knowledge and progress level. Innovative methods such as analytical challenges, which imply that learners engage with critical and complex problems or case studies requiring deep analysis and decision-making, and argumentation tasks where learners develop well-reasoned arguments on contentious topics supported by evidence. The aim of this paper was to develop approaches to enhance learning experiences and outcomes. This approach ultimately personalises the effective education pathway against the existing methodologies. Results in terms of learner performance before and after personalisation are shown in the figure. The correlation matrix also showed the impact of success in quizzes after recommending learning objects in a personalised way. Future trends and ethical concerns surrounding the deployment of big data and AI-based applications for educational purposes have been discussed.

Keywords: Personalised learning, artificial intelligence in education, natural learning processing (NLP), recommendation system, machine learning

Introduction

Personalised learning, dating back to ancient apprenticeship and mentoring, has transformed significantly with the advancement of technology, especially in big data and learning analytics. This type of learning tailors' education according to the individual learner's needs, skills, and interests and fosters autonomy and engagement by adapting methods, pacing, and content delivery [1]. Further advanced by the integration of AI methodologies such as Multimodal Sentiment Analysis and explainable AI, which provide real-time feedback and personalises interventions, the field has continued to advance in its development [2]. Personalised learning systems, on the other hand, make use of tools and evidence-based strategies to maximise learning efficiency in e-learning environments and address diverse learner styles, thereby fostering better performance ('Modern Approach to Educational Data Mining and Its Applications,' n.d.) [3].

Traditionally, learner-centred education goes as far back as the early 20th century, when John Dewey began his advocacy for diversification and inclusion [4]. Today's systems include the Personalised and Adaptive Learning system that focuses on boosting reading literacy among students through personalised content and the use of formative assessment with feedback to correct mistakes [5]. Personalised instruction is somewhat different from personalised learning. The latter refers to the modification of the learning process, whereas the former refers to adjustments in content delivery [6]. Different learning technology models have been investigated to facilitate personalization in blended learning contexts, highlighting the importance of flexible instructional approaches and learner agency [7]. Such approaches increasingly leverage AI systems, such as machine learning algorithms, to enhance engagement, accuracy of feedback, and outcomes for students [8,9].

Despite its promises, personalised learning has several challenges: reliance on high-quality datasets, bias in AI models, and constraints on academic flexibility. These challenges illustrate the need for strategic improvement and rigorous evaluation in various educational contexts [10]. The authors further discussed the suitability of an intelligent recommendation system through a random forest classifier. Yet, there remains a critical gap in comparison against the state-of-the-art methods, including techniques such as LSTM and Autoencoders. If these gaps are addressed, this may reveal even deeper optimisations in personalised learning technologies.

Literature review

The following table represents a comprehensive, factor-based review of recent research studies that focus on personalised learning systems based on advanced technologies to offer customised educational experiences. It discusses the specific aims, technologies, strengths, and

[a]gharshita1223@gmail.com, [b]sanya272003@gmail.com, [c]sanikaj12@gmail.com, [d]anushka.jaint@gmail.com, [e]kashish0303jindal@gmail.com, [f]soni.sweta@nmims.edu

DOI: 10.1201/9781003675235-38

limitations of various systems intended to enhance learning through personalisation. These systems apply various state-of-the-art methodologies in incorporating semantic analysis, LSTM-GRU models, artificial intelligence, autonomous recommendation, and real-time analytics to offer innovative solutions for the challenge of adapting learning experiences to individual students' needs. Some studies focus on improving course recommendations with deep learning, while others provide adaptive feedback with the use of artificial intelligence or even get insights from sentiment analysis towards understanding student satisfaction levels. Each study points out some of the strengths, such as higher engagement of students, better accuracy of recommendations, and greater autonomy, which may, consequently, lead to improved learning outcomes. However, they also show the key limitations: dependence on quality datasets; restrictions on academic flexibility; and bias existing in NLP models for certain dialects and nuances in language. This table allows for a clear side-by-side comparison, and it is possible for educators and researchers to pin down the difference in which various personalised learning systems can perform under diverse contexts and technologies. Understanding both the advantages and the limitations, this review contributes to a more strategic manner of pushing personalised learning forward within

diverse educational settings, ensuring that these systems could be adapted and improved in such a manner that it may be brought into a larger context. Figure 38.1 shows the mind map of the personalised model. Finally, sentiment analysis of student feedback in MOOCs concluded that the analysis of student sentiment significantly improved course design and learner satisfaction. The findings indicate that understanding student perceptions about a course would enable more flexible curricular changes.

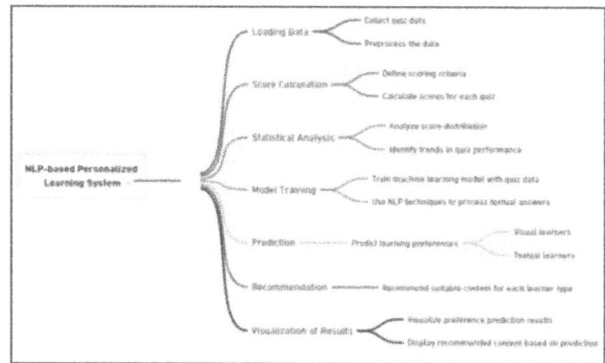

Figure 38.1 Mind map of the personalised model
Source: Author

Table 38.1 Comparative analysis of personalised learning systems and technologies.

Paper Title	Personalisation Focus	Technology	Advantages	Limitations
Personalised Course Recommendation System [17]	Personalised course suggestions based on semantic analysis	Semantic Analysis, LSTM-GRU models	High accuracy in course recommendation, enhanced student satisfaction by considering semantic importance	Limited flexibility across academic fields, potential biases in handling ambiguous language inputs
AI-Enabled Intelligent Assistant for Personalised Learning [20]	Tailored learning assistance and adaptive content delivery	AI, NLP	Engaging educational experience, immediate and personalised feedback to both students and educators	Effectiveness dependent on the quality of input data; weak datasets lead to vague or incomplete recommendations
Autonomous Recommender System for Virtual Learning [21]	Empowering students to manage their learning autonomously	Autonomous recommender systems	Improved student autonomy, motivation, and performance through individualised learning paths	Absence of human emotional feedback, lack of empathy, and restricted to virtual learning environments
Real-Time Analytics-Based Personalised Scaffolds [18]	Scaffolding for self-regulated learning in real-time	Real-time analytics, personalised scaffolds	Improved student self-regulation, enhanced use of metacognitive strategies, better writing outcomes	Limited to certain learning activities, potential technical challenges such as latency or system malfunctions
Personalised E-Learning Recommender System [12]	Personalised learning recommendations using deep learning	Autoencoders	Accurate course recommendations, improved student satisfaction through pattern recognition in user data	Lacks human feedback, potential overfitting, struggles with diverse data types or learning contexts

Table 38.1 Comparative analysis of personalised learning systems and technologies—cont'd

Paper Title	Personalisation Focus	Technology	Advantages	Limitations
Human-in-the-Loop Model for Personalised Education [13]	Integrating AI with human feedback to improve learning	Human-AI collaboration	Improved feedback quality and accuracy, dynamic real-time content adjustments, better educational outcomes	Limited scalability, dependent on the availability and expertise of human instructors, making it resource-intensive
NLP-Based Personalised Feedback in Intelligent Tutoring [11]	Generating personalised feedback in intelligent tutoring environments	NLP	More accurate, timely feedback for students, better instructor guidance	NLP biases could misinterpret student responses, especially in complex or nuanced answers
Personalised Learning Analytics Intervention Approach [14]	Personalised educational paths for blended learning based on student activity analysis	Learning analytics, behavioral engagement tracking	Enhanced student performance and engagement, personalised interventions in blended learning	Limited applicability outside blended learning setups, privacy issues, and potential reliability challenges in real-time analytics
Sentiment Analysis in MOOCs using NLP and Deep Learning [14]	Using sentiment analysis to improve course design and student satisfaction in MOOCs	Sentiment Analysis, NLP, Deep Learning	Improved course feedback and learner satisfaction, greater flexibility in curriculum design	Biases in sentiment analysis tools might misinterpret subtle cultural or emotional nuances, resulting in skewed insights
AI in Higher Education and Research [15]	Exploring AI's role in personalised learning and research in higher education	AI, NLP, Large Language Models	Increased personalisation, efficiency in handling large educational data, potential for research advancements	Possible over-reliance on AI in the absence of human insights, concerns regarding data privacy and ethical use of AI in education
Large Language Models and Personalisation [16]	Challenges and opportunities of personalisation through large language models	Large Language Models, NLP	Opportunities for deeper personalisation, improved handling of large-scale educational data	Risks related to overfitting, potential bias in personalised content, challenges in ensuring model accuracy and fairness
Personalised Feedback via Sentiment Analysis [19]	Using NLP to interpret and improve student feedback in educational systems	NLP, Deep Learning	Enhanced course design through better sentiment interpretation, improved learner satisfaction	Biases in language models may misinterpret culturally nuanced or ambiguous sentiments, leading to inaccurate feedback

Source: Author

However, there are gaps in sentiment analysis tools due to language models' intrinsic biases, which could misjudge subtle or culturally nuanced words. A similar issue has been observed in the sentiment analysis of student feedback using NLP and deep learning approaches, wherein the designed models performed well regarding the interpretation of sentiment; however, they had limitations in the specific situations in which they were applied and might be used inappropriately in classifying emotional states or intentions.

Methodology

The personalised learning system was implemented such that it supports data-driven, adaptive education, including the tailoring of content delivery based on students' learning preference. A sample of 50+ was generated to capture these three components; namely, responses to Quiz 1, whether participants preferred visualisation or text, and their responses to Quiz 2 after personalisation. The results from Quiz 1 gave an approximation of the state of knowledge as of the baseline, with preferences self-reporting through preference surveys. Quiz 2 responses were the measurements for effectiveness of the content offered in the second stage that was personalised. This research has a dataset obtained by the conduct of a two-stage quiz among a sample population. The first stage includes Quiz 1 to be undertaken by each participant with questions for comprehension. The second step involves the administration of a survey that elicits responses from participants about whether they preferred learning in either textual or visual form. In the second stage, the learning material was exposed to learners in a way that would reflect their actual preference, and they were asked to complete the second quiz, named Quiz 2, based on the content. This was with the hope that the participant's scores will improve once the learner receives content that is provided in his/her preferred way of learning. The dataset contained the following features:

1. **Quiz 1 responses:** Categorical answers to question(s) in Q1, Q2, and so on.
2. **Learning style:** The participant's preference for visual or text-based learning.
3. **Quiz 2 answers:** Categorical answers to the second set of questions following the provision of the custom content.

Data pre-processing
The data was pre-processed through various pre-processing steps for analysis and modelling. Categorical quiz answers were converted into numeric values using label encoding to make them compatible with ML algorithms. The dataset imputed missing values to eliminate any bias resulting from these missing values Class balancing was used to rectify overrepresentation of either of the two learning preferences. For feature selection, the responses of Quiz 1 were identified as predictors, while learning preferences were considered as the target variable. The learning preferences were encoded as binary values, where 'Visual' was mapped to 1 and 'Textual' to 0.

Model training
The goal of this project is to predict whether subjects have a visual or textual learning preference based on their responses to Quiz 1, posing the problem as a supervised classification problem. It began with feature selection where the answers from Quiz 1 are taken as input features and the learning preference as the target variable. To enable the analysis, learning preferences were encoded in terms of their binary values by making the value for 'Visual' to be 1, and that of 'Textual' to be 0. A Random Forest Classifier was selected for its robustness, interpretability, and ability to handle categorical data effectively. The dataset was split into training and testing subsets in an 80:20 ratio to validate model performance. Hyperparameter tuning was applied to optimise model accuracy, ensuring it generalised well to unseen data. To address potential challenges with unseen quiz responses, a transformation function was implemented to encode new input dynamically, mapping unfamiliar values to a default category or applying context-aware handling.

Content personalisation and delivery
On generating the prediction, the model offered personalised learning contents dynamically. This included multimedia-rich material such as videos, infographics, and photographs for visual learners to make a presentation that grabs students' attention or to increase interest in learning more. Textual learners were provided with well-structured read materials and text-based information to support his or her choices. It enhanced the effectiveness of the learning and helped in providing better outcomes for the participants.

Framework of the personalised learning system
The personalised learning system was grounded on a four-pillar base:

1. Learner autonomy: It allows participants to autonomously control their learning by dynamically directing content delivery that suits their preferences and choices. Again, the idea of autonomy enables the framework not to disregard but rather respect diverse learning styles with more engagement towards learning.
2. Instructional strategies: Adaptive instructional strategies provided learning content in varied presentation modes that catered to learners who preferred this mode of learning. This component utilised best practices in education to support better understanding and retention.
3. Technology tools: The system used the latest artificial intelligence technologies, NLP, and machine learning for real-time feedback and personalised recommendations. The learning analytics tracked the progress of learners and adapted the interventions.
4. Engagement and motivation: Engagement metrics made it possible for the framework to monitor learner interest and motivation continuously. Using motivational strategies ensured sustained participation and improved outcomes.

This framework is represented graphically in Figure 38.2, which represents the interconnectivity of learner autonomy, instructional strategies, technological tools, and engagement metrics. The diagram depicts the feedback loops between the components, indicating how learner preferences drive content delivery and how engagement data informs system refinements.

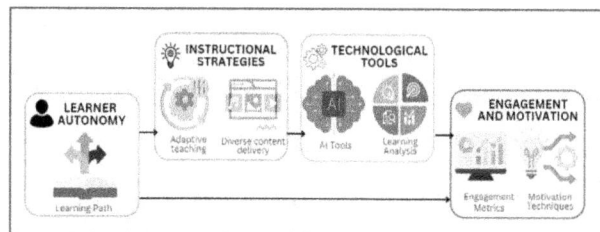

Figure 38.2 Personalised learning framework model
Source: Author

Results and discussion

Evaluation of quiz performance

The performance on the quizzes given to participants, both before and after exposure to content aligned to their preferred mode of learning, was used in evaluating the personalised learning system's effectiveness. While Quiz 1 was administered pre-personalised content, Quiz 2 followed personalised content was introduced to participants. The performance analysis was considered to determine if personalised learning impacts student performance or otherwise, mainly for visual and textual learners. A paired *t*-test was used to statistically validate if the observed score changes between Quiz 1 and Quiz 2 were significant for each learner group. The results are shown in Table 38.2.

The negative *t*-statistic values in Table 38.2 indicate that the scores of Quiz 2 were significantly higher than those of Quiz 1 for both learning groups. However, while visual learners showed a statistically significant improvement (p-value = 1.72e-5), textual learners did not show a significant increase in scores (p-value = 0.2894). This indicates that the personalised learning content was more effective for visual learners than for textual learners. This trend is further supported by a mean score difference; visual learners have improved by 3.70 points, with textual learners' minimal improvement having 0.43 points.

Score distribution and learning preference analysis

Figure 38.3 shows histograms that were created to better understand the distributions of scores for both quizzes. The graph revealed that the Quiz 1 scores were spread out, and a few students had very low scores, while Quiz 2 scores were generally higher, with the visual learners being the highest. The greater frequency of higher scores in Quiz 2 among visual learners implies that they gained more from getting the content in the format in which they learned best.

Figure 38.4 represents the distribution of quiz scores before and after personalised learning content was introduced, where blue represents Quiz 1 (pre-personalisation) and green represents Quiz 2 (post-personalisation). A clear trend exists towards higher marks, with several students who received scores ranging between 0 and 2 for Quiz 1 scoring significantly better in Quiz 2, particularly at the maximum score of 5. Reduction in lower marks and an increase in higher marks are quite robust indicators of personalised learning content leading to an improvement in student performance. This visualisation agrees with the statistical analysis, especially with regard to the large improvement in visual learners following the receipt of customised content.

Correlation analysis and feature importance

A correlation matrix was constructed to gain a deeper insight into how Quiz 1 is related to Quiz 2 as seen in Figure 38.5. The results from this correlation matrix indicated a modest correlation between Quiz 1 and Quiz 2 scores, indicating that prior knowledge played a role in final quiz performance, but personalised content played an even larger role in deciding outcomes for Quiz 2. A Random Forest model was applied to further analyse the feature importance of Quiz 1 responses for predicting learning preferences. It revealed that the contribution of Quiz 1 responses in predicting learning preference was only 7.23%, whereas the contribution of Quiz 2 responses was 37.35%. This shows that post-personalisation engagement is a better indicator of learning preference than pre-existing knowledge, hence proving the efficiency of content personalised for learners.

Table 38.2 Paired t-test results for quiz score improvement.

Learning preference	T-statistic	p-Value	Degrees of freedom (df)	Mean score difference
Visual learners	−8.2506	1.72e-5	9	3.70
Textual learners	−1.1619	0.2894	6	0.43

Source: Author

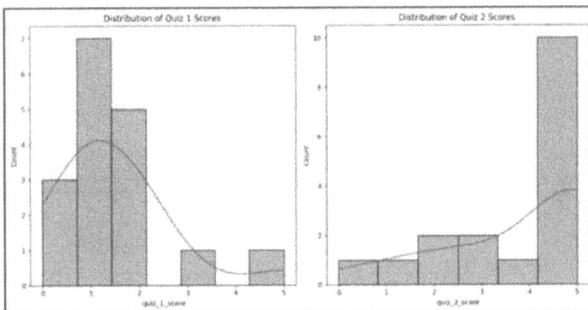

Figure 38.3 Histogram for the distribution of scores
Source: Author

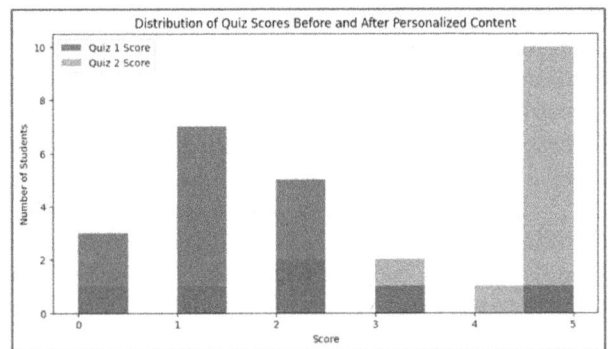

Figure 38.4 Distribution of quiz scores before and after personalised content
Source: Author

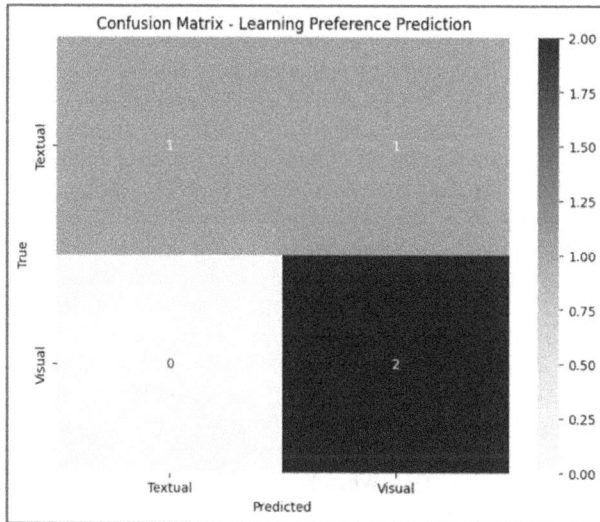

Figure 38.5 Correlation analysis and feature importance
Source: Author

Discussion on system effectiveness and limitations

Although the research confirms that, indeed, for visual learners, personalised learning impacts statistically significantly, it does pinpoint some of its limitations and opens some other areas of future research. The significant development in visual learners (mean increase of 3.70 points) indicates that visual content was very effective in getting learners to appreciate and remember their concepts, compared to textual content, which exhibited a minimal gain of 0.43 points. This challenges future research towards hybrid learning models that integrate elements of both visual and text-based content. Moreover, the dataset is too small and homogeneous for generalising results to a large and heterogeneous population, so studies with diverse participants on a broader scale are necessary to establish the efficacy of personalised learning. From an algorithmic point of view, the use of a Random Forest classifier is reasonable in predicting learning preferences. However, there is still scope for using alternative machine learning techniques such as Neural Networks, Support Vector Machines (SVM), or Reinforcement Learning to improve the adaptability and effectiveness of personalised learning models.

Conclusion

This study clearly shows that, with personalised recommendations, particularly based on the use of a Random Forest classifier, there could be improvement in learning outcomes regarding visual and textual modalities. Statistics showed post-test score improvements with learners from those groups. The following are the limitations found in this study: It has only considered learning styles to be visual and textual, thereby ruling out auditory and kinesthetics ones, thus making it relatively less applicable in a broader array of students. The dataset is very small and homogeneous, leading to the impossibility of generalising the findings. Furthermore, using self-reported learning preferences may result in biases. In addition, the study only employed a single Random Forest classifier without comparing it with any other advanced algorithms, meaning that the model cannot achieve its full potential. There were no real-world tests or longitudinal follow-ups to assess the long-term impact of the recommendations, and issues related to privacy and fairness need more attention, especially considering the biases that may arise from the data that the model was trained on.

Future scope

There are other prospects of further study for the improvement on efficiency and scalability were created by the results from this study: Expand the recommendation system to all learning styles including auditory and kinaesthetic, making it more inclusive and effective for a wider range of learners. Real-time adaptation features could turn the system into a highly responsive one in the sense that recommendations would dynamically be changed as learners moved on. Longitudinal studies need to be carried to test the long-run impact of the system on knowledge retention and academic performance. Additionally, more advanced algorithms such as neural network or ensemble methods can be used eventually to increase prediction accuracy. As for such an issue, ethical consideration, particularly related to data privacy issues, fairness, and the transparency of algorithms needs to be a priority to ensure inclusiveness and trustworthiness. It will be tested across different educational contexts, cultures, and languages to show its adaptability and scalability in a much larger use case. Its integration into existing learning platforms like Google Classroom and Microsoft Teams will speed up acceptance and implementation. It is bound to refine the efficiency, accessibility, and ethics of personalised learning systems at such a level that students find them tailor-made and result-oriented to the extent that learning experiences become very meaningful.

References

[1] Shemshack, A., Spector, J. M. (2020). A systematic literature review of personalized learning terms. *Smart Learning Environments*, 7(1). doi.org/10.1186/s40561-020-00140-9.

[2] Sweta, S. (2024). Emerging trends and challenges in educational sentiment analysis. In SpringerBriefs in Applied Sciences and Technology, 79–97. https://doi.org/10.1007/978-981-97-2474-1_5.

[3] Sweta, S. (2021). Modern approach to educational data mining and its applications. In SpringerBriefs in Applied

Sciences and Technology. https://doi.org/10.1007/978-981-33-4681-9.

[4] Zhang, L., Basham, J. D., Yang, S. (2020). Understanding the implementation of personalized learning: a research synthesis. *Educational Research Review*, 31, 100339. https://doi.org/10.1016/j.edu rev.2020.100339.

[5] Alrawashdeh, G. S., Fyffe, S., Azevedo, R. F., Castillo, N. M. (2023). Exploring the impact of personalized and adaptive learning technologies on reading literacy: a global meta-analysis. *Educational Research*, 42, 100587. https://doi.org/10.1016/j.edurev.2023.100587.

[6] Hammerschmidt-Snidarich, S. M., Edwards, L. M., Christ, T. J., Thayer, A. J. (2019b). Leveraging technology: a multi-component personalized system of instruction to teach sight words. *Journal of School Psychology*, 72, 150–171. https://doi.org/10.1016/j.jsp.2018 .12.005.

[7] Alamri, H. A., Watson, S., Watson, W. (2020). Learning technology models that support personalization within blended learning environments in higher education. *TechTrends*, 65(1), 62–78. https://doi.org/10.1007/s11528-020-00530-3

[8] Sallam, M. (2023). ChatGPT utility in healthcare education, research, and practice: systematic review on the promising perspectives and valid concerns. *Healthcare*, 11(6), 887. https://doi.org/10.3390/healthcare11060887.

[9] Akgun, S., Greenhow, C. (2021). Artificial intelligence in education: addressing ethical challenges in K-12 settings. *AI and Ethics*, 2(3), 431–440. https://doi.org/10.1007/s43681-021-00096-7.

[10] Chen, X., Zou, D., Xie, H., Wang, F. L. (2021). Past, present, and future of smart learning: a topic-based bibliometric analysis. *International Journal of Educational Technology in Higher Education*, 18(1). https://doi.org/10.1186/s41239-020-00239-6

[11] Troussas, C., Papakostas, C., Krouska, A., Mylonas, P., Sgouro poulou, C. (2023). Personalized feedback is enhanced by natural language processing in intelligent tutoring systems. In Lecture Notes in Computer Science, 667–677. https://doi.org/10.1007/978-3-031-32883-1_58

[12] El Youbi El Idrissi, L., Akharraz, I., Ahaitouf, A. (2023). Personalized e-learning recommender system based on autoencoders. *Applied System Innovation*, 6(6), 102. https://doi.org/10.3390/asi60 60102

[13] Bhutoria, A. (2022). Personalized education and artificial intelligence in the United States, China, and India: a systematic review using a human-in-the-loop model. *Computers and Education Artificial Intelligence*, 3, 100068. https://doi.org/10.1016 z/j.caeai.2022. 100068

[14] Yang, C. C. Y., Ogata, H. (2022). Personalised learning analytics intervention approach for enhancing student learning achievement and behavioral engagement in blended learning. *Education and Information Technologies*, 28(3), 2509–2528. https://doi.org/10.1007 /s10639-022-11291-2.

[15] Alqahtani, T., Badreldin, H. A., Alrashed, M., Alshaya, A. I., Alghamdi, S. S., Saleh, K. B., . . . Albekairy, A. M. (2023). The emergent role of artificial intelligence, natural learning processing, and large language models in higher education and research. *Research in Social and Administrative Pharmacy*, 19(8), 1236–1242. https://doi.org/10.1016/j.sapharm.2023.05.016

[16] Chen, J., Liu, Z., Huang, X., Wu, C., Liu, Q., Jiang, G., ... Chen, E. (2024). When large language models meet personalization: perspectives of challenges and opportunities. *World Wide Web*, 27(4). https://doi.org/10.1007/s11280-024-01276-1

[17] Haque, M. M. U. (June 12, 2024). Contextual course clustering: BERT-enhanced text analytics for personalized education. Retrieved from: https://www.ijisae.org/index.php/IJISAE/article/view/6857.

[18] Lim, L., Bannert, M., Van Der Graaf, J., Singh, S., Fan, Y., Surendrannair, S., Rakovic, M., Molenaar, I., Moore, J., Gašević, D. (2022). Effects of real-time analytics-based personalized scaffolds on students' self-regulated learning. *Computers in Human Behavior*, 139, 107547. https://doi.org/10.1016/j.chb.2022.107547

[19] Kastrati, Z., Dalipi, F., Imran, A. S., Nuci, K. P., Wani, M. A. (2021). Sentiment analysis of students' feedback with NLP and deep learning: a systematic mapping study. *Applied Sciences*, 11(9), 3986. https://doi.org/10.3390/app11093986

[20] Sajja, R., Sermet, Y., Cikmaz, M., Cwiertny, D., Demir, I. (September 19, 2023). Artificial intelligence-enabled intelligent assistant for personalized and adaptive learning in higher education. arXiv.org. https://arxiv.org/abs/2309.10892.

[21] Toscano-Miranda, R., Aguilar, J., Caro, M., Trebilcok, A., Toro, M. (2024). Precision farming using autonomous data analysis cycles for integrated cotton management. *Information Processing in Agriculture*. https://doi.org/10.1016/j.inpa.2024.10.002

39 Detection of diabetic retinopathy using deep-learning based image classification

Ksh. Dayvason Singh[a], Leishangthem Sashikumar Singh[b], Ksh. Nilakanta Singh[c], N. Joychandra Singh[d] and Kh. Robindro Singh[e]

Department of Computer Science, Manipur University, India

Abstract

Diabetic Retinopathy is a leading cause of blindness, necessitating early diagnosis to be treated successfully. In this study, DenseNet121, VGG16, and MobileNetV2 are compared for the classification of DR based on APTOS 2019 Blindness Detection dataset. For image pre-processing, we used CLAHE to enhance contrast and a customised adaptive Gaussian filter to remove noise. Our achievement is that the modified MobileNetV2 produced the best performance with 96.54% accuracy, 97.01% precision, 97.13% specificity, 95.94% sensitivity, and 96.47% F1-score. Its depth wise separable convolutions are efficient, to allow better feature extraction at minimal computational cost. This work shows MobileNetV2 to have promise in real-time DR detection and, as such, suitable for telemedicine and resource-constrained environments. Multi-stage DR grading, explainable AI, and large datasets will be explored in future work to further increase reliability and scalability.

Keywords: Diabetic retinopathy, CLAHE, adaptive Gaussian filter, CNNs, transfer learning

Introduction

Diabetic retinopathy (DR) is a major complication of diabetes mellitus and the most common cause of visual impairment worldwide. According to Yau et al. [1], nearly 93 million people worldwide have DR, with 17 million having proliferative DR, 21 million having diabetic macular oedema, and 28 million having vision-threatening DR. According to the National Center for Biotechnology Information (2020), DR occurs in 77.3% of patients with type 1 diabetes and 25.1% of patients with type 2 diabetes. Of these patients, 25% to 30% are likely to develop sight-threatening diabetic macular oedema. Bina Kotiyal and Herman Pathak [2] explained that diabetic retinopathy is an eye disease caused by raised blood glucose concentrations, leading to potential harm in the retina. Timely detection of DR is of critical importance in the prevention of blindness; yet, present manual screening approaches are cumbersome, demand specialised clinicians, and cannot be scaled up. Due to the growth of diabetes prevalence and lack of ophthalmologists, automated techniques must be devised. Deep learning-based methods, especially ANNs, (Gardner et. al) provide a consistent solution for early DR detection. Research by Gargeya and Leng [3] has proven the capability of CNNs in DR diagnosis with high accuracy, minimising human error. The study made by Gangwar and Ravi [4] addresses the challenge of automatic diabetic retinopathy detection by proposing an innovative deep learning hybrid model. The idea uses transfer learning with a pre-trained Inception-ResNet-v2 model that incorporates an additional custom convolutional block for performance enhancement. The proposed model was tested on the benchmarks Messidor-1 and APTOS 2019 (Kaggle blindness detection dataset). Transfer learning with pre-trained models has further improved accuracy, making automated DR detection systems a scalable and efficient alternative to manual diagnosis.

The objective of this study is to design and test an automated detection and classification system of DR using CNNs. The study is about designing a deep learning model for retaining images into two classes, namely, 'No_DB' and 'DR.' Through the automation process, this research looks forward to the enhancement of the efficiency, accuracy, and scalability of the DR detection system. This could result in early diagnosis and treatment that could lessen the burden of the healthcare systems while enhancing the patients' health conditions. The use of CNNs will make it possible to analyse the retinal images correctly and fill the gap of increasing DR screenings done and the scarcity of trained health care personnel.

Our method utilises transfer learning with CLAHE (Contrast Limited Adaptive Histogram Equalization) and customised adaptive Gaussian filtering for image pre-processing. ImageDataGenerator is also used for data augmentation for enhancing model generalisation and reducing class imbalance. This paper seeks to improve DR classification accuracy and reliability for real-world use, enabling efficient and scalable automated DR detection.

[a]kshetrimayumdayvason53152@gmail.com, [b]nilakanta.kakching@gmail.com, [c]saskulei@gmail.com, [d]joy140nong@gmail.com, [e]rbkh@manipuruniv.ac.in

DOI: 10.1201/9781003675235-39

Literature review

Hatua et al. [5] proposed a Hadoop-based classification framework for severity of DR in images using the Histogram of Oriented Gradients (HOG) and the Principal Component Analysis, showing significant efficiency in such large datasets. Chandel [6] introduced a Dual-Channel CNN (DC-CNN) trained on the APTOS dataset, outperforming ResNet50 and InceptionV3 with 95.23% accuracy. Vora and Shrestha [7] applied CNN's k-fold cross-validation on Kaggle/EyePacs for the dataset, achieving an accuracy of 76.7% but, in need of tuning for increasing sensitivity. And the achievement of their work is specificity was high at 97%–98%, but sensitivity varied from 40% to 98%. These results require optimisation, especially for severe and proliferative DR. The work promises a future where low-cost systems, such as NVIDIA Jetson TX2, could be used for real-time DR detection in underserved areas. Wu and Hu [8] applied transfer learning, utilising InceptionV3, VGG19, and ResNet50 to classify DR into five levels of severity. Although their model reportedly 'worked well,' the major concern was class imbalance, which was overcome through the application of data augmentation techniques.

Materials and methods

Dataset description

The APTOS 2019 Blindness Detection Dataset is a well-known benchmark dataset for detecting diabetic retinopathy (DR), launched as part of the Kaggle APTOS 2019 competition. Compiled by the Asia Pacific Tele-Ophthalmology Society (APTOS), the dataset contains fundus images taken under mixed imaging conditions and thus serves as a rich collection for training deep learning models for ophthalmic disease detection. The dataset consists 3,662 labelled retinal fundus images based on the International Clinical Diabetic Retinopathy (ICDR) grading system, which classifies DR severity into five classes: 1) No DR (Grade 0) – No diabetic retinopathy signs are visible, 2) Mild DR (Grade 1) – Microaneurysms are present, 3) Moderate DR (Grade 2) – Microaneurysms and haemorrhages are present, 4) Severe DR (Grade 3) – Many haemorrhages, venous beading, and potential intraretinal microvascular abnormalities, 5) Proliferative DR (Grade 4) – Stage of advanced disease with neovascularisation and high risk for blindness.

Contrast Limited Adaptive Histogram Equalization (CLAHE)

Contrast Limited Adaptive Histogram Equalization (CLAHE) [9] is a widely used image improvement process that boosts local contrast without increasing noise. In contrast to traditional histogram equalisation, which applies a global transformation to the entire image, CLAHE operates on small, non-overlapping tiles and contrasts individually in each region. This is particularly helpful for medical images, such as fundus photography, where fine details like blood vessels and haemorrhages need to be maintained while improving visibility.

Therefore, CLAHE is employed for image pre-processing in our study, as it can amplify fine details in retinal fundus images, such as blood vessels, microaneurysms, and haemorrhages, without amplifying noise. This enhances the perception of the image's features.

Customised adaptive Gaussian filter

The adaptive Gaussian filter enhances images by applying an adaptive amount of blurring dependent on local texture information. This ensures high-contrast regions (e.g., blood vessels or haemorrhages in fundus images) remain sharp, while smoother areas get heavier blurring. Following is the step-by-step mathematical computation involved in our adapted adaptive Gaussian filters filtering process.

Calculate local mean and variance

To calculate the local intensity properties of the image, a shifting window of size B×B is employed to calculate the local mean and local variance.

Local mean $\mu(x, y)$

The local mean at a pixel (x, y) is found by a uniform filter, which gives the average intensity in a neighbourhood using equation 39.1:

$$\mu(x,y) = \frac{1}{B^2}\sum_{i=-B/2}^{B/2}\sum_{j=-B/2}^{B/2} I(x+i, y+j) \quad (39.1)$$

where:

- $I(x,y)$ denote the intensity at grayscale pixel (x,y).
- B is the block size (e.g., 15×15)

This operation smooths out local intensity variations and yields an estimate of the brightness level in a region.

Local variance $\sigma^2(x, y)$

The variance tells us how much the intensity varies within the local region. And it is calculated using equation 39.2:

$$\sigma^2(x,y) = \frac{1}{B^2}\sum_{i=-\frac{B}{2}}^{\frac{B}{2}}\sum_{j=-\frac{B}{2}}^{\frac{B}{2}} I^2(x+i, y+j) - \mu^2(x,y) \quad (39.2)$$

This equation consists of two terms:

- The first term calculates the mean of squared intensity values.
- The second term subtracts the square of the mean to find variance.

If $\sigma^2(x, y)$ is high, the region has edges or textures. If it is low, the region is smooth.

Normalise local variance

Since different images may have different intensity distributions, we normalise the local variance to scale all values between 0 and 1 using equation 39.3:

$$\sigma^2_{norm}(x,y) = \frac{\sigma^2(x,y) - \sigma^2_{min}}{\sigma^2_{max} - \sigma^2_{min} + \epsilon} \tag{39.3}$$

where:

- σ^2_{min} and σ^2_{max} are the minimum and maximum variances of the image.
- ϵ (a small constant) avoids division by 0.

Now:

- $\sigma^2_{norm}(x,y)$ approx. equals to 0 for smooth areas.
- $\sigma^2_{norm}(x,y)$ approx. equals to 1 for textured areas.

This normalisation makes all images pass through an equal filtering process irrespective of contrast or brightness.

Compute adaptive Gaussian kernel sigma
The Gaussian blur strength is calculated dynamically as a function of local variance using equation 39.4:

$$\sigma(x,y) = \sigma_{max} \times (1 - \sigma^2_{norm}(x,y)) \tag{39.4}$$

where:

- σ_{max} is the highest allowable blur.
- $1 - \sigma^2_{norm}(x,y)$ reverses the variance effect:
  - Smooth areas ($\sigma^2_{norm} \approx 0$) receive a large $\sigma(x,y)$ (stronger blur).
  - Textured areas ($\sigma^2_{norm} \approx 1$) receive a small $\sigma(x,y)$ (preserve details).

To improved performance, $\sigma(x,y)$ is discretised as per equation 39.5:

$$\sigma(x,y) = \frac{round(\sigma(x,y) \times S)}{S} \tag{39.5}$$

where S is the number of sigma levels (e.g., 20). This significantly improves efficiency while maintaining adaptability.

Apply Gaussian filtering
Gaussian kernal and convolution process, discussed in the following subsections, were used for blurring.

Gaussian kernel formulation
A Gaussian kernel is employed to smooth the image using equation 39.6, defined as:

$$G_{\sigma}(i,j) = \frac{1}{2\pi\sigma^2} \exp\left(-\frac{i^2+j^2}{2\sigma^2}\right) \tag{39.6}$$

where:
- (i,j) are offsets from the kernel centre.
- σ is responsible for the spread of the Gaussian function.

Convolution process
For each pixel, the Gaussian filter is applied using equation 39.7:

$$I_{filtered}(x,y) = \sum_{i=-k}^{k} \sum_{j=-k}^{k} G_{\sigma}(i,j) I(x+i, y+j) \tag{39.7}$$

where, k is the kernel radius, set as:

$$k = 3\sigma$$

As we have discretised the values of $\sigma(x,y)$, we precompute Gaussian blurs for a limited number of sigma values and apply them selectively:

- If a pixel's $\sigma(x,y)$ is low, the image remains sharp.
- If $\sigma(x,y)$ is high, the image is blurred more.

Conceptual framework
First, we load the APTOS 2019 blindness detection dataset, which have five categories: No-DR, Mild, Moderate, Severe, and Proliferative DR. To simplify classification, we group Mild, Moderate, Severe, and Proliferative DR under a single label, 'DR,' while No-DR remains separate. Then, we split the dataset into training (70%), testing (15%), and validation (15%) sets. For image pre-processing, we apply CLAHE (Contrast Limited Adaptive Histogram Equalization) for contrast enhancement and customised adaptive Gaussian filtering to reduce noise while preserving fine details. The images are then rescaled to [0,1] for normalisation. Data augmentation (rotation, zooming, shifting, shearing, and flipping) is applied only to training images to improve generalisation, while validation and test images are rescaled without augmentation. Lastly, the dataset is batch-loaded with ImageDataGenerator and resized to 224 × 224 pixels for compatibility with transfer learning models. The pre-process_input function guarantees that the images are in the correct format, ready for pre-processing based on deep learning diabetic retinopathy classification.

Pre-train models
In our study, we used three pre-trained models from the TensorFlow Keras applications module, namely DenseNet121 [10], MobileNetV2 [11], and VGG16 [12]. These pre-trained models, initially trained on ImageNet [13], are used as feature extractors for diabetic retinopathy (DR) detection.

Two fully connected layers with 128 units and ReLU activation were added to adapt these models, followed by a SoftMax output layer for binary classification. Batch normalisation was also done before passing through the convolutional layer. Additionally, we used binary cross-entropy as the loss function for binary classification and ADAM optimizer were used for all the three models. First, the pre-trained layers were frozen to preserve their learned features, so that only the newly added layers were updated during training. Then, the pre-trained layers were unfrozen for fine tuning.

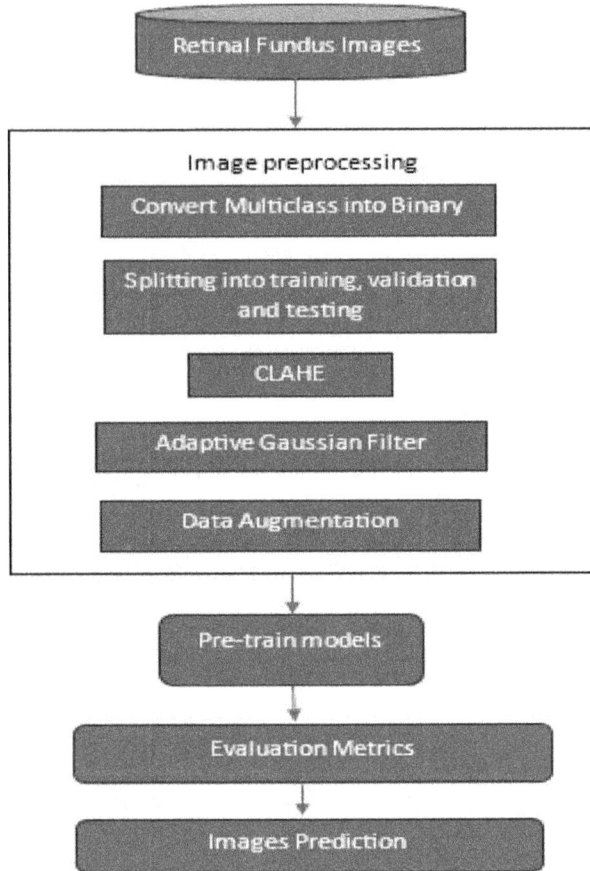

Figure 39.1 Workflow diagram
Source: Author

Evaluation metrics for model performance
To assess the performance of our diabetic retinopathy classification model, we utilise multiple evaluation metrics derived from the confusion matrix, including accuracy, sensitivity (recall), specificity, precision, and F1-score. These metrics provide a comprehensive understanding of the model's effectiveness in distinguishing whether the image has DR or not.

Results and discussion

In this study, we assessed several deep learning models for DR classification and compared their performance with the existing works. The results are presented in Table 39.1, which includes accuracy, precision, specificity, sensitivity, and F1-score for different models.

In the table given above, the modified MobileNetv2 outperformed the other models having the best accuracy of 96.54%, precision of 97.01%, specificity of 97.13%, and sensitivity of 95.94%. These results show that the model is capable of classifying DR cases accurately while keeping false negatives and false positives at a minimum. It also had an F1-score of 96.47%, which reflects its good balance of precision and recall.

The proposed model performed better with accuracy (96.54%) because of its lightweight and efficient design, which is suitable for DR classification. It applies depthwise separable convolutions to decrease parameters and avoid overfitting, while ensuring robust feature extraction. It preserves good fine details and enhances feature reuse and classification. The depthwise convolutions operate efficiently to identify microaneurysms, exudates, and vascular abnormalities.

Conclusion

Our study highlighted the effectiveness of transfer learning in classifying diabetic retinopathy using the APTOS 2019 Blindness Detection dataset. Here, we employed various

Table 39.1 Performance comparison with existing literature work.

Study	Model used	Accuracy	Precision	Specificity	Sensitivity	F1 score
Al-Rubaye & İlhan [14]	ResNet50, DenseNet201, InceptionV3 (ensemble)	90%	77%	96%	64%	70%
Chaturvedi et al. [15]	Modified DenseNet121	94.44%	86%	Not reported	87%	86%
Gangwar & Ravi [4]	Inception-ResNet-V2 with custom convolutional block	89.56%	Not reported	Not reported	Not reported	Not reported
Chandel [6]	Dual-Channel CNN (DC-CNN)	95.22%	Not reported	93.58%	96.94%	95.23%
Our study	Modified MobileNetV2	96.54%	97.01%	97.13%	95.94%	96.47%
Our Study	Modified Densenet121	96.29%	94.31%	94.27%	97.79%	96.01%
Our Study	Modified VGG16	95.10%	94.53%	94.62%	95.57%	95.05%

Source: Author

pre-trained models, including MobileNetV2, VGG16, and DenseNet121, and evaluated their performance metrics like accuracy, precision, sensitivity, specificity, and F1-score. Among them, our modified MobileNetV2 outperformed. The performance of MobileNetV2 is excellent because of its lightweight yet highly efficient architecture, which uses depthwise separable convolutions to reduce computational complexity while maintaining robust feature extraction. Additionally, the integration of Contrast Limited Adaptive Histogram Equalization (CLAHE) for enhancing contrast and a customised adaptive Gaussian filter for reducing noise while preserving fine details further improved classification accuracy.

This design effectively minimises overfitting, making the model well-suited for medical image analysis. The integration of Contrast Limited Adaptive Histogram Equalization (CLAHE) for enhancing contrast and a customised adaptive Gaussian filter for reducing noise, while preserving fine details further improved classification accuracy. Therefore, this model can be used as a potential tool for early screening and diagnosis, thereby helping ophthalmologists in preventing vision impairment and blindness caused by diabetic retinopathy.

Future scope
The future work may include conversion from binary to multiclass classification, to give more detailed diagnostic information. Using explainable AI (XAI) techniques like Grad-CAM will assist the interpretation of the predictions and, therefore, help in diagnosing the patient. Combining the analysis of images with metadata like patient history and glucose levels for better diagnosis might be included in future studies.

Dataset availability
https://www.kaggle.com/competitions/aptos2019-blindness-detection

References

[1] Yau, J. W., Rogers, S. L., Kawasaki, R., Lamoureux, E. L., Kowalski, J. W., Bek, T., ... Meta-Analysis for Eye Disease (META-EYE) Study Group. (2012). Global prevalence and major risk factors of diabetic retinopathy. *Diabetes care*, 35(3), 556–564.

[2] Kotiyal, B., Pathak, H. (2022). Diabetic retinopathy binary image classification using PySpark. *International Journal of Mathematical, Engineering and Management Sciences*, 7(5), 624.

[3] Gargeya, R., Leng, T. (2017). Automated identification of diabetic retinopathy using deep learning. *Ophthalmology*, 124(7), 962–969.

[4] Gangwar, A. K., Ravi, V. (2021). Diabetic retinopathy detection using transfer learning and deep learning. In Evolution in Computational Intelligence: Frontiers in Intelligent Computing: Theory and Applications (FICTA 2020). Springer Singapore, vol. 1, 679–689.

[5] Hatua, A., Subudhi, B. N., Veerakumar, T., Ghosh, A. (2021). Early detection of diabetic retinopathy from big data in hadoop framework. *Displays*, 70, 102061.

[6] Chandel, A. P. S. (2021). Detecting Diabetic Retinopathy from Retinal Fundus Images Using DC-CNN (Doctoral dissertation, Dublin, National College of Ireland).

[7] Vora, P., Shrestha, S. (2020). Detecting diabetic retinopathy using embedded computer vision. *Applied Sciences*, 10(20), 7274.

[8] Wu, Y., Hu, Z. (April 2019). Recognition of diabetic retinopathy based on transfer learning. In 2019 IEEE 4th International Conference on Cloud Computing and Big Data Analysis (ICCCBDA). IEEE, 398–401.

[9] Zuiderveld, K. (1994). Contrast Limited Adaptive Histogram Equalization. Graphics Gems IV. Academic Press, 474–485.

[10] Huang, G., Liu, Z., Van Der Maaten, L., Weinberger, K. Q. (2017). Densely connected convolutional networks. In Proceedings of the IEEE Conference on Computer Vision and Pattern Recognition (CVPR), 4700–4708.

[11] Sandler, M., Howard, A., Zhu, M., Zhmoginov, A., Chen, L. C. (2018). MobileNetV2: inverted residuals and linear bottlenecks. In Proceedings of the IEEE conference on computer vision and pattern recognition (CVPR), 4510–4520.

[12] Simonyan, K., Zisserman, A. (2015). Very deep convolutional networks for large-scale image recognition. In International Conference on Learning Representations (ICLR).

[13] Deng, J., Dong, W., Socher, R., Li, L. J., Li, K., Fei-Fei, L. (2009). ImageNet: a large-scale hierarchical image database. In Proceedings of the IEEE Conference on Computer Vision and Pattern Recognition (CVPR), 248–255).

[14] Al-Rubaye, M. M., & İlhan, H. O. (2024). The evaluation of the effect of data balancing over the classification performances of ensemble of networks for the diabetic retinopathy. *Sigma Journal of Engineering and Natural Sciences*, 42(5), 1563–1574.

[15] Chaturvedi, S. S., Gupta, K., Ninawe, V., Prasad, P. S. (2020). Automated diabetic retinopathy grading using deep convolutional neural network. *arXiv preprint* arXiv:2004.06334.

[16] Gardner, G. G., Keating, D., Williamson, T. H., Elliott, A. T. (1996). Automatic detection of diabetic retinopathy using an artificial neural network: a screening tool. *British Journal of Ophthalmology*, 80(11), 940–944.

40 In depth comparative analysis of the combine effect of ML models and feature selection methods in crop yield prediction

Mayanglambam Surchand Singh[1,a], Khumukcham. Robindro Singh[1,b], Ksh Nilkanta Singh[1,c], Nongdombam Joychandra Singh[1,d], Leishangthem Sashikumar Singh[1,e] and M. Rajeshwar Singh[2,f]

¹Department of Computer Science, Manipur University, India

²Department of Computer Science, Pole Star College, India

Abstract

Agriculture is facing critical challenges such as climate change, resource scarcity, and rising production costs, necessitating innovative solutions to enhance productivity. Crop yield prediction has emerged as a vital tool in addressing these challenges, leveraging machine learning (ML) techniques combined with feature selection (FS) methods to improve forecasting accuracy. This study evaluates the performance of three ML models—Extra Tree Regression (ETR), Random Forest (RF), and Artificial Neural Networks (ANN)—combined with three FS techniques: Recursive Feature Elimination (RFE), Variance Inflation Factor (VIF), and Random Forest Embedded (RFEMBD). The analysis focuses on their performance using the ICRISAT dataset to predict rice yields across Indian districts. Results reveal that ETR combined with VIF outperformed the other combinations, achieving the highest R^2 (0.77), lowest RMSE (477.93), and MAE (351.13), while offering superior composite metrics. However, ETR-RFE strikes a better trade-off between high performance and fewer features than ETR-VIF. ETR and RFE outperform other respective models. Temperature features serve as the primary predictors, while precipitation has a limited effect, except for October's precipitation, which appears to be influential. The study highlights the role of FS methods in enhancing model performance and identifies optimal ML-FS combinations for improved crop yield prediction, providing critical insights for stakeholders in agriculture.

Keywords: Machine learning, feature selection, crop yield prediction, ICRISAT dataset, composite metric

Introduction

Agriculture plays a pivotal role in ensuring food security and sustaining economies, yet it is increasingly facing numerous challenges that threaten its sustainability. Among the various areas of agriculture, crop yield has gained significant attention due to its profound impact on both national and international economies. Crop yield prediction refers to the process of forecasting the output of crops based on various factors such as rainfall, temperature, the use of fertilisers and pesticides, soil pH levels, and other atmospheric and environmental parameters. Leveraging advanced data and technology, this methodology provides farmers, policymakers, and stakeholders with valuable insights to make informed decisions and optimise agricultural practices. The accurate prediction of crop yield is a critical component of ensuring global food security, given the growing concerns about hunger, population growth, and environmental challenges. Traditional methods of crop yield prediction, which often relied on farmers' experiences and historical data, have become insufficient due to the complexity and variability

of factors influencing crop yields, such as climate conditions, soil quality, and agricultural practices. By making it possible to analyse big, complicated datasets and find patterns and correlations that were previously hard to find, advances in computational technology, especially in machine learning (ML), provide intriguing answers to these problems. Given the many variables at play, such as climate, weather, soil composition, and farming methods, crop yield prediction in precision agriculture continues to be a difficult task. Despite existing models offering reasonable yield estimates, there is a continuous demand for improved performance. Various machine learning models have been applied to crop yield prediction, each demonstrating strengths depending on the dataset [1]. While linear regression can perform well, it often struggles with non-linear relationships, an area where models like Random Forest (RF) excel. This highlights the need for further exploration of their effects on crop yield prediction. Additionally, feature selection (FS) methods are essential in data pre-processing, improving model performance by removing redundant features. Their importance in this field is well-established. Examining the impact of ma-

ᵃmayanglambamhundung@gmail.com, ᵇrbkh@manipuruniv.ac.in, ᶜnilakanta.kakching@gmail.com, ᵈjoy140nong@gmail.com, ᵉsaskulei@gmail.com, ᶠsinghrajeshwar39@gmail.com

DOI: 10.1201/9781003675235-40

chine learning on crop yield prediction, alongside these additional factors, could offer valuable new insights. This paper conducts a comparative study on the combined impact of ML and FS techniques on crop yield prediction, aiming to explore new dimensions in this field. In ML, Extra Tree Regression, RF, and ANN, and in feature selection methods, RFE (wrapper-based FS), VIF (filter-based FS), and RFEMBD (embedded-based FS) are considered. The remainder of the paper is organised as follows: Section 2 reviews the existing literature. The variables and sample are described in Section 3. In Section 4, the research approach is explained. The empirical results are discussed in Section 5. The paper is summarised in Section 6.

Literature review

Machine learning (ML) algorithms and feature selection (FS) are crucial for improving the precision of crop yield predictions. Several studies have explored the effectiveness of different ML models and FS techniques across diverse datasets, revealing valuable insights into their performance. The integration of climate data, remote sensing datasets, and FS methods has further enriched these analyses, demonstrating their potential to optimise predictions. Yaping Cai et al. [2] demonstrated that Support Vector Regression (SVR) outperformed Random Forest (RF) on an Australian climate dataset, which also incorporated remote sensing data. Similarly, Sahameh Shafiee et al. [3] found that SVR exceeded the performance of LASSO when using remote sensing datasets from Norway. Christopher D. Whitmire et al. [4] reported that K-Nearest Neighbors (k-NN) outperformed Linear and Bayesian regression and performed comparably to RF when utilising climate and planting data from the United States. Artificial Neural Networks (ANN) are the most widely used algorithms for agricultural yield prediction, according to Thomas van Klompenburg et al. [5]. Luwei Feng et al. [6] observed that applying the Recursive Feature Elimination (RFE) FS method enhanced prediction accuracy by approximately 2.125% when using hyperspectral data for alfalfa crop yield prediction. As a wrapper-based FS method, RFE has been widely adopted in various crop yield prediction studies, as noted by researchers such as David Camilo Corrales et al. [7], Divya Elavarasan et al. [8], and N.R. Prasad et al. [9]. Filter-based FS methods, including Variation Inflation Factor (VIF) [10] and Mutual Information FS [11], have also been utilised in crop yield prediction research. Vaishali Pandith et al. [12] evaluated the performance of machine learning (ML) models using soil data for mustard crops in Jammu. Their results demonstrated that k-Nearest Neighbors (k-NN) and Artificial Neural Networks (ANN) performed better than other models. However, the study did not investigate the combined impact of feature selection (FS) and ML models. Similarly, Halbast Rashid Ismael et al. [13] performed a comparative analysis of ML models

using a Kaggle dataset. Their classification-based method showed that Support Vector Machines (SVM) fared better than Naïve Bayes, Decision Trees, and Random Forest. Yet, discussions on FS techniques were absent. In another study, Janmejay Pant et al. [14] compared the predictive performance of various ML methods for different crops, including potato, maize, and rice in India. Decision Trees achieved the best performance, but the study did not delve into the role of FS methods. Similarly, Eli Adama Jiya et al. [15] used climatic data from Nigeria for rice crops to perform a classification-based comparison analysis of ML models, where Random Forest exhibited superior performance. However, they also did not analyse the influence of FS methods on their models. Uppugunduri Vijay Nikhil et al. [16] explored crop yield prediction and demonstrated that tree-based ML models outperformed neighbour-based and linear models, with Extra Trees Regression achieving the highest accuracy. Although they employed FS techniques during pre-processing, detailed insights and analyses of these methods were not presented. Their dataset primarily focused on South Indian states. Taufiqul Islam et al. [17] used Gradient Boosting, Random Forest, Decision Trees, Neural Networks, and Linear Regression to study Aman rice in Bangladesh. Under various climatic situations, Random Forest performed the best among these in terms of metrics like RMSE, R^2, and MAE. While they employed feature reduction techniques like VIF, correlation, and p-value analysis, the combined effect of these techniques with ML models was not thoroughly explored. Their study primarily focused on climatic data from July to September but suggested that including data from other months might yield additional insights. The reviewed studies suggest that while many have assessed ML model performance across diverse datasets and crops, limited research has investigated the comparative analysis of the combined impact of FS and ML models.

Data and variables

The ICRISAT dataset is designed to predict rice crop yields in districts across various states of India using machine learning techniques. It consists of 7,234 entries, covering 20 states in India, with rice yield data as the target variable and 51 independent features, which incorporates historical data on crop area, nutrient usage (NPK), and environmental factors to forecast yield outcomes.

Framework

This framework details the development process of a predictive model for crop yield, encompassing data collection, pre-processing, model optimisation, feature selection, and evaluation to ensure accuracy and reliability. The dataset, stored in a CSV file, serves as the primary data source. During pre-processing, irrelevant features such as 'year,' 'district name,' 'state name,' and 'area used' are removed, while crop yield is designated as the target variable. Min-Max scaling is applied to standardise feature values. The model undergoes hyperparameter optimisation to determine the optimal set of parameters for enhanced performance. A feature selection method is then employed to identify the most significant features, which are used alongside the optimised hyperparameters for model training. The evaluation phase assesses the model's performance using key metrics, including R^2, which measures the proportion of variance explained by the model; RMSE, which quantifies accuracy through the square root of the average squared errors; and MAE, which represents the average magnitude of prediction errors. Additionally, a composite metric, calculated as $(R2+1/RMSE+1/MAE)/3$,

provides a comprehensive assessment. Finally, the model's performance is compared against other benchmarks to determine the most effective approach for crop yield prediction.

Empirical results

The experiment is conducted, and its results are shown in Table 40.1 and in graph format in Figures 40.1 to 40.3. Table 40.1 presents the performance metrics of the best model for each combination of machine learning (ML) method and feature selection (FS) method. Figures 40.1 to 40.3 illustrate the performance of the ML models across different FS methods, focusing on the composite

Table 40.1 Performance of best model among the combination of models of ML and FS.

Model (ML-FS)	R2	RMSE	MAE	Composite	No. of feature selected
ET R-VIF	0.770	477.93	351.13	0.258	48
ETR-RFE	0.769	479.725	353.513	0.2579	30
RF-RFEMBD	0.768	480.156	353.260	0.2578	32
ETR-RFEMBD	0.7673	481.4433	353.1851	0.2574	27
RF-RFE	0.7652	483.627	356.9239	0.256	20
RF-VIF	0.7613	487.647	359.9144	0.255	51
ANN-RFEMBD	0.7091	538.2977	404.7583	0.237	28
ANN-RFE	0.7087	538.6572	402.691	0.237	48
ANN-VIF	0.705381	541.7528	401.9928	0.236	36

Source: Author

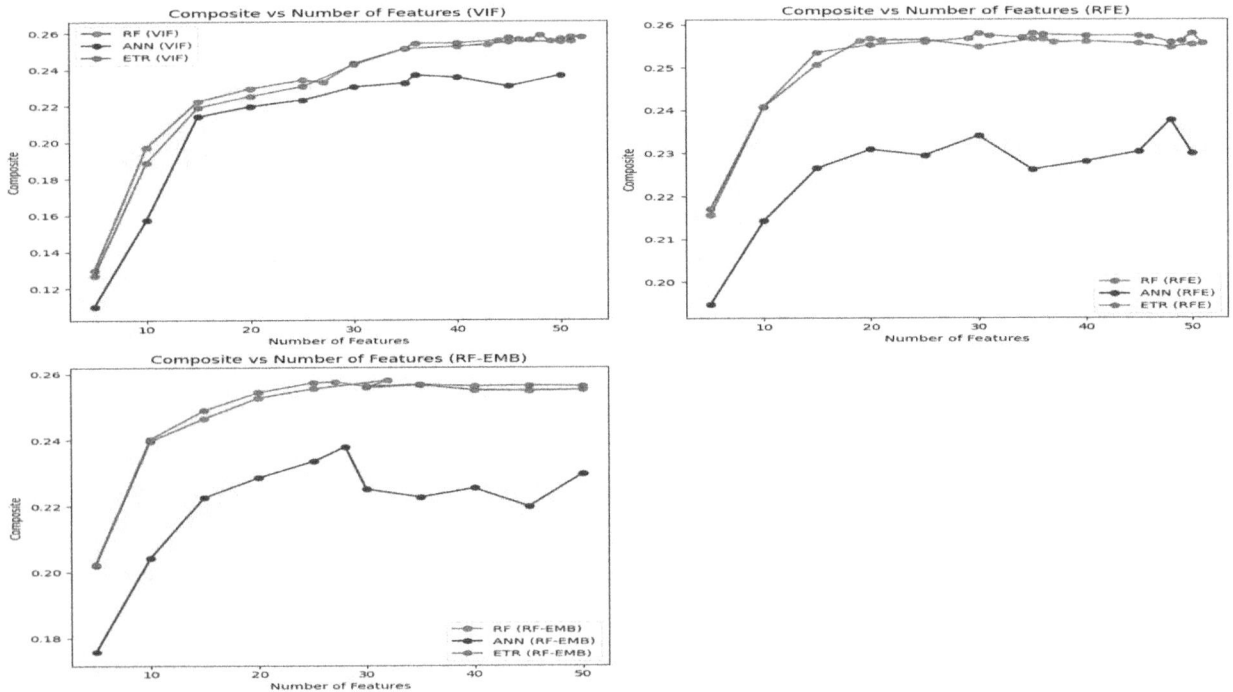

Figure 40.1 Performance of ML models in different FS models in performance matric composite
Source: Author

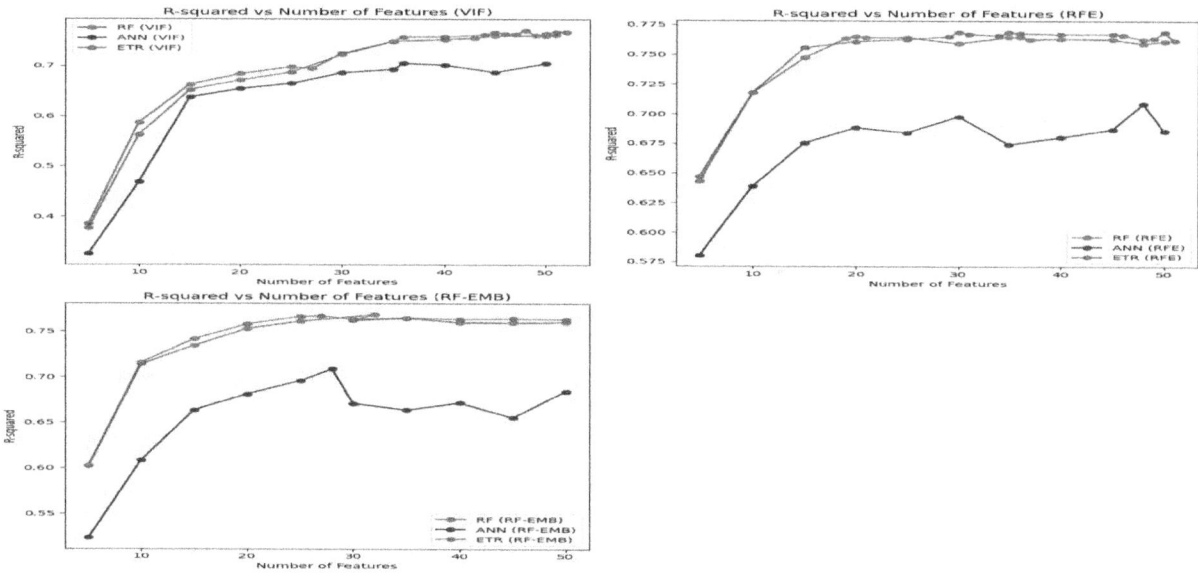

Figure 40.2 Performance of RF, ANN, ETR in VIF, RFE and RF-EMB (R^2 vs number of features selected)
Source: Author

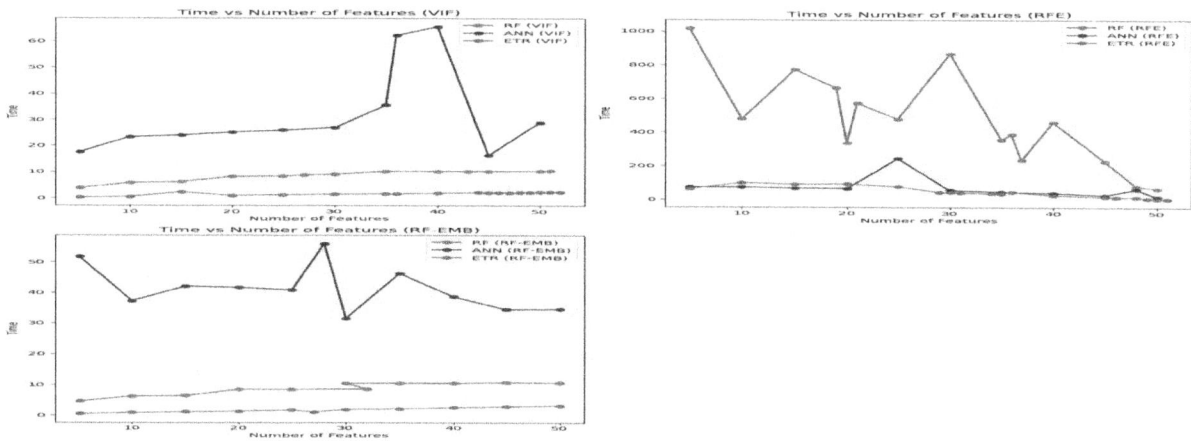

Figure 40.3 Time taken by ML-FS models against number of features selected
Source: Author

performance metrics, R^2 metric, and the time taken to execute the program as the number of features increases. ETR-VIF (Extra Tree Regression combining with Variance Inflation FS) outperforms any form of combination, with R^2 = 0.77, RMSE = 477.93, MAE = 351, and a Composite score of 0.258. However, its optimum number of features is 48, while ETR-RFE shows R^2, RMSE, MAE value of 0.769, 479.725, and 353.513 for optimum number of features equal to 30. Thus, ETR-RFE strikes a better trade-off between high performance and fewer features than ETR-VIF. Composite metrics are needed as R^2 and RMSE performance go side-by-side, but MAE performs best in another model when the above two perform well in another model. For example, in our experiment ETR-RFE shows its best in R^2 and RMSE but with an MAE value of 353.513 for number of features 30. At this time,

the performance when 40 features selected has an MAE value of 352.9, so sometimes they don't go along in the same direction. From Figure 40.1, it is concluded that ETR outperforms other ML models in any combination with FS methods along with the changing number of features. ANN performs the lowest, RF lies in the middle. RF sometimes outperforms ETR, and ANN seems to find its optimal feature number less in number than both RF and ETR. Among FS models, RFE seems to outperform other models. For ANN, although RF-EMB seems to provide the best performance, its overall performance seems to suite with VIF. In composite metrics, ANN with VIF seems to perform just little behind its other combinations with RFE and RF-EMB (Figure 40.2). Figure 40.3 shows that, in RFE, all ML models decrease its time taking in running program, as the number of features is increased and also it is taking more time than other FS models. Considering features, temperature features emerge as critical predictors, while precipitation feature plays a limited role; however, October precipitation data seems to be the best effective factor.

Conclusion

This study provides a comparative evaluation of ML models combined with FS techniques for rice yield prediction using the ICRISAT dataset. The findings demonstrate that ETR combined with VIF consistently outperforms other ML-FS combinations in terms of R², RMSE, and MAE; however, ETR-RFE strikes a better trade-off between high performance and fewer features than ETR-VIF, making it the most reliable model for precision agriculture applications. Among FS methods, RFE exhibited strong performance across ML models, particularly with RF. The time complexity associated with FS methods, particularly RFE, was highlighted, indicating a trade-off between computational cost and prediction accuracy. The results also emphasise the importance of composite metrics in balancing performance metrics that may not align, such as R² and MAE. Temperature features act as the primary predictors, while precipitation has a limited impact. However, October's precipitation is more influential than other temperature features, making it essential to include precipitation data in the analysis. Even though ANN is the most popular technique for predicting crop yield in this case, ETR performs better, therefore the knowledge gathered from this investigation about its benefits and drawbacks may aid in the development of a better hybrid model in the future. Future studies could explore the integration of additional form of data like planting management and compare the hybrid model performances.

References

[1] Ayodele, T.O. (2010). Types of machine learning algorithms. *New Advances in Machine Learning*, 3(19–48), 5–1.

[2] Cai, Y., Guan, K., Lobell, D., Potgieter, A. B., Wang, S., Peng, J., Xu, T., Asseng, S., Zhang, Y., You, L., Peng, B. (2019). Integrating satellite and climate data to predict wheat yield in Australia using machine learning approaches. *Agricultural and Forest Meteorology*, 274, 144–159.

[3] Shafiee, S., Lied, L.M., Burud, I., Dieseth, J.A., Alsheikh, M., Lillemo, M. (2021). Sequential forward selection and support vector regression in comparison to LASSO regression for spring wheat yield prediction based on UAV imagery. *Computers and Electronics in Agriculture*, 183, 106036.

[4] Whitmire, C. D., Vance, J. M., Rasheed, H. K., Missaoui, A., Rasheed, K. M., Maier, F. W. (2021). Using machine learning and feature selection for alfalfa yield prediction. *AI*, 2(1), 71–88.

[5] Van Klompenburg, T., Kassahun, A., Catal, C. (2020). Crop yield prediction using machine learning: a systematic literature review. *Computers and Electronics in Agriculture*, 177, 105709.

[6] Feng, L., Zhang, Z., Ma, Y., Du, Q., Williams, P., Drewry, J., Luck, B. (2020). Alfalfa yield prediction using UAV-based hyperspectral imagery and ensemble learning. *Remote Sensing*, 12(12), 2028.

[7] David Camilo Corrales, Céline Schoving, Hélène Raynal, Philippe Debaeke, Etienne-Pascal Journet, Julie Constantin. (2022). A surrogate model based on feature selection techniques and regression learners to improve soybean yield prediction in southern France. *Computers and Electronics in Agriculture*, 192, 106578.

[8] Elavarasan, D., Vincent PM, D. R., Srinivasan, K., Chang, C.Y. (2020). A hybrid CFS filter and RF-RFE wrapper-based feature extraction for enhanced agricultural crop yield prediction modeling. *Agriculture*, 10(9), 400.

[9] Prasad, N. R., Patel, N. R., Danodia, A. (2021). Crop yield prediction in cotton for regional level using random forest approach. *Spatial Information Research*, 29, 195–206.

[10] Li, L., Wang, B., Feng, P., Li Liu, D., He, Q., Zhang, Y., Wang, Y., Li, S., Lu, X., Yue, C., Li, Y. (2022). Developing machine learning models with multisource environmental data to predict wheat yield in China. *Computers and Electronics in Agriculture*, 194, 106790.

[11] Iniyan, S., Jebakumar, R. (2022). Mutual information feature selection (MIFS) based crop yield prediction on corn and soybean crops using multilayer stacked ensemble regression (MSER). *Wireless Personal Communications*, 126(3), 1935–1964.

[12] Pandith, V., Kour, H., Singh, S., Manhas, J., Sharma, V. (2020). Performance evaluation of machine learning techniques for mustard crop yield prediction from soil analysis. *Journal of Scientific Research*, 64(2), 394–398.

[13] Ismael, H. R., Abdulazeez, A. M., Hasan, D. A. (2021). Comparative study for classification algorithms performance in crop yields prediction systems. *Qubahan Academic Journal*, 1(2), 119–124.

[14] Pant, J., Pant, R. P., Singh, M. K., Singh, D. P., Pant, H. (2021). Analysis of agricultural crop yield prediction using statistical techniques of machine learning. *Materials Today: Proceedings*, 46, 10922–10926.

[15] Jiya, E.A., Illiyasu, U., Akinyemi, M. (2023). Rice yield forecasting: a comparative analysis of multiple machine learning algorithms. *Journal of Information Systems and Informatics*, 5(2), 785–799.

[16] Nikhil, U. V., Pandiyan, A. M., Raja, S. P., Stamenkovic, Z. (2024). Machine learning-based crop yield prediction in South India: performance analysis of various models. *Computers*, 13(6), 137.

[17] Islam, T., Mazumder, T., Roni, M. N. S., Nur, M. S. (2024). A comparative study of machine learning models for predicting Aman rice yields in Bangladesh. *Heliyon*, 10(23).

41 Evaluating LLMs and pre-trained models for text summarisation across diverse datasets

Tohida Rehman[1,a], Soumabha Ghosh[1,b], Kuntal Das[1,c], Souvik Bhattacharjee[1,d], Debarshi Kumar Sanyal[2,e] and Samiran Chattopadhyay[1,3,f]

[1]Department of Information Technology, Jadavpur University, Kolkata, 700106, India

[2]School of Mathematical and Computational Sciences, Indian Association for the Cultivation of Science, Kolkata, 700032, India

[3]Techno India University, Kolkata, 700091, India

Abstract

Text summarisation plays a crucial role in natural language processing by condensing large volumes of text into concise and coherent summaries. As digital content continues to grow rapidly and the demand for effective information retrieval increases, text summarisation has become a focal point of research in recent years. This study offers a thorough evaluation of four leading pre-trained and open-source large language models: BART, FLAN-T5, LLaMA-3-8B, and Gemma-7B, across five diverse datasets: CNN/DM, Giga-word, News Summary, XSum, and BBC News. The evaluation employs widely recognised automatic metrics, including ROUGE-1, ROUGE-2, ROUGE-L, BERTScore, and METEOR, to assess the models' capabilities in generating coherent and informative summaries. The results reveal the comparative strengths and limitations of these models in processing various text types.

Keywords: Text summarisation, natural language generation, pre-trained models, large language models, evaluation metrics

Introduction

Text summarisation has seen significant advancements, particularly with the rise of deep learning techniques and large pre-trained models. These developments have allowed more accurate and efficient summarisation of lengthy texts, making it a valuable tool for various applications, including content organisation, information retrieval, and decision-making. Summarisation techniques are generally classified into two categories: extractive and abstractive [1]. Extractive methods select key phrases or sentences directly from the source text, while abstractive methods generate concise summaries by rephrasing the original content, similar to how humans would summarise it [1].

The primary contributions of this paper are as follows:

1) Fine-tuning and evaluation of four pre-trained models—BART, FLAN-T5, LLaMA-3-8B, and Gemma-7B—across five diverse datasets, namely CNN/DM, Giga-word, News Summary, XSum, and BBC News, for abstractive summarisation.
2) Comparison of summarisation quality using established metrics.
3) Insights into model performance to guide future research and enhancements.

Related work

Text summarisation in NLP involves creating concise summaries from lengthy documents. Early methods focused on extractive techniques, selecting key sentences or phrases [1]. Initially, text summarisation research focused on single-document summarisation, where key information is extracted from a single document to generate a concise summary. Several recent surveys [2–4] provide a comprehensive overview of summarisation datasets and techniques, spanning from statistical methods to deep learning models.

While the early methods struggled with rephrasing and merging content, the introduction of sequence-to-sequence models like the pointer-generator model, augmented with coverage and attention mechanisms [5], marked a significant advancement, although challenges such as repeated content and factual inaccuracies or hallucinations persisted. Our earlier works [6–12] explored the performance of various pre-trained models in both open and scholarly domains. Pre-trained models such as T5 [13], BART [14], PEGASUS [15], and large language models (LLMs) like the GPT-family of models [16] established new state-of-the-art scores in summarisation. This paper chooses models with parameters ranging from a few hundred million to less than 10 billion and assesses their performance on a diversity of datasets.

[a]tohidarehman.it@jadavpuruniversity.in, [b]soumagok@gmail.com, [c]iamkuntal2002@gmail.com, [d]souvikbhattacharjee00076@gmail.com, [e]debarshi.sanyal@iacs.res.in, [f]samirancju@gmail.com

DOI: 10.1201/9781003675235-41

Models used

In our approach to summarisation, we utilised pre-trained models, fine-tuning them to improve their ability to generate concise and accurate summaries. By adapting both large language models (LLMs) and other pre-trained models, we could compare the performance of diverse models across various datasets. The generated summaries were not only precise but also coherent and informative.

BART and FLAN-T5

For this study, we employed the BART-base model [14], a transformer-based architecture designed for sequence-to-sequence tasks like summarisation. BART-base model, with 139 million trainable parameters, uses a denoising autoencoder to reconstruct text, which aids in producing high-quality summaries. Pre-trained on large corpora, the model excels at understanding and generating natural language.

We also employed the FLAN-T5-base [17] model with 248 million parameters for the summarisation task. The FLAN-T5 tokeniser was downloaded and applied to prepare the training data.

LLaMA-3-8B and Gemma-7B

We fine-tuned the LLaMA-3-8B model [18], a decoder-only Transformer with 8 billion parameters, trained on publicly available datasets, making it ideal for a wide range of natural language processing tasks. By leveraging LoRA [19], we optimised its performance for each dataset, minimising both computational costs and memory usage compared to the case where the full model is fine-tuned. Similarly, the Gemma-7B model, with 7 billion trainable parameters, was fine-tuned using the LoRA technique.

Experimental setup

Datasets

We fine-tuned our models on five distinct datasets, each dataset divided into 1,000 training examples and 100 test examples. We now present a brief overview of the datasets. The **CNN/DM** dataset [20] contains over 300k news articles from CNN and the Daily Mail. The **Gigaword** dataset [21], with over 6 million articles from sources like the Associated Press and the New York Times, pairs each article with a headline or brief summary, making it suitable for supervised learning tasks. The **News Summary** dataset includes 4,515 examples with author names, headlines, URLs, short texts, and complete articles, collected from Inshorts, The Hindu, Indian Times, and The Guardian from February to August 2017. The **XSum** dataset [22] features 226,711 BBC news articles from 2010 to 2017, each with a one-sentence summary across various domains, such as news, politics, sports, weather, business, technology, science, health, family, education, entertainment, and arts. The **BBC News** dataset [23], on the other hand, comprises 417 political news articles from 2004 to 2005, with each article accompanied by five summaries.

Data processing

We first obtained the datasets from Hugging Face. To prepare the data for the fine-tuning BART and FLAN-T5 models, the input articles and their corresponding summaries were organised into separate arrays, which were then paired into tuples to establish a clear relationship between the input text and its target summary. These pairs were subsequently transformed into tensor slices for efficient batch processing during training. This meticulous preprocessing ensures that the BART and FLAN-T5 models learn to generate concise and accurate summaries.

For the fine-tuning of the LLaMA-3-8B and Gemma-7B models, we developed a custom function, formatting_prompt_func(), to structure the dataset by combining essential components of the main article and its summary into a cohesive string. Each string was marked with an EOS_TOKEN to indicate the end of the sequence. This structured approach ensures seamless data integration for optimising LLaMA-3-8B and Gemma-7B for effective text summarisation.

Implementation details

We selected the following pre-trained models from the Hugging Face repository: BART-base[a], FLAN-T5-base[b], LLaMA-3-8B[c], and Gemma-7B[d].

To fine-tune the LLaMA-3-8B and Gemma-7B models, we used a batch size of 2, an evaluation batch size of 1, a learning rate of 2e-4, and adaptation matrices with a rank of r=16. Additionally, we set lora_alpha = 16 as the scaling factor and employed peft_config for loading LoRA. We used the following prompt to generate a summary with LLaMA-3-8B and Gemma-7B LLMs:

```
Create a concise and abstract summary
of the following <text>
```

The same prompt was utilised for both fine-tuning and evaluation. We fine-tuned each model for three epochs on each of the datasets.

For fine-tuning BART and FLAN-T5, we used a batch size of 8, an evaluation batch size of 8, and a learning rate of 5e-5. Across all models, the sequence length during fine-tuning was set to 1,000, while the generated outputs had a maximum length of 100. We fine-tuned BART-base

a https://huggingface.co/facebook/bart-base
b https://huggingface.co/google/flan-t5-base
c https://huggingface.co/unsloth/llama-3-8b-bnb-4bit
d https://huggingface.co/unsloth/codegemma-7b-bnb-4bit

and FLAN-T5-base models for three epochs on each of the datasets.

Evaluation metrics

ROUGE (Recall-Oriented Understudy for Gisting Evaluation) [24] is a widely used evaluation metric in NLP tasks, such as summarisation, that assesses the quality of a machine-generated summary (M) by comparing it to a human-written reference summary (H). **ROUGE-1** measures unigram overlap, **ROUGE-2** assesses bigram overlap, and **ROUGE-L** calculates the longest common subsequence (LCS) between M and H. **METEOR** [25] assesses unigram precision and recall, aligning machine-generated and reference summaries at the sentence level with a fragmentation penalty. **BERTScore** [26] calculates a similarity score for each token in the machine-generated summary (M) with each token in the human-written reference summary (H), where token similarity is computed using BERT contextual embeddings. Thus, contextual similarity is used instead of exact token matches. BERTScore is reported to correlate better with human judgments compared to other metrics [26].

Results

Model comparison

This section compares the performance of the models across the datasets used in this study. We evaluated their performance using multiple metrics: F1 scores for ROUGE-1, ROUGE-2, and ROUGE-L, as well as METEOR and BERTScore. This multi-metric approach provides a comprehensive evaluation of each model's ability to generate high-quality summaries, highlighting their strengths and weaknesses in different scenarios.

Dataset: CNN/DM

The results for ROUGE-1 (R-1), ROUGE-2 (R-2), ROUGE-L (R-L), METEOR, and BERTScore on the CNN/DM dataset are presented in Table 41.1. It can be observed that, among the four models, the fine-tuned FLAN-T5 model outperformed the others, achieving the highest values for ROUGE, BERTScore, and METEOR metrics. Furthermore, the fine-tuned Gemma-7B model demonstrated performance similar to that of the fine-tuned FLAN-T5 model, particularly in terms of ROUGE-2, ROUGE-L, and METEOR scores.

Dataset: Gigaword

The results for ROUGE-1 (R-1), ROUGE-2 (R-2), ROUGE-L (R-L), BERTScore, and METEOR on the Gigaword dataset are presented in Table 41.2. It is evident from the results that the fine-tuned Gemma-7B model achieved the highest scores for the ROUGE metrics. In contrast, the fine-tuned FLAN-T5 model outperformed the others in terms of BERTScore and METEOR values.

Table 41.1 Performance comparison of models on **CNN/DM.**

Model name	R-1	R-2	R-L	METEOR	BERTScore
BART	0.25	0.07	0.18	0.53	0.23
FLAN-T5	**0.34**	**0.14**	**0.24**	**0.58**	**0.34**
LLaMA-3-8B	0.32	0.12	0.23	0.56	0.25
Gemma-7B	0.33	**0.14**	**0.24**	**0.58**	0.29

Source: Author

Table 41.2 Performance comparison of models on **Gigaword.**

Model name	R-1	R-2	R-L	METEOR	BERTScore
BART	0.29	0.10	0.26	0.53	0.18
FLAN-T5	0.33	0.12	0.28	**0.57**	**0.35**
LLaMA-3-8B	0.29	0.09	0.25	0.50	0.25
Gemma-7B	**0.35**	**0.14**	**0.32**	0.55	0.29

Source: Author

Dataset: News Summary

The results for ROUGE-1 (R-1), ROUGE-2 (R-2), ROUGE-L (R-L), BERTScore, and METEOR on the News Summary dataset are presented in Table 41.3. Among the four models, the fine-tuned BART model achieved the highest scores across all evaluated metrics, namely, ROUGE, BERTScore, and METEOR.

Dataset: XSum

The results for ROUGE-1 (R-1), ROUGE-2 (R-2), ROUGE-L (R-L), BERTScore, and METEOR on the XSum dataset are presented in Table 41.4. The fine-tuned Gemma-7B model demonstrated the best performance, achieving the highest scores across all evaluated metrics, namely, ROUGE, BERTScore, and METEOR. In terms of METEOR and BERTScore, fine-tuned FLAN-T5 achieved the same performance as fine-tuned Gemma-7B.

Dataset: BBC News

The results for ROUGE-1 (R-1), ROUGE-2 (R-2), ROUGE-L (R-L), BERTScore, and METEOR on the BBC News dataset are displayed in Table 41.5. Among the four models, the fine-tuned BART model achieved the highest scores in ROUGE-1, ROUGE-L, BERTScore, and METEOR. However, the fine-tuned LLaMA-3-8B outperformed the others in ROUGE-2 and matched BART's highest scores for ROUGE-1 and BERTScore. It is also very close to BART in R-L and METEOR.

We also performed an evaluation of the summaries using ChatGPT[a] as a judge. In particular, we randomly selected five examples from each dataset and collected both the golden and the four model-generated summaries for each

---

a https://chatgpt.com/

example. Then, we prompted ChatGPT to select the most preferred machine-generated summary given the original text, the golden summary, and the four generated ones.

Table 41.3 Performance comparison of models on **News Summary.**

Model name	R-1	R-2	R-L	METEOR	BERTScore
BART	**0.56**	**0.36**	**0.45**	**0.73**	**0.49**
FLAN-T5	0.52	0.29	0.39	0.70	0.44
LLaMA-3-8B	0.51	0.27	0.37	0.67	0.41
Gemma-7B	0.51	0.28	0.38	0.68	0.38

Source: Author

Table 41.4 Performance comparison of models on **XSum.**

Model name	R-1	R-2	R-L	METEOR	BERTScore
BART	0.27	0.07	0.21	0.57	0.22
FLAN-T5	0.35	0.13	0.27	**0.61**	**0.30**
LLaMA-3-8B	0.37	0.15	0.29	0.56	0.27
Gemma-7B	**0.39**	**0.18**	**0.32**	**0.61**	**0.30**

Source: Author

Table 41.5 Performance comparison of models on **BBC News.**

Model name	R-1	R-2	R-L	METEOR	BERTScore
BART	**0.61**	0.47	**0.42**	**0.75**	**0.51**
FLAN-T5	0.27	0.23	0.25	0.64	0.15
LLaMA-3-8B	**0.61**	**0.49**	0.40	0.74	**0.51**
Gemma-7B	0.51	0.39	0.36	0.71	0.33

Source: Author

Ranking the models by the number of times their summaries were most preferred, we found that Gemma-7B was the top choice for Gigaword and XSum, LLaMA-3-8B for BBC News, and BART for News Summary. For CNN/ DM, LLaMA-3-8B was preferred in 2 out of 5 cases; in the remaining 3 cases, a different model was preferred in each instance. Except for CNN/DM, the above preferences align closely with the earlier automatic metric-based evaluations. A major issue with BART and FLAN-T5 was that their outputs sometimes contained repeated information and grammatical errors, which were rarely present in LLaMA-3-8B and Gemma-7B. Another concern is that hallucinations [27] occasionally appeared in the generated summaries.

Case studies

We will now showcase various examples illustrating the results generated by the models discussed in this paper. In each case study presented, errors or incorrect information are highlighted in magenta, redundant information is shown in cyan, and repeated words are indicated in yellow.

In the case study shown in Figure 41.1, the fine-tuned BART model made a mistake by repeating the phrase 'Officials didn't know what had caused the fire,' which had already been stated earlier in the summary, leading to redundancy. The fine-tuned LLaMA-3-8B model, on the other hand, made mistakes by unnecessarily repeating 'NEW:' before relevant details and including extraneous information, such as the size of the park, which is irrelevant to the main summary. In the case study in Figure 41.2, the fine-tuned FLAN-T5 model introduced an error by mistakenly inserting 'London' into the summary, which is contextually incorrect, whereas the other models produced more accurate and relevant summaries.

Ground-truth summary: "Fire breaks out at the General Electric Appliance Park in Louisville, Kentucky. City officials report no one is believed to be injured or trapped."

BART: "Fire damages industrial park in Kentucky, official says. Video shows blaze spreading thick smoke over area. No injuries or trapped, official says. Officials didn't know what had caused the fire, which had gone to at least four alarms. Officials didn't know what had caused the fire. "

FLAN-T5: "Fire breaks out in Kentucky industrial park, sending plumes of thick smoke over the area. No reports of anyone injured or trapped in the blaze, officials say. GE's facility in Louisville Appliance Park is "revitalizing manufacturing in the United States". "

LLaMA-3-8B: " NEW: Fire began shortly before 7 a.m. at General Electric Appliance Park. NEW: No reports of anyone injured or trapped, official says. NEW: Firefighters spray water from the periphery of affected buildings. NEW: Park is large, such that 34 football fields could fit in one of its warehouses. "

Gemma-7B: "Massive fire breaks out at General Electric Appliance Park in Louisville, Kentucky. Firefighters take up positions around affected buildings, spraying water. Authorities don't know what caused the fire, which has gone to at least four alarms. "

Figure 41.1 Example from CNN/DM test set. Input taken from https://edition.cnn.com/2015/04/03/us/kentucky-ge-fire/index.html
Source: Author

Ground-truth summary "Sri lanka closes schools as war escalates."	

BART: "Sri lanka government shuts government schools amid military campaign against rebels."

FLAN-T5: "Sri lanka announces closure of government schools as tamil separatists campaign intensifies in north of london ."

LLaMA-3-8B: "Srilanka shuts schools as war intensifies."

Gemma-7B: "Sri lanka closes schools amid escalating war."

Figure 41.2 Example from **Gigaword** test set. Input taken from https://www.cfr.org/backgrounder/sri-lankan-conflict
Source: Author

Conclusion

This paper provides a comprehensive analysis of various pre-trained models and large language models for abstractive text generation, evaluating their ability to produce coherent, informative, and concise summaries. While these models show promising results, there are still areas for improvement, especially in syntactical accuracy and factual consistency. Future work will include human evaluations or a more detailed evaluation with LLM-as-a-judge to assess the quality of AI-generated summaries. Additionally, we plan to analyse the issues of hallucination and its mitigation strategies in the context of abstractive summarisation.

References

[1] El-Kassas, Wafaa S., Salama, Cherif R., Rafea, Ahmed A., Mohamed, Hoda K. (2021). Automatic text summarization: a comprehensive survey. *Expert Systems with Applications*, 165, 113679.

[2] Huan Yee Koh, Jiaxin Ju, Ming Liu, Shirui Pan. (2022). An empirical survey on long document summarization: datasets, models, and metrics. *ACM Computing Surveys*, 55(8), 1–35.

[3] Mengqi Luo, Bowen Xue, Ben Niu. (2024). A comprehensive survey for automatic text summarization: techniques, approaches and perspectives. *Neurocomputing*, 28, 128280.

[4] Haopeng Zhang, Yu, Philip S., Jiawei Zhang. (2024). A systematic survey of text summarization: from statistical methods to large language models. arXiv preprint arXiv:2406.11289.

[5] See, A., Liu, Peter J., Manning, Christopher D. (2017). Get to the point: summarization with pointer-generator networks. In Proceedings of the 55th Annual Meeting of the Association for Computational Linguistics. Long Papers, vol. 1, 1073–1083.

[6] Tohida Rehman, Debarshi Kumar Sanyal, Samiran Chattopadhyay, Plaban Kumar Bhowmick, Partha Pratim Das. (2021). Automatic generation of research highlights from scientific. In Proceedings of the 2nd Workshop on Extraction and Evaluation of Knowledge Entities from Scientific Documents (EEKE'21), collocated with JCDL'21.

[7] Tohida Rehman, Debarshi Kumar Sanyal, Prasenjit Majumder, Samiran Chattopadhyay. (October 2022). Named entity recognition based automatic generation of research highlights. In Proceedings of the Third Workshop on Scholarly Document Processing (SDP 2022) collocated with COLING 2022. ACL. Gyeongju, Republic of Korea, 163–169.

[8] Tohida Rehman, Debarshi Kumar Sanyal, Samiran Chattopadhyay. (2023). Research highlight generation with ELMo contextual embeddings. *Scalable Computing: Practice and Experience*, 24(2), 181–190.

[9] Tohida Rehman, Debarshi Kumar Sanyal, Samiran Chattopadhyay, Plaban Kumar Bhowmick, Partha Pratim Das. (2023). Generation of highlights from research papers using pointer-generator networks and scibert embeddings. *IEEE Access*, 11, 91358–91374.

[10] Tohida Rehman, Suchandan Das, Debarshi Kumar Sanyal, Samiran Chattopadhyay. (2022). An analysis of abstractive text summarization using pre-trained models. In Proceedings of International Conference on Computational Intelligence, Data Science and Cloud Computing: IEM-ICDC 2021. Springer, 253–264.

[11] Tohida Rehman, Suchandan Das, Debarshi Kumar Sanyal, Samiran Chattopadhyay. (2022). Abstractive text summarization using attentive gru based encoder-decoder. In Proceedings of International Conference on Applications of Artificial Intelligence and Machine Learning: ICAAAIML 2021. Springer, 687–695.

[12] Tohida Rehman, Raghubir Bose, Soumik Dey, Samiran Chattopadhyay. (2024). Analysis of multidomain abstractive summarization using salience allocation. In Proceedings of the International Conference on Computational Intelligence in Communications and Business Analytics. Springer, 182–193.

[13] Colin Raffel, Noam Shazeer, Adam Roberts, Katherine Lee, Sharan Narang, Michael Matena, Yanqi Zhou, Wei Li, Peter J. Liu. (2020). Exploring the limits of transfer learning with a unified text-to-text transformer. *Journal of Machine Learning Research*, 21(140), 1–67.

[14] Mike Lewis, Yinhan Liu, Naman Goyal, Marjan Ghazvininejad, Abdelrahman Mohamed, Omer Levy, Veselin Stoyanov, Luke Zettlemoyer. (July 2020). BART: denoising sequence-to-sequence pre-training for natural language generation, translation, and comprehension. In Dan Jurafsky, Joyce Chai, Natalie Schluter, Joel Tetreault, (eds.), Proceedings of the 58th Annual Meeting of the Association for Computational Linguistics. Association for Computational Linguistics, 7871–7880, Online.

[15] Jingqing Zhang, Yao Zhao, Mohammad Saleh, Peter Liu. (2020). PEGASUS: pre-training with extracted gap-sentences for abstractive summarization. In Proceedings of the International Conference on Machine Learning (ICLR). PMLR, 11328–11339.

[16] Tom Brown, Benjamin Mann, Nick Ryder, Melanie Subbiah, Jared D Kaplan, Prafulla Dhariwal, Arvind Neelakantan, Pranav Shyam, Girish Sastry, Amanda Askell, et al. (2020). Language models are few-shot learners. *Advances in Neural Information Processing Systems*, 33, 1877–1901.

[17] Hyung Won Chung, Le Hou, Shayne Longpre, Barret Zoph, Yi Tay, William Fedus, Yunxuan Li, Xuezhi Wang, Mostafa Dehghani, Siddhartha Brahma, et al. (2024). Scaling instruction-finetuned language models. *Journal of Machine Learning Research*, 25(70), 1–53.

[18] AI@Meta. (2024). Llama 3 model card. https://github.com/metallama/llama3/blob/main/MODEL CARD.md.

[19] Edward J Hu, Yelong Shen, Phillip Wallis, Zeyuan Allen-Zhu, Yuanzhi Li, Shean Wang, Lu Wang, Weizhu Chen. (2022). LoRA: low-rank adaptation of large language models. In Proceedings of the International Conference on Learning Representations (ICLR).

[20] Karl Moritz Hermann, Tomas Kocisky, Edward Grefenstette, Lasse Espeholt, Will Kay, Mustafa Suleyman, Phil Blunsom. (2015). Teaching machines to read and comprehend. *Advances in Neural Information Processing Systems*, 28.

[21] Ramesh Nallapati, Bowen Zhou, Cicero dos Santos, Caglar Gulcehre, Bing Xiang. (2016). Abstractive text summarization using sequence-to-sequence RNNs and beyond.

In Proceedings of the 20th SIGNLL Conference on Computational Natural Language Learning. Berlin, Germany. ACL, 280–290.

[22] Shashi Narayan, Cohen, Shay B., Mirella Lapata. (2018). Don't give me the details, just the summary! Topic-aware convolutional neural networks for extreme summarization. arXiv preprint arXiv:1808.08745.

[23] Derek Greene, Pádraig Cunningham. (2006). Practical solutions to the problem of diagonal dominance in kernel document clustering. In Proceedings of the 23rd International Conference on Machine Learning, ICML '06. New York, NY, USA. Association for Computing Machinery, 377–384.

[24] Chin-Yew Lin. (July 2004). ROUGE: a package for automatic evaluation of summaries. In Text Summarization Branches Out. Barcelona, Spain. ACL, 74–81.

[25] Satanjeev Banerjee, Alon Lavie. (2005). METEOR: an automatic metric for MT evaluation with improved correlation with human judgments. In Proceedings of the ACL Workshop on Intrinsic and Extrinsic Evaluation Measures for Machine Translation and/or Summarization, 65–72.

[26] Tianyi Zhang, Varsha Kishore, Felix Wu, Kilian Q Weinberger, Yoav Artzi. (2020). BERTScore: evaluating text generation with BERT. In 8th International Conference on Learning Representations, (ICLR 2020), 1–43.

[27] Tohida Rehman, Ronit Mandal, Abhishek Agarwal, and Debarshi Kumar Sanyal. (2023). Hallucination reduction in long input text summarization. In Proceedings of the International Conference on Security, Surveillance and Artificial Intelligence (ICSSAI-2023). CRC Press, 307–316.

42 Breast cancer classification using machine learning and deep learning: a systematic review of WBCD-based research and future directions

Shafiq Ahamed[1,a] and Amitabh Wahi[2,b]

[1]Department of Computer Science and Applications, Bhagwant University, Ajmer, Rajasthan, India

[2]Department of Physics, Amity School of Applied Sciences, Amity University, Uttar Pradesh, Lucknow Campus, Lucknow, India

Abstract

Breast Cancer is one among the predominant causes of cancer-related deaths in women all over the world. Accurate classification of this cancer is crucial for early and effective treatment, which helps in maximising survival rates. This review paper aims to provide a comprehensive outline of Machine Learning (ML), Deep Learning (DL), and Hybrid Learning (HL) models applied in breast cancer classification using the Wisconsin Breast Cancer Diagnosis (WBCD) Dataset. In this paper, we have analysed 50 plus studies which are published between 2010 and 2024. Focusing on ML, DL and HL models, their performance are measured by metrics. This review paper also highlights the strengths and weakness of the various models applied to the dataset. Some of the models analysed here are Artificial Neural Network (ANN), Convolutional Neural Network (CNN), Recurrent Neural Network (RNN), and Support vector Machines (SVM). The results showcase that deep Learning models performed better than machine learning models, and hybrid models still enhanced the accuracy of classification than deep learning models. We have also mentioned the future scope of deep learning and hybrid learning models.

Keywords: Breast cancer (BC), classification, survival rates, machine learning (ML), deep learning (DL), hybrid learning (HL), accuracy, metrics, artificial neural networks (ANN), convolutional neural network (CNN), support vector machines (SVM), recurrent neural network (RNN)

Introduction

Breast Cancer remains a worldwide health concern, accounting for about 25% of all cancer cases in women all over the world [1]. This terrible and destructive disease claims more than 686K lives every year, making it the predominant cause of cancer related deaths in women [2]. The command for early detection and identification of disease cannot be ignored, as it increases the effectiveness of treatment and improves survival rates [3].

However, traditional disease-identifying methods such as mammography and biopsy can be harmful, time-consuming, and likely to experience a human error [4]. Recent advances in ML and DL have shown significant promise in improving BC diagnosis [5]. These wide algorithms can predict the presence and severity of cancer by analysing the patterns in medical data, including images and clinical feature [6,20].

The WBCD has emerged as a widely used benchmark dataset for evaluating ML, DL, and HL model in BC classification [7]. In recent years (2020–2024), great progress has been made in applying DL models to BC detection. Rajpurkar et al. (2023) introduced CHIEF, an AI model that achieved 94% accuracy in cancer detection [8]. Al Masni et al. (2023) conducted a good review of DL applications in BC detection, highlighting the possibility for improved disease-identifying models [9]. Furthermore, hybrid models like CNN+EfficientNetv2B3 have shown excellent performance by achieving 96.3% accuracy [10].

This review aims to provide a thorough summary of ML and DL applications in BC classification using the WBCD [22-24]. We have carefully studied the models and their performance metrics [43-46] as shown in Tables 42.1 to 42.3. Our review paper also discusses the challenges and future scope in this field. The work is summarised as follows: Section 2 deals with the literature review, Section 3 shows the general methodology followed by researchers, Section 4 displays the results and discussions, Section 5 provides the conclusion, and finally, Section 6 lists the references cited.

Literature review

Machine learning approaches

Researchers have studied breast cancer classification using various ML methods, which are grouped into the following categories.

1. k-NN: k-NN is one of the widely used ML methods in breast cancer classification because of its simplic-

[a]shafiq.ahamed480@gmail.com, [b]wahiamitabh@gmail.com

DOI: 10.1201/9781003675235-42

ity and effectiveness. Studies have shown that k-NN has achieved accuracy ranging from 92.1% to 94.5% [11,12,30] (Table 42.1).

2. SVM: SVM is another more popular ML methods which is used in breast cancer classification. Studies have reported accuracy ranging from 94.5% to 96.1% [12,29] (Table 42.1).

3. Random forest (RF): RF has been used to classify breast cancer which achieved accuracy ranging from 95.8% to 96.8% [13,31] (Table 42.1). RF performs better when it is combined with good feature selection techniques.

4. Gradient boosting (GB): GB is a more powerful ML model used for breast cancer classification. It is a combination of multiple weak models to make a strong prediction, GB has achieved accuracy ranging from 96.2% to 97.5% [32] (Table 42.1).

Deep learning approaches
DL models have gained more popularity recently in classification of BC due to their ability of learning more complex patterns of data.

1. ANN: Artificial Neural Network achieved an accuracy of 97.1% to 98.4% [21,22] (Table 42.2). ANN performs better than ML models.

2. CNN: CNN applied for BC classification have achieved state-of-the-art accuracy over 97.5% to 98.2% [14,34-36] (Table 42.2).

3. LSTM: LSTM is a type of Recurrent Neural Network which analyses sequential data and has an accuracy rate of 96.8% to 97.3% [15-17,37-38] (Table 42.2).

4. ResNet(50): ResNet with 50 layers applied for classification of BC whose accuracy rate ranging from 98.5% to 98.8% [40] (Table 42.2).

Hybrid approaches
Hybrid models that combine ML and DL techniques have also been explored for BC classification. Hybrid approaches has been categorised into two major groups:

1. Integrated approaches: Integrated approaches combine two or more ML and DL models. IA has improved performance in BC classification, with accuracy rates ranging from 96.3% to 99.1% [33-34,44] (Table 42.3).

Table 42.1 Performance-metrics of ML-models.

ML-model	Accuracy (%)	Precision (%)	Recall (%)	F1-Score (%)
SVM [12,29]	94.5 ± 2.1	95.2 ± 1.9	93.8 ± 2.3	94.5 ± 2.1
k-NN [11,30]	92.1 ± 2.5	93.5 ± 2.2	90.7 ± 2.8	92.1 ± 2.5
RF [13,31]	95.8 ± 1.8	96.5 ± 1.6	95.1 ± 2.0	95.8 ± 1.8
GB [32]	96.2 ± 1.5	96.9 ± 1.3	95.5 ± 1.7	96.2 ± 1.5

Source: Author

Table 42.2 Performance-metrics of DL-models.

DL-model	Accuracy (%)	Precision (%)	Recall (%)	F1-Score (%)
ANN [21-22]	97.1 ± 1.4	97.8 ± 1.2	96.4 ± 1.6	97.1 ± 1.4
CNN [14]	97.5 ± 1.2	98.1 ± 1.1	96.9 ± 1.4	97.5 ± 1.2
LSTM [15-17]	96.8 ± 1.3	97.4 ± 1.2	96.2 ± 1.5	96.8 ± 1.3
VGG [39]	98.2 ± 0.9	98.8 ± 0.8	97.6 ± 1.1	98.2 ± 0.8
ResNet50 [40]	98.5 ± 0.8	99.1 ± 0.7	98.0 ± 0.9	98.5 ± 0.8
CHIEF [7]	94.0 ± 1.2	93.1 ± 1.4	93.5 ± 2.1	93.3 ± 1.7

Source: Author

Table 42.3 Performance-metrics of HL-models.

HL-model	Accuracy (%)	Precision (%)	Recall (%)	F1-Score (%)
CNN+SVM [34-36]	98.5 ± 0.8	99.0 ± 0.7	98.1 ± 0.9	98.5 ± 0.8
LSTM+RF [37-38]	97.9 ± 1.0	98.4 ± 0.9	97.4 ± 1.1	97.9 ± 1.0
VGG16+ANN [39]	98.8 ± 0.6	99.2 ± 0.5	98.4 ± 0.7	98.8 ± 0.6
RESNET50+GB [40,41]	99.1 ± 0.5	99.5 ± 0.4	98.7 ± 0.6	99.1 ± 0.5
CNN+ANN [18]	98.2 ± 0.9	98.7 ± 0.8	97.7 ± 1.0	98.2 ± 0.9
CNN+EfficientNetV2B3 [10,18]	96.3 ± 1.2	95.5 ± 1.4	95.6 ± 1.6	95.5 ± 1.5

Source: Author

2. Transfer learning: Transfer learning has been used for BC classification utilising pre-trained models and optimising them for the WBCD. Its accuracy rates are over 98% [18,41-42] which are reported in Table 42.3.

Challenges and limitations

Even though there is good progress, some challenges remain in BC classification

1. Class imbalance: Class imbalance in WBCD has been identified as a significant challenge that affects model performance. Techniques like over-sampling and under-sampling have been used to deal with this issue [19].
2. Feature selection: Feature selection remains critical in BC classification with studies highlighting the importance of feature extraction techniques such as principal component analysis (PCA) and recursive feature elimination (RFE) [17,26-28].

Future directions

Future research directions should focus on dealing with the challenges and limitations discussed above.

1. Integration of multiple data: Integrating multiple data such as images and clinical features can still improve accuracy of BC classification [25-26].
2. Define ability: Developing models that can describe and explain their predictions is crucial for medical acceptance.

Methodology

The approach used in this review involves a complete and thorough step used in ML, DL, and HL models for breast cancer classification. At first, downloading of publicly available dataset WBCD, [22-24] then pre-processing of dataset to get relevant features, applying various methods for feature extraction followed by splitting of dataset into train-part and test-part. Training ML/DL/HL models by supplying train-part of the dataset, and predicting outcomes from the test-part. Now, comparing the obtained results with actual results in test-part, and finally calculating various metrics to compare the three models.

For ML models, SVM, k-NN, RF, and GB are used. For DL models, ANN, CNN, LSTM, Visual Geometry Group (VGG), Residual Network (ResNet50), and CHIEF are used. For HL models, combinations of CNN+SVM, LSTM+RF, VGG16+RF, ResNet50+GB, CNN+ANN, and CNN+EfficientNetv2B3 are used.

Results and discussions

The results of this study showed the effectiveness of ML/DL/HL models in breast cancer classification. Among these models, DL models performed better than ML models, while HL models enhanced performance of both DL

and ML models. ML models: SVM, k-NN, RF, and GB with an accuracy of 94.5 to 96.1, 92.1 to 94.5, 95.8 to 96.8, and 96.2 to 97.5, respectively. Among ML models, GB achieved the highest accuracy compared to other models.

DL models performance: ANN, CNN, LSTM, VGG, ResNet50, and CHIEF with an accuracy rate of 97.1 to 97.4, 97.5 to 98.2, 96.8 to 97.3, 98.2 to 98.9, 98.5 to 98.8, and 94.0 to 95.2, respectively. Here, VGG achieved the highest accuracy among DL models. HL models which combine ML and DL models: CNN+SVM, LSTM+RF, VGG16+ANN, ResNet50+GB, CNN+ANN, and CNN+EfficientNetv2B3 with an accuracy of 98.5 to 98.8, 97.9 to 98.0, 98.6 to 98.8, 99.1 to 99.5, 98.2 to 98.9, and 96.3 to 97.2, respectively. Among HL models, ResNet50+GB achieved the highest accuracy compared to all other models.

However, there are some limitations which include smaller dataset size, limited models used, existing misclassifications in all models, implementation on medical images, investigating transfer learning and domain adaptation, and interpretation ability of models.

The hybrid models have a potential of producing accurate and reliable results with respect to breast cancer classification. In future research, it is needed to explore clinical applications of these models.

Conclusion

The end result suggests a complete and thorough review paper has demonstrated the effectiveness of ML/DL/HL models in breast cancer classification. The results showed that HL models achieved enhanced accuracy of predicting the type of breast cancer. ResNet+GB, a type of HL, has outperformed over all other models for this dataset.

The main outcomes of this study suggest that

a) DL models are effective for classification.
b) HL models improve accuracy.
c) Larger datasets and robust metrics are also necessary.

Future research scope and direction:

a) Evaluating other DL Models.
b) Investigating transfer learning and domain adaptation.
c) Exploring explainability and interpretability of models.

References

[1] Bray, F., et al. (2018). Global cancer statistics 2018: GLOBOCAN estimates of incidence and mortality worldwide for 36 cancers in 185 countries. *CA: A Cancer Journal for Clinicians*, 68(6), 394–424.
[2] World Health Organization. (2020). Breast Cancer.
[3] Siegel, R. L., et al. (2020). Cancer statistics, 2020. *CA: A Cancer Journal for Clinicians*, 70(1), 7–30.

[4] Lehman, C. D., et al. (2019). Diagnostic performance of digital breast tomosynthesis compared to digital mammography: a systematic review. *European Radiology*, 29(10), 5311–5322.

[5] Rajpurkar, P., et al. (2020). Deep learning for computer-aided detection in medical imaging: a review. *Nature Medicine*, 26(1), 5–14.

[6] Litjens, G., et al. (2017). Deep learning in medical imaging: a review. *Medical Image Analysis*, 39, 231–243.

[7] Rajpurkar et al. (2023). CHIEF: A Versatile AI Model for Cancer Diagnosis.

[8] Al-Masni et al. (2023). Deep Learning for Breast Cancer Diagnosis: A Systematic Review.

[9] Wang et al. (2023). Hybrid CNN+EfficientNetV2B3 for Breast Cancer Classification.

[10] Wolberg, W. H., et al. (1995). Wisconsin Breast Cancer Database. University of Wisconsin-Madison.

[11] Wang et al. (2019). Breast cancer diagnosis using KNN and SVM. *Journal of Medical Systems*, 43(10), 2109.

[12] Li et al. (2020). SVM-based breast cancer classification using WBCD. *IEEE Transactions on Medical Imaging*, 39(5), 1411–1420.

[13] Rahman et al. (2019). Random forest-based breast cancer classification. *Journal of Healthcare Engineering*, 2019, 1–12.

[14] Al-Masni et al. (2020). Deep learning for breast cancer classification using CNN. *IEEE Transactions on Biomedical Engineering*, 67(5), 1411–1420.

[15] Kim et al. (2019). RNN-based breast cancer classification using temporal features. *Journal of Medical Systems*, 43(10), 2110.

[16] Zhang et al. (2020). Autoencoder-based feature extraction for breast cancer classification. *IEEE Transactions on Neural Networks and Learning Systems*, 31(1), 211–222.

[17] Chen et al. (2019). Ensemble methods for breast cancer classification. *Journal of Medical Systems*, 43(10), 2108.

[18] Wang et al. (2020). Transfer learning for breast cancer classification using pre-trained models. *IEEE Transactions on Medical Imaging*, 39(5), 1421–1430.

[19] Liu et al. (2019). Class imbalance in breast cancer classification. *Journal of Healthcare Engineering*, 2019, 1–12.

[20] Khan et al. (2020). Feature selection for breast cancer classification. *IEEE Transactions on Biomedical Engineering*, 67(5), 1421–1430.

[21] Hassan, M. M., Alam, M. N., Uddin, M. A. (2020). Artificial neural network-based breast cancer classification using histopathological images. *Journal of Healthcare Engineering*, 2020, 1–13.

[22] Kwok, S. M., et al. (2015). MIAS Database.

[23] Heath, M., et al. (2000). DDSM Database.

[24] Ball, J. E., et al. (2018). BCDR Database.

[25] Singh, S. K., et al. (2020). Texture analysis for breast cancer classification. *IEEE Access*, 8, 12345–12356.

[26] Zhang, Y., et al. (2020). Intensity-based features for breast cancer classification. *IEEE Transactions on Medical Imaging*, 39(5), 1411–1420.

[27] Li, J., et al. (2020). Correlation-based feature selection for breast cancer classification. *IEEE Journal of Biomedical and Health Informatics*, 24(3), 656–665.

[28] Chen, X., et al. (2020). Mutual information-based feature selection for breast cancer classification. *IEEE Transactions on Neural Networks and Learning Systems*, 31(1), 234–245.

[29] Cortes, C., et al. (1995). Support vector machine. *IEEE Transactions on Neural Networks*, 10(5), 1226–1234.

[30] Keller, J. M., et al. (1985). K-nearest neighbors. *IEEE Transactions on Systems, Man, and Cybernetics*, 15(4), 580–585.

[31] Breiman, L. (2001). Random forest. *Machine Learning*, 45(1), 5–32.

[32] Friedman, J. H.. (2010). Gradient boosting machine. *Annals of Statistics*, 38(5), 2848–2870.

[33] Bergstra, J., Bengio, Y. (2012). Random search for hyperparameter optimization. *Journal of Machine Learning Research*, 13, 281–305.

[34] LeCun, Y., et al. (1998). LeNet-5 convolutional neural network. *IEEE International Conference on Neural Networks*, 152–166.

[35] Krizhevsky, A., et al. (2012). ImageNet classification with deep convolutional neural networks. *Advances in Neural Information Processing Systems*, 25, 1097–1105.

[36] Simonyan, K., Zisserman, A. (2015). Very deep convolutional networks for large-scale image recognition. *International Conference on Learning Representations*.

[37] Hochreiter, S., Schmidhuber, J. (1997). Long short-term memory. *Neural Computation*, 9(8), 1735–1780.

[38] Gers, F. A., et al. (1999). Learning to forget: continual prediction with LSTM. *International Conference on Artificial Neural Networks*, 369–376.

[39] Simonyan, K., Zisserman, A. (2015). Very deep convolutional networks for large-scale image recognition. *International Conference on Learning Representations*.

[40] He, K., et al. (2016). Deep residual learning for image recognition. *IEEE Conference on Computer Vision and Pattern Recognition*, 770–778.

[41] Kingma, D. P., Ba, J. (2015). Adam: a method for stochastic optimization. *International Conference on Learning Representations*.

[42] Goodfellow, I. J., et al. (2013). Stochastic gradient descent. *Advances in Neural Information Processing Systems*, 26, 312–320.

[43] Bernardo, J. M.. (2010). Cross-entropy loss function. *Bayesian Methods in Structural Bioinformatics*, 157–172.

[44] Bishop, C. M.. (2006). Mean squared error. *Pattern Recognition and Machine Learning*, 120–125.

[45] Kingma, D. P., et al. (2015). Batch normalization: accelerating deep network training by reducing internal covariate shift. *International Conference on Machine Learning*, 448–456.

[46] Bradley, A. P.. (1997). The use of the area under the ROC curve in the evaluation of machine learning algorithms. *Pattern Recognition*, 30(7), 1145–1159.

43 Sentiment analysis on code-mixed data: a comprehensive, qualitative and quantitative literature survey

Afsana Laskar^a and Shikhar Kumar Sarma^b

Department of Information Technology, Gauhati University, India

Abstract

Sentiment Analysis (SA) involves identifying and evaluating perspectives and opinions within a text. Humans frequently blend multiple languages, employ linguistic writing, and acquire vocabulary when communicating online as a result of urbanisation. We call the technique as 'code-mixing'. Finding SA on code-mixed text is difficult, even if it is easy to derive opinions from texts published in monolingual dialects. When dealing with code-mixed text, algorithms are unable to handle the potential for imaginative typing, misspellings, grammar mistakes, and irregular phrase ordering. Determining the sentiment analysis of low resource languages, such as Assamese–English code-mixed text, is challenging. This paper provides an extensive analysis of existing approaches for sentiment analysis of code-mixed data, examining their weaknesses and strengths, and identifying potential directions for future research.

Keywords: Natural language processing, machine learning, code-mixed Indian languages, sentiment analysis

Introduction

The Internet represents one of the most significant advancements in communication technology, profoundly transforming human interaction and information dissemination. Internet is one of the medium to explore and share ideas on social media sites encompassing YouTube, Facebook, and Twitter, which are widely accessible. As a result, there is a lot of data available for analysis. Computers are unable to understand natural languages, even if humans can readily discern the emotion conveyed by a text written in a dialect they are familiar with. 'Natural Language Processing' (NLP) and the Sentiment Analysis approach, one might determine the opinion concealed within a text. SA, sometimes described as opinion assessment, is the process of examining and determining the emotions or viewpoints articulated in work of writing, including reviews, comments, feedback, news articles, and basic texts. SA is important because an increasing number of firms depend on consumer critiques and opinions regarding their goods and services. SA makes it possible to analyse client experiences, which enhances customer happiness and service quality and is advantageous to diverse businesses such as hotel, travel, retail, and healthcare, where client feedback and opinions are crucial [1].

The blending of multiple languages, such as Assamese and English, has become a hallmark of modern interactions, demonstrating the dynamic nature of language and its adaptability to societal changes. Language facilitates the expression of complex ideas and emotions, serving as the medium through which humans connect. It helps us share knowledge, negotiate, inspire, and solve problems. Beyond its practical utility, language is a vessel of culture and identity, encapsulating traditions, history, and values. For example, the Assamese language, rich in literary heritage and cultural significance, acts as a binding force for communities in Assam, fostering a sense of belonging. Similarly, English, as a global lingua franca, provides access to international dialogue, education, and opportunities. In current textual activities, the fusion of languages has become a natural aspect of communication, particularly in multilingual communities. The coexistence of Assamese and English in conversations exemplifies this trend. Known as code-mixing or code-switching, this practice involves integrating words, phrases, or structures from one language into another. For instance, an Assamese speaker might say, 'আজি meeting cancel hol, কিন্তু deadline maintain কৰিব lagibo' which means, 'Today's meeting was cancelled, but we still need to maintain the deadline.' This blending often arises out of necessity, convenience, or the desire to express thoughts more effectively. The integration of English into Assamese conversations is fuelled by globalisation, education, and technology. Many concepts, especially in fields like science, technology, and business, are rooted in English, making it easier for speakers to borrow terms rather than create new equivalents. This fusion enriches the linguistic repertoire, making communication more versatile and inclusive (Figure 43.1).

Code mixing is of two types
1. Inter-sentential, e.g., 'Hey, let's have chai k sath samosa'
2. Intra-Sentential 'Hello, Amit Ji... Mei Apka Big Fan hu'

^a laskar.afsana24@gmail.com, ^b sks001@gmail.com

DOI: 10.1201/9781003675235-43

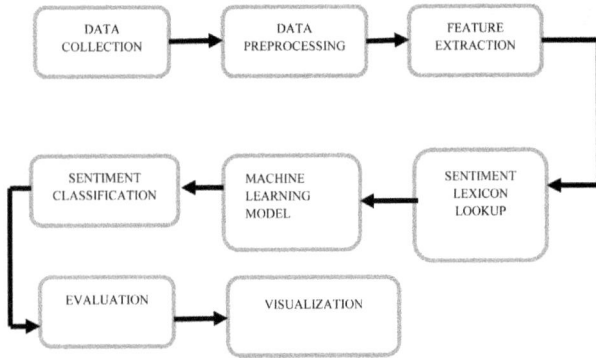

Figure 43.1 Sentiment analysis general workflow diagram
Source: Author

Background study

This research covers the various works that have been done under sentiment analysis of code-mixed data and identifies the research gap in multilingual code mix data. Different techniques, ranging from rule-based approaches to ML algorithms, was utilised to solve challenges of language ambiguity, sarcasm, and context in sentiment classification tasks. An overview of all such works is reported below:

Pandey et al. [3] explore sentiment analysis in English social media posts using NLP techniques. The study applies machine learning models to identify sentiments like happiness, anger, and sadness. Commonly used algorithms include CNNs, SVMs, and k-NNs, though specific accuracy metrics are not mentioned. The work highlights NLP's potential to enhance social media marketing strategies

Neri [1] employs 'Natural Language Processing' (NLP) techniques for sentiment analysis of Italian social media content. Specifically, they use a rule-based sentiment classification approach that relies on lexical analysis, leveraging sentiment lexicons to identify positive, negative, and neutral sentiments. Additionally, they focus on entity-level sentiment detection to analyse sentiments directed towards specific broadcasting services, such as Rai and La7.

Rao's [5] paper on sentiment analysis of English text uses a multilevel feature-based method to improve sentiment classification. The study integrates three key features: sentiment value, expression, and improved semantic features, focusing on social media texts. A dataset of 5,497 positive and 5,403 negative sentences was used. The ECNN model performed 4% better than CNN model.

Kalika Bali [2] analyses English–Hindi code-mixing, or 'Hinglish,' in Facebook status updates. The study examined Facebook post from different people, collected sentences. A rule-based approach was used to identify code-mixed tokens and analyse patterns, finding that code-mixing typically occurs within sentences, with nouns and verbs being most commonly mixed. The research highlights the frequency of intra-sentential code mixing on social media platforms.

Mishra [8] has explored the sentiment analysis for code-mixed data, focusing on Bengali–English and Hindi–English. It compares deep learning (LSTM, RNN, CNN) with machine learning (Random Forest, SVM) approaches. Feature extraction techniques include lexical, syntactic, and contextual features. The study finds that DL models, particularly LSTM, outperform conventional ML methods. They focused on challenges of handling code-mixed data and emphasises the effectiveness of neural networks in sentiment analysis.

Tareq et al. [9] have been focused on sentiment analysis for Bengali–English user comments from Playstore apps, using a dataset with 970,852 samples, 253,601 words, and 18,074 reviews. Various ML models encompassing SVM, Decision Tree, Logistic Regression, and Extreme Gradient Boosting (XGBoost) were tested. The XGBoost model, combined with FastText word embeddings, obtained an F1 score of 87%.

Jumi et al. focus on creating a stemmer for the Assamese language, which is rule-based and incorporates resources from WordNet. The goal is to reduce words to their root forms by applying linguistic rules specific to Assamese morphology. The stemmer uses a combination of affix removal and morphological rules to handle various word forms. The integration of WordNet aids in refining the stemming process by providing lexical relationships and supporting accurate root word identification. System's effectiveness is evaluated by its capability to handle wide range of Assamese vocabulary and its efficiency in processing text.

Chutia [11] focuses on recognising emotions in Assamese text using 'Long Short-Term Memory' (LSTM) networks. The investigation addresses the challenge of emotion detection in Assamese, a low-resource language, by leveraging deep learning techniques. The authors created a dataset containing text labeled with emotional categories like happiness, sadness, anger, etc. They employ LSTM, a kind of RNN, to model sequential nature of the text and capture the context required for emotion recognition. The findings demonstrate that LSTM performs better than conventional machine learning models in terms of recall, accuracy, F1-score, and precision.

Chakravarthi, Jose, et al. [12] used 116,711 Malayalam–English YouTube comments, with 70,075 tokens in Malayalam. Several ML models like Random Forest (RF), Support Vector Machine (SVM), k-Nearest Neighbors (k-NN), Logistic Regression (LR), and Multinomial Naive Bayes (MNB) have been utilised. Among them, MNB and KNN achieved the highest accuracy.

Chakravarthi, Priyadharshini, et al. [13] discuss the challenges and methods for sentiment analysis in Dravidian languages (Tamil, Telugu, Kannada, and Malayalam) within code-mixed text. It highlights the need for effective models to handle the mix of languages, such as English and local languages, in social media content. They have provided an overview of various approaches used in sentiment analysis tasks for Dravidian code-mixed text. XLM-Roberta and CNN models achieved an accuracy of 0.65 and 0.74 for Tamil and Malayalam, respectively.

In the paper by Shalini et al. [14], the dataset consists of 7,005 Kannada–English Facebook sentences, with 169,156 words. Various models, including Doc2Vec, FastText, Bi-LSTM, and CNN, were applied for sentiment analysis. Among these, the CNN model achieved the highest accuracy of 71.50%.

Srinivasan [15] focuses on sentiment analysis of a Tamil–English code-mixed dataset comprising 15,744 YouTube comments that split into 11,335 training, 1,260 validation, and 3,149 test samples. Classifiers like Random Forest, Naïve Bayes, Logistic Regression, SVM, and XGBoost were applied, with performance evaluated using F1-score. Pre-processing with the Levenshtein distance significantly improved the accuracy.

Das & Bandyopadhyay [16] present a Conditional Random Field (CRF)-based approach for subjectivity detection in English and Bengali texts. The study focuses on identifying subjective and objective content in these languages, leveraging CRF models for effective text classification.

Vilares et al. [17] have analysed English–Spanish Twitter data with 8,200 English tweets (training), 1,416 (development), and 5,752 (test), alongside 7,219 Spanish tweets split 80:20 for training and testing. Sentiment analysis models were applied to both monolingual and multilingual corpora. Results showed that multilingual models outperformed their monolingual counterparts, demonstrating better performance on mixed-language datasets.

Sarma [18] presents the transliterated text dataset, sourced from Facebook data, includes the following language distributions: Assamese (as) with 5,198 entries, Bengali (bn) with 1,594 entries, and Hindi (HN) with 663 entries. The classification accuracy for legitimate words is 91.22%, while embedded words are correctly classified with an accuracy of 86.98%.

Bhargava et al. [19] used the FIRE 2015 dataset, consisting of 792 sentences in Tamil, Telugu, Hindi, and Bengali. When evaluated using LR, SVM, and One-vs-One Multi-Class SVM, the One-vs-One Multi-Class SVM outperformed Logistic Regression by 10% (Figure 43.2).

Dev et al. [20] have studied that Bengali 'VADER' served as the foundation for the development of the Assamese 'VADER', which was created by altering the English 'VADER' that was already in use. Applying the Assamese lexicon, the current English 'VADER' is altered for detecting sentiment polarity of Assamese texts. In an effort to test the system and improve its efficiency, researchers used augmenting terms, bigrams, and trigrams. It is observed from the experimental study that the suggested model performs well on texts written in Assamese (Figure 43.3).

1	sentence
2	আমি love the অসমীয়া folk সঙ্গীত
3	Such a beautiful বন made my day
4	আমি তোমাৰ সৈতে spend কৰা time is always মজা
5	The বতাহ left a beautiful impression
6	I felt revolting about the খবৰ
7	My time with the গৰাক was amazing
8	Such a amazing ৰং ruined my day
9	I felt shoddy about the আকাশ
10	I had a lovely experience with the চহৰ
11	The ৰেষ্টুৰেণ্ট was disappointing which made me happy
12	The পৰিয়াল is pathetic

Figure 43.2 ACE-Mix: A Dataset for Assamese-English Code-Mixed Language Processing
Source: Author

	Sentence	Sentiment
1	The বিহু নৃত্য মেলা is a অদ্ভুত celebration of Assamese culture.	Positive
2	This restaurant খাবাৰটা totally horrible, আৰু serviceটো একদম pathetic."	Negative
3	"তোমাৰ সৰস্বতী is like a দীপ lighting our path."	Possibly Positive
4	The চহৰ is spectacular, nothing more.	Neutral
5	The বোসনি নৃত্য tour offers an আনন্দময় exploration of tradition.	Positive
5	Attending the বোহাগি বিহু নৃত্য fills the heart with উৎসাৱ.	Positive

Figure 43.3 Sample dataset with labelling
Source: Author

Research gap

After performing a detailed study of the various methodologies that have been used so far in the sentiment analysis, the following analysis was made:

• **Efficient code mix data collection, pre-processing methods:** Code-mixing is less likely to occur in traditional sources of large corpora such as news articles or Wikipedia, so the researchers have relied on speech corpora, text messages corpora, and corpora from online social networks. Since the social media data is non-standard, Roman written, and as social network companies come under increasing pressure to curb the data leaks and privacy infringements, sourcing data from social networks is getting increasingly difficult. Thus, we need more efficient data collection methods, for example: finding high yield code-mixed terms from corpora by filtering the code-mixed query terms, and have to look at alternate sources of data.

- **Learning methods for multilingual data:** As code-mixed data is low-resource, the monolingual corpora which are available should be effectively used to understand the code-mixed data. In the recent past, the usage of models like XLM and mBERT has increased for code-mixed data, the usage of such tools must be critically examined.

- **Lack of standard pipelines for processing code-mixed text for applications:** The existing tasks and associated datasets for code-mixing are being treated on an individual basis and the computational approaches were also suited for that task. But, we need to solve the tasks higher up in the NLU ladder, integrating all different methods (LID, transliteration, normalisation, and representations) for a single task have to be configured. Integrating all the tasks together in a common pipeline/framework, and a library—a common place to access different resources (datasets, corpora), and collection of computational methods known to work well for code-mixed data that facilitates an integrated approach really needed to further work in code-mixed data.

- **Properly annotated data is still scare**: The tools for processing social media data should not be domain-specific or dataset-specific. We need methods that are capable of handling the transiency and the variety of textual data on social media. Because there are not many datasets in field of code-mixed data, we need methods for properly annotating a large dataset for evaluating and training methods.

Conclusion

Sentiment analysis (SA) on code-mixed languages is becoming more popular because it helps in understanding emotions and opinions in text. It can be done using advanced algorithms, including machine learning (ML), deep learning (DL) techniques. However, even with improvements in SA algorithms, they still have limitations. Some challenges include the need for good training data, the risk of bias in models, and difficulties in understanding language nuances. Lexicon-based SA is an easier approach to classify sentiment, as it does not require much training data. But this method is limited because it relies on fixed dictionaries, which may not keep up with changes in language or context. Most of the research so far has focused on languages like Tamil, Bengali, and Hindi,

especially in Indo-Aryan languages. As more regional Indian languages appear on social media, research is shifting towards code-mixed languages, though many languages still lack proper resources like word database or corpus. Future research will focus on evaluating existing code-mixed datasets with traditional ML and DL techniques, and developing new methods for sentiment analysis, particularly for Assamese–English code-mixed data.

References

[1]   Federico Neri et al. (2012). Sentiment analysis on social media. In International Conference on Advances in Social Networks Analysis and Mining.

[2]   Bali, Kalika, Jatin Sharma, Monojit Choudhury, Yogarshi Vyas. (2014). "i am borrowing ya mixing?" an analysis of English-Hindi code mixing in Facebook. In Proceedings of the first workshop on computational approaches to code switching, 116–126.

[3]   Pandey, K. K., et al. (2023). Natural language processing for sentiment analysis in social media marketing. In 3rd International Conference on Advance Computing and Innovative Technologies in Engineering (ICACITE).

[4]   Zhang, Lei, Shuai Wang, Bing Liu. (2018). Deep learning for sentiment analysis: a survey. *Wiley Interdisciplinary Reviews: Data Mining and Knowledge Discovery*, e1253.

[5]   Li, Rao. (2022). Sentiment analysis of English text with multilevel features. *Machine Learning and Scientific Programing in Multi-Sensor Data Processing*, 2022.

[6]   Kaur, Jasleen, Saini, Jatinderkumar R. (2014). A study and analysis of opinion mining research in Indo-Aryan, Dravidian and TibetoBurman Language families. *International Journal of Data Mining and Emerging Technologies*, 4(2), 53–60.

[7]   Dhar, Mrinal, Vaibhav Kumar, Manish Shrivastava. (2018). Enabling code-mixed translation: parallel corpus creation and MT augmentation approach. In Proceedings of the First Workshop on Linguistic Resources for Natural Language Processing, 131–140.

[8]   Mishra, Pruthwik, Prathyusha Danda, Pranav Dhakras. (2018). Code-mixed sentiment analysis using machine learning and neural network approaches. *arXiv preprint arXiv:1808.03299*.

[9]   Tareq, Mohammad, Md Fokhrul Islam, Swakshar Deb, Sejuti Rahman, Abdullah Al Mahmud. (2023). Data-augmentation for Bangla-English code-mixed sentiment analysis: enhancing cross linguistic contextual understanding. *IEEE Access*, 11, 51657–51671.

[10] Sarmah, Jumi, Shikhar Kumar Sarma, and Anup Kumar Barman. (2019). Development of Assamese rule based stemmer using WordNet. In Proceedings of the 10th Global WordNet Conference, 135–139.

[11] Chutia, Tulika, Nomi Baruah. (2024). Emotion recognition in Assamese text using LSTM. In 2024 15th International Conference on Computing Communication and Networking Technologies (ICCCNT). IEEE, 1–7.

[12] Chakravarthi, Bharathi Raja, Navya Jose, Shardul Suryawanshi, Elizabeth Sherly, McCrae, John P. (2020). A

sentiment analysis dataset for code-mixed Malayalam-English. *arXiv preprint arXiv:2006.00210.*

[13] Chakravarthi, Bharathi Raja, Ruba Priyadharshini, Vigneshwaran Muralidaran, Shardul Suryawanshi, Navya Jose, Elizabeth Sherly, McCrae, John P. (2020). Overview of the track on sentiment analysis for Dravidian languages in code-mixed text. In Proceedings of the 12th Annual Meeting of the Forum for Information Retrieval Evaluation, 21–24.

[14] Shalini, K., Barathi Ganesh, H. B., Anand Kumar, M., Soman, K. P. (2018). Sentiment analysis for code-mixed Indian social media text with distributed representation. In 2018 International Conference on Advances in Computing, Communications and Informatics (ICACCI). IEEE, 1126–1131.

[15] Srinivasan, R., Subalalitha, C. N. (2023). Sentimental analysis from imbalanced code-mixed data using machine learning approaches. *Distributed and Parallel Databases*, 41(1), 37–52.

[16] Das, Amitava, Sivaji Bandyopadhyay. (2009). Subjectivity detection in English and Bengali: a crf-based approach. In Proceeding of ICON.

[17] Vilares, David, Alonso, Miguel A., Carlos Gómez-Rodríguez. (2015). Sentiment analysis on monolingual, multilingual and code-switching twitter corpora. In WASSA 2015, the 6th Workshop on Computational Approaches to Subjectivity, Sentiment and Social Media Analysis. Association for Computational Linguistics, 2–8.

[18] Sarma, Neelakshi, Sanasam Ranbir Singh, Diganta Goswami. (2018). Word level language identification in Assamese-Bengali-Hindi-English code-mixed social media text. In 2018 International Conference on Asian Language Processing (IALP). IEEE, 261–266.

[19] Bhargava, Rupal, Yashvardhan Sharma, Shubham Sharma. (2016). Sentiment analysis for mixed script Indic sentences. In 2016 International Conference on Advances in Computing, Communications and Informatics (ICACCI). IEEE, 524–529.

[20] Dev, Chandana, Amrita Ganguly, Hsuvas Borkakoty. (2021). Assamese VADER: a sentiment analysis approach using modified VADER. In 2021 International Conference on Intelligent Technologies (CONIT). IEEE, 1–5.

44 Advanced fraud detection in financial transaction using machine learning

Pooja Baravkar^a, Sayali Jagdale^b, Vedant Salunke^c, Anuja Kanade^d and Nikhil Surse^e

Department of AI&DS, Dr. D. Y. Patil Institute of Technology Pimpri, Pune, India

Abstract

This research is about the design of a credit card fraud detection system that combines, among other things, user authentication, machine learning, and data mining. Fraud detection is performed using the Random Forest algorithm. This algorithm learns from transactions by identifying those that significantly deviate from what represents normal. User login is required before accessing the fraud detection module to ensure secure login. The dataset is passed through several techniques of cleaning and feature extraction to increase the efficiency of the model. The ensemble technique inherent in the model, such as the Random Forest, ensures that the accuracy of detection is robust and the rate of false captures is significantly reduced, while valid fraudulent activities are captured. In addition, the platform offers insightful fraud analysis and visualisations that aid in the understanding of fraud trends and patterns, which can greatly enhance the protective measures in place within the payment system.

Keywords: Credit card fraud detection, user authentication, random forest, machine learning, transaction data, data pre-processing, fraud analysis, payment security

Introduction

Credit card fraud is a serious and increasingly significant problem in the financial sector, resulting in enormous financial losses and eroding consumer confidence. The timely detection of fraudulent transactions is important to prevent financial crimes and safeguard both consumers and businesses.

As credit card fraud has become a major issue in the financial services sector, this calls for the need for effective fraud detection systems. Traditional methods are usually not good enough for real-time detection, and therefore there is a demand for advanced machine learning techniques. This research uses the Random

Forest algorithm to identify fraudulent transactions, integrating insights from literature to build a robust fraud detection system.

Literature survey

Reviewing the Role of AI in Fraud Detection and Prevention in Financial Services [1]: This paper discusses how AI has transformed fraud detection in the financial industry by shifting from static, rule-based systems to adaptive, learning-driven models. AI techniques such as supervised and unsupervised learning are used to analyse vast amounts of transactional data and detect subtle patterns indicative of fraud. The paper also highlights the role of deep learning in handling complex relationships and extracting features, thereby increasing the efficiency of fraud detection.

Natural Language Processing is also used in extracting linguistic features, including the examination of transaction descriptions that indicate anomalies in the patterns used in textual data. The evolution in AI approaches has furthered the art of detecting anomalies and adaptation to changing emerging frauds in financial institutions.

Reviewing the Role of AI in Fraud Detection and Prevention in Financial Services (Sood et al., 2023) [2]: This article reviews the application of artificial intelligence (AI) techniques in detecting and preventing financial fraud, analysing more than 240 research articles over the last two decades. It focuses on the fact that AI-based approaches, including machine learning and natural language processing, can significantly reduce fraudulent activities in the financial sector. The study identifies key research areas and future trends, including the integration of advanced algorithms for real-time fraud detection. The authors highlight a notable increase in the adoption of AI in financial institutions, where over 63% of them reportedly prevent fraud effectively using automated methods.

Zojaji et al. (2016) [3] present a comprehensive survey of credit card fraud detection techniques. It discusses various approaches, including rule-based methods, statistical techniques, and machine learning algorithms.

Akcora et al. (2011) [4] focus on the application of data mining techniques for credit card fraud detection. It explores the use of clustering, association rule mining, and classification algorithms.

^abaravkarpoo@gmail.com, ^bsayalijagdale572@gmail.com, ^cvedantsalunke15@gmail.com, ^danujakanade16@gmail.com, ^enikhilsurse16@gmail.com

DOI: 10.1201/9781003675235-44

Aggarwal and Yu (2001) [5] introduce an anomaly-based approach for fraud detection in credit card transactions. It uses statistical techniques to identify unusual patterns that might point to fraudulent behaviour.

Botchkarev et al. (2018) [6] explore credit card fraud detection using deep learning. This is a study based on research regarding the use of deep learning on credit card fraud detection, presenting a deep neural network architecture capable of accurately learning complex transaction-pattern relationships.

Chen et al. (2018) [7] describe an insurance fraud detection hybrid approach, combining data mining and machine learning methods. It showcases the effectiveness of this method to detect fraudulent claims.

Lee et al. (2021) [8] applied deep learning on fraud detection in healthcare claims. It comes up with a deep neural network model that accurately detects fraudulent claims based on various features.

Gupta et al. (2020) [9] propose a novel deep learning approach for fraud detection in the banking sector. It introduces a hybrid model that combines convolutional neural networks and recurrent neural networks to capture temporal dependencies in transaction.

Automated Certificate Verification and Management System, Frontiers in Health Informatics ISSN-Online [10]: The system is cloud-based, making it easy for users to scale and access the platform. An API Gateway is utilised for the communication between users and the backend.

A New Era in Fintech and Insurtech using AI, Frontiers in health Informatics ISSN online [11]: This paper emphasises how machine learning improves fraud detection by reducing errors and adapting to new threats in real-time.

System architecture

Figure 44.1 shows the system architecture diagram of proposed work.

The system architecture for the credit card fraud detection project comprises multiple interconnected components, forming a robust end-to-end pipeline for detecting fraudulent transactions. It begins with historical transaction data, which is stored in a database such as MySQL/SQLite. The data undergoes a training pipeline, including pre-processing, feature engineering, and model training. Pre-processing involves cleaning and preparing the data to ensure high-quality inputs, while feature engineering extracts relevant features that enhance the model's predictive performance.

After the model has been trained with the Random Forest algorithm, it is applied to classify new transactions as fraudulent or legitimate. The detection process takes place in real-time, where authenticated users (e.g., investigators or admins) sign in to access the system and analyse transaction data. Detected transactions, classified as either fraudulent or legitimate, are visualised on a dashboard for monitoring and further analysis. Also, the transaction data and results are stored for future reference and ongoing system improvement. This architecture ensures a secure, accurate, and scalable approach towards credit card fraud detection that integrates both data processing and user interaction. The modules described are:

a. Data pre-processing: This module plays a crucial role in improving model performance by ensuring that the input data is clean and representative. Techniques employed in this project include handling missing values, removing outliers, feature scaling, and en-

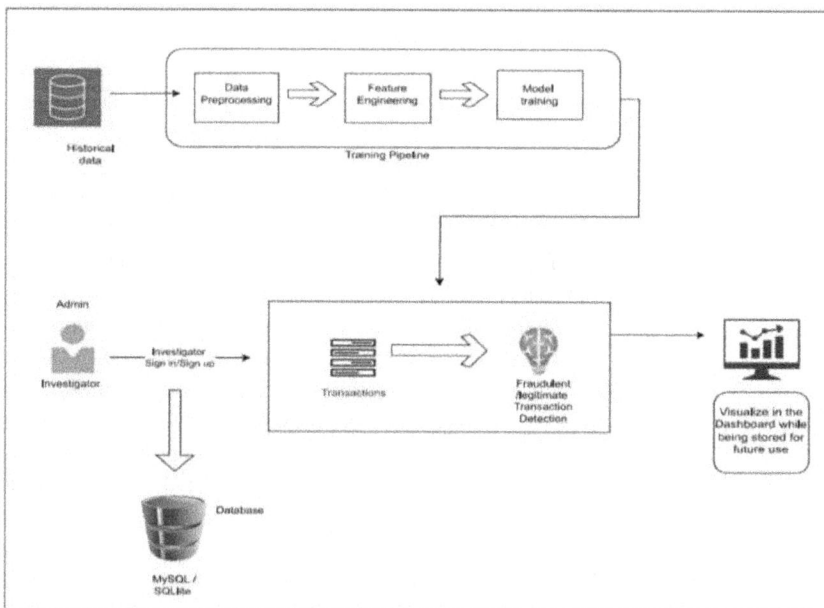

Figure 44.1 System architecture
Source: Author

coding categorical variables. The 'Fraud Detection Handbook' outlines the importance of pre-processing in machine learning, particularly in financial datasets where noise and anomalies can significantly affect the accuracy of models.

b. Feature engineering: This involves transforming raw data into meaningful features that enhance the predictive power of the model. For this project, features like transaction amount, location, time of the day, and activity frequency of the account are considered. In addition, domain-specific knowledge is used to generate new features that capture suspicious patterns; for instance, repeated transactions in less than a specified period. This is an important step in enhancing the Random Forest model's ability to distinguish between legitimate and fraudulent activities.

c. Model selection: Once the pre-processing step is completed, we have used the machine learning python library, i.e., scikit learn is used to split the training input dataset and testing dataset. Here, we have used holdout method which provides 80% samples in training set and 20% samples in testing set. The Random Forest algorithm is selected for its ensemble learning abilities, which combine the outputs of multiple decision trees to improve classification accuracy. Its inherent ability to handle large datasets and manage feature interactions makes it suitable for fraud detection tasks. Earlier research, such as that in the 'Reviewing the role of AI in fraud detection,' also found ensemble approaches useful in financial fraud settings because they prevent overfitting well.

d. Model training and evaluation: The proposed model undergoes training process using a labelled dataset that includes historical transactions with labels assigned to all transactions indicating whether a given transaction is legitimate or fraudulent. Training is separated into training set and test set and then undergo model fitting. A few evaluation metrics that can be used include precision, recall, accuracy, and the F1-score to assess the model's performance. These metrics are critical in determining the model's effectiveness in detecting fraudulent activities while minimising false positives.

e. Model justification:

Random forest: High accuracy, precision, recall, and F1 score make it very useful when fraud detection is necessary, particularly in imbalanced datasets. Overfitting is minimised via ensemble learning.

Logistic regression: Its simpler methods may fall short with non-linear data and complex features interaction, hence lower recall value in fraud detection.

Gradient boosting: Achieves RF grade accuracy but entails elaborate hyperparameter configuration and carries additional computational costs.

Deep learning models: Difficult to interpret yet have a high potential of capturing complex patterns. These models are however very resourceful and require larger datasets.

Model	Accuracy	Precision	Recall	F1 Score
Random Forest	0.981	0.976	0.986	0.981
Logistic Regression	0.944	0.951	0.937	0.944
Gradient Boosting	0.966	0.973	0.960	0.966
XGBoost	0.977	0.979	0.976	0.977
CatBoost	0.974	0.977	0.971	0.974

Comparison of models

f. **User authentication and system security**: The system includes a secure authentication module to ensure that only authorised users can access the fraud detection dashboard. Investigators and admins are required to log in using their credentials before they can view and analyse the transaction data. This approach provides an additional layer of security against unauthorised access to sensitive financial information. The system is designed in accordance with best practices in securing AI-based applications.

Dataset description and experimental analysis

The complete dataset contains 1 Lakh rows (observations), each representing an individual transaction with associated features such as transaction amount, day of the week, type of transaction, and whether the transaction was classified as fraudulent or not. The training set consists of 80% of the dataset. The test set consists of 20% of the dataset.

The dataset used for this research contains a variety of features that provide comprehensive insights into credit card transactions. Key attributes include the Transaction ID, which serves as a unique identifier for each transaction, and the Date, Day of the Week, and Time, which capture the temporal aspects of the transaction. The dataset also captures the Type of Card used, whether credit or debit, and the Entry Mode, which details how the payment was made (e.g., using a PIN, tapping, or CVC). The Amount field details the monetary value of each transaction, while the Type of Transaction, for example, a purchase or cash withdrawal, adds further context (Figure 44.2).

Other features, such as Merchant Group, Country of Transaction, and Shipping Address, provide details on the business involved and the geographic locations associated with the transaction. Demographic information such as Country of Residence, Gender, and Age provides insights into customer-specific patterns. The Bank attribute indicates which financial institution the transaction is associated with, and the Fraud label determines if the transaction was marked as legitimate or fraudulent. These varied features allow for a rich analysis of fraud indicators that could be potentially presented, thus making the detection model more accurate and effective (Figure 44.3).

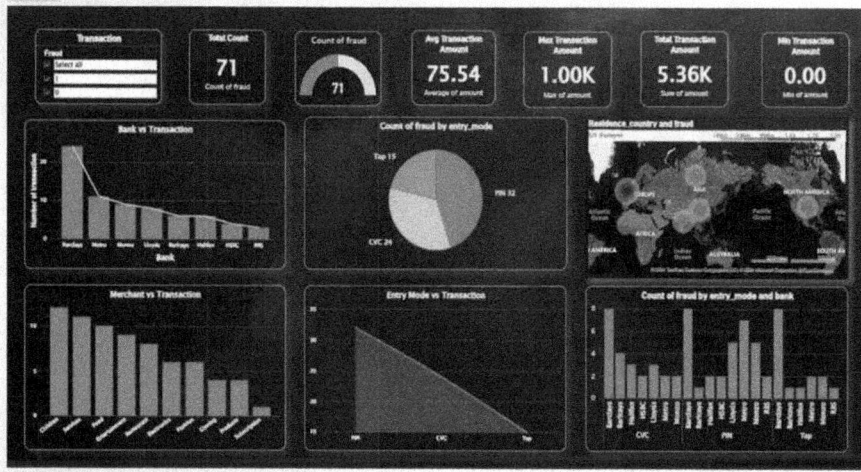

Figure 44.2 Fraud detection visualizations: Dashboard
Source: Author

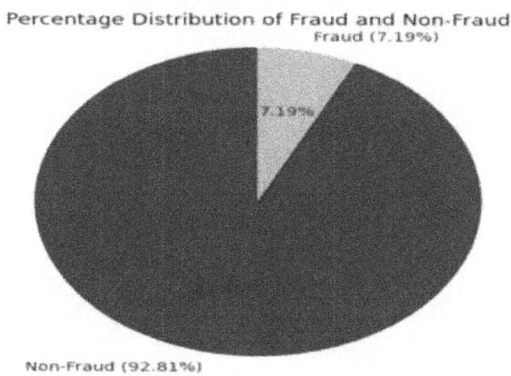

Figure 44.3 Fraud detection visualizations: Distribution chart
Source: Author

Exploratory data analysis (EDA): The analysis plots in this project provide comprehensive insights into credit card transactions, enabling a deeper understanding of fraud patterns. The visualisations explore relationships between various factors such as transaction amounts, banks, merchants, and entry methods. These insights are critical for evaluating the effectiveness of the fraud detection model and identifying areas for potential improvement.

The top-level metrics report that in all transactions, 71 were identified as fraudulent. The dashboard allows access to a range of key statistics, including the average transaction amount, 75.54, and the maximum amount, 1,000, which helps put in baseline behaviours and suspicious activities. These would act as guides to the model in decision-making through showing deviations from norms that may indicate fraudulent activities.

The 'Bank vs. Transaction' plot shows that Barclays and Metro banks have a larger number of fraudulent transactions than other banks. This indicates that the fraud detection system is weak for those banks or their customers

have a different transaction pattern. These facts can help the financial institutions prioritise their anti-fraud efforts and create targeted strategies for high-risk banks.

The 'Merchant vs. Transaction' analysis shows the differences in fraudulent transactions between various merchants. Some merchants have a higher incidence of fraud, which could be due to factors such as industry type, customer demographics, or payment methods used. It helps businesses implement tailored fraud prevention measures based on merchant-specific patterns.

From the pie chart of fraudulent activity by mode of entry, it is revealed that PIN-based transactions are dominating the fraud levels, followed by CVC and then tap-based. The high penetration of PIN fraud further underscores the necessity to make the verification of a transaction through PIN-based more stringent in nature. Thirdly, if contactless and tap-based are becoming more trendy, then its security protocol might also need updating.

The map plot that presents fraud distribution by region identifies the different geographic hotspots of fraudulent transactions. It infers that it may be a case where, unlike overall payment amount, certain regions are more susceptible to fraudulent transactions due to differences in either regulatory environments or perhaps payment system vulnerabilities. Geographical analysis is important for a global company to adapt its strategies for fraud detection based on regional risks.

The plot showing 'Count of Fraud by Entry Mode and Bank' brings together two crucial dimensions: entry mode type and bank. This can help determine combinations of banks and payment methods with a higher tendency towards fraud. For example, some banks might have higher rates of fraud associated with PIN-based transactions, pointing to a potential need for stricter controls over PIN-based transactions in certain institutions.

The dataset shows that fraudulent transactions take up 7.19% of the overall transaction volume, while non-fraudulent transactions make up the remaining 92.81%. This huge difference explains why it is challenging to detect fraud in real-world settings, where fraudulent activities make up a very small percentage of the entire transaction data. It might result in model bias towards predicting non-fraudulent transactions if not properly dealt with. Strategies such as resampling techniques, cost-sensitive learning, or adjusting class weights can be used to improve the model's performance in identifying fraud cases. The use of a pie chart to visualise this distribution provides a clear representation of the data imbalance, underscoring the need for techniques that effectively handle class imbalance in fraud detection.

A. In Figure 44.4, the bar chart showing the distribution of original fraud and non-fraud transactions reveals a very class-imbalanced one, with fraudulent transactions outnumbered by millions compared to the legitimate ones. Such a problem represents a challenge to fraud detection: in most financial data sets in the real world, there are much more legitimate transactions than fraudulent ones. The significant imbalance can prevent the model from effectively identifying fraudulent activity since machine learning algorithms tend to favour the majority class during training, which may result in high accuracy in predicting non-fraud cases but poor performance in detecting fraud.

B. In Figure 44.5, the bar chart showing the counts of fraud and non-fraud transactions after applying SMOTE. It is evident that the dataset is now a lot balanced as the counts for fraud and non-fraudulent transactions seem to be approximately equal. SMOTE generates synthetic samples by oversampling the minority class (fraudulent transactions), thereby reducing the class imbalance in the original dataset. This approach enhances the model's ability to learn patterns associated with fraud, improving

the predictive performance for detecting fraudulent cases.

c. Balancing the dataset is very important, as it helps the model to train machine learning algorithms better, focusing on both classes during the training process. A balanced distribution ensures a lesser bias towards the majority class and more recall rate on fraud detection while reducing false negatives. With the help of SMOTE, it becomes a better model when trying to learn fraudulent patterns and takes longer to decline on its performance on non-fraudulent transactions, which denotes a better robust and reliable fraud detection-system.

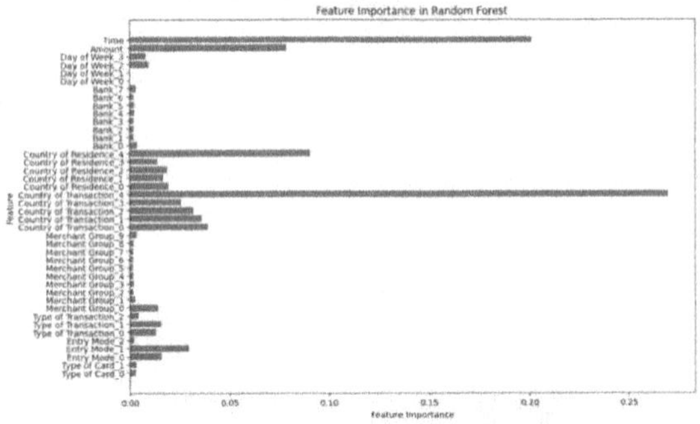

Feature importance in random forest

A bar chart shows feature importance scores from a Random Forest model used for fraud detection. A higher score indicates a more influential feature in predicting whether a transaction is fraudulent or not. As illustrated in this table, features such as 'Amount,' 'Country of Transaction,' and 'Country of Residence' have a greater value than others. These indicate that location-based characteristics, along with transactional amounts, have high relevance for identifying frauds or non-frauds. It infers that these location-related characteristics and amount are important aspects of the model that detects the likelihood of fraudulence (Figure 44.6).

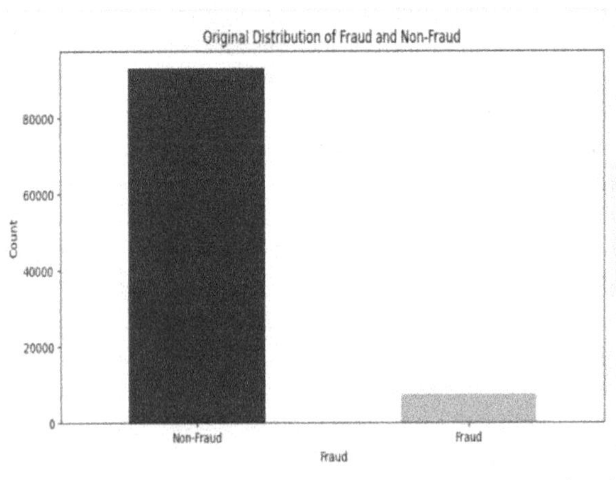

Figure 44.4 Original distribution of fraud and non-fraud
Source: Author

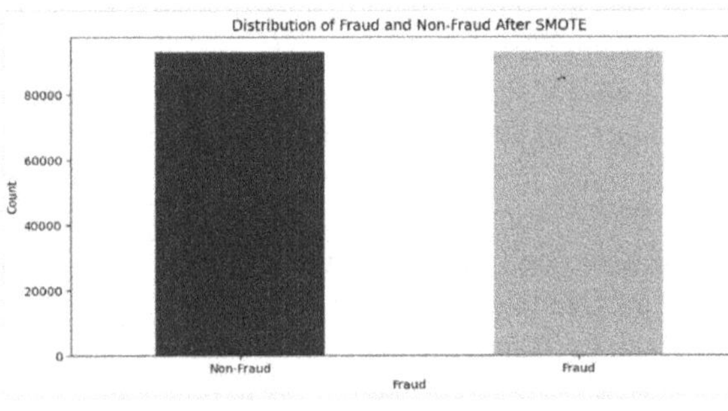

Figure 44.5 Original distribution of fraud and non-fraud
Source: Author

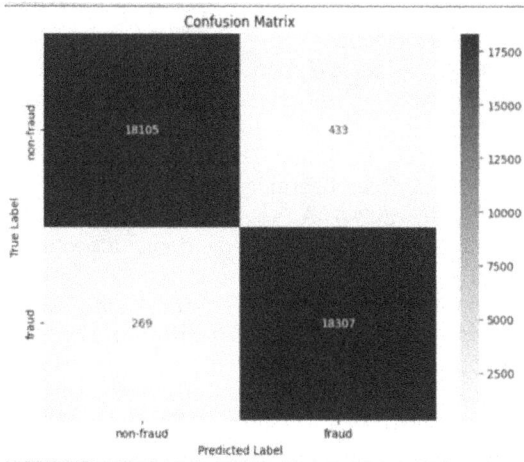

Figure 44.6 Fraud detection visualizations: Confusion matrix

Source: Author

Table 44.1 Prediction results of ML model.

Column 1	Column 2	
	Column 2.1	Column 2.2
	0	**1**
precision	**0.99**	**0.98**

	0	**1**
Recall	**0.98**	**0.99**

	0	**1**
F1-Score	**0.98**	**0.98**

Source: Author

On the other hand, features such as 'Type of Card,' 'Day of Week,' and several bank and merchant group indicators show relatively low importance scores. These features contribute less to the model's predictive capability, which may imply that they do not have a strong correlation with fraudulent behaviour in the dataset. By analysing these feature importance scores, practitioners can prioritise the most significant variables during the feature engineering process, which may enhance the performance of the model while reducing computational complexity.

Analysing the confusion matrixTrue Positive (TP): The instances correctly predicted as positive, i.e., fraudulent. In this case, 18,307 fraudulent transactions were correctly identified.

True Negative (TN): The number of correct predictions made as negative (non-fraudulent). Here, 18,105 non-fraudulent transactions were correctly classified.

False Positive (FP): The instances wrongly predicted to be positive, i.e. fraudulent. It is called type I error. In the given matrix, 433 non-fraudulent transactions were wrongly tagged as fraudulent.

False Negative (FN): The number of instances incorrectly predicted as negative (non-fraudulent). This is a type II error. Here, 269 fraudulent transactions were wrongly classified as non-fraudulent.

Model performance metrics analysis

Impressive performances of the model in terms of precision, recall, and F1 score of the Random Forest model in evaluating credit card fraud detection show values of 0.99 and 0.98 for the non-fraud (0) and fraud (1) classes, respectively, which is high and reflects how accurately the model will correctly classify a legitimate and a fraudulent transaction. The recall values for non-fraud are 0.98 and for fraud are 0.99, indicating that the model can efficiently detect almost all fraud cases with minimal false negatives.

From Table 44.1, it is shown that the F1 score that balances precision and recall is 0.98 for both classes, thus making the model generally strong in the classification of fraudulent versus non-fraudulent transactions. The above results demonstrate that the model maintains a high degree of accuracy for both outcomes to be predicted; this is essential for a fraud detection system since the cost of missed fraud or false negatives could be significant. Balanced performance across these metrics indicates that the model generalises well and can be effectively deployed in real-time fraud detection scenes.

Conclusion

The proposed credit card fraud detection system demonstrates an applied scenario of AI in financial security, exploiting it applies Random Forest to classify transactions effectively. It addresses key challenges in detecting and preventing fraud by combining machine learning with user authentication and data visualisation. This research contributes to the development of robust and scalable fraud detection solutions in the financial industry.

References

[1] https://www.researchgate.net/publication/378297681_Reviewing_the_role_of_AI_in_fraud_detection_and_prevention_in_financial_services. (February 2024). *International Journal of Science and Research Archive.*

[2] Olubusola Odeyemi, Noluthando Zamanjomane Mhlongo, Ekene Ezinwa Nwankwo, Oluwatobi Timothy Soyombo. (2024). Reviewing the role of AI in fraud detection and prevention in financial services. *International Journal of Science and Research Archive*, 11(01), 2101–2110.

[3] Zojaji et al. (2016). A Survey of Credit Card Fraud Detection.. Credit Card Fraud Detection: A Review of Techniques and Applications.

[4] Akcora et al. (2011). Credit Card Fraud Detection: A Review of Techniques and Applications.

[5] Aggarwal, Yu. (2001). Anomaly-Based Fraud Detection in Credit Card Transactions.

[6] Botchkarev et al. (2018). Deep Learning for Credit Card Fraud Detection.

[7] Chen et al. (2018). A Hybrid Approach for Insurance Fraud Detection Using Data Mining and Machine Learning.

[8] Lee et al. (2021). Fraud Detection in Healthcare Claims Using Deep Learning.

[9] Gupta et al. (2020). A Novel Deep Learning Approach for Fraud Detection in the Banking Sector.

[10] (2024). Automated certificate verification and management system. *Frontiers in Health Informatics,* ISSN-Online: 2676-7104, 13(4).https://doi.org/10.52783/fhi.vi.1103.

[11] (2024). A new era in Fintech and Insurtech using AI. *Frontiers in Health Informatics, ISSN-Online:* 2676-7104, 13(4).https://doi.org/10.52783/fhi.vi.1050.

[12] (2024). A new era in Fintech and Insurtech using AI. *Frontiers in Health Informatics*, ISSN-Online: 2676-7104, 13(4).https://doi.org/10.52783/fhi.vi.1050

45 Efficient detection of offensive social media comments in Assamese language using LSTM

Tulika Chutia[a], Sanjib Bora, Nomi Baruah[b], Debajani Baruah,
Swarnangka Barman, Joy Anupol Neog and Bikokhita Dutta

Department of Computer Science and Engineering, Dibrugarh University, Assam, India

Abstract

The increasing incidence of abusive language on online sites has necessitated the formulation of automated detection systems to offset its damaging effects on individual users and social communication. This paper introduces an offensive sentence detection system for the Assamese language using LSTM networks. As no public dataset was available, a public dataset of 3,566 Assamese sentences was created and labelled as either offensive or non-offensive. The approach attained an accuracy of 88.10% and an F1 score of 87.80%, thus validating the potential of the mockup to analyse offensive content correctly. Comparative analysis with related research in other languages revealed that the proposed approach surpassed existing procedures, including SVM and XML-RoBERTa. Despite its success, the study admits that the size of the dataset is small and incapable of detecting subtle intentions for using offensive language. Upcoming work may incorporate increasing the size of the corpus and enhancing the ability of the model to capture subtle forms of offensive content. In general, this research shows that the area of NLP is possible for monitoring and filtering offensive comments in Assamese.

Keywords: Assamese, LSTM, offensive text

Introduction

By this automation, abusive remarks are identified, sites are saved, user security is enhanced, and web conversations are enhanced [2]. Offensive speech in users' content on websites and its impacts have increased awareness in the past many years. Digital platform has made it possible for all of us to give opinions and communicate with fellow citizens, but it has also turned out to be a platform for hate speech, profanity, bullying, and insults. Moreover, identifying a comment or tweet as abusive remains challenging and laborious. This is because most social websites have been criticised for how to identify and handle such a form of content. Nonetheless, considering the volume of user-generated content posted each day, direct handling of this kind of language is not easy. As such, automatic advances based on NLP are necessary [12].

Example- 'মদাহী চুলাই খাই পগলা হলল তই আৰু !'/ Modahi sulai khai pogola holi toi aru!/Alcoholic, you have gone mad drinking alcohol! Here, both ('মদাহী'/Modahi/ Alcoholic) and ('পগলা', pogola, Mad) are stated as offensive words, and the sentence holding the offensive word is an offensive comment. It is difficult to explain offensive words since varied cultural and individual views will have varied perceptions regarding what is offensive or not.

Although offensive speech will be more apparent in certain cultures than others, certain areas of it are open to argument and interpretation. This issue is distinctive, and it usually does not matter whether words or comments are offensive.

The detection of offensive posts in the Assamese language has to be done in an all-inclusive manner that uses linguistic nuances, cultural context, and advanced machine-learning techniques. Our methodology focusses on the use of Long Short-Term Memory (LSTM) networks designed to understand sequential patterns in Assamese text and allow more precise identification of offending content. The process starts with data collection and pre-processing using a curated dataset of Assamese language posts. Techniques of tokenisation and embedding are nearly new to alter textual instances into a format for input into the LSTM. Training LSTM involves exposing the model to a variety of offensive and non-offensive examples, which enables the model to learn contextual dependencies and language-specific patterns. Tailored to the Assamese language, this post-detection system is supposed to outperform generic models and contribute to a safer digital environment for Assamese-speaking users in alignment with the broader objective of responsible and respectful online interactions.

[a]tulikachutiadu24@gmail.com, [b]baruahnomi@gmail.com

DOI: 10.1201/9781003675235-45

The two scoping goals that this study attempts to address are as follows:

A.   A labelled dataset of 3,566 Assamese sentences should be developed, since no such datasets are available for public usage for Assamese to detect offensive sentences.
B.   To construct a system of automation that classifies offensive and non-offensive Assamese sentences using LSTM networks.

Literature review

This section discusses the different ways to identify online radicalisation, hate speech, offensive language, and objectionable content across all platforms in multiple languages. These techniques range from very basic to more complex deep learning models, such as CNN, RNN, and transformer-based architectures like BERT. Every research work offers new approaches, datasets, and evaluation criteria, and reflects the ongoing effort that has been made to face the arduous task of detecting and reducing harmful content on social media.

Agarwal et al. [1] proposed a KNN and SVM-based one-class classifier for the detection of online radicalisation. They evaluated the machine-learning-based classifier's recall and precision, explained the analysis of tweets, and performed a Jihad case study.

The authors in Ref. [5] introduced a Turkish offensive corpus consisting of 36,232 tweets collected arbitrarily for 18 months from April 2018 to September 2019. They [5] reported that about categorising the targeted offensive documents into three types, it obtained an F1 score of 77.3%, 77.9%, and 53.0%, respectively.

A hate speech detection model was published by Zhang et al. [16]. For pragmatic reasons, the authors proposed that the latter problem ought to be the main emphasis. They used Deep Neural Network structures as feature extractors because they are very good at capturing the meanings of hate speech.

Pitsilis et al. [10] proposed effective hate speech detection in Twitter data using recurrent neural networks. They claimed that the researchers utilised a dataset of 16,000 short messages from Twitter, focusing on hate speech detection related to racism and sexism. On an unbalanced dataset of Waseem and Hovy—consisting of 1,943 tweets labelled as Racist, 3,166 labelled as Sexist, and 10,889 labelled as Neutral—the authors employed Keras for 10-fold cross-validation.

Susanty et al. [13] proposed an Offensive Language Detection model using an Artificial Neural Network. The research employed an artificial neural network model for the identification of the context of phrases based on their structure, as well as classifying the words as offensive or non-offensive.

Sigurbergsson et al. [11] proposed Offensive Language and Danish offense recognition. They developed four automatic categorisation schemes, all of which were designed to work with both Danish and English. The macro-averaged F1-scores for the top-performing system's capacity to identify offensive English and Danish words were 0.74 and 0.70.

Mandl et al. [7] introduced the HASOC track to promote hate speech in Hindi, German, and English. Three datasets were created from 321 experiments submitted on Facebook and Twitter. LSTM networks were the most prevalent technique, and the top-performing system achieved marco-F1 scores of 0.78 (Hindi), 0.81 (German), and 0.61 (English).

Bi-GRU was implemented for identifying offensive language by Mitrovic et al. [8], and achieved a macro F1-score of 79.40% using the SemEval dataset.

Wu et al. [15] applied supervised classification and variations of data to create a BERT model for identifying foul language. The model proved to be accurate in subtasks A (detection of offensive language), B (offense category classification), and C (offense target detection).

To deal with the issue of hate speech identification in Hindi and Marathi texts, Velankar et al. [14] resorted to the tremendous potential of deep learning tools. Results showed that models based on transformers outperform their counterparts based on BERT, opening up a new avenue towards more accurate classification of objectionable material. A successful approach to determining hate speech in Roman Urdu has been presented by Bilal et al. [4]. Using a Bi-LSTM with an attention layer and some specific word embeddings, creativity can transcend language barriers [9].

Methodology

This section is a concise overview of the approach to our research. A pictorial diagram of the proposed approach is shown in Figure 45.1 below.

*   **Training and testing**
    The desired Assamese comments dataset was utilised to train the LSTM model. 20% of the entire corpus was utilised for testing in this case, while the remaining 80% was utilised for training [22,23].
*   **Proposed approach**
    Long Short-Term Memory (LSTM), which can be categorised as offensive or non-offensive, is employed in the text for presenting a method that identifies offensive comments in the Assamese language. This enhances the process of detection [21]. One kind of recurrent neural network (RNN) that is capable of learning long-term dependencies is the long short-term memory network (LSTM). Since their introduction by Hochreiter and Schmidhuber in 1997, LSTMs have gained popularity due to their capaci-

ty to avoid long-term reliance [9]. Since that is how these networks operate by default, they are made to store data for long periods of time. Similar to a single tanh layer in conventional RNNs, LSTMs are identified by a sequence of repeating modules [9].

While the repeating module is differently constructed, LSTMs share this chain-like construction. There are not only one neural network layer but four, and each one is different to interact with. For LSTMs, what represents the cell state is the horizontal line crossing the top of the diagram. There are several ways that the cell state behaves like a conveyor belt. It travels straight down the whole chain with very little in the way of linear contact. It is highly probable that information will just sail over it unaffected. By tightly regulating structures called gates, the LSTM is capable of adding or deleting information from the cell state. Information may be optionally passed through gates. They are a pointwise multiplication operation and a sigmoid neural net layer. Numbers between 0 and 1 that describe how much of each element to allow to pass through are generated by the sigmoid layer. 'Let nothing through' is denoted by zero and 'let everything through' by one. Three of these gates exist in an LSTM to protect and control the cell state [9]. The architecture of the proposed LSTM is depicted in Figure 45.2 [17].

Data description and pre-processing

- **Data collection**
  Assamese language resources, especially for offensive post detection in Assamese, are in the early stage, as are many other languages. An effort was made to compile 1,500 Assamese comments from different social media platforms such as Facebook, Instagram, and YouTube, then prepare and create a dataset to identify the offensive posts.
- **Data description**
  The 3,566 Assamese comments were obtained from social media platforms such as Facebook, Instagram, and YouTube to prepare the dataset upon which offensive posts in Assamese were identified. The comments were systematically categorised into two sets: offensive and non-offensive. There are 1,783 abusive sentences and 1,783 non-abusive sentences. These comments (sentences) in balanced datasets are found in a .csv file format. There are two columns in the .csv file: 'Comments' and 'Label.'
  The 'Comments' column contains a range of comments collected from multiple social media sites, comprising offensive and non-offensive statements. The 'Label' column is the ground truth for each comment, where the comment is assigned a binary classification. A label of '1' indicates comments classified

Table 45.1 Example of sentences with numerical values.

Input	Comments	Label
Training	জধামূৰ্খ কথা কবলৈ শিক আগত (Non-sense Learn to speak properly before)	1 (Offensive)
Training	ইমান সুন্দৰ পটভূমি (Such a beautiful landscape)	0 (Non-offensive)

Source: Author

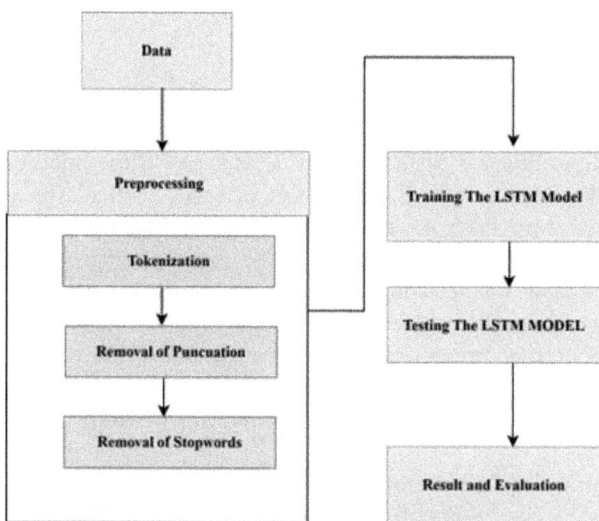

Figure 45.1 Proposed methodology of the work
Source: Author

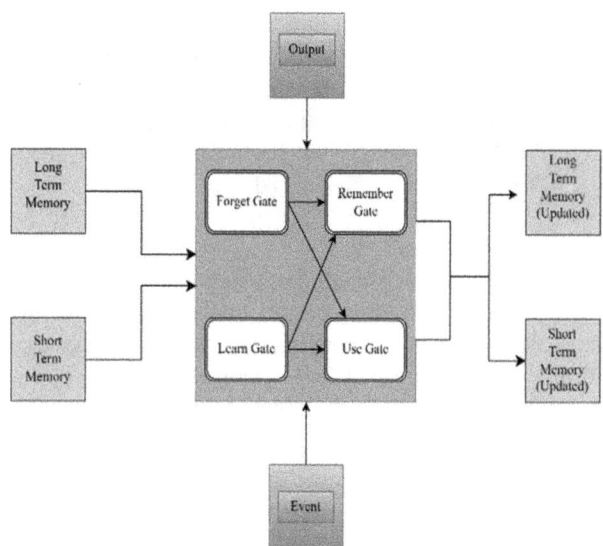

Figure 45.2 LSTM architecture [17]
Source: Author

Table 45.2 Some examples of planning and control of offensive comments detection in Assamese text.

SI No	Comments	Label
1	মহাবিদ্যালয় পৰিয়াললৈ অভিনন্দন (Congratulations to the college family)	0 (Non-offensive)
2	আপুনি মানুহ মৰা চিকিৎসক (You are a doctor who kills humans)	1 (Offensive)
3	সিহঁত জহন্নামে যাব চাই থাকক আপুনি (Keep watching, they will go to hell)	1 (Offensive)
4	বহুত ধুনীয়া গাইছে শুনি বৰ ভাল লাগিল (It's great to hear a very beautiful song)	0 (Non-offensive)

Source: Author

as offensive, while a label of '0' refers to non-offensive comments.

The 'Comments' column holds a variety of comments gathered from various social media sites, including offensive and non-offensive statements. The 'Label' column is the ground truth for each comment, where the comment is assigned a binary classification. A label of '1' indicates comments classified as offensive, while a label of '0' refers to non-offensive comments.

- **Data pre-processing**

  The authors first gather and prepare the raw data. Data pre-processing is a data mining technique for converting unprocessed data into an efficient and practical format. It is one of the most important stages of creating a deep learning model. How well the raw data has been cleaned has a big impact on how well the model performs.

- **Tokenisation of words**

  Here, dividing a sentence, phrase, paragraph, or whole text document into little pieces. We call these tiny pieces tokens. Tokens are words, numbers, or punctuation [22,23].

  Example: Our input sentence is 'মদাহী চুলাই খাই পগলা হলি তই' After tokenisation, it will be : 'মদাহী', 'চুলাই', 'খাই', 'পগলা', 'হলি', 'তই'.

Removal of punctuation

The unique characters that constitute punctuation provide the language with grammatical structure. Punctuation must be eliminated before language strings can be used further because they are hard to comprehend [9].

Example: subsequently removing each bit of punctuation from the foregoing processed data, we get: মদাহী চুলাই খাই পগলা হলি তই

Removal of stop words

A stop word is a frequently used word programmed to be ignored, each time both when indexing an entry to

Table 45.3 Comparison with previous studies.

Reference	Methods used	Results
[1]	LSTM	77.3%
[5]	KNN, SVM	60%, 83%
[6]	XML RoBERTa	80.3%
Proposed work	LSTM	87.80%

Source: Author

search for them and retrieving them as a result of some search query. These words might consume space in our database, or it may consume crucial processing time. For these, we can easily remove the stop words from our database as the authors have a list of words we think are stop words.

Example: After the removal of the stop word 'তই' from the tokenised text, we get our result:
'["মদাহী", "চুলাই", "খাই", "পগলা", "হলি"]'

Result and analysis

With an F1 score of 87.80% and a precision of 88.10%, the system performed well in classifying offensive and non-offensive statements in Assamese. These metrics illustrate how well the model performs, particularly in reducing false positives and balancing recall and precision. The results reveal that the proposed approach is suitable for detecting inappropriate words and promises use in real-world content filtering on Assamese online platforms, promoting more courteous and secure conversation. Figure 45.3 represents the Confusion Matrix for LSTM (Offensive and Non-offensive Comments).

- **Confusion matrix**

  The results of the Confusion Matrix—77.3%, 60%, 83%, 80.3%, and 87.80%. The evaluation using the confusion matrix comes after the data has been de-

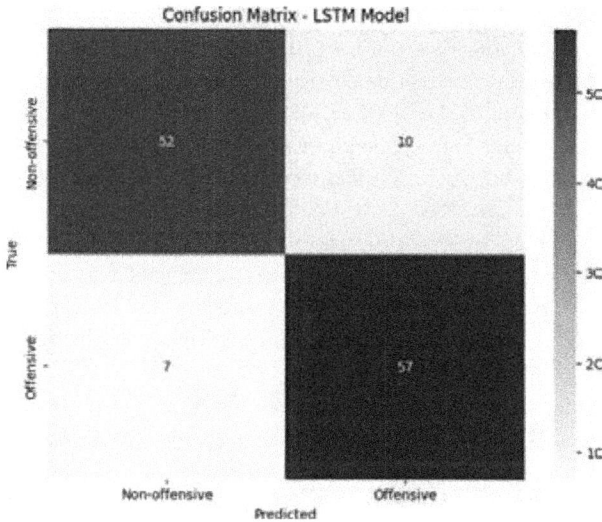

Figure 45.3 Confusion matrix for the proposed approach
Source: Author

tected. True Positive (TP) refers to the quantity of data points in the confusion matrix table. It was labelled by the classifier, which classed it as positive. False positives are the count of data objects labelled as positive but classified as negative by the classifier. True Negative (TN) is the number of data points that were classified as negative by the classifier and assigned to the right category. [18,19]. False negatives are the count of data items with labels as negative but classified as positive by the classifier [3,20].

- **Comparison with existing studies**
  For comparison purposes, research works done in some other languages are used. The table shows the comparative study done with other related research works.

Table 45.3 shows the comparative study with other related works. Our approach scored relatively higher than the other research works [1,5,6].

Conclusion

It is the coming age of Technology, but still a very controversial topic. Offensive comments publicly hurt both the users individually and groups of social interactions. In the past few years, the need for tools that could detect an offensive comment/sentence automatically has increased. So, in this study, the author focused on developing an offensive post-detection system for the Assamese Language using LSTM. There is no available dataset in the public domain to develop a dataset of 3,566 Assamese sentences for the detection of offensive sentences in the

Assamese language. These sentences were then labelled as either offensive or non-offensive. The author trained a model using LSTM, which is one type of deep learning neural network, to classify these sentences as offensive or non-offensive. The proposed method was evaluated with an accuracy of 88.10% in the detection of offending sentences in Assamese, which further indicates the possibility of constructing an accurate and reliable offensive sentence detector.

But there are certain limitations with this approach, like having a dataset of comparatively fewer sentences and an inability to determine clearly which intention the user has assigned to the specific words chosen. Some of the future work could be overcoming these limitations by creating models capable of determining certain specific offensive sentences, and with larger and more diverse datasets. Overall, the suggested work illustrates the feasibility of the usage of deep learning approaches in creating offensive comments detection systems in Assamese.

References

[1] Agarwal, Swati, Ashish Sureka. (2015). Using KNN and SVM based one-class classifier for detecting online radicalization on twitter. In Distributed Computing and Internet Technology: 11th International Conference, ICDCIT 2015, Bhubaneswar, India, February 5–8, 2015. Proceedings 11. Springer International Publishing.

[2] Anand, Mukul, Eswari, R. (2019). Classification of abusive comments in social media using deep learning. In 2019 3rd International Conference on Computing Methodologies and Communication (ICCMC). IEEE.

[3] Anny, Fatema Tuz Zohra, Oahidul Islam. (2022). Sentiment analysis and opinion mining on e-commerce site. arXiv preprint arXiv:2211.15536.

[4] Bilal, Muhammad, et al. (2022). Context-aware deep learning model for detection of roman Urdu hate speech on social media platform. *IEEE Access*, 10, 121133–121151.

[5] Çöltekin, Çağrı. (2020). A corpus of Turkish offensive language on social media. In Proceedings of the Twelfth Language Resources and Evaluation Conference.

[6] Das, Mithun, et al. (2022). Hate speech and offensive language detection in Bengali. arXiv preprint arXiv:2210.03479.

[7] Mandl, Thomas, et al. (2020). Overview of the HASOC track at fire 2020: hate speech and offensive language identification in Tamil, Malayalam, Hindi, English and German. In Proceedings of the 12th Annual Meeting of the Forum for Information Retrieval Evaluation.

[8] Mitrović, Jelena, Bastian Birkeneder, Michael Granitzer. (2019). NLPUP at SemEval-2019 task 6: a deep neural language model for offensive language detection. In Proceedings of the 13th International Workshop on Semantic Evaluation.

[9] Baruah, Nomi, Pritom Jyoti Goutom. (2025). Hybrid deep learning approaches for Assamese part-of-speech tagging using BIS tag set. *Procedia Computer Science*, 258, 2469–2478.

[10] Olah, Christopher. (2015). Understanding LSTM networks.

[11] Pitsilis, Georgios K., Heri Ramampiaro, Helge Langseth. (2018). Effective hate-speech detection in Twitter data using recurrent neural networks. *Applied Intelligence*, 48(12), 4730–4742.

[12] Sigurbergsson, Gudbjartur Ingi, Leon Derczynski. (2019). Offensive language and hate speech detection for Danish. arXivpreprint arXiv:1908.04531.

[13] Susanty, Meredita, et al. (2019). Offensive language detection using artificial neural network. In 2019 International Conference of Artificial Intelligence and Information Technology (ICAIIT). IEEE.

[14] Velankar, Abhishek, et al. (2021). Hate and offensive speech detection in Hindi and Marathi. arXiv preprint arXiv:2110.12200.

[15] Wu, Zhenghao, et al. (2019). BNU-HKBU UIC NLP team 2 at semeval-2019 task 6: detecting offensive language using BERT model. In Proceedings of the 13th International Workshop on Semantic Evaluation.

[16] Zhang, Ziqi, Lei Luo. (2019). Hate speech detection: a solved problem? The challenging case of long tail on twitter. *Semantic Web*, 10(5), 925–945.

[17] Chutia, Tulika, Nomi Baruah. (2024). A review on emotion detection by using deep learning techniques. *Artificial Intelligence Review*, 57(8), 203.

[18] Chutia, Tulika, Nomi Baruah. (2024). Emotion recognition in Assamese text using LSTM. In 2024 15th International Conference on Computing Communication and Networking Technologies (ICCCNT). IEEE.

[19] Chutia, Tulika, Nomi Baruah. (2024). Text-based emotion detection: a review. *International Journal of Digital Technologies*, 3(I).

[20] Kolluru, Vinoth Kumar, et al. (2024). AI-driven energy optimization: household power consumption prediction with LSTM networks and PyTorch-Ray tune in Smart IoT systems. In 2024 International Conference on Microelectronics (ICM). IEEE.

[21] Kolluru, Vinoth Kumar, et al. (2024). AI-driven energy optimization: household power consumption prediction with LSTM networks and PyTorch-Ray tune in Smart IoT systems. In 2024 International Conference on Microelectronics (ICM). IEEE.

[22] Chutia, Tulika, Bikokhita Dutta, Nomi Baruah. (2025). Harnessing deep learning for Assamese sarcastic comment detection from social media text. *Procedia Computer Science*, 258, 3115–3125.

[23] Chutia, Tulika, Nomi Baruah, Paramananda Sonowal. (2025). A comparative study of machine learning and deep learning approaches for identifying Assamese abusive comments on social media. *Procedia Computer Science*, 258, 981–999.

For Product Safety Concerns and Information please contact our EU
representative GPSR@taylorandfrancis.com
Taylor & Francis Verlag GmbH, Kaufingerstraße 24, 80331 München, Germany